Genetic Engineering Handbook

Genetic Engineering Handbook

Edited by David Rhodes

SYRAWOOD
PUBLISHING HOUSE

New York

Published by Syrawood Publishing House,
750 Third Avenue, 9th Floor,
New York, NY 10017, USA
www.syrawoodpublishinghouse.com

Genetic Engineering Handbook
Edited by David Rhodes

International Standard Book Number: 978-1-68286-454-8 (Hardback)

Cataloging-in-publication Data

Genetic engineering handbook / edited by David Rhodes.
    p. cm.
Includes bibliographical references and index.
ISBN 978-1-68286-454-8
1. Genetic engineering. 2. Biotechnology. I. Rhodes, David.
QH442 .G45 2017
660.65--dc23

Printed in the United States of America.

TABLE OF CONTENTS

PREFACE

Genetic engineering is the science of using biotechnology to modify and improve organisms and enhance their characteristics. This field produces genetically modified organisms (GMOs), genetically modified food and genetically modified crops. It incorporates techniques like DNA sequence, gene transfer, genome editing, gene therapy, etc. This book includes contributions of experts and scientists which will provide innovative insights into this field. It also provides interesting topics for research which readers can take up. Different approaches, evaluations, methodologies and advanced studies on genetic engineering have been included in it. Scientists and students actively engaged in this subject will find this text full of crucial and unexplored concepts.

After months of intensive research and writing, this book is the end result of all who devoted their time and efforts in the initiation and progress of this book. It will surely be a source of reference in enhancing the required knowledge of the new developments in the area. During the course of developing this book, certain measures such as accuracy, authenticity and research focused analytical studies were given preference in order to produce a comprehensive book in the area of study.

This book would not have been possible without the efforts of the authors and the publisher. I extend my sincere thanks to them. Secondly, I express my gratitude to my family and well-wishers. And most importantly, I thank my students for constantly expressing their willingness and curiosity in enhancing their knowledge in the field, which encourages me to take up further research projects for the advancement of the area.

Editor

Stable Expression of *mtlD* Gene Imparts Multiple Stress Tolerance in Finger Millet

Ramanna Hema[1], Ramu S. Vemanna[1], Shivakumar Sreeramulu[1¤], Chandrasekhara P. Reddy[1], Muthappa Senthil-Kumar[1,2]*, Makarla Udayakumar[1]

1 Department of Crop Physiology, University of Agricultural Sciences, GKVK, Bangalore, India, 2 National Institute of Plant Genome Research (NIPGR), Aruna Asaf Ali Marg, New Delhi, India

Abstract

Finger millet is susceptible to abiotic stresses, especially drought and salinity stress, in the field during seed germination and early stages of seedling development. Therefore developing stress tolerant finger millet plants combating drought, salinity and associated oxidative stress in these two growth stages is important. Cellular protection through osmotic adjustment and efficient free radical scavenging ability during abiotic stress are important components of stress tolerance mechanisms in plants. Mannitol, an osmolyte, is known to scavenge hydroxyl radicals generated during various abiotic stresses and thereby minimize stress damage in several plant species. In this study transgenic finger millet plants expressing the mannitol biosynthetic pathway gene from bacteria, mannitol-1-phosphate dehydrogenase (*mtlD*), were developed through *Agrobacterium tumefaciens*-mediated genetic transformation. *mtlD* gene integration in the putative transgenic plants was confirmed by Southern blot. Further, performance of transgenic finger millet under drought, salinity and oxidative stress was studied at plant level in T_1 generation and in T_1 and T_2 generation seedlings. Results from these experiments showed that transgenic finger millet had better growth under drought and salinity stress compared to wild-type. At plant level, transgenic plants showed better osmotic adjustment and chlorophyll retention under drought stress compared to the wild-type. However, the overall increase in stress tolerance of transgenics for the three stresses, especially for oxidative stress, was only marginal compared to other *mtlD* gene expressing plant species reported in the literature. Moreover, the *Agrobacterium*-mediated genetic transformation protocol developed for finger millet in this study can be used to introduce diverse traits of agronomic importance in finger millet.

Editor: Sunghun Park, Kansas State University, United States of America

Funding: This work was supported by grants from Indian council of agricultural research, New Delhi. HR acknowledges the financial assistance from department of crop physiology towards her PhD research. MS-K thanks the council of scientific and industrial research, New Delhi for the senior research fellowship award (No. 9/271(86)/2004/EMR-1) for his PhD research and also international foundation for science (IFS), Sweden (C/4066-1/Ac/17341R). The funders had no role in study design, data collection and analysis, decision to publish, or preparation of the manuscript.

Competing Interests: The authors have declared that no competing interests exist.

* E-mail: skmuthappa@nipgr.ac.in

¤ Current address: Metahelix Life Sciences Limited, KIADB 4th Phase, Bommasandra, Bangalore, India

Introduction

Finger millet (*Eleusine coracana*) constitutes ~12% of the global millet area and is cultivated in more than 25 countries in Africa and Asia. It is an important food crop in India, Nepal and several African countries [1–3]. It is an under-utilized but an important cereal crop for livelihood in rural areas in many of these countries [4]. It has nutritional qualities better than that of wheat and other prominent cereal crops [5–7]. Over the years several approaches, including utilization of somaclonal variations [8] and conventional plant breeding strategies, were used to develop high-yielding finger millet varieties like GPU28 and Indaf 9 [9–11]. However, since the major area of cultivation of these varieties is rain-fed, the potential yields are affected by severe drought stress [12–15] and drought-induced oxidative stress [16]. In addition, salinity stress also reduces potential yields [17,18]. More specifically, seed germination and seedling establishment stages are highly susceptible to drought and salinity stress [12,18]. Various agronomic practices during sowing and plant establishment in the field [19–21] have not improved the productivity of finger millet [12]. Hence, developing plants having higher intrinsic stress tolerance is necessary for yield improvement.

Osmolytes have been attributed to enhance abiotic stress tolerance in plants [22,23]. Relevance of diverse range of osmolyte compounds including carbohydrate-, sugar alcohol- and amino acid-derived compounds under abiotic stress have been shown [22–25]. Developing finger millet plants that can accumulate novel osmolytes is useful for improving stress tolerance. From this perspective, utilizing the existing genetic variability among finger millet germplasm for increased osmolyte production and genetic engineering for a particular osmolyte are approaches that have been attempted in finger millet [16,26–29]. However, incorporating traits from a stress tolerant germplasm line into cultivated variety by conventional breeding is time consuming [30]. Hence, developing transgenic plants expressing novel stress responsive genes has relevance in improving crop performance under abiotic stresses like drought and salinity. To date three research groups have demonstrated genetic transformation in finger millet [17,31–35], however, genetic engineering in this crop has not been routine and not widely employed as a method for crop improvement. One of the reasons for this is lack of efficient and easy genetic transformation protocol. In this direction one focus of this study was to optimize *Agrobacterium*-mediated transformation and plant regeneration from callus in finger millet.

Mannitol, a sugar alcohol, is an important osmolyte and scavenger of reactive oxygen species (ROS). This compound has been reported to impart tolerance to abiotic stresses in different plant species [36–39]. Also, apart from protection against free radicals, transgenic Arabidopsis plants producing mannitol have been shown to express various abiotic stress tolerance related genes [40] leading to enhanced protection during several abiotic stresses. A mannitol biosynthetic gene, mannitol-1-phosphate dehydrogenase (*mtlD*), from *Escherichia coli* is known to catalyze mannitol production in various transgenic plants engineered with this gene and also these plants showed enhanced abiotic stress tolerance [41–47]. Stress tolerance of these transgenic plants has been shown to be due to either by enhanced osmoprotection or through protection against free radicals that otherwise cause oxidative stress. For example, the transgenic tobacco plants expressing *mtlD* gene targeted to chloroplast showed increased hydroxyl radical scavenging capacity and retention of chlorophyll in leaf tissue under methyl viologen-induced oxidative stress [36].

Finger millet is known to accumulate proline under stress [48,49], but, mannitol production has not been reported in this plant. However, like tobacco [41,50,51], potato [46], petunia [43], eggplant [52], sorghum [53], rice [54] and other plants, the substrate for mtlD enzyme action and pathway for production of mannitol is expected to present in finger millet. This study demonstrates a protocol for *Agrobacterium tumefaciens* (here after referred as *Agrobacterium*)-mediated transformation in finger millet. Also, by using this protocol *mtlD* gene expressing transgenic plants were developed and their tolerance under drought, salinity and oxidative stresses was studied.

Materials and Methods

Plant material and growth conditions

Finger millet. The finger millet (var. GPU 28) seeds were obtained from all India coordinated research project (small millets), University of Agricultural Sciences, GKVK Bangalore India. For the seedling level experiments, seeds were germinated on moist filter paper at 30°C and 80% relative humidity. Seedlings (1.5 to 2 cm length) were used for stress experiments. For plant level experiments, plants were grown in plastic pots with the potting mixture consisting red soil: sand: vermiculite in the ratio of 3:1:1 by volume (22% water holding capacity). GPU28 was used for all experiments described in this manuscript except for optimization of hormonal concentrations for callus induction and plant regeneration in MS medium.

Tobacco. Tobacco (var. KST 19) plants were grown in the greenhouse and ~30–35 day old plants were used for stress experiments unless otherwise specified.

For both plant species, recommended fertilizers including micro and macronutrients were provided. Prophylactic measures were taken to maintain the plants disease and pest free. Plants were grown in the greenhouse at 22–27°C with 60% relative humidity and a 12 h photoperiod with mild-day light intensity of ~500 μmol m^{-2} s^{-1}.

Bacterial strain and binary vectors

The binary vectors pCAMBIA 1301 having beta-glucuronidase (*UidA*) [55] and pCAMBIA 1380 having *mtlD* [52,54] (NCBI accession ×51359), both driven by CaMV35S promoter were used in this study. Both constructs have the plant selectable antibiotic gene, hygromycin phosphotransferase II (*HptII*). These constructs were mobilized into *Agrobacterium* strain EHA105 and used for plant transformation experiments.

Agrobacterium-mediated genetic transformation in finger millet

Callus induction. Seeds were surface sterilized with 0.1% mercuric chloride for 5 min and then rinsed for four times in sterile de-ionized water. Sterilized seeds were placed on Murashige and Skoog (MS) medium [56] with 3.0 mg/L auxin (2, 4-D) and 0.5 mg/L benzyl adenine (BA; Himedia laboratories, Mumbai, India) and incubated in dark for 30 days at 29°C. Calli obtained from the scutellum portion of the seeds were sub-cultured. Some seeds developed into seedlings and were periodically removed from the plate. Also, brown calli were discarded before sub-culturing and only embryogenic calli were used for transformation.

Transformation. *Agrobacterium* strains EHA105 harboring the binary constructs were grown on AB minimal medium with 50 mg/L kanamycin for 16 h at 28°C. Cultures (OD$_{600}$ = 0.1) were resuspended in liquid co-cultivation medium supplemented with 100 μM acetosyringone (Sigma, Aldrich, India). Calli (45 day old) were incubated with bacterial (carrying *mtlD* or *UidA* gene construct) suspension medium for 10 min. They were then blotted dry on sterile filter paper to remove excess *Agrobacterium* and co-cultivated for 48 h at 25°C in dark. Subsequently, they were washed using sterile water containing 300 mg/L cefotaxime for four to five times, blotted on sterile filter paper and inoculated on antibiotic selection medium (MS+0.5 mg/L BA+3 mg/L 2, 4-D+ 30 mg/L hygromycin and 300 mg/L cefotaxime). This step was repeated if *Agrobacterium* growth occurred.

Plant regeneration. After three to four weeks of incubation the surviving calli were transferred to regeneration medium (MS with 0.5 mg/L BA) with hygromycin for plant-let growth and maintained in light for 20 days. Later, the shoot-lets were sub-cultured to rooting medium (MS with 0.5 mg/L BA and 0.1 mg/ L NAA) for two weeks. Rooted plants were then transplanted in to soil mixture for further growth. Overall steps involved in mtlD transgenic plant development are described in figure S1.

Stress imposition in finger millet seedlings

Finger millet seeds from T1 generation were germinated on moist filter paper at 30°C for 36 h and uniform seedlings were acclimated with respective stress (−0.2 MPa PEG, 50 mM NaCl, and 1 mM menadione) for 8 h. Then, the seedlings were transferred to Petri plates with filter paper containing polyethylene glycol (PEG-5000; −1.4 or −1.5 MPa) or NaCl (300 and 400 mM) or the naptho-quinone compound, menadione (Sigma chemicals, Bangalore India; 4 and 5 mM), solution. PEG, NaCl and menadione induces osmotic, salinity and oxidative stresses respectively and solutions were prepared as described in previous literatures [27,57,58]. Menadione undergoes rapid auto-oxidation with the regeneration of the parent quinone and thus produces superoxide radicals and hydrogen peroxide [59,60]. After subjecting the seedlings to respective severe stress, seedlings were allowed to recover at 30°C for 72 h on the filter paper with water in the Petri plates. At the end of recovery, the root and shoot length (cm) were recorded. In all the experiments 50 seedlings were maintained per treatment. Seedlings and/or plants that are maintained at 30°C (room temperature) throughout the experimental period were considered as absolute controls [27].

Drought stress imposition in finger millet plants

Fifteen day old plants were planted in pots filled with potting mixture of known weight. Pots were irrigated until all the soil macro and micro pores were filled and excess water was drained. Based on water holding capacity for this soil mixture, total weight of pot with soil mix for 100% field capacity (FC) was arrived. Plants were maintained at 100% FC (control) until imposition of drought stress. Drought stress was imposed by following gravimetric method as described previously [23,61]. Briefly, acclimation was done by gradually with-holding irrigation until the soil water content reached to 50% FC over one week period. Immediately after acclimation, plants were exposed to severe stress (50% or 20% FC) by further withholding the irrigation and the plants were maintained at these stress levels for one week. At the end of the stress period, stress responses were assessed in the leaves. Soil moisture regimes were monitored by weighing the pot with plant once a day and replenished the weight lost through transpiration by adding water. Pot mixture without plant was used to know the evaporative soil water loss (blank) at the beginning of the experiment. Soil water potential was measured using WP4 dew-point potentiometer (Decagon Devices Inc, Washington, DC, USA) and the value at 100% FC was −0.03.

Oxidative stress imposition in finger millet and tobacco plants

Leaf segments (1 cm diameter) were taken from 45 day old transgenic and wild-type tobacco or finger millet plants grown under non-stress condition. The leaf segments were initially incubated in lower concentration (0.5 μM) of methyl viologen (Sigma chemicals, Bangalore India) with 100 μmol m^{-2} s^{-1} light for 2 h for stress acclimation. They were then exposed to high light (700–800 μmol m^{-2} s^{-1}) with 2 μM methyl viologen for 6 h. Methyl viologen accepts electron from the last step of photosynthetic electron transport chain and produces free radicals [62] and has been extensively used to create oxidative stress in plants and algae [55,57,63]. At the end of stress period the leaf segments were used for total chlorophyll estimation or cell viability estimation by 2,3,5-Triphenyltetrazolium chloride (TTC) assay. Superoxide radicals were quantified during the stress period by 2,3-bis-(2-methoxy-4-nitro-5-sulfophenyl)-2H-tetrazolium-5-carboxanilide (XTT) assay.

Molecular confirmation of putative transgenic finger millet plants by PCR

The genomic DNA was isolated from the leaves of putative transformed and wild-type plants. The transformed plants were confirmed by PCR using primers specific to *UidA* gene 5′-GGG CAG GCC AGC GTA TCG TG-3′ and 5′-GTC CCG CTA GTG CCT TGT CCA GTT 3′ and primer pairs 5′-ATC CGC TGC CCC TCT CT-3′ and 5′-ACA TGG CCT TTC AGC TGC GTG GTA-3′ that amplify 500 bp fragment of the *mtlD* gene. Primer pairs 5′- GAG GCT ATT CGG CTA TGA CTG-3′ and 5′-ATC GCG AGG GGC GAT ACC GTA-3′ were used to amplify 750 bp fragment of *hptII* gene. Further, primer pairs 5′-TCC ATA ATG AAG TGT GAT GT-3′ and 5′-GGA CCT GAC TCG TCA TAC TC-3′ were used to amplify 300 bp fragment of *Actin* gene.

Southern blot

Putative finger millet transgenic plants having *UidA* or *mtlD* genes were analyzed for integration and copy number of the transgene. Genomic DNA from the transgenic and wild-type plants were restrict digested (XbaI), resolved on 0.8% agarose gel and transferred to positively charged nylon membrane. The membrane was exposed to UV (1200 μJ for 60 s) for cross linking. The blot was probed with the respective radio labeled (^{32}P dCTP 3000 Ci mmol) inserts. Pre-hybridization was done at 42°C for 2 h and hybridization was done at 55°C overnight with blocking solution (0.5 M Na phosphate buffer, 1 mM EDTA, and 7% SDS, pH 7.2). The hybridized blot was washed in the following sequence with 6× saline sodium citrate (SSC) for 10 min at 37°C, 4× SSC for 10 min at 37°C; 2× SSC for 10 min at 37°C; 2× SSC for 15 min at 55°C; 0.2× SSC for 10 min at 55°C. The hybridization signals were detected following exposure of Kodak X-ray film to the membrane at −70°C for 24 h and developed by autoradiography.

Dot blot

Genomic DNA (50 ng) from finger millet plants was spotted on the nitrocellulose membrane and fixed by exposing to UV (1200 μJ for 60 s) in an UV cross linker (Herolab CL-1, Gemini BV Laboratory, The Netherlands). Probe (*Hpt*) was prepared by labeling it with ^{32}P dCTP as described under Southern blot. The blot was hybridized using the protocol described under Southern blot and developed by autoradiography.

RT-PCR

Total RNA was extracted according to the protocol described by Datta et al. (1989) and first strand cDNA was synthesized by oligo (dT) primers using Molony Murine Leukaemia Virus reverse transcriptase (MMLV-RT; MBI Fermentas, Hanover, MD, USA) according to manufacturer's instructions. The cDNA pool was used as a template to perform RT- PCR analysis using the following primers *Actin-F tccataatgaagtgtgatgt*, *Actin-R ggacctgactcgtca-tactc*.and mtlD F- caagcatgcggcgtacatcc, mtlD R- gcggatcatctt-cactgcggaa. PCR conditions were 94°C for 2 min, 25 cycles of 94°C for 45 s, 52–58°C for 30 s, 72°C for 30 s and a final extension of 72°C for 10 min.

GUS histochemical assay

GUS activity in putative transformed callus was assayed as described in the literature [64]. Hygromycin resistant calli and a portion of ear head (T$_0$ generation) from the PCR positive transgenic plants were incubated at 37°C in GUS staining solution [containing 50 mM phosphate buffer, pH 7.0 and 1 mM 5-Bromo 4-chloro 3-indolyl glucuronide (X-gluc; Sigma, Aldrich, India)] for 8 h. After incubation they were transferred to 80% (v/v) ethanol/water solution to remove the chlorophyll, and later photographed. Similarly, seedlings (T$_1$ generation) were stained with GUS solution and photographed.

Quantification of osmotic adjustment

From the drought exposed (gravimetrically) finger millet plants 5th leaf was excised from stressed and well-watered plants and sampled at midday, quickly sealed in plastic bags, and kept on ice. These leaves were used to determine the relative water content (RWC) and solute potential.

Relative water content. After determining the fresh weight, the leaf segments were floated on deionized water for 5 h to determine their turgid weight. The dry weight was determined after oven-drying to a constant weight. The RWC was calculated using the formula:

$RWC\% =$

$$100 \times [(Fresh{\cdot}weight - dry{\cdot}weight)/(turgid{\cdot}weight - dry{\cdot}weight)]$$

Solute potential of cells in leaf (Ψs). Leaf samples were frozen in liquid nitrogen, thawed, and centrifuged for 5 min at 20 000 g. The Ψs of the extracted sap was measured by VAPRO vapor pressure osmometer (Wescor Inc., Logan, UT, USA).

$$\Psi_{s100} = (\Psi s \times RWC)/100$$

Osmotic adjustment. From the values of RWC and solute potential of control and stress-grown plants, the osmotic adjustment was calculated using the formula: Osmotic adjustment (OA) = drought leaf Ψ_{S100} - control leaf Ψ_{S100}

Estimation of chlorophyll content

Chlorophyll was extracted from 100 mg of leaf tissue in 1 ml of acetone and dimethyl sulphoxide (1:1, volume/volume) mix. The absorbance was recorded at 663 and 645 nm using UV–visible spectrophotometer Model DU800 (Shimadzu Corporation, Kyoto, Japan). Total chlorophyll was estimated as described previously [65] and expressed as percent reduction relative to the corresponding control.

Mannitol estimation

The standard graph was developed using Mannitol (Himedia laboratories, Mumbai, India) using the previously described method [43,66] with below mentioned modifications. Surface-sterilized seeds were ground in ethanol and centrifuged at 7000 g for 5 min at room temperature. Similarly the surface sterilized germinating seedlings was ground in ethanol and centrifuged. Mannitol (Sigma Aldrich, Bangalore, India) chemical was used to prepare standard. Distilled water (2 ml) was used as reagent blanks. One blank (oxidized) was carried through the procedure as described below. The other (unoxidized) was treated with stannous chloride (Himedia laboratories, Mumbai, India) prior to addition of periodic acid. Periodic acid reagent (Himedia laboratories, Mumbai, India) (0.5 ml) was added to the oxidized blank and sample tubes (2 ml of supernatant). Later, the contents were mixed and incubated at room temperature for 8 to 10 min. To these tubes 0.5 ml of stannous chloride was added. The stannous chloride is known to be oxidized by periodate into stannic acid, which appears as a milky precipitate in the tube. In order to solubilize the precipitate, 5 ml of chromotropic acid reagent was added rapidly and the mix was vigorously vortexed. The tube was placed in a boiling water bath for 30 min and then cooled and volume was made up to 25 ml with distilled water. Absorbance at 570 nm was measured using spectrophotometer [67] (Spectramax plus, Spinco Bioteck, Bangalore India).

TTC assay

Cell viability was measured by 2,-3,-5-triphenyltetrazolium chloride (TTC; Sigma, Aldrich, India) as described in the previous literature [68]. Briefly, TTC assay solution (0.1%) was prepared by dissolving TTC in sodium phosphate buffer (pH 7.4). Segments (1 cm diameter each having dry weight 0.27 mg) from stressed and control finger millet leaves were initially washed in sterile water and incubated in the TTC solution at room temperature for 5 h in dark. Subsequently, they were boiled with 5 ml of 2-methoxy

ethanol until dryness to extract bound formazan. To this 5 ml of 2-methoxy ethanol was again added and absorbance was measured at 485 nm using UV–visible spectrophotometer Model DU800.

XTT assay

Leaf segments (1 cm diameter each having dry weight (0.27 mg) were incubated in 1 ml of K-phosphate buffer (20 mM, pH 6.0) containing 500 µM of XTT (cell proliferation kit II, Roche Diagnostics Corporation, Basel, Switzerland) along with methyl viologen and exposed to high light stress. XTT is a sodium 3′[1-(phenylamine-carbonyl-3, 4-tetrazolium)-bis [4 methoxy-6-nitro] benzene sulfonic acid hydrate]. Later, increase in absorbance at 470 nm of incubation medium was measured using a micro titer plate reader (Sunrise-Magellan, Tecan, Australia) as described in the previous literature [69].

Statistical analysis

Analysis of variance (anova) was performed using MS Excel software, and data points were analyzed for Fisher's least significant difference (LSD, $P \leq 0.05$) or Duncan's multiple range test or students t-test. Statistical significance of values in graphs was indicated either as asterisks or as letters.

Results

Callus induction and plant regeneration in finger millet

Fully matured finger millet (var. GPU28) seeds were used as explant (Figure 1A). Among 11 different hormonal treatments two of them namely, MS medium containing 0.5 mg/L BA & 2 mg/L 2, 4-D and MS medium containing 0.5 mg/L BA & 3 mg/L 2, 4-D showed higher callus growth ranging from 200–245 mg/seed callus (Table S1). MS medium with 0.5 mg/L BA & 3 mg/L 2, 4-D eventually produced better quality embryogenic calli. Hence, this concentration was used for callus induction in this experiment. Seven days after incubation of sterilized seeds in callus induction medium, callus initiation was observed from the scutellum portion of the seeds. After sub-culturing two times in ~15 days, embryogenic calli were observed indicating the hormone concentrations used in callus induction medium was successful for callus induction (Figure 1B). Further, to achieve efficient regeneration of shoot-lets, ~30 day old hard calli were sub-cultured on to regeneration medium. Hormonal concentration for optimum regeneration of shoots was initially optimized and MS medium with 0.5 mg/L BA was found to produce green and good quality shoot-lets (Table S2). Results showed that hard friable calli produced better morphogenesis and shoot-lets than the soft watery calli. Green tissues and subsequent initiation of shoots from theses calli was seen after 15 days. Later, these shoot-lets were sub-cultured and multiple shoots were obtained (Figure 1C). Further, individual shoots were transferred to fresh regeneration medium for root initiation (Figure 1D) and plants were allowed to grow till 4–5 leaves stage. MS medium with 0.5 mg/L BA and 0.1 mg/L NAA provided better rooting compared to MS medium with 0.5 mg/L BA. Plants were then transferred to styrofoam disposable cups having sterile vermiculite, covered with polythene bags and maintained in growth room for three weeks (Figure 1E). Later, hardened plants were transferred to soil mixture and grown in greenhouse (Figure 1F). These plants flowered and produced seeds normally, demonstrating the successful generation of finger millet plants from scutellum derived calli.

Figure 1. Callus induction and regeneration from scutellum portion of matured seeds in finger millet. Seeds were incubated on the callus induction medium for 30 days (A), and the calli developed from scutellum portion of seeds was photographed (B). Further, these calli were transferred to regeneration medium and fifteen days later shoot initiation was photographed (C). Later, these shoot-lets were transferred to root induction medium (D). After this step, plants were transferred to potting mixture for hardening (E) and were transferred to greenhouse. Inset in (B) shows enlarged callus derived from single seed.

Development of transgenic finger millet expressing *UidA* gene through *Agrobacterium*-mediated transformation

Embryogenic calli infected with *Agrobacterium* strain EHA105 harboring pCAMBIA 1301/*UidA* constructs were co-cultivated for 48 h. The *Agrobacterium* culture was incubated with 100 μM of acetosyringone for virulence gene induction [70]. Previously standardized hygromycin concentration (30 mg/L) was used for selecting the transformed calli [71,72]. After co-cultivation, the infected calli were incubated on the selection medium (MS+ 0.5 mg/L BA+3 mg/L 2, 4-D+30 mg/L hygromycin and 300 mg/L cefotaxime) under light (100 μmol m^{-2} s^{-1}) for 10 days. The surviving calli were then transferred to regeneration medium (MS+0.5 mg/L BA+30 mg/L hygromycin and 300 mg/L cefotaxime). Later, as per previously standardized protocol (Figure 1C), they were sub-cultured into shooting medium with hygromycin (Figure 2A). Hygromycin resistant independent shoots were developed in to plants.

PCR analysis using *UidA* and *HptII* gene specific primers showed presence of *UidA* gene in putative transgenic plants (T$_0$ generation; Figure 2B). Further, the genomic dot blot analysis using *HptII* gene specific probe showed the signals in the lane loaded with DNA from transgenic lines and no signal was observed in the wild-type (Figure 2C) suggesting the integration of construct. To identify the copy number of *UidA* gene, Southern blot was performed using *UidA* gene specific probe. The genomic DNA was digested with *Xba*I which cuts once in T-DNA. The hybridization signals in blot for the two independent transgenic lines (UidA-1 and UidA-5) showed single copy integration of *UidA* gene in the finger millet genome (Figure 2D).

GUS expression

To further confirm *UidA* gene expression at protein level, GUS-histochemical staining in callus and a portion of ear head was performed. The transgenic callus stained blue with X-gluc indicating the GUS activity, whereas wild-type callus did not stain (Figure 2E). Similarly, ear head from the transformed line (UidA-1) showed intense staining (Figure 2E). Further, 3 day old seedlings (T$_1$ generation line, UidA-1) selected on the hygromycin (Figure S2) were also subjected to histochemical staining. Dark blue staining of shoots, young leaves and roots further confirmed the stability of the transgene expression in T$_1$ generation (Figure 2F).

Further, transformation efficiency was computed based on the number of calli infected and number of calli regenerated into shoots in the selection medium. Results showed an average of 6% transformation efficiency (Table S3). Taken together, these results demonstrated the establishment of regeneration and *Agrobacterium*-mediated transformation protocols in finger millet.

Development of transgenic finger millet plants expressing *mtlD* gene by *Agrobacterium*-mediated transformation

With the aim to improve drought, oxidative and salinity tolerance in finger millet during early seedling growth, the pCAMBIA 1380/mtlD construct was used for mannitol production in finger millet. Adapting the regeneration and transformation protocols developed in this study, the finger millet (var. GPU28) putative transgenic plants expressing mtlD were developed. Hygromycin resistant putative transgenic finger millet seedlings (Figure S3) from independent transgenic events were analyzed by PCR, genomic dot blot and Southern blot. PCR analysis using the *mtlD* gene specific primers showed amplification in all the selected putative transgenic plants whereas no amplification was observed in wild-type plants (Figure 3A). Further, genomic dot blot analysis also confirmed the presence of *HptII* gene in the transgenic lines (Figure 3B). Further, copy number of *mtlD* gene integrated in the genome was assessed by Southern blot (Figure 3C). The expression of *mtlD* gene in fingermillet was confirmed by RT-PCR (Figure 3D). The segregation analysis [73] revealed that the transgenic events are likely to have monogenic inheritance

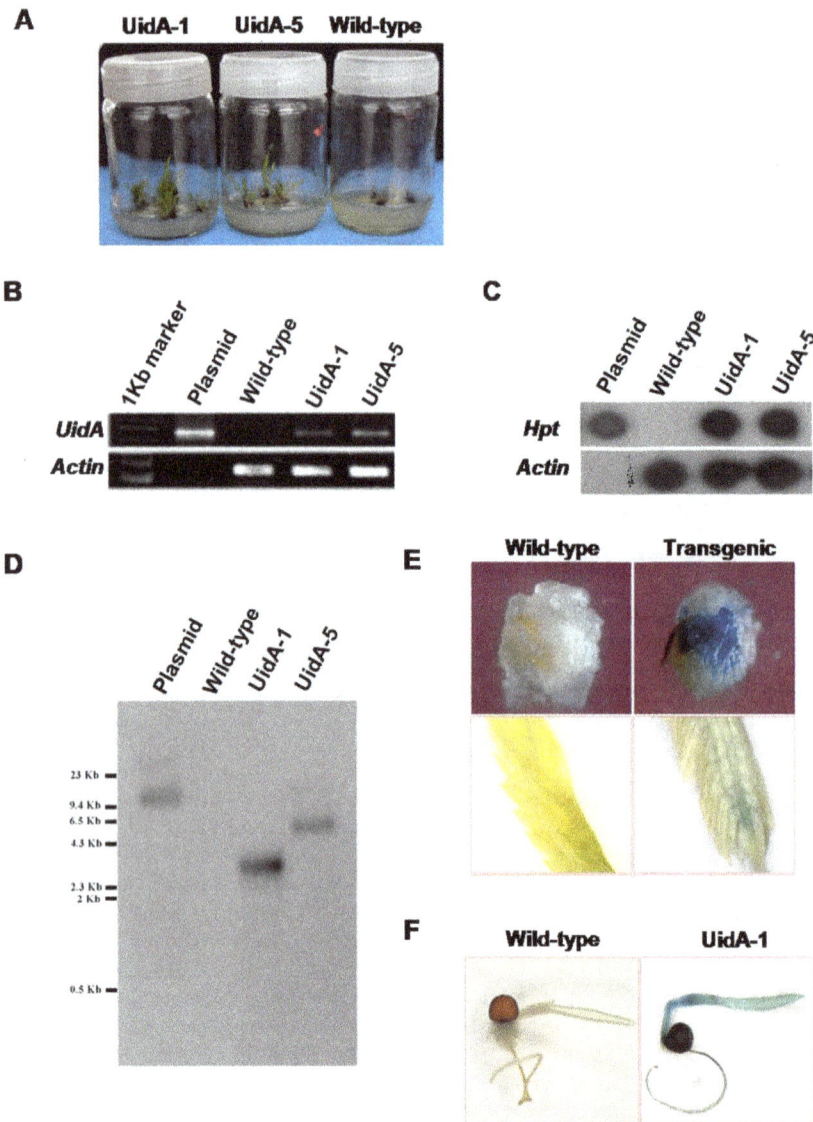

Figure 2. *Agrobacterium*-mediated *UidA* gene transformation in finger millet. Initially calli were infected with *Agrobacterium* harboring *UidA* gene construct and co-cultivated for two days and transferred to the hygromycin selection medium. Photograph of survived calli on selection medium was taken after 30 days (A). Plants (T_0) obtained from the survived calli were analyzed for the presence of *UidA* gene by PCR (B) and for *HptII* gene by genomic dot blot (C). Also, the copy number of *UidA* gene was assessed by Southern blot (D). GUS histochemical assay was performed in transformed calli (T_0 generation; upper panel), T_0 generation finger-let from ear head (UidA-1 T_0; lower panel) (E) and 3 day old T_1 generation seedlings (UidA-1) and the GUS expression were photographed (F). UidA-1 and UidA-2 indicate two independent transgenic lines.

(Table 1). Unlike reported in a previous study [74] where wheat plants accumulating higher mannitol showed stunted growth, finger millet transgenic plants showed normal growth. For example, 15 day old wild-type and transgenic plants maintained ~10 leaves per plant with similar leaf size and overall plant height.

Performance of *mtlD* expressing transgenic finger millet plants under methyl viologen-induced oxidative stress

Drought stress and salinity stress induces higher levels of free radical production in finger millet [16]. Mannitol is known to reduce injury to the plants from ROS generated during drought and salinity stress [36,74,75]. Hence, in order to select stress tolerant plants among the putative transgenic plants methyl viologen-induced oxidative stress was imposed. At the end of stress

period cell viability (TTC assay) was assessed. Transgenic lines maintained higher cell viability compared to wild-type plants under MV stress (Figure 4A). During the stress superoxide radical production was estimated by XTT assay. The transgenic lines mtlD-3 and mtlD-5 showed lower levels of XTT reduction compared to wild-type indicating lower levels of superoxide radicals in the transgenic lines (Figure 4B). Seeds from mtlD-1, mtlD-3 and mtlD-5 lines were used for further experiments.

Assessing the drought stress tolerance of mtlD expressing transgenic finger millet plants

Transgenic seedlings were subjected to PEG, NaCl and menadione stress and their growth were measured. Transgenic seedlings from all three lines (mtlD-1 T_1, mtlD-3 T_1 and mtlD-5

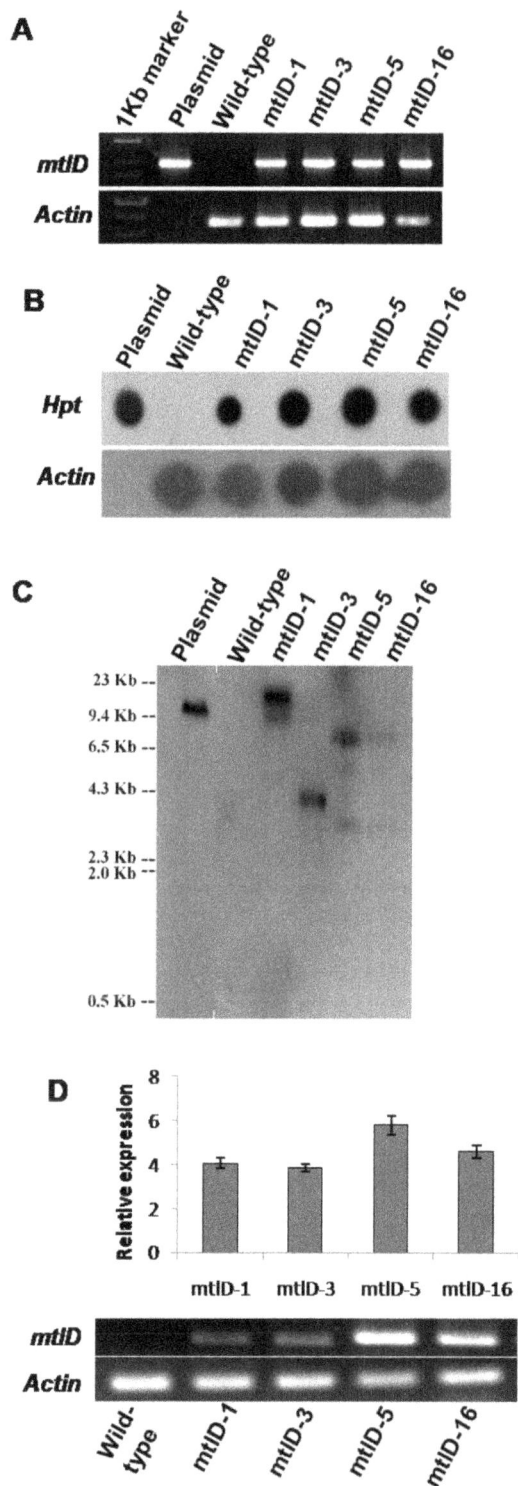

Figure 3. Molecular characterization of finger millet transgenic plants expressing *mtID* gene. Putative finger millet transgenic lines (T_0 generation) transformed with pCAMBIA 1380-mtID construct were analyzed for the presence of *mtID* gene by PCR (A) and *HptII* gene by dot blot analysis (B). Also, Southern blot was performed to assess the copy number of *mtID* gene in these transgenic plants (C). Transcript expression of transgene was analyzed in transgenic plants by semi quantitative RT-PCR. Quantity one (Biorad) software was used to measure the band intensity. The corresponding increase in the relative density of bands over wild-type is presented in the histogram (D).

T_1) showed moderate level of stress tolerance under PEG-induced osmotic stress (Figure S4A). However, only seedlings from mtID-1 T_1 showed increased salinity tolerance compared to wild-type (Figure S4B). Also, only the mtID-5 T_1 seedlings showed increased growth under oxidative stress compared to wild-type (Figure S4C).

Further, transgenic plants (T_1 generation) were grown for 15 days in pots and used for testing their drought tolerance. Initially the plants were subjected to gradual acclimation stress and later maintained at severe stress levels for seven days. Phenotypic observations were made from the stressed plants (mtID-1 T_1) and wild-type. At 50% FC stress for seven days, wild-type and the transgenic plants did not show visible wilting phenotype. At much higher stress levels (20% FC) the transgenic plants showed improved tolerance to drought stress than wild-type plants (Figure 5A). Stressed plants reduced relative water content compared to non-stressed plants (Figure S4D). Transgenic plants (mtID-1, mtID-3 T_1 and mtID-5 T_1) exposed to 50% FC stress maintained higher osmotic adjustment and ~10% less chlorophyll degradation in all the three transgenic lines than wild-type plants (Figure 5B and 5C).

Assessing the osmotic, salinity and oxidative stress tolerance of finger millet *mtID* transgenic seedlings

The T_2 generation transgenic seedlings (1.5–2.0 cm length) germinated on hygromycin medium (Figure S5) were transferred to PEG or NaCl or menadione stress and recovery growth was assessed. Transgenic and wild-type seedlings exposed to PEG, NaCl and menadione showed growth reduction compared to their no-stress control. However, mtID-1-2 T_2 seedlings showed no significant improvement in stress tolerance under PEG stress (Figure 6A). Similarly, mtID-5-1 T_2 seedlings showed only slight improvement in growth under PEG-induced osmotic stress (Figure 6A). The transgenic seedlings exposed to 300 mM NaCl showed slight improvement in growth compared to wild-type. However, at 400 mM the growth differences of transgenic seedlings were not significantly different from wild-type (Figure 6B). Stress tolerance of all the three lines was equal to wild-type under menadione-induced oxidative stress (Figure 6C). In order to ensure that these transgenic seedlings indeed accumulated mannitol, seeds obtained from transgenic plants (T_2 generation) were used for mannitol quantification. Results showed that mannitol content was high in all the tested seed lots (Figure 7). Similarly the mannitol content was higher in germinated seedlings of transgenics (Figure S6). This calorimetric assay also detected ~30 mg of mannitol per gram of wild-type seeds. This indicates finger millet may naturally accumulate mannitol in seeds. However, at this time, we do not have experimental evidence for the pathway involved in mannitol synthesis in finger millet. Also, BLAST search for identifying homologs of mannitol biosynthetic genes of celery in foxtail millet, phylogenetically close plant species to finger millet, did not reveal any hits. Nevertheless, our results indicated that mannitol accumulation has marginally improved stress tolerance in finger millet.

In order to understand the reason for only marginal improvement in oxidative stress tolerance levels in transgenic finger millet, transgenic tobacco plants expressing mtID was used and compared their stress tolerance increase with finger millet plants (mtID-5 T_1). Wild-type tobacco plants are more susceptible to stress compared to wild-type finger millet [76]. The stress tolerance of tobacco and finger millet transgenic plants expressing mtID was analyzed by subjecting the leaf segments to methyl viologen coupled with high light-induced oxidative stress. The transgenic tobacco plants showed ~10% increase in chlorophyll

Table 1. Segregation analysis for mtlD finger millet transgenic seedlings.

	Number of seedlings incubated	Number of seedlings survived*	Chi square value**
mtlD-1	71	53	(3:1) 0.731429
mtlD-3	77	51	(3:1) 1.958748
mtlD-5	68	50	(3:1) 0.732168
mtlD-16	71	50	(3:1) 1.047471

*one day old transgenic (T_1 generation) seedlings were incubated on 60 mg/L of hygromycin. 5 days after incubation they were scored for hygromycin resistance based on their survival.
**The chi-square test was performed using MS Excel POPTOOLS.

maintenance under stress compared to its wild-type (Figure S7). However, the finger millet transgenic plants showed only ~4% increase in chlorophyll maintenance compared to wild-type plants (Figure S7).

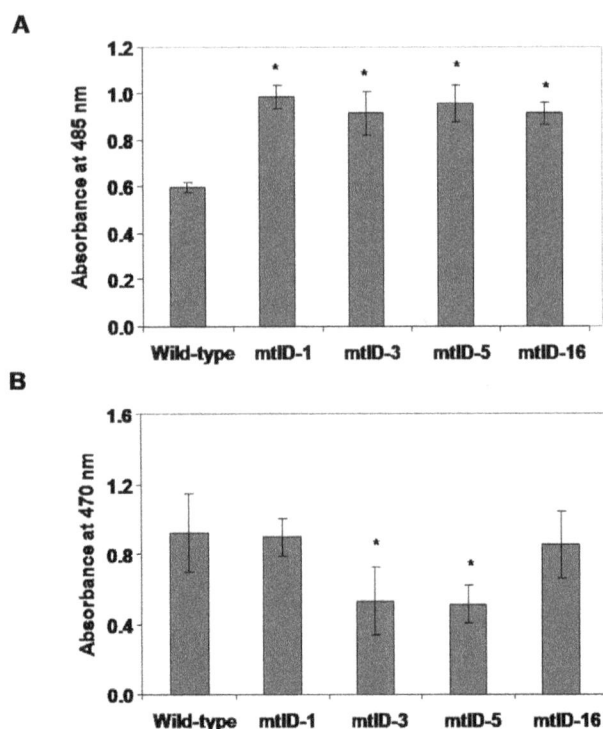

Figure 4. Cell viability and superoxide radical accumulation in the finger millet transgenic plants expressing *mtlD* gene under methyl viologen coupled with high light-induced oxidative stress. Excised leaf segments from T_1 generation finger millet transgenic plants were exposed to oxidative stress as described in material and methods section. At the end of stress, a sub set of leaf segments were incubated in TTC solution and cell viability was calorimetrically estimated by measuring the extent of reduction of TTC into formazan (A). Another sub set of leaf segments were used for superoxide radical estimation during stress. XTT solution was added to the incubation medium and XTT reduction during stress was measured by calorimetric method (B). Each bar represents the mean of standard error values (n = 10). Experiments were repeated twice. Result from one experiment is presented here and results for another independent experiment were similar. Asterisks indicate values are statistically significant (student's t test; $p < 0.05$) versus corresponding wild-type.

Discussion

Genetic engineering of finger millet for improved abiotic stress tolerance is stymied because of the lack of an efficient method for genetic transformation. Here we discuss development of plant regeneration from scutellum derived embryogenic callus and also a reproducible protocol for genetic transformation by *Agrobacterium*-mediated method.

Callus induction and plant regeneration in finger millet has been reported previously using shoot apex [77–80]. In this study scutellum derived callus production was standardized. This protocol has two major advantages. First, scutellum derived callus possess higher regeneration efficiency compared to callus induction from apical meristem. Second, since this protocol uses seeds as explants, limitation due to explant generation during large scale studies can be overcome. For example, seed can be directly used for callus induction, whereas, for obtaining shoot apex explant at least few additional weeks are needed for growing sterile seedlings from seeds [17,32]. Despite these studies showing callus induction and regeneration, only a few studies have demonstrated genetic engineering of finger millet for agronomically important traits. Recent preliminary studies in our lab that were aimed to develop transgenic finger millet plants for improved grain zinc content [81]and salinity stress tolerance [82] using the method developed in this study demonstrated reproducibility of the method described in this study.

Particle gun-mediated genetic transformation has been one of the methods used for transforming finger millet [17,32]. However, this method is expensive and usually results in multiple insertions of the transgene in the genome. Multiple insertions negatively impact both laboratory research and commercial release of transgenic plants [83]. Contrary to particle gun-mediated method, *Agrobacterium*-mediated transformation is simple, permits large scale experiments with less cost and has higher reproducibility. Hence, protocol described in this study and two other recent studies [31,84] used *Agrobacterium*-mediated transformation. These transformation protocols are expected to widen the scope for genetic engineering in finger millet in future (Figure S8).

By using the *Agrobacterium*-mediated method finger millet plants expressing *mtlD* gene were developed. The transgenic plants exhibited tolerance under drought, salinity and oxidative stress. The transgenic plants marginally performed better compared to the corresponding wild-type plants under both drought and salinity stress. For example, *mtlD* gene expressing transgenic plants showed enhanced osmotic adjustment under drought stress and also transgenic seedlings showed better growth under osmotic stress and salinity. It is possible that mannitol accumulated in finger millet transgenic plants acted as osmoprotectant and protected cells from free radicals in addition to inducing several other stress tolerance mechanisms as inferred in previous studies

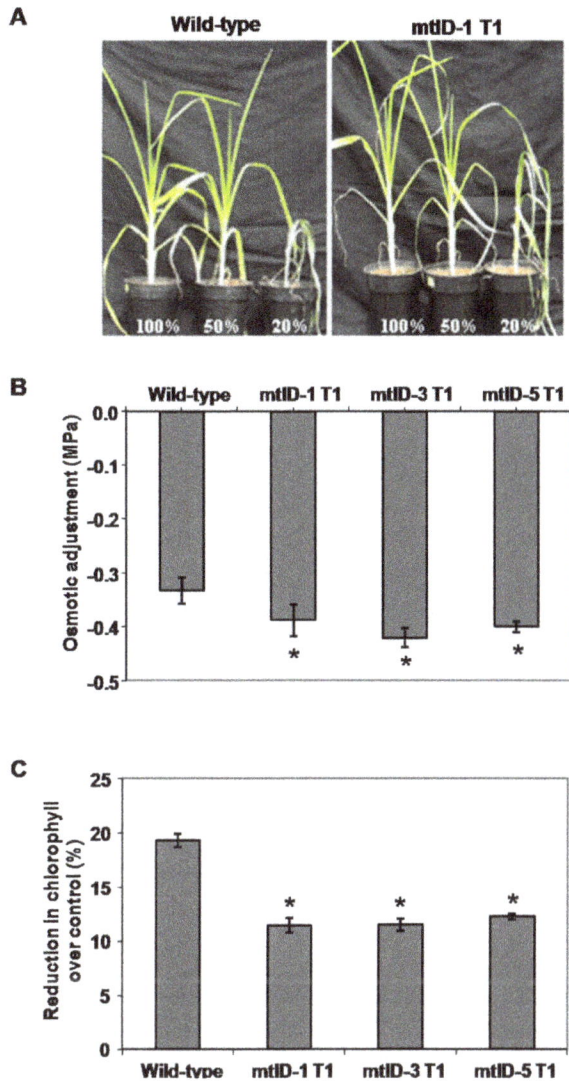

Figure 5. Effect of drought stress on the growth, osmotic adjustment and chlorophyll content of *mtID* transgenic finger millet plants. Drought stress was imposed to 15 day old plants (T$_1$ generation) by gradually withholding irrigation in such a way that the soil moisture reaches to 50% FC in one week. Later, plants were exposed to severe stress (50% or 20% FC) for 7 days. Photographs (line mtID-1 T$_1$) were taken at the end of the stress (A). Stress tolerance of the transgenic plants exposed to 50% FC was analyzed by measuring osmotic adjustment (B) and percent reduction in chlorophyll over respective controls (C). Each bar represents the mean of standard error values (n = 5). Asterisks indicate values are statistically significant (student's t test; $p < 0.05$) versus corresponding wild-type. Values mentioned in the picture represent respective filed capacity.

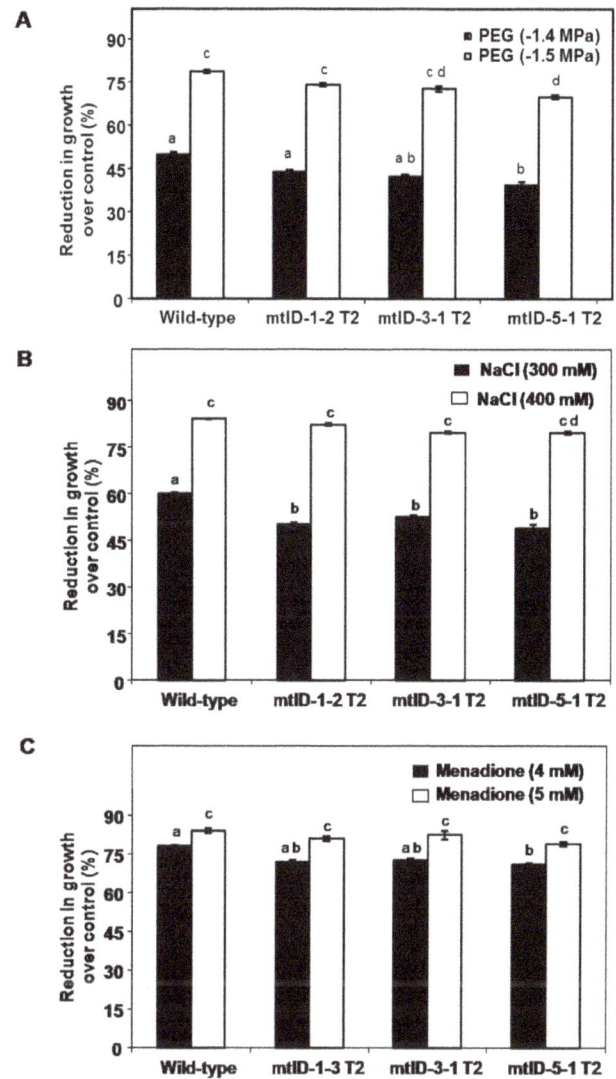

Figure 6. Performance of transgenic finger millet seedlings expressing *mtID* gene under osmotic, salinity and oxidative stress. Finger millet transgenic (T$_2$ generation) and wild-type seedlings (1.5 cm length) were initially acclimated with lower concentration of corresponding stress (−0.2 MPa PEG, 50 mM NaCl, and 1 mM menadione) for 8 h and further subjected to indicated concentrations of respective severe stress levels for 48 h. Seedlings were allowed to recover for two days and recovery growth after osmotic stress (A), salinity stress (B) and menadione induced oxidative stress (C) were measured and percent reduction in growth over corresponding control was calculated. Each bar represents the mean of standard error values (n = 20). Experiments were repeated twice. Result from one experiment is presented here and results for another independent experiment were similar. Alphabets above bar indicates the statistical significance (ANNOVA). Same alphabets indicate no significant difference ($p < 0.05$).

[36,40,53,74] and hence leading to increase in their tolerance compared to wild-type. However, the fold improvement in finger millet (a C4 plant) stress tolerance, especially the oxidative stress tolerance, compared to wild-type was much lesser than the increase in tolerance of several C3 plants expressing *mtID* gene under several abiotic stresses (Table S4). For example, increase in seedling growth in eggplants expressing *mtID* gene was up to 80% over their wild-type under both PEG and NaCl stress (Table S4). However, finger millet transgenic seedlings showed only an average increase of ~10% over their wild-type. Consistently, transgenic sorghum (C4) plants expressing *mtID* gene showed only

up to 23% increase in biomass under salinity stress compared their wild-type plants [53], whereas, C3 plants like rice and wheat transgenic plants showed an average of 80% and 50% increase over their corresponding wild-type [54,74].

In order to confirm these observations that are based on previous literature information, we developed tobacco plants expressing *mtID* gene and analyzed their performance under oxidative stress. Interestingly the tobacco transgenic plants showed higher level (~13%) of chlorophyll maintenance over wild-type

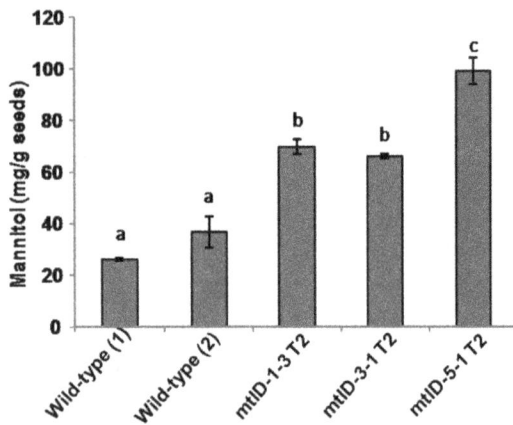

Figure 7. Mannitol accumulation in *mtlD* gene expressing finger millet. Mannitol content in the seeds obtained from T_2 generation plants were estimated for the transgenic lines. Alphabets above bar indicates the statistical significance (Duncan's multiple range test). Same alphabets indicate no significant difference ($p<0.05$). Experiments were repeated twice. Result from one experiment is presented here and results for another independent experiment were similar. WT (1) and WT (2) are two independent replicate plant seeds.

under oxidative stress (Figure S7). However, finger millet transgenic seedlings expressing *mtlD* gene showed only slight improvement ($<3\%$) in oxidative stress tolerance (Figure 6). Similarly, at plant level, increase in chlorophyll stability in *mtlD* finger millet lines over wild-type was only 7%. This indicated that *mtlD* gene expression in finger millet imparted only marginal level of tolerance over wild-type. Consistent with these results the ascorbate peroxidase gene, *EcApx1*, cloned from finger millet has been attributed to impart better oxidative stress tolerance in the wild-type finger millet [85]. Tobacco plants expressing a NAC class of transcription factor, *EcNAC1*, from finger millet showed better chlorophyll stability under oxidative and salinity stresses [63]. Also, finger millet [86] and maize have been shown to maintain higher membrane stability and also overall plant growth under oxidative stress compared to tomato and beans [87]. Taken together, the tolerance levels of finger millet mtlD expressing transgenic plants are only marginal because wild-type plants may have higher intrinsic basal tolerance [63]. Hence, we speculate that an additional scavenging system brought out by mannitol have not dramatically contributed for stress protection in transgenic plants over wild-type.

The mechanism for only marginal increase in stress tolerance of finger millet could be due to mannitol-mediated osmotic regulation and free radical scavenging [88]. Although mannitol accumulated in finger millet mtlD expressing transgenic plants have contributed to cellular protection through its osmolytic properties, high levels of endogenous accumulation of proline [49,89] in both wild-type and mtlD expressing transgenic plants countered the net contribution due to mannitol alone. Similarly, the germinating seeds also showed higher accumulation of mannitol (figure S6) indicating contribution of mannitol biosynthesis in reducing the free radical generation. In the transgenic plants also the scavenging ability have not been high because the wild-type plants already has efficient free radical scavenging ability [85]. It is also possible that wild-type finger millet plants better managed oxidative stress compared to tobacco because of their unique adaptation. For example, finger millet has better re-positioning of chloroplast when exposed light stress compared to tobacco [90]. Faster relocation of chloroplasts away from the high

light exposed regions protects photosynthetic machinery in C4 species, finger millet [90]. However, the exact mechanism of mtlD-mediated mannitol production in plants and the fate of mannitol produced, including its internal transport during stress [91], is not yet understood.

Conclusions

Regeneration and *Agrobacterium*-mediated transformation protocols in finger millet are demonstrated in this study. Finger millet genetic transformation described in previous study using shoot apex as explant [84] and this study are useful for this crop improvement. Salient features of this study include (1) use of commercially important finger millet (var. GPU28); (2) efficient explant type (mature seeds). Our results also reiterate the earlier notion that finger millet may have better adaptation to abiotic stress. Hence, other novel strategies should be developed for further improving stress tolerance in finger millet.

Supporting Information

Figure S1   Steps involved in development of finger millet transgenic plants expressing *mtlD* gene. The T_0 transgenic plants (var. GPU28) which were confirmed by PCR and Southern blot were grown in greenhouse and seeds were collected. For further experiments, out of 20 putative transgenic events, four T_0 transgenic events were used. These four events were designated as mtlD-1, mtlD-3, mtlD-5, mtlD-16 and taken forward. Seeds obtained from three (T_0) events (mtlD-1, mtlD-3 and mtlD-5) were used for further experiments. Seeds from each event were bulked and used for seedling experiment (named as mtlD-1 T_1, mtlD-3 T_1 and mtlD-5 T_1) or plant level experiment. Plants obtained from individual seeds representing each (T_1) event were assessed for presence of *mtlD* gene by PCR and then seeds were collected from these plants. These seeds were designated as mtlD-1-2 T_2, mtlD-3-1 T_2 and mtlD-5-1 T_2 and used for mannitol quantification. A sub-set of seedlings obtained from these seeds were used for stress experiments.

Figure S2   Germination of seeds obtained from *UidA* gene expressing finger millet transgenic plants on hygromycin. Wild-type (var. GPU28) finger millet seeds were germinated in petri dishes on filter paper with different concentration of hygromycin for 5 days at 30°C with 70% relative humidity in dark and germination percentage was recorded at the end of treatment period (A). Similarly, seeds obtained from the GUS positive transgenic finger millet plants (UidA-1; T1 generation) and wild-type (var. GPU28) were germinated on hygromycin (60 mg/L) and number of seedlings survived was recorded (B). Asterisks indicate values are statistically significant (student's t test; $p<0.05$) versus corresponding wild-type. Each bar represents the mean of standard error values (n = 10).

Figure S3   Growth of mtlD expressing finger millet transgenic (T_1 generation) and wild-type seedlings on hygromycin medium. Seeds from mtlD transgenic finger millet (T1; mtlD-1 T1 generation) and wild-type (WT, var. GPU28) were inoculated on the MS medium containing hygromycin (20 mg/L or 40 mg/L) for five days. Seedling growth on antibiotic medium was photographed at the end of treatment period.

Figure S4   Performance of mtlD expressing finger millet transgenic seedlings under osmotic, salinity and oxida-

tive stress. Finger millet transgenic (T_1 generation) and wild-type (var. GPU28) seedlings (1.5 cm length) were initially acclimated with lower concentration of corresponding stresses (−0.2 MPa PEG, 50 mM NaCl, and 1 mM menadione) for 8 h and then subjected to indicated concentrations of respective severe stress levels for 48 h. Seedlings were allowed to recover for two days and recovery growth of osmotic stress (A), salinity stress (B) and menadione induced oxidative stress (C) were measured and percent reduction in growth over corresponding control was calculated. The percent relative water content was measured from the plants exposed to 100%, 50% and 30% field capacity (D). Each bar represents the mean of standard error values (n = 20). Experiments were repeated twice. Alphabets above bar indicates the statistical significance (ANNOVA). Same alphabets indicate no significant difference ($p<0.05$).

Figure S5 Growth of mtID expressing finger millet transgenic (T_2 generation) and wild-type seedlings on hygromycin medium. Seeds from mtID expressing transgenic finger millet (mtID-1-2 T_2) and wild-type (WT, var. GPU28) were incubated on the MS medium containing hygromycin (20 mg/L or 40 mg/L) for five days. Seedling growth on antibiotic medium was photographed at the end of treatment period.

Figure S6 Mannitol accumulation in *mtID* gene expressing finger millet germinating seedlings. Mannitol content in the seedlings obtained from T_2 generation plants were estimated as described in material and methods. Alphabets above bar indicates the statistical significance (Duncan's multiple range test). Same alphabets indicate no significant difference ($p<0.05$).

Figure S7 Chlorophyll retention in tobacco and finger millet plants expressing *mtID* gene under methyl viologen coupled with high light-induced oxidative stress. Leaf segments were taken from transgenic tobacco, finger millet (mtID-5) and corresponding wild-type plants grown under non-stress condition. These leaf segments were exposed to high light (800 µmol m^{-2} s^{-1}) stress with methyl viologen as described in materials and methods section. At the end of stress period, total chlorophyll was measured and percent reduction in total chlorophyll over their corresponding non-stress control was calculated. Values are mean of three replications and the error bar represents standard error. Alphabets above bar indicates the statistical significance

(ANNOVA). Same alphabets indicate no significant difference ($p<0.05$).

Figure S8 Comparison of different features of genetic transformation protocols in finger millet reported in previous literature. In the literature, three research groups have demonstrated genetic transformation in finger millet. First, particle gun-mediated protocol was developed by Latha and co-workers. Second, *Agrobacterium*-mediated protocol was developed by Antony Ceasar and Ignacimuthu. Both protocol uses shoot apex as explants. Third, Sharma et al., and Jagga-Chugh et al., have demonstrated *UidA* gene expression. In this pictorial representation, different features of these protocols are compared with the protocols developed from this current study.

Table S1 Standardization of hormonal concentrations for efficient callus induction in finger millet (var. Indaf 9).

Table S2 Standardization of hormonal concentrations for efficient shoot induction from the callus in finger millet (var. Indaf 9).

Table S3 Efficiency of *Agrobacterium*-mediated finger millet transformation.

Table S4 Comparison of mtID gene expressing plant species reported in the literature with their corresponding wild-type plants for their abiotic stress tolerance.

Acknowledgments

Binary vector constructs were obtained from Dr. M. V. Rajam, Delhi University (South campus), New Delhi, India. Authors thank Dr. Ramegowda Venkategowda for critical reading of the manuscript. Authors also thank Mr. Narayansamy and Mr. Ramanji for plant care during the research period.

Author Contributions

Conceived and designed the experiments: RH MS-K MU. Performed the experiments: RH RSV SS MS-K. Analyzed the data: RH RSV CPR MS-K. Contributed reagents/materials/analysis tools: CPR MS-K MU. Wrote the paper: MS-K.

References

1. Gowda BTS, Halaswamy BH, Seetharam A, Virk DS, Witcombe JR (2000) Participatory approach in varietal improvement: A case study in finger millet in India. Current Science 79: 366.
2. Adhikari RK (2012) Economics of finger millet (*Elecusine coracana* G.) production and marketing in peri urban area of Pokhara valley of Nepal. Journal of Development and Agricultural Economics 4: 151–157.
3. Hamza NB (2009) Progress and prospects of millet research in Sudan;Tadele Z, editor. 260–260 p.
4. Vijayalakshmi D, Gowda KN, Jamuna KV, Ray BRM, Sajjan JT (2012) Empowerment of self help group women through value addition of finger millet. Journal of Dairying, Foods and Home Sciences 31: 223–226.
5. Mbithi-Mwikya S, Van Camp J, Mamiro PRS, Ooghe W, Kolsteren P, et al. (2002) Evaluation of the nutritional characteristics of a finger millet based complementary food. Journal of Agricultural and Food Chemistry 50: 3030–3036.
6. Chandrashekar A (2010) Finger Millet: *Eleusine coracana*. In: Steve LT, editor. Advances in food and nutrition research: Academic Press. pp. 215–262.
7. Oghbaei M, Prakash J (2012) Bioaccessible nutrients and bioactive components from fortified products prepared using finger millet (*Eleusine coracana*). Journal of the Science of Food and Agriculture 92: 2281–2290.
8. Radchuk V, Radchuk R, Pirko Y, Vankova R, Gaudinova A, et al. (2012) A somaclonal line SE7 of finger millet (*Eleusine coracana*) exhibits modified cytokinin homeostasis and increased grain yield. Journal of Experimental Botany 63: 5497–5506.
9. Nandini B, Ravishankar CR, Mahesha B, Shailaja H, Murthy KNK (2010) Study of correlation and path analysis in F 2 population of finger millet. International Journal of Plant Sciences 5: 602–605.
10. Shet RM, Gireesh C, Jagadeesha N, Lokesh GY, Jayarame G (2009) Genetic variability in segregating generation of interspecific hybrids of finger millet (*Eleusine coracana* (L.) Gaertn.). Environment and Ecology 27: 1013–1016.
11. Nagaraja A, Jayarame G, Krishnappa M, Gowda KTK (2008) GPU 28: a finger millet variety with durable blast resistance. Journal of Mycopathological Research 46: 109–111.
12. Hebbar SS, Nanjappa HV, Ramachandrappa BK (1992) Performance of finger millet after harvest of early kharif crops under rainfed conditions. Crop Research 5: 92–97.
13. Sashidhar VR, Gurumurthy BR, Prasad TG, Udayakumar M, Seetharam A, et al. (1986) Genotypic variation in carbon exchange rate functional leaf area and productivity in finger millet (*Eleusine coracana*) an approach to identify desirable plant types for higher water use efficiency under rainfed conditions. Field Crops Research 13: 133–146.

14. Reddy VC, Patil VS (1983) Influence of soil moisture stress on yield and its components of finger millet. Mysore Journal of Agricultural Sciences 17: 325–328.

15. Shivkumar BG, Yadahalli YH (1996) Performance of fingermillet genotypes under late sown dryland conditions as influenced by intercrops and intercropping systems. Indian Journal of Agricultural Research 30: 173–178.

16. Bhatt D, Negi M, Sharma P, Saxena SC, Dobriyal AK, et al. (2011) Responses to drought induced oxidative stress in five finger millet varieties differing in their geographical distribution. Physiology and Molecular Biology of Plants 17: 347–353.

17. Mahalakshmi S, Christopher GSB, Reddy TP, Rao KV, Reddy VD (2006) Isolation of a cDNA clone (PcSrp) encoding serine-rich-protein from Porteresia coarctata T. and its expression in yeast and finger millet (Eleusine coracana L.) affording salt tolerance. Planta 224: 347–359.

18. Shailaja HB, Thirumeni S (2007) Evaluation of salt-tolerance in finger millet (Eleusine coracana) genotypes at seedling stage. The Indian Journal of Agricultural Sciences 77: 672–674

19. Kalarani MK, Sivakumar R, Mallika V, Sujatha KB (2001) Effect of seed soaking treatments in late sown finger millet under rainfed condition. Madras Agricultural Journal 88: 509–511.

20. Kalarani MK, Thangaraj M, Sivakumar R, Vanangamudi M, Srinivasan PS (2001) Ameliorants for late sown finger millet under rainfed condition. Indian Journal of Plant Physiology 6: 435–437.

21. Yadav DS, Goyal AK, Vats BK (1999) Effect of potassium in Eleusine coracana (L.) Gaertn. under moisture stress conditions. Journal of Potassium Research 15: 131–134.

22. Hare PD, Cress WA, Staden JV (1998) Dissecting the roles of osmolyte accumulation during stress. Plant, Cell & Environment 21: 535–553.

23. Kathuria H, Giri J, Nataraja KN, Murata N, Udayakumar M, et al. (2009) Glycinebetaine-induced water-stress tolerance in codA-expressing transgenic indica rice is associated with up-regulation of several stress responsive genes. Plant Biotechnology Journal 7: 512–526.

24. Serraj R, Sinclair TR (2002) Osmolyte accumulation: can it really help increase crop yield under drought conditions? Plant, Cell & Environment 25: 333–341.

25. Hayat S, Hayat Q, Alyemeni MN, Wani AS, Pichtel J, et al. (2012) Role of proline under changing environments: A review. Plant Signaling & Behavior 7: 1456–1466.

26. Rai S, Gaur AK, Guru SK, Agrawal S, Arora S (2009) In vitro evaluation of relative tolerance of Eleucine coracana genotypes against water deficit stress. Indian Journal of Plant Physiology 14: 82–87.

27. Uma S, Prasad TG, Udayakumar M (1995) Genetic variability in recovery growth and synthesis of stress proteins in response to polyethylene glycol and salt stress in finger millet. Annals of Botany 76: 43–49.

28. Dinesh-Kumar SP, Sashidhar VR, Prasad TG, Udayakumar M, Seetharam A (1987) Solute accumulation, solute potential, germinability and seedling vigour of seeds of finger millet (Eleusine coracana Gaertn.) raised under rain-fed conditions and under irrigation. Plant, Cell & Environment 10: 661–665.

29. Sashidhar VR, Gurumurthy BR, Prasad TG, Udayakumar M, Seetharam A, et al. (1986) Genotypic variation in carbon exchange rate functional leaf area and productivity in finger millet Eleusine coracana an approach to identify desirable plant types for higher water use efficiency under rainfed conditions. Field Crops Research 13: 133–146.

30. Jayaprakash TL, Sashidhar VR, Prasad TG, Udaykumar M (1992) Photoperiodic response of advanced breeding lines and cultivars of finger millet under irrigated and rainfed conditions. Plant Physiology & Biochemistry 19: 49–54.

31. Antony Ceasar S, Ignacimuthu S (2011) Agrobacterium-mediated transformation of finger millet (Eleusine coracana (L.) Gaertn.) using shoot apex explants. Plant Cell Reports 30: 1759–1770.

32. Latha AM, Rao KV, Reddy VD (2005) Production of transgenic plants resistant to leaf blast disease in finger millet (Eleusine coracana (L.) Gaertn.). Plant Science 169: 657–667.

33. Gupta P, Raghuvanshi S, K Tyagi A (2001) Assessment of the efficiency of various gene promoters via biolistics in leaf and regenerating seed callus of millets, Eleusine coracana and Echinochloa crusgalli. Plant Biotechnology 18: 275–282.

34. Jagga-Chugh S, Kachhwaha S, Sharma M, Kothari-Chajer A, Kothari SL (2012) Optimization of factors influencing microprojectile bombardment-mediated genetic transformation of seed-derived callus and regeneration of transgenic plants in Eleusine coracana (L.) Gaertn. Plant Cell Tissue and Organ Culture 109: 401–410.

35. Sharma M, Kothari-Chajer A, Jagga-Chugh S, Kothari SL (2011) Factors influencing Agrobacterium tumefaciens-mediated genetic transformation of Eleusine coracana (L.) Gaertn. Plant Cell Tissue and Organ Culture 105: 93–104.

36. Shen B, Jensen RG, Bohnert HJ (1997) Mannitol protects against oxidation by hydroxyl radicals. Plant Physiology 115: 527–532.

37. Smirnoff N, Cumbes QJ (1989) Hydroxyl radical scavenging activity of compatible solutes. Phytochemistry 28: 1057–1060.

38. Seckin B, Sekmen AH, Turkan I (2009) An enhancing effect of exogenous mannitol on the antioxidant enzyme activities in roots of wheat under salt stress. Journal of Plant Growth Regulation 28: 12–20.

39. Zhifang G, Loescher WH (2003) Expression of a celery mannose 6-phosphate reductase in Arabidopsis thaliana enhances salt tolerance and induces biosynthesis of both mannitol and a glucosyl-mannitol dimer. Plant, Cell & Environment 26: 275–283.

40. Chan Z, Grumet R, Loescher W (2011) Global gene expression analysis of transgenic, mannitol-producing, and salt-tolerant Arabidopsis thaliana indicates widespread changes in abiotic and biotic stress-related genes. Journal of Experimental Botany 62: 4787–4803.

41. Tarczynski MC, Jensen RG, Bohnert HJ (1992) Expression of a bacterial mtlD gene in transgenic tobacco leads to production and accumulation of mannitol. Proceedings of the National Academy of Sciences 89: 2600–2604.

42. Thomas JC, Sepahi M, Arendall B, Bohnert HJ (1995) Enhancement of seed germination in high salinity by engineering mannitol expression in Arabidopsis thaliana. Plant, Cell & Environment 18: 801–806.

43. Chiang YJ, Stushnoff C, McSay AE (2005) Overexpression of mannitol-1-phosphate dehydrogenase increases mannitol accumulation and adds protection against chilling injury in petunia. Journal of the American Society for Horticultural Science 130: 605–610.

44. Hu L, Lu H, Liu QL, Chen XM, Jiang XN (2005) Overexpression of mtlD gene in transgenic Populus tomentosa improves salt tolerance through accumulation of mannitol. Tree Physiology 25: 1273–1281.

45. Khare N, Goyary D, Singh NK, Shah P, Rathore M, et al. (2010) Transgenic tomato cv. Pusa Uphar expressing a bacterial mannitol-1-phosphate dehydrogenase gene confers abiotic stress tolerance. Plant Cell Tissue and Organ Culture 103: 267–277.

46. Rahnama H, Vakilian H, Fahimi H, Ghareyazie B (2011) Enhanced salt stress tolerance in transgenic potato plants (Solanum tuberosum L.) expressing a bacterial mtlD gene. Acta Physiologiae Plantarum 33: 1521–1532.

47. Prabhavathi V, Yadav JS, Kumar PA, Rajam MV (2002) Abiotic stress tolerance in transgenic eggplant (Solanum melongena L.) by introduction of bacterial mannitol phosphodehydrogenase gene. Molecular Breeding 9: 137–147.

48. Kandpal RP, Rao NA (1985) Changes in the levels of polyamines in ragi (Eleusine coracana) seedlings during water stress. Biochemistry International 11: 365–370.

49. Prasad SV, Rao GG, Rao GR (1980) Studies on salt tolerance of ragi (Eleusine coracana Gaertn): Germination and free proline accumulation. Proceedings of the Indian Academy of Sciences 89: 481–484.

50. Sheveleva EV, Jensen RG, Bohnert HJ (2000) Disturbance in the allocation of carbohydrates to regenerative organs in transgenic Nicotiana tabacum L. Journal of Experimental Botany 51: 115–122.

51. Liu J, Huang S, Peng X, Liu W, Wang H (1995) Studies on high salt tolerance of transgenic tobacco. Chinese Journal of Biotechnology 11: 275–280.

52. Prabhavathi V, Rajam MV (2007) Mannitol-accumulating transgenic eggplants exhibit enhanced resistance to fungal wilts. Plant Science 173: 50–54.

53. Maheswari M, Varalaxmi Y, Vijayalakshmi A, Yadav SK, Sharmila P, et al. (2010) Metabolic engineering using mtlD gene enhances tolerance to water deficit and salinity in sorghum. Biologia Plantarum 54: 647–652.

54. Pujni D, Chaudhary A, Rajam MV (2007) Increased tolerance to salinity and drought in transgenic indica rice by mannitol accumulation. Journal of Plant Biochemistry and Biotechnology 16: 01–07.

55. Hema R, Senthil-Kumar M, Shivakumar S, Chandrasekhara Reddy P, Udayakumar M (2007) Chlamydomonas reinhardtii, a model system for functional validation of abiotic stress responsive genes. Planta 226: 655–670.

56. Murashige T, Skoog F (1962) A revised medium for rapid growth and bioassays with tobacco tissue cultures. Physiologia Plantarum 13: 473–497.

57. Senthil-Kumar M, Srikanthbabu V, Mohan Raju B, Ganeshkumar, Shivaprakash N, et al. (2003) Screening of inbred lines to develop a thermotolerant sunflower hybrid using the temperature induction response (TIR) technique: a novel approach by exploiting residual variability. Journal of Experimental Botany 54: 2569–2578.

58. Aarati P, Krishnaprasad BT, Ganeshkumar, Savitha M, Gopalakrishna R, et al. (2003) Expression of an ABA responsive 21 kDa protein in finger millet (Eleusine coracana Gaertn.) under stress and its relevance in stress tolerance. Plant Science 164: 25–34.

59. Shi MM, Kugelman A, Iwamoto T, Tian L, Forman HJ (1994) Quinone-induced oxidative stress elevates glutathione and induces gamma-glutamylcysteine synthetase activity in rat lung epithelial L2 cells. Journal of Biological Chemistry 269: 26512–26517.

60. Borges AA, Cools HJ, Lucas JA (2003) Menadione sodium bisulphite: a novel plant defence activator which enhances local and systemic resistance to infection by Leptosphaeria maculans in oilseed rape. Plant Pathology 52: 429–436.

61. Karaba A, Dixit S, Greco R, Aharoni A, Trijatmiko KR, et al. (2007) Improvement of water use efficiency in rice by expression of HARDY, an Arabidopsis drought and salt tolerance gene. Proceedings of the National Academy of Sciences 104: 15270–15275.

62. Dodge AD, Harris N (1970) The mode of action of paraquat and diquat. Biochemical Journal 118: 43–44.

63. Ramegowda V, Senthil-Kumar M, Nataraja KN, Reddy MK, Mysore KS, et al. (2012) Expression of a finger millet transcription factor, EcNAC1, in tobacco confers abiotic stress-tolerance. PloS one 7: e40397–e40397.

64. Jefferson RA, Kavanagh TA, Bevan MW (1987) GUS fusions beta glucuronidase as a sensitive and versatile gene fusion marker in higher plants. EMBO Journal 6: 3901–3908.

65. Hiscox JD, Israelstam GF (1979) A method for the extraction of chlorophyll from leaf tissue without maceration. Canadian Journal of Botany 57: 1332–1334.

66. Abebe T, Guenzi AC, Martin B, Cushman JC (2003) Tolerance of mannitol-accumulating transgenic wheat to water stress and salinity. Plant Physiology 131: 1748–1755.

67. Corcoran AC, Page IH (1947) A method for the determination of mannitol in plasma and urine. Journal of Biological Chemistry 170: 165–171.
68. Towill LE, Mazur P (1975) Studies on the reduction of 2,3,5-triphenyl tetrazolium chloride as a viability assay for plant tissue cultures. Canadian Journal of Botany 53: 1097–1102.
69. Schopfer P, Plachy C, Frahry G (2001) Release of reactive oxygen intermediates (superoxide radicals, hydrogen peroxide, and hydroxyl radicals) and peroxidase in germinating radish seeds controlled by light, gibberellin, and abscisic acid. Plant Physiology 125: 1591–1602.
70. Gelvin S (2006) Agrobacterium virulence gene induction. In: Wang K, editor. Agrobacterium Protocols: Humana Press. pp. 77–85.
71. Hema R (2001) Agrobacterium mediated gene transfer for osmolyte production in finger millet (*Eleusine coracana* (L.) Gaertn) [MSc]. Bangalore: University of Agricultural Sciences. 100 p.
72. Shivakumar S (2006) Overexpression of P5CS in finger millet (*Eluesine coracana* L. Gaertn) to improve the seed germination and seedling establishment under water deficit stress [Ph.D]. Bangalore: University of Agricultural Sciences.
73. James VA, Avart C, Worland B, Snape JW, Vain P (2002) The relationship between homozygous and hemizygous transgene expression levels over generations in populations of transgenic rice plants. Theoretical and Applied Genetics 104: 553–561.
74. Abebe T, Guenzi AC, Martin B, Cushman JC (2003) Tolerance of mannitol accumulating transgenic wheat to water stress and salinity. Plant Physiology 131: 1748–1755.
75. Macaluso L, Lo Bianco R, Rieger M (2007) Mannitol-producing tobacco exposed to varying levels of water, light, temperature and paraquat. Journal of Horticultural Science & Biotechnology 82: 979–985.
76. Hema R (2006) Relevance of mannitol in dehydration stress tolerance: development and analysis of the transgenics expressing mtlD in model systems and finger millet (*Eleusine coracana* (l.) gaertn) [PhD]. Bangalore: University of Agricultural Sciences. 197 p.
77. Eapen S, George L (1989) High frequency plant regeneration through somatic embryogenesis in finger millet (*Eleusine coracana* Gaertn). Plant Science 61: 127–130.
78. Prasanna S, Kothari SL, Chandra N (1990) High frequency embryoid and plantlet formation from tissue cultures of the finger millet *Eleusine coracana* (L.) Gaertn. Plant Cell Reports 9: 93–96.
79. Kumar S AK, Kothari SL. (2001) *In vitro* induction and enlargement of apical domes and formation of multiple shoots in millet, *Eleusine coracana* (L.) Gaertn. and crowfoot grass, E. indica (L.) Gaertn. Current Science 81: 1482–1485.
80. Anjaneyulu E, Attitalla IH, Hemalatha S, Raj SB, Balaji M (2011) An efficient protocol for callus induction and plant regeneration in finger millet (*Eleusine coracana* L.). World Applied Sciences Journal 12: 919–923.
81. Yamunarani BR (2009) Molecular characterization of *Eleusine coracana* (L.) Gaertn. genotypes for variability in zinc content and development of transgenics with high grain zinc. University of Agricultural Sciences, Bangalore, India.
82. Vasantha KM (2013) Targeting genes for ion homeostasis and salt tolerance in fingermillet (*Eleusine coracana* (L.) gaertn.): over-expressing genes for salt compartmentation and proton gradient generation and their significance in salt tolerance. Bangalore, India: University of Agricultural Sciences.
83. Oltmanns H, Frame B, Lee L-Y, Johnson S, Li B, et al. (2010) Generation of backbone-free, low transgene copy plants by launching T-DNA from the Agrobacterium chromosome. Plant Physiology 152: 1158–1166.
84. Ignacimuthu S, Ceasar SA (2012) Development of transgenic finger millet (*Eleusine coracana* (L.) Gaertn.) resistant to leaf blast disease. Journal of Biosciences 37: 135–147.
85. Bhatt D, Saxena S, Jain S, Dobriyal A, Majee M, et al. (2013) Cloning, expression and functional validation of drought inducible ascorbate peroxidase (*Ec-apx1*) from *Eleusine coracana*. Molecular Biology Reports 40: 1155–1165.
86. Govind G, Vokkaliga ThammeGowda H, Jayaker Kalaiarasi P, Iyer D, Senthil-Kumar M, et al. (2009) Identification and functional validation of a unique set of drought induced genes preferentially expressed in response to gradual water stress in peanut. Molecular Genetics and Genomics 281: 591–605.
87. Gopalakrishna R, Kumar G, KrishnaPrasad BT, Mathew MK, Udayakumar M (2001) A stress-responsive gene from groundnut, Gdi-15, is homologous to flavonol 3-O-glucosyltransferase involved in anthocyanin biosynthesis. Biochemical and Biophysical Research Communications 284: 574–579.
88. Rathinasabapathi B (2000) Metabolic engineering for stress tolerance: installing osmoprotectant synthesis pathways. Annals of Botany 86: 709–716.
89. Kandpal RP, Rao NA (1985) Alterations in the biosynthesis of proteins and nucleic-acids in finger millet *Eleusine coracana* seedlings during water stress and the effect of proline on protein biosynthesis. Plant Science 40: 73–80.
90. Yamada M, Kawasaki M, Sugiyama T, Miyake H, Taniguchi M (2009) Differential positioning of C4 mesophyll and bundle sheath chloroplasts: aggregative movement of C4 mesophyll chloroplasts in response to environmental stresses. Plant and Cell Physiology 50: 1736–1749.
91. Conde A, Silva P, Agasse A, Conde C, Gerós H (2011) Mannitol transport and mannitol dehydrogenase activities are coordinated in *Olea europaea* under salt and osmotic stresses. Plant and Cell Physiology 52: 1766–1775.

2

Efficient CRISPR/Cas9-Mediated Gene Editing in *Arabidopsis thaliana* and Inheritance of Modified Genes in the T2 and T3 Generations

WenZhi Jiang[1], Bing Yang[2], Donald P. Weeks[1]*

1 Department of Biochemistry, University of Nebraska, Lincoln, Nebraska, United States of America, **2** Department of Genetics, Development and Cell Biology, Iowa State University, Ames, Iowa, United States of America

Abstract

The newly developed CRISPR/Cas9 system for targeted gene knockout or editing has recently been shown to function in plants in both transient expression systems as well as in primary T1 transgenic plants. However, stable transmission of genes modified by the Cas9/single guide RNA (sgRNA) system to the T2 generation and beyond has not been demonstrated. Here we provide extensive data demonstrating the efficiency of Cas9/sgRNA in causing modification of a chromosomally integrated target reporter gene during early development of transgenic Arabidopsis plants and inheritance of the modified gene in T2 and T3 progeny. Efficient conversion of a nonfunctional, out-of-frame *GFP* gene to a functional *GFP* gene was confirmed in T1 plants by the observation of green fluorescent signals in leaf tissues as well as the presence of mutagenized DNA sequences at the sgRNA target site within the *GFP* gene. All GFP-positive T1 transgenic plants and nearly all GFP-negative plants examined contained mutagenized *GFP* genes. Analyses of 42 individual T2 generation plants derived from 6 different T1 progenitor plants showed that 50% of T2 plants inherited a single T-DNA insert. The efficiency of the Cas9/sgRNA system and stable inheritance of edited genes point to the promise of this system for facile editing of plant genes.

Editor: Frederik Börnke, Leibniz-Institute for Vegetable and Ornamental Plants, Germany

Funding: Funding for this research was provided, in part, by the National Science Foundation (MCB-0952533 and EPSCoR-1004094 to D.P.W.) and the Department of Energy (DOE DE-EE0001052 and DOE CAB-COMM DOE DE-EE0003373 to D.P.W.) and Iowa State University Plant Science Institute Innovation Grant (to B.Y.). The funders had no role in study design, data collection and analysis, decision to publish, or preparation of the manuscript.

Competing Interests: The authors have declared that no competing interests exist.

* E-mail: dweeks1@unl.edu

Introduction

In recent years, zinc finger nuclease (ZFN) technology [1] and TAL Effector Nuclease (TALEN) technology [2–5] have become powerful gene editing tools for targeted gene modification in human cells, fruit flies, zebrafish, nematodes and plants. For both ZFNs and TALENS, engineered sequence-specific DNA binding domains are fused with a subunit of the nonspecific DNA nuclease, Fok1. As a result, pairs of ZFNs and TALENS targeting adjacent DNA target sites generate double strand breaks (DSBs) at or near the target site. Repair of the DSB by error-prone nonhomologous end joining (NHEJ) or homologous recombination (HR) often lead to gene sequence modification, including gene knockout. Within the past year, another highly promising system for gene editing, the clustered regulatory interspersed short palindromic repeat (CRISPR)/CRISPR-associated protein (Cas) system, has evolved from studies of bacterial defense systems that provide protection against invading viruses or plasmid DNAs [6–8]. CRISPR loci are variable short spacers separated by short repeats and are transcribed and processed into short non-coding RNAs. These short RNAs can form a functional complex with Cas proteins and guide the complex to cleave complementary foreign DNAs. The type II CRISPR/Cas system derived from *Streptococcus pyogenes* is the most widely used for gene editing [6–9]. It has the marked advantage of possessing a required PAM recognition sequence of only two nucleotides (GG). Development of single guide RNAs

(sgRNAs) that are fusions of essential portions of tracrRNA with the "guide RNA" of crRNAs was an important improvement in facilitating rapid adoption of the CRISPR technology for targeted gene modification in eukaryotic cells [6,9]. To obtain a functional RNA-guided gene disruption in a host cell, one needs only to transform the cell with the *Cas9* gene and a gene (generally driven by an RNA polymerase III-dependent promoter) encoding a sgRNA that contains a 20 bp sequence complementary to the segment of DNA in the host cell that is the target for disruption by a DSB. Once a host cell containing the *Cas9* gene has been established, subsequent modification of a target gene requires only transformation with a sgRNA gene producing a sgRNA complementary to the target gene. Simultaneous modification of two or more genes simply requires transformation of cells with two or more appropriately targeted sgRNA genes [e.g., 10]. The Cas9/sgRNA system has been used successfully for gene disruption, gene activation/repression and various other kinds of genome editing in several types of cells and organisms [e.g., 8–19] – with a number of other gene editing functions possible in the near future [20].

Very recently, there were reports of successful expression of the CRISPR/Cas9 system in higher plant tissue culture systems and during transient expression in *Agrobacterium* inoculated plant cells and tissues [16,21–26]. In this report we demonstrate that the Cas9/sgRNA system delivered by *Agrobacterium tumefaciens* is fully functional when delivered to Arabidopsis by the floral dip method.

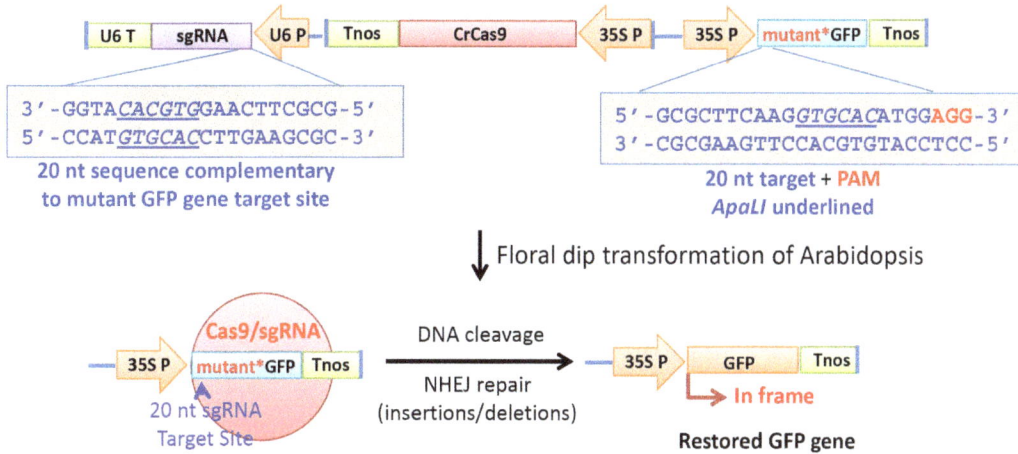

Figure 1. Design of a Cas9/sgRNA system for mutagenesis and restoration of activity of a non-functional (out-of-frame) mutant *GFP* gene in Arabidopsis.

Results using T1 transgenic Arabidopsis are presented showing efficient targeting of specific DNA sequences for DNA cleavage and error-prone repair by NHEJ (i.e., successful conversion of an out-of-frame mutant *GFP* gene to a functional *GFP* gene that provides a visual demonstration and verification of Cas9/sgRNA activity). Results with progeny from several independent T1 plants have been used to demonstrate successful inheritance of modified genes in the T2 and T3 generations and to suggest potential silencing of the Cas9/sgRNA beyond the T1 generation in Arabidopsis.

Results and Discussion

Strategy for Detection of Cas9 and sgRNA Induced Mutagenesis in Arabidopsis

Double strand DNA breaks (DSBs) at the target site in a non-functional *GFP* gene caused by ZFN, TALEN and Cas9/sgRNA expression are most often repaired by the non-homologous end joining (NHEJ) DNA repair mechanism. This repair is often accompanied by small deletions or insertions of nucleotides at the site of repair. We previously took advantage of this mechanism to develop a reporter system for detecting transient expression of Cas9/sgRNA activity in Arabidopsis leaves transformed with two *Agrobacterium* lines, one carrying binary vectors containing the Cas9 and sgRNA genes and another line carrying a nonfunctional, out-of-frame, *GFP* gene [16]. For present studies aimed at generating stably transformed Arabidopsis plants expressing the Cas9/sgRNA genes, we used the same GFP reporter system (outlined in Figure 1) that places all three required genes on a single binary vector instead of the two vectors previously employed. This reporter system involves use of a nonfunctional mutant version of a *GFP* gene containing a small 20 nucleotide insertion (plus a 3 nucleotide extension that includes a GG PAM recognition sequence) immediately downstream of the ATG start codon of the *GFP* gene. This insert was designed to create a shift in the reading frame of the gene so the products lack the ability to produce a fluorescent signal in transgenic plant cells. A single guide RNA (sgRNA) gene construct containing a 20 bp sequence complementary to the 20 nucleotide target in the mutant GFP gene was engineered to be driven by the Arabidopsis U6 gene promoter. Incorporation of the sgRNA produced from this gene into the Cas9 protein should guide the Cas9/sgRNA complex to the 20 bp

target sequence in the mutant *GFP* gene and cause subsequent double strand cleavage of the *GFP* gene target sequence. NHEJ DNA repair of the DSB involving insertion or deletion of nucleotides at the cleavage site would, in approximately one-third of the cases, reestablish a correct reading for the *GFP* gene. In such cases, the mutation should allow production and visualization of green fluorescent protein in transgenic plant cells. Important advantages of this system are that it generates a positive GFP signal that is easily detected in even small clusters of plant cells and that expression of GFP has no negative influence on cell or plant physiology - as is the case for a number Cas9/sgRNA targets used in previous studies with plants or plant cells [21–26]. In addition to our previous success in using restoration of GFP signals from a mutant GFP gene in Arabidopsis and tobacco leaf tissues transiently transformed with Cas9/sgRNA genes [16], others [22] have successfully used a similar reporter system (a nonfunctional split YFP target gene) to obtain Cas9/sgRNA-dependent transient expression of YFP in Arabidopsis protoplast cells and detection of the positive signal via flow cytometry. Unlike earlier studies with plant cells and tissues [21–26], we sought both to gauge the efficiency with which Cas9/sgRNA can mutagenize a target gene and to demonstrate that such a gene can be stably inherited in second and third generation transgenic plants.

As outlined in Figure 2, Arabidopsis plants were transformed with *Agrobacteria* using a floral dip method [27]. The treated plants were allowed to develop and resulting T1 generation seeds were germinated on plates containing solid MS medium and hygromycin. First or second true leaves of hygromycin resistant T1 seedlings were analyzed by confocal microscopy for the presence or absence of green fluorescent signals. PCR primers flanking the *GFP* gene target site were used to amplify the target region. The amplified target regions were genotyped by a PCR/restriction enzyme digestion protocol that allowed detection of mutagenized sgRNA target sites by loss of an *ApaLI* restriction site. The *ApaLI* resistant fragments were cloned and mutagenesis of the sgRNA target sites in several T1 Arabidopsis plants was verified using DNA sequencing of the target sequence. Finally, T2 and T3 progeny of several T1 plants were analyzed to confirm inheritance of mutagenized *GFP* genes and to assess the presence or absence of Cas9/sgRNA activity in the T2 and T3 generation.

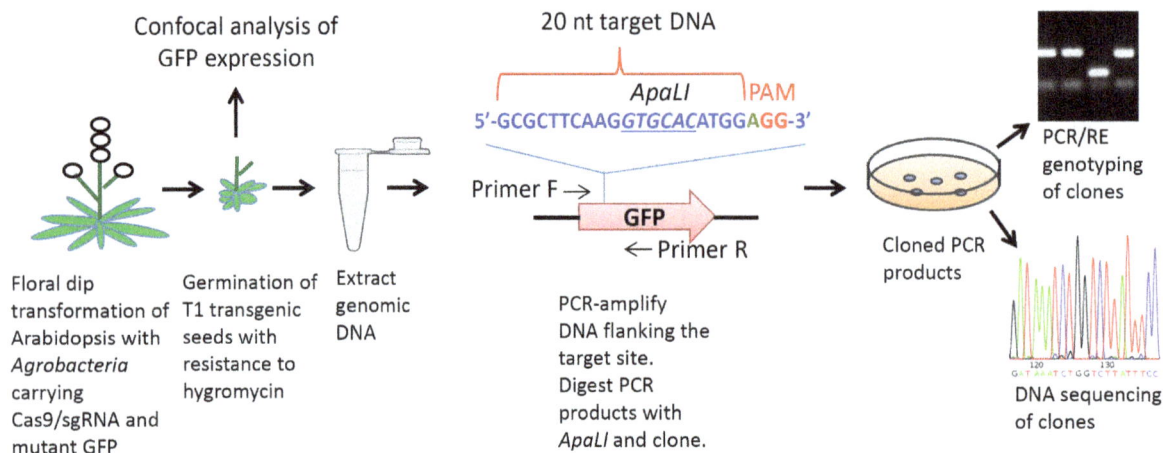

Figure 2. Cas9/sgRNA-mediated mutagenesis in Arabidopsis. Scheme for Cas9/sgRNA-mediated mutagenesis of a non-functional (out-of-frame) mutant *GFP* gene, detection of T1 Arabidopsis leaves with restored *GFP* gene function by confocal microscopy and documentation of target site DNA mutagenesis.

Evidence for High-level Expression of Cas9 and sgRNA Genes in T1 Arabidopsis Plants

Of the seeds collected from T0 parental plants approximately 1% germinated in the presence of hygromycin. This typical recovery rate for transformants using the floral dip method indicated no marked negative effect of the Cas9/sgRNA system on the efficiency of Arabidopsis transformation or seed development. Examination by confocal fluorescence microscopy of leaves of resulting hygromycin-resistant TI plants showed widespread presence of functional *GFP* genes in leaf cells (Figure 3). Such fluorescence was observed with leaves of 35 of 60 T1 Arabidopsis plants obtained by germinating T1 seeds on plates containing levels of hygromycin (20 μg mL^{-1}) lethal to germinating seedlings of wild type plants. As a negative control, Arabidopsis inflorescences were inoculated with *Agrobacteria* containing a binary vector identical to the one described above but carrying a sgRNA gene producing a randomly selected 20 nucleotide targeting sequence plus an AGG PAM sequence (GGATAACATGGCCATCAT-CAAGG) that was not complementary to the sgRNA target site in the nonfunctional *GFP* gene or to any Arabidopsis genome sequence. None of the 61 hygromycin resistant T1 plants obtained in this experiment displayed green fluorescence in their leaf cells or other plant parts (data not shown). These results demonstrated that neither the floral dip transformation procedure nor the Cas9 gene alone, per se, were responsible for mutagenesis of the out-of-frame *GFP* gene (i.e., successful mutagenesis required a sgRNA gene encoding a properly targeted sgRNA sequence as well as the Cas9 gene).

Two notes should be made in regard to our choice of the conversion of a nonfunctional GFP gene to an active GFP as an indication of Cas9/sgRNA activity. First, the visual signals created in leaf tissues was not meant to be used as an accurate measure of mutagenesis rates – rather we have relied on more accurate measures offered by restriction enzyme analyses of PCR products and DNA sequencing (described below). Second, we wished to determine when during seed and plant development the Cas9/sgRNA complex was active in causing targeted gene disruptions. The later aim was accomplished when it was observed that in essentially all of the transgenic plants expressing GFP, there was a patchy pattern of GFP expression seen across the surface of individual leaves (Figure 3A and 3C). The patchiness of GFP expression contrasts sharply with the uniform distribution of

chlorophyll fluorescence captured in the same photographic frames (Figure 3 B and 3D). The simplest explanation for the mosaic patterns observed in plants expressing the Cas9 and sgRNA genes is that while the T-DNA region carrying the Cas9, sgRNA and nonfunctional *GFP* genes is inserted into the chromosome of an ovule progenitor cell prior to fertilization [28], the action of the Cas9 and sgRNA complex in binding and mutagenizing the nonfunctional *GFP* gene and converting it to a functional *GFP* gene is a stochastic process that can occur in somatic cells anytime during seed development and, perhaps, leaf development. The presence in a single leaf of large numbers of different mutagenized DNA sequences in the target region of the *GFP* gene (along with nonmutagenized sequences) (documented and discussed below) strongly suggests that this explanation is correct.

Another form of variation in expression of GFP across the surface of the leaf was also evident when comparing green fluorescence intensities emanating from stomatal guard cells with surrounding epidermal cells (Figure 4). This differential expression pattern was observed consistently among all the GFP-positive leaves examined and is similar to the preferential expression of *GFP* genes driven by the 35S CaMV promoter in guard cells relative to expression in surrounding cells observed by others [29].

Analysis of Relative Abundance of Mutagenized and Nonmutagenized *GFP* Genes in Leaves of T1 Arabidopsis Plants

A combination of PCR amplification of the sgRNA target region within the mutant *GFP* gene and restriction digestion using the enzyme *Apa*LI was used to obtain a rough estimate of the efficiency of Cas9/sgRNA-mediated mutagenesis. The aim of this analysis was to determine the proportion of nonfunctional, out-of-frame, *GFP* genes that were not mutagenized by the Cas9/sgRNA system in T1 plant cells relative to the proportion of genes in which the Cas9/sgRNA complex was successful in altering the *Apa*LI recognition site within the nonfunctional *GFP* gene. To accomplish this, total DNA was extracted from a single leaf of 12 different hygromycin-resistant, GFP-positive, T1 Arabidopsis seedlings. These DNA samples were subjected to PCR amplification with DNA primers complementary to sites approximately 125 bp on either side of the *Apa*LI cut site within the 20 bp sgRNA

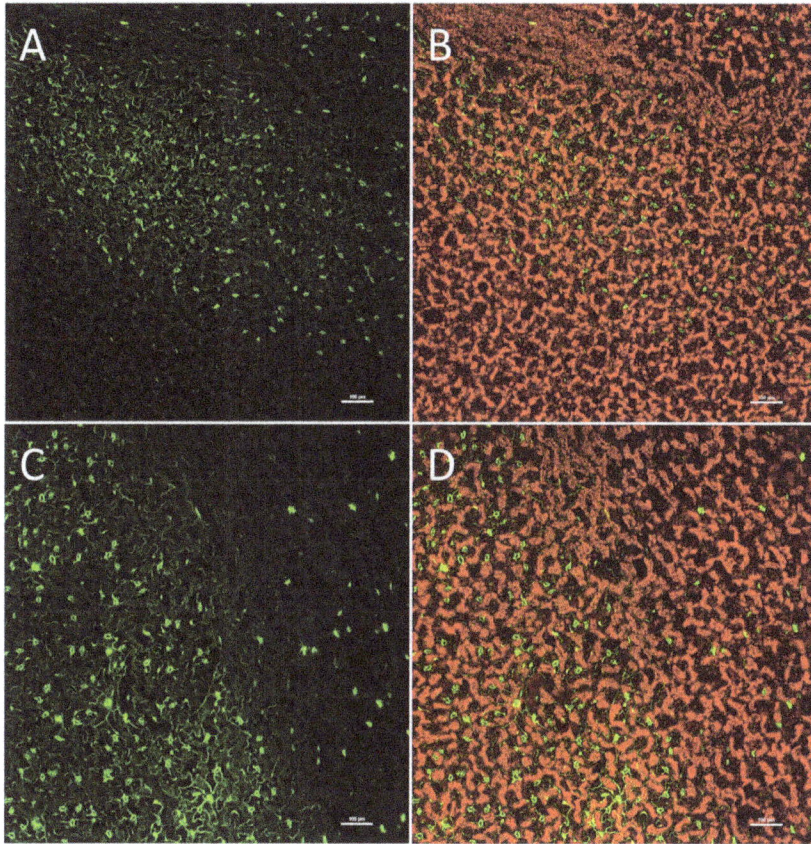

Figure 3. Cas9/sgRNA-mediated mutagenesis of nonfunctional mutant *GFP* genes in Arabidopsis leaves. Expression in T1 Arabidopsis leaf cells of functional *GFP* genes produced by Cas9/sgRNA-mediated mutagenesis of nonfunctional mutant *GFP* genes introduced along with the Cas9 and sgRNA genes using *A. tumefaciens* and a floral dip transformation protocol. A and C) Detection of green fluorescence protein signals in transgenic leaves. B and D) Merged image of red chlorophyll fluorescence and GFP fluorescence. Leaves from hygromycin resistant seedlings were photographed ten days (A and B) and twenty days (C and D) after seed germination. Bar, 100 μm.

target site of the mutant *GFP* gene (Figure 2). Digestion of the approximately 250 bp PCR product from amplification of a cloned nonfuctional, out-of-frame, *GFP* gene with *Apa*LI resulted in the expected production of two DNA fragments each of approximately 125 bp in length (Figure 5, lane NG). The *Apa*LI digestion patterns of PCR-amplified DNA from the 12 individual T1 plants (all displaying GFP positive leaf segments) revealed that all (as expected) contained a 250 bp, *Apa*LI-resistant DNA product

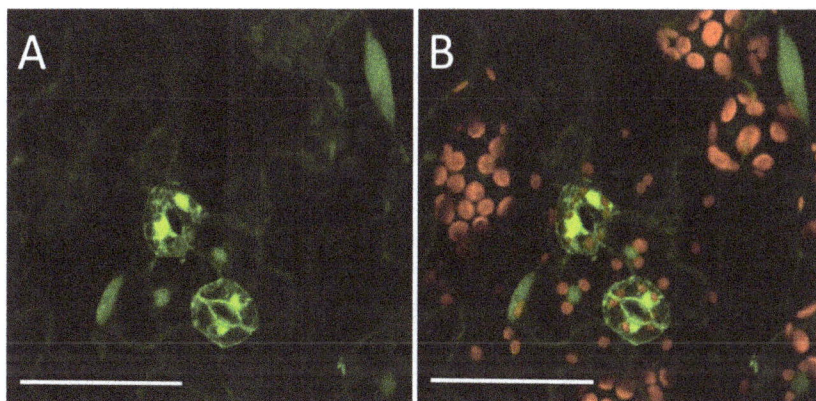

Figure 4. Robust expression in guard cells of functional *GFP* genes created by Cas9/sgRNA-mediated mutagenesis. A) Detection of green fluorescence protein signals in guard cells of transgenic leaf stomata with lesser expression in surrounding leaf epidermal cells of T1 transgenic Arabidopsis plants. B) Merged image of red chlorophyll fluorescence and GFP fluorescence. Bar, 50 μm. Photographed twenty days after seed germination.

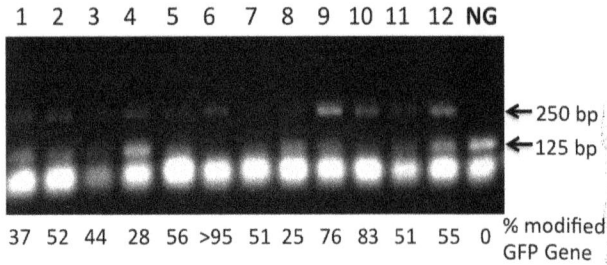

Figure 5. Efficiency of Cas9/sgRNA mutagenesis in Arabidopsis plants. PCR/Restriction Enzyme (PCR/RE) analysis of total DNA extracts from individual hygromycin resistant T1 Arabidopsis plants (lanes 1–12) showing the relative proportion of nonfunctional *GFP* genes mutagenized by Cas9/sgRNA activity. Bottom arrow indicates the expected ~125 bp DNA fragments resulting from *Apa*LI cleavage of the ~250 bp PCR product amplified from nonfunctional, out-of-frame, nonmutagenized *GFP* genes that contains an intact *Apa*LI cleavage site in the sgRNA target region. Top arrow indicates the expected ~250 bp size of PCR products from *GFP* genes mutagenized by the Cas9/sgRNA system in such a manner that they are no longer are susceptible to cleavage by *Apa*LI. NG (Nonfunctional *GFP* Gene), the PCR products amplified from a sgRNA target site of a cloned nonfunctional, out-of-frame, *GFP* gene digested with *Apa*LI. % modified *GFP* Gene = (pixels in 250 bp band)/ (pixels in 250 bp band+pixels in 125 bp band) ×100.

and that the portion of target DNA sequences mutagenized by the Cas9/sgRNA complex ranged from 37% to greater than 95% (Figure 5, lanes 1–12). These data demonstrate a high level of Cas9/sgRNA-driven DNA cleavage and resulting gene mutagenesis in somatic cells of Arabidopsis T1 plants.

As an extension of the experiments described above, we also examined 12 out of 25 GFP-negative T1 Arabidopsis plants to determine if they contained mutagenized target *GFP* genes as ascertained by PCR/restriction enzyme analysis (Figure S1) similar to those depicted in Figure 5. Interestingly, 11 out of 12 GFP-negative plants examined were shown to have at least one mutagenized target site (data not shown). This again demonstrated a high frequency of Cas9/sgRNA-induced mutagenesis. The large proportion of GFP negative T1 plants (11 out of 12) containing at least one mutagenized target region illustrates, as expected, that only a fraction (~one-third) of mutagenized target sites restore a proper reading frame to the coding region of the nonfunctional, out-of-frame, *GFP* gene and allow the associated green fluorescence phenotype.

Confirmation of Cas9/sgRNA-directed Mutagenesis using DNA Sequencing

To confirm Cas9/sgRNA-directed mutagenesis of nonfunctional *GFP* genes and to determine the nature of the mutagenesis, we conducted a separate experiment in which we sequenced five independent clones of PCR products amplified from target gene DNAs isolated from each of five different T1 plants (#21–#25), three of which were GFP positive and two of which were GFP negative. Prior to PCR amplification, DNA isolated from each plant had been treated with *Apa*LI to prevent amplification of nonmutagenized target region DNAs (i.e., those containing an intact *Apa*LI restriction site) and to enrich for DNA regions carrying target sites with altered *Apa*LI recognition sequences.

All five T1 plants, whether GFP-positive or GFP-negative, contained multiple PCR products displaying site-specific nucleotide insertions and deletions - hallmarks of Cas9/sgRNA-catalyzed DNA cleavage and NHEJ DNA repair (Figure 6). As expected for DNA repair events performed by the NHEJ system, most of the

repairs involved insertion or deletions of only a few nucleotides, but with occasional creation of sizable deletions of 20 or more nucleotides. Also as expected for Cas9/sgRNA DNA cleavages that occur two or three base pairs upstream of the PAM site, all of the observed insertions/deletions were located in close proximity to the expected cleavage site.

The DNA sequence analyses shown in Figure 6 demonstrate the presence of multiple different mutagenized *GFP* target sequences within single leaves from five T1 Arabidopsis plants. These observations suggest that independent cas9/sgRNA-driven mutagenesis events occurred in somatic cells during seed or seedling development. This conjecture is consistent with the observation of mosaic patterns of GFP expression in T1 Arabidopsis plant leaves (Figure 3) in which some patches of cells display green fluorescence and neighboring patches do not. Together, these observations strongly suggest that even though Cas9 and sgRNA genes possibly become active soon after insertion of T-DNA into a host chromosome following floral dip transformation of Arabidopsis, the action of the Cas9/sgRNA complex in targeting the 20 bp sequence in the nonfunctional *GFP* gene is often not immediate, but occurs perhaps more or less randomly in cells during portions of seed or seedling development. These observations also demonstrate that the Cas9/sgRNA complex acted to modify nonfunctional *GFP* genes that were part of a T-DNA region stably integrated into a chromosomal location. Thus, the findings of the present study in regard to inheritance of *GFP* genes should also pertain equally to inheritance of endogenous Arabidopsis genes modified by the Cas9/sgRNA complex.

Evidence for Inheritance of Cas9/sgRNA Derived Mutant Phenotypes and Genotypes in T2 Plants

All 24 transgenic T1 plants examined for this portion of our studies produced normal levels of seeds with generally high levels of viability. That is, there was no evidence for an effect of Cas9 and sgRNA gene expression on the ability of T1 plants to produce viable progeny. Indeed, examination of the Arabidopsis genome for sequences with sufficient homology to the 20 bp sgRNA target sequence to cause potential off-site targeting [e.g., 26, 30, 31] failed to reveal any sites of concern [19].

Germination and seedling survival rates of 70% to 80% in the presence of hygromycin were obtained for T2 generation seeds from the individual T1 plants tested suggesting that, in most cases, the hygromycin resistance gene (and accompanying Cas9, sgRNA and *GFP* genes) was present in a single copy in T2 plants. Leaves from T2 seedlings derived from 6 different T1 plants (T1 Plants #3 to #8) were excised and examined for green fluorescence using a confocal fluorescence microscope. T2 seedlings from all but one of the 6 T1 progenitor plants displayed no green fluorescence. All of the T2 seedlings derived from T1 Plant #6 had leaves displaying uniform green fluorescence (Figure S2).

To determine if only a single gene was inherited in the T2 progeny of the 6 T1 plants examined above (T1 Plants #3 to #8), DNA was extracted from 7 hygromycin-resistant T2 plants derived from each of the 6 T1 progenitors. The 250 bp region containing the target site for Cas9/sgRNA cleavage was PCR amplified from each DNA preparation and subjected (without cloning) to DNA sequence analysis. Only one DNA sequence was obtained for each of the 7 individual T2 plants in each of the 6 groups (Figure 7A and B). For T2 plants derived from T1 generation plants #4, #5 and #6, the presence of a single mutagenized GFP gene sequence provided compelling evidence that, in each case, only a single *GFP* gene had been inherited from the T1 progenitor plant. For T2 plants derived from T1 plants #3, #7, and #8, the gene number could not be determined because only the original *GFP* gene

Plant #21   In/Del

CATGGAGCGCTTCAAGGTGCACATGGAGGACTAGTAAAGGAGAAGAAC

CATGGAGCGCTTCAAGGTGCACAATGGAGGACTAGTAAAGGAGAAGAAC   +1

CATGGAGCGCTTCAAGGTG..CATGGAGGACTAGTAAAGGAGAAGAAC   -2

CATGGAGCGCTTCAAGGTG..CATGGAGGACTAGTAAAGGAGAAGAAC   -2

CATGGAGCGCTT...........TGGAGGACTAGTAAAGGAGAAGAAC   -11

CATGGAGCGCTT................GACTAGTAAAGGAGAAGAAC   -16

Plant #22

CATGGAGCGCTTCAAGGTGCACATGGAGGACTAGTAAAGGAGAAGAAC

CATGGAGCGCTTCAAGGTGAACTTTTCACTGAGGACTCTTGACA[ ]AC   +23

CATGGAGCGCTTCAAGGTGACCATGGAGGACTAGTAAAGGAGAAGAAC   0 (+2,-2)

CATGGAGCGCTTCAAGGTG..CATGGAGGACTAGTAAAGGAGAAGAAC   -2

CATGGAGCGCTTCAAGGTG..CATGGAGGACTAGTAAAGGAGAAGAAC   -2

CATGGAGCGCTTCAAGGTGCA....GAGGACTAGTAAAGGAGAAGAAC   -4

Plant #23

CATGGAGCGCTTCAAGGTGCACATGGAGGACTAGTAAAGGAGAAGAAC

CATGGAGCGCGCTTCATGATATCTCCACTGACCATATGGAGGACTA[ ]AC   +11 (+19,-8)

CATGGAGCGCTTCTAGGTGCAA..GGAGGACTAGTAAAGGAGAAGAAC   -2 (+9,-11)

CATGGAGCGCTTCAAGGTG.....GGAGGACTAGTAAAGGAGAAGAAC   -5

CATGGAGCGCTTCAAGGTG.....GGAGGACTAGTAAAGGAGAAGAAC   -5

CATGGAGCGCTTCAAGGTGCA.........................   -33

Plant #24

CATGGAGCGCTTCAAGGTGCACATGGAGGACTAGTAAAGGAGAAGAAC

CATGGAGCGCTTCAAGGTGCACAATGGAGGACTAGTAAAGGAGAAGAAC   +1

CATGGAGCGCTTCAAGGTGCA..TGGAGGACTAGTAAAGGAGAAGAAC   -2

CATGGAGCGCTTCAAGGTGC...TGGAGGACTAGTAAAGGAGAAGAAC   -3

CATGGA.....................GGACTAGTAAAGGAGAAGAAC   -21

CATGGAG....................GACTAGTAAAGGAGAAGAAC   -21

Plant #25

CATGGAGCGCTTCAAGGTGCACATGGAGGACTAGTAAAGGAGAAGAAC

CATGGAGCGCTTCAAGGTG..CATGGAGGACTAGTAAAGGAGAAGAAC   -2

CATGGAGCGCTTCAAGGTG..CATGGAGGACTAGTAAAGGAGAAGAAC   -2

CATGGAGCGCTTCAAGGTG....TGGAGGACTAGTAAAGGAGAAGAAC   -4

CATGGAGGAGGCTAGTAG.................AGGAGAAGAAC   -24 (+11, -35)

CATGGAGCGCTTCAAGGTG......................AAGAAC   -21

Figure 6. Confirmation by DNA sequencing of Cas9/sgRNA-mediated mutagenesis of the sgRNA target site within the nonfuctional, out-of-frame, *GFP* gene. Five cloned DNA fragments of 250 bp containing DNA from PCR amplified sgRNA target regions of (previously) nonfunctional *GFP* genes from three different GFP-positive T1 Arabidopsis plants (Plants 21, 22 and 25) and two different GFP-negative plants (Plants 23 and 24) were subjected to DNA sequencing. DNA sequences of a segment of 48 nucleotides surrounding the sgRNA target site are shown for each clone with the sequence of the nonmutagenized DNA region shown as the top line of each group. The 20 nucleotide target sequence for the Cas9/sgRNA complex is depicted in blue, the PAM site in red and the *Apa*LI recognition site is underlined in blue. Brackets ([]) denote nondisplayed DNA sequences. For the Cas9/sgRNA-mutagenized DNA sequences, deleted nucleotides are depicted as red dots and inserted nucleotides are shown in green. The net length of insertions and/or deletions (In/Del) are presented in the column to the right.

sequence found in the transforming T-DNA were present. The sequencing data for T2 progeny of T1 Plant #6 revealed restoration of a proper reading frame in the *GFP* gene in these plants due to deletion (in the genetically competent cells of the T1 progenitor plant) of a single A nucleotide and the consequential expression of the green fluorescence phenotype noted above (Figure S2). The presence of only a single *GFP* gene in each of the T2 plants was further confirmed by careful examination of the DNA sequencing trace from one individual plant from each of the 6 groups of T2 plants. In all 6 cases (in which *GFP* gene was mutagenized or not) this examination showed uniform distribution of peaks with no indication of minor overlapping peaks (Figure 7C through F). As an example of a pattern that might be expected if more than one species of Cas9/sgRNA-altered *GFP* gene was present (or if a copy of the original nonfunctional *GFP* gene and a copy of a Cas9/sgRNA-altered *GFP* gene were both present),

Figure 7G displays a DNA sequencing trace of the same region after PCR amplification of DNA isolated from leaves of T1 Plant #1– a plant that was shown earlier to contain multiple altered and nonaltered *GFP* genes by *Apa*LI restriction enzyme digestion of PCR amplified *GFP* genes (Figure 5). In this case, the uniform pattern of DNA sequence peaks upstream of the sgRNA target site becomes overlapping and lower in height as the sequencing trace proceeds through the area of the genes containing deletions and insertions caused by Cas9/sgRNA-mediated DNA cleavage and NHEJ DNA repair. From this set of experiments, we conclude that when coupled with the segregation ratios of hygromycin-resistant to hygromycin-sensitive T2 seedling (~3:1), the DNA sequence data provided in Figure 7 for T2 progeny of T1 generation plants #4, #5 and #6 provide compelling evidence for inheritance of only single *GFP* genes in the T2 progeny of each of the three different T1 progenitors – implying the presence of only a single

A. T2 Plants Containing a Mutagenized GFP Gene

T1 Plant #4 Progeny

```
Original   GA GCGCTTCAAGGTGCACATGGAGGACTAGT
T1 —       GA GCGCTTCAAGGTGCACAATGGAGGACTAG
           GA GCGCTTCAAGGTGCACAATGGAGGACTAG
           GA GCGCTTCAAGGTGCACAATGGAGGACTAG
T2         GA GCGCTTCAAGGTGCACAATGGAGGACTAG
Mut+, GF-  GA GCGCTTCAAGGTGCACAATGGAGGACTAG
           GA GCGCTTCAAGGTGCACAATGGAGGACTAG
           GA GCGCTTCAAGGTGCACAATGGAGGACTAG
           GA GCGCTTCAAGGTGCACAATGGAGGACTAG
```

T1 Plant #5 Progeny

```
Original   GA GCGCTTCAAGGTGCACATGGAGGACTAGT
T1 —       GA GCGCTTCAAGGTGCACATTGGAGGACTAG
           GA GCGCTTCAAGGTGCACATTGGAGGACTAG
           GA GCGCTTCAAGGTGCACATTGGAGGACTAG
T2         GA GCGCTTCAAGGTGCACATTGGAGGACTAG
Mut+, GF-  GA GCGCTTCAAGGTGCACATTGGAGGACTAG
           GA GCGCTTCAAGGTGCACATTGGAGGACTAG
           GA GCGCTTCAAGGTGCACATTGGAGGACTAG
           GA GCGCTTCAAGGTGCACATTGGAGGACTAG
```

T1 Plant #6 Progeny

```
Original   GA GCGCTTCAAGGTGCACATGGAGGACTAGT
T1 —       GA GCGCTTCAAGGTGCAC-TGGAGGACTAG
           GA GCGCTTCAAGGTGCAC-TGGAGGACTAG
           GA GCGCTTCAAGGTGCAC-TGGAGGACTAG
           GA GCGCTTCAAGGTGCAC-TGGAGGACTAG
T2         GA GCGCTTCAAGGTGCAC-TGGAGGACTAG
Mut+, GF+  GA GCGCTTCAAGGTGCAC-TGGAGGACTAG
           GA GCGCTTCAAGGTGCAC-TGGAGGACTAG
           GA GCGCTTCAAGGTGCAC-TGGAGGACTAG
```

B. T2 Plants Lacking a Mutagenized GFP Gene

T1 Plant #3 Progeny

```
Original   GA GCGCTTCAAGGTGCACATGGAGGACTAGT
T1 —       GA GCGCTTCAAGGTGCA--TGGAGGACTAGT
           GA GCGCTTCAAGGTGCACATGGAGGACTAGT
           GA GCGCTTCAAGGTGCACATGGAGGACTAGT
           GA GCGCTTCAAGGTGCACATGGAGGACTAGT
T2         GA GCGCTTCAAGGTGCACATGGAGGACTAGT
Mut-, GF-  GA GCGCTTCAAGGTGCACATGGAGGACTAGT
           GA GCGCTTCAAGGTGCACATGGAGGACTAGT
           GA GCGCTTCAAGGTGCACATGGAGGACTAGT
```

T1 Plant #7 Progeny

```
Original   GA GCGCTTCAAGGTGCACATGGAGGACTAGT
T1 —       GA GCGCTTCAAGGTGCA----GAGGACTAGT
           GA GCGCTTCAAGGTGCACATGGAGGACTAGT
           GA GCGCTTCAAGGTGCACATGGAGGACTAGT
           GA GCGCTTCAAGGTGCACATGGAGGACTAGT
T2         GA GCGCTTCAAGGTGCACATGGAGGACTAGT
Mut-, GF-  GA GCGCTTCAAGGTGCACATGGAGGACTAGT
           GA GCGCTTCAAGGTGCACATGGAGGACTAGT
           GA GCGCTTCAAGGTGCACATGGAGGACTAGT
```

T1 Plant #8 Progeny

```
Original   GA GCGCTTCAAGGTGCACATGGAGGACTAGT
T1 —       GA GCGCTTCAAGGTGCACAATGGAGGACTAG
           GA GCGCTTCAAGGTGCACATGGAGGACTAGT
           GA GCGCTTCAAGGTGCACATGGAGGACTAGT
           GA GCGCTTCAAGGTGCACATGGAGGACTAGT
T2         GA GCGCTTCAAGGTGCACATGGAGGACTAGT
Mut-, GF-  GA GCGCTTCAAGGTGCACATGGAGGACTAGT
           GA GCGCTTCAAGGTGCACATGGAGGACTAGT
           GA GCGCTTCAAGGTGCACATGGAGGACTAGT
```

C. T2Progeny of T1 Plant #4 (GF-)

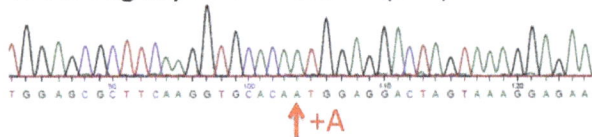

TGGAGCGCTTCAAGGTGCACAATGGAGGACTAGTAAAGGAGAA
↑ +A

D. T2 Progeny of T1 Plant #5 (GF-)

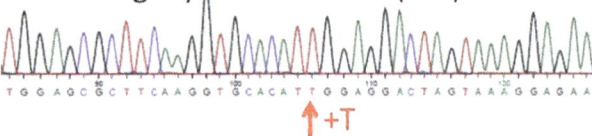

TGGAGCGCTTCAAGGTGCACATTGGAGGACTAGTAAAGGAGAA
↑ +T

E. T2Progeny of T1 Plant #6 (GF+)

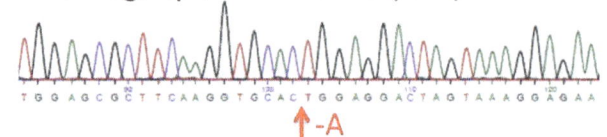

TGGAGCGCTTCAAGGTGCACTGGAGGACTAGTAAAGGAGAA
↑ -A

F. T2 Progeny of T1 Plant #3 (GF-)

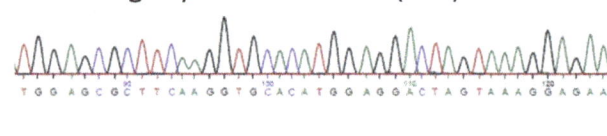

TGGAGCGCTTCAAGGTGCACATGGAGGACTAGTAAAGGAGAA

G. T1 Plant #1 (GF+)

TGGAGCGCTTCAAGGTGCAC........CT..AA.GAGA

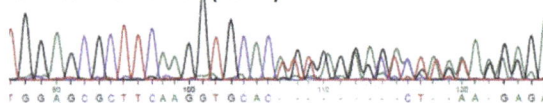

Figure 7. Confirmation of inheritance of a single modified or nonmodified *GFP* gene in each of 7 T2 progeny from 6 individual T1 generation plants – and evidence of Cas9 gene and/or sgRNA gene silencing in T2 progeny. DNA was isolated from each of 7 T2 progeny from each of 6 different progenitor T1 plants (T1 Plants #3 to #8). PCR was used to amplify a 250 bp DNA fragment containing the sgRNA target region of the nonfunctional, out-of-frame, *GFP* gene. DNA sequencing of the fragment provided the sequence of the 31 bp region displayed for each of the 42 T2 plants. The DNA sequence of the original *GFP* gene is provide as the top line in each column along with the sequence of one mutagenized *GFP* gene found in a leaf of the original progenitor T1 plant. A) DNA sequences of three groups of T2 plants in which there has been a Cas9/sgRNA-mediated gene modification including insertion of an A nucleotide (Plant #4 progeny), a T nucleotide (Plant #5 progeny), or deletion of an A nucleotide (Plant #6 progeny) that restored a proper reading frame and resulted in T2 progeny displaying a green fluorescence phenotype, B) DNA sequences of three groups of T2 plants (progeny of T1 Plants #3, #7 and #8) in which there was no inherited Cas9/sgRNA-mediated gene

modification. (GF−), No green fluorescence phenotype; (GF+), Green fluorescence phenotype; (Mut+), Inherited mutagenized *GFP* gene; (Mut-), No inherited mutagenized *GFP* gene. C, D, E, and F) DNA sequencing traces from sequencing of PCR amplified Cas9/sgRNA target sites from a single leaf of an individual T2 progeny from T1 progenitor Plants #3, #7 and #8, respectively. G) A DNA sequencing trace from sequencing of the PCR amplified Cas9/sgRNA target sites isolated from a single leaf of T1 Plant #1 showing multiple overlapping DNA peaks caused by the presence of multiple different DNA sequences in the separate mutagenized *GFP* genes present in different patches of cells scattered throughout the leaf.

T-DNA region in the genetically-competent cells of T1 plants #4, #5 and #6 used to produce the T2 and T3 generation plants analyzed in this study.

The data of Figure 7 also revealed a potentially important and unexpected finding – the probable silencing of the Cas9 gene and/or the sgRNA gene in progeny of T1 plants. Remembering the large amount of DNA sequencing information described earlier (Figure 6) that showed multiple different mutagenized *GFP* genes in single leaves from individual T1 plants, it is clear that the Cas9 and sgRNA genes were active in many different cells during T1 seed development (and, potentially, leaf development). Thus, the lack of mutagenesis of *GFP* gene sequences observed in leaves of three different groups of T2 plants (those produced from T1 plants #3, #7 and #8, Figure 7) and the lack of expression of a functional GFP in any leaf of these plants suggest that the Cas9 and/or sgRNA genes in these T2 plants may have been silenced in the genetically competent cells of progenitor T1 plants or during early T2 seed development. Confirmation of silencing and its timing relative to seed and plant development will require future studies. Because of the possibility that Cas9 and sgRNA gene silencing can occur even during early development of T1 generation seeds and is likely not uniform in timing between different cells and tissues during seed and/or leaf development, attempts to accurately define the timing of gene silencing likely will be difficult. That is because such studies may require highly sensitive RT-PCR or RNA-seq procedures to detect the presence or absence of Cas9 and sgRNA transcripts from minute samples of various tissues taken during the entire course of development of multiple T1 generation Arabidopsis seeds, seedlings and plants – a challenging, but potentially revealing, task.

As an extension of our experiments showing stable inheritance of genes modified by the Cas9/sgRNA system in the T2 generation, we extracted DNA from several T3 generation progeny of T2 plants representing the three sets of T2 plants containing mutagenized GFP gene sequences examined in Figure 7. As expected, all plants displayed a single mutagenized GFP gene sequence that was identical to the one found in the T2 plant from which they were derived (Figure S3) - providing verification of stable inheritance of the mutagenized GFP genes in the T3 generation. A summary of the inheritance data from the present study is provided in Table S1.

Together, results of the present study point to the ease and high efficiency in producing T1 Arabidopsis plants that successfully express the Cas9/sgRNA system for targeted gene modification. These studies also verify stable inheritance of newly acquired genes/traits in T2 and T3 plants. The unexpected discovery of an apparent lack of Cas9/sgRNA activity in T2 and T3 plants suggests a potential role for gene silencing in dictating the extent and duration of Cas9/sgRNA-mediated mutagenesis during plant development.

An important virtue of the Cas9/sgRNA system for plant biotechnology applications is that once a desired gene modification has been achieved, the T-DNA region carrying Cas9, sgRNA and other transgenes can be eliminated entirely from progeny by simple genetic crosses – as we have demonstrated earlier for TALEN-modified rice plants (3). The lack of transgenes may allow such genetically enhanced crop plants to avoid present regulatory

restraints and, thus, make their way into the marketplace faster than is presently the case for transgenic plants. Presuming that future studies demonstrate activity of the Cas9/sgRNA system in a wide range of plants, this system for precise and efficient gene modification would appear to have significant promise for greatly speeding scientific advances in plant biology and for engineering crop plants with improved traits for productivity and nutrition.

After this article was submitted for publication and distributed for review, a manuscript by Feng et al. [32] appeared that reported inheritance of Cas9/sgRNA-generated mutant genes in T2 generation Arabidopsis plants and other results that support the major conclusions of the present article.

Materials and Methods

Preparation of Plant Materials

Arabidopsis thaliana ecotype (Col-0) was used in all experiments. Seeds were sown on soil and were stratified for 3 days at 4°C. Seeds were geminated and plants were grown within a controlled environment chamber at 22°C with a 12 hr light/12 hr dark photoperiod and under 75% relative humidity till early flowering stage about 4 weeks from germination. For selection of transgenic Arabidopsis that are resistant to antibiotic hygromycin, seeds were surface sterilized by exposure to 100% ethanol for one minute and then by treatment with 50% commercial bleach for 5 minutes. After three thorough washes with water, the seeds were sown on 1% agarose plates containing MS medium with a final concentration of 20 μg mL^{-1} hygromycin and 100 μg mL^{-1} carbenicillin. Plates containing the seeds were incubated in the controlled growth chamber described above. Germinated seedlings with resistance to hygromycin were transferred to soil ten days after germination.

Construction of Plant Expression Vectors

Design and construction of the Cas9 gene and sgRNA gene from *Streptococcus pyogenes* [6,9], the nonfunctional, out-of-frame *GFP* gene, and the *Agrobacterium tumefaciens* binary vector containing these gene constructs have been described earlier [16]. A map of the binary vector used in the present studies as well as the complete DNA sequence of the T-DNA region of the binary vector are provided as Figures S3 and S4, respectively.

Floral Dip-*Agrobacterium*-mediated Arabidopsis Transformation

The transformation of Arabidopsis was performed as described earlier [27]. In brief, *Agrobacterium tumefaciens*, strain C58cc, carrying the Cas9 gene, sgRNA gene, mutant *GFP* gene and a hygromycin resistance gene was grown overnight at 27°C with shaking at 200 RPM in 5 mL of Luria–Bertani (LB) medium supplemented with appropriate antibiotics. The next afternoon, 1.5 mL of the saturated bacterial culture was transferred to 500 mL of LB medium containing appropriate antibiotics and grown with shaking overnight at 27°C. Cells were harvested the next day by centrifugation and diluted in 5% sucrose buffer to a final OD$_{600}$ of 0.8. Silwet L-77 was added to a final concentration of 0.01%. Inflorescence clusters of Arabidopsis plants were dipped into the bacterial culture [27]. Inoculated plants were maintained in growth chambers until seeds were fully mature. T1 generation seeds were collected from individual T0 plants and germinated on plates containing solid MS medium with 20 μg/mL hygromycin to select successfully transformed T1 plants and to determine rates of transformation obtained in the floral dip transformation protocol.

Enrichment of Mutagenized Target Sites using Restriction Enzyme Digestion of PCR Amplified sgRNA Target Site Regions Present in Mutagenized *GFP* Genes

PCR amplification of the sgRNA target sites in DNA from GFP-positive T1 Arabidopsis plants and subsequent restriction enzyme digestion were used to verify Cas9/sgRNA-stimulated cleavage and subsequent erroneous DNA repair by NHEJ. DNA was extracted from young T1 Arabidopsis leaves using a NucleoSpin Plant II kit (MACHEREY-NAGEL GmbH and Co.KG, Germany). To eliminate PCR amplification of most DNA sequences from the original nonfunctional *GFP* gene, the extracted DNA was cleaved at the resident *Apa*LI cut site that is located within the 20 bp sgRNA target site near the PAM site. In this way, subsequent PCR amplification favored production of mutagenized DNA. For PCR amplification, primers 125 bp upstream and 125 bp downstream of the *Apa*LI restriction site were employed. PCR amplified and *Apa*LI-digested DNA was agarose gel-purified and the resulting ~250 bp DNA fragments were cloned into a pBlueScript vector. DNA sequencing of the 250 bp fragments was used to determine the types of Cas9/sgRNA/NHEJ-mediated mutations obtained.

DNA Sequence Analysis of sgRNA Target Regions of 42 Individual T2 Plants

Total DNA was isolated from an individual leaf of each of 7 T2 progeny from 6 different T1 progenitor plants (i.e., 42 separate leaf samples). The same primers as described above were used to PCR amplify the ~250 bp DNA region containing the sgRNA target site in DNA from each leaf. This PCR amplified DNA was sequenced directly (i.e., no cloning of fragments was involved) to determine if one or more than one modified or nonmodified *GFP* gene was present.

Fluorescence Confocal Microscopy of Arabidopsis Leaf Cells

Second or third true leaves of T1 and T2 seedlings were cut and analyzed for GFP fluorescence using a Nikon ECLIPSE 90i system confocal fluorescence microscope at 100× and 600× magnification. The excitation and detection wavelengths were set at 448 nm and 500–550 nm, respectively, for GFP fluorescence and at 641 nm and 662–737 nm for chlorophyll auto-fluorescence to minimize cross talk between the two fluorescence channels.

Supporting Information

Figure S1 Analyses of mutagenized sgRNA target sites in nonfunctional *GFP* genes by PCR/RE in DNAs from 24 T1 transgenic Arabidopsis plants.

Figure S2 Expression of a functional *GFP* gene in a leaf of a T2 progeny of T1 generation Plant #6.

Figure S3 Confirmation of inheritance of a single modified *GFP* gene in each of 7 T3 progeny from 3 individual T2 generation plants.

Figure S4 Map of plant expression vector containing Cas9, sgRNA and mutant *GFP* genes.

Figure S5 Sequences of plant expression vector containing Cas9, sgRNA and mutant *GFP* genes from LB to RB.

Table S1 Summary of expression of Cas9/sgRNA-induced mutations of the targeted nonfunctional GFP gene in somatic tissue of T1 generation plants and inheritance of the mutagenized GFP gene in T2 and T3 generation Arabidopsis plants.

Acknowledgments

The authors thank Christian Elowski and Dr. Joe Zhou for assistance with confocal fluorescence miscroscopy, Dr. Istvan Ladunga for assistance in running the CasOT program and Dr. David Marks for helpful discussions.

Author Contributions

Conceived and designed the experiments: WJ BY DPW. Performed the experiments: WJ. Analyzed the data: WJ BY DPW. Contributed reagents/materials/analysis tools: WJ. Wrote the paper: WJ BY DPW.

References

1. Beerli RR, Barbas CF (2002) engineering polydactyl zinc finger transcription factors. Nat Biotechnol 20: 135–141.
2. Li T, Huang S, Zhao X, Wright DA, Carpenter S, et al. (2011) modularly assembled designer TAL effector nucleases for targeted gene knockout and gene replacement in eukaryotes. Nucleic Acids Res 14: 6315–6325.
3. Li T, Liu B, Spalding MH, Weeks DP, Yang B (2012) High-efficiency TALEN-based gene editing produces disease-resistant rice. Nat Biotechnol 30: 390–392.
4. Gaj T, Gersbach CA, Barbas CF 3rd (2013) ZFN, TALEN, and CRISPR/Cas-based methods for genome engineering. Trends Biotechnol 31: 397–405.
5. Chen K, Gao C (2013) TALENs: customizable molecular DNA scissors for genome engineering of plants. J Genet Genomics 40: 271–279.
6. Jinek M, Chylinski K, Fonfara I, Hauer M, Doudna JA, et al. (2012) a programmable dual-RNA-guided DNA endonuclease in adaptive bacterial immunity. Science 6096: 816–821.
7. Wiedenheft B, Sternberg SH, Doudna JA. (2012) RNA-guided genetic silencing systems in bacteria and archaea. Nature 7385: 331–338.
8. Cong L, Ran FA, Cox D, Lin S, Barretto R, et al. (2013) multiplex genome engineering using CRISPR/Cas systems. Science 6121: 819–823.
9. Mali P, Yang L, Esvelt KM, Aach J, Guell M, et al. (2013) RNA-guided human genome engineering via Cas9. Science 6121: 823–826.
10. Wang H, Yang H, Shivalila CS, Dawlaty MM, Cheng AW, et al. (2013) one-step generation of mice carrying mutations in multiple genes by CRISPR/Cas-mediated genome engineering. Cell 4: 910–918.
11. Chang N, Sun C, Gao L, Zhu D, Xu X, et al. (2013) genome editing with RNA-guided Cas9 nuclease in zebrafish embryos. Cell Res 4: 465–472.
12. Cho SW, Kim S, Kim JM, Kim JS. (2013) targeted genome engineering in human cells with the Cas9 RNA-guided endonuclease. Nat Biotechnol 3: 230–232.
13. DiCarlo JE, Norville JE, Mali P, Rios X, Aach J, et al. (2013) genome engineering in Saccharomyces cerevisiae using CRISPR-Cas systems. Nucleic Acids Res 7: 4336–4343.
14. Gratz SJ, Cummings AM, Nguyen JN, Hamm DC, Donohue LK, et al. (2013) genome engineering of Drosophila with the CRISPR RNA-guided Cas9 nuclease. Genetics [Epub ahead of print].
15. Hwang WY, Fu Y, Reyon D, Maeder ML, Tsai SQ, et al. (2013) Efficient genome editing in zebrafish using a CRISPR-Cas system. Nat Biotechnol 3: 227–229.
16. Jiang W, Zhou H, Bi H, Fromm F, Yang B, et al. (2013) demonstration of CRISPR/CAS9/sgRNA-mediated targeted gene modification in Arabidopsis, tobacco, sorghum and rice. Nucleic Acids Res 41:e188.
17. Miao J, Guo D, Zhang J, Huang Q, Qin G, et al. (2013) Targeted mutagenesis in rice using CRISPR-Cas system. Cell Res 23: 1233–1236.

18. Qi LS, Larson MH, Gilbert LA, Doudna JA, Weissman JS, et al. (2013) repurposing CRISPR as an RNA-guided platform for sequence-specific control of gene expression. Cell 5: 1173–1183.
19. Xiao A, Cheng Z, Kong L, Zhu Z, Linm S, et al. (2014) CasOT: a genome-wide Cas9/gRNA off-target searching tool. Bioinformatics Jan 2. [Epub ahead of print].
20. Mali P, Esvelt KM, Church GM. (2013) Cas9 as a versatile tool for engineering biology. Nat Methods 10: 957–963.
21. Feng Z, Zhang B, Ding W, Liu X, Yang DL, et al. (2013) efficient genome editing in plants using a CRISPR/Cas system. Cell Res 23: 1229–1232.
22. Li J, Norville JE, Aach J, McCormack M, Zhang D, Bush J, et al. (2013) multiplex and homologous recombination–mediated genome editing in Arabidopsis and Nicotiana benthamiana using guide RNA and Cas9. Nat Biotechnol 31: 688–691.
23. Mao Y, Zhang H, Xu N, Zhang B, Gou F, et al. (2013) application of CRISPR-cas system for efficient genome engineering in plants. Mol. Plant 6: 2008–2011.
24. Nekrasov V, Staskawicz B, Weigel D, Jones JDG, Kamoun S (2013) targeted mutagenesis in the model plant Nicotiana benthamiana using Cas9 RNA-guided endonuclease. Nat Biotechnol 31: 691–693.
25. Shan Q, Wang Y, Li J, Zhang Y, Chen K, et al. (2013) targeted genome modification of crop plants using a CRISPR-Cas system. Nat Biotechnol 31: 686–688.
26. Xie K, Yang Y (2013) RNA-guided genome editing in plants using A CRISPR-Cas System. Molecular Plant [Epub ahead of print].
27. Zhang X, Henriques R, Lin S, Niu Q, Chua N. (2006) agrobacterium-mediated transformation of the Arabidopsis thaliana using the floral dip method. Nature Protocol 1: 1–6.
28. Desfeux C, Clough SJ, Bent AF (2000) Female reproductive tissues are the primary target of agrobacterium-mediated transformation by the Arabidopsis floral-dip method. Plant Physiol 123: 895–904.
29. Yang Y, Costa A, Leonhardt N, Siegel RS, Schroder JI (2008) Isolation of a strong Arabidopsis guard cell promoter and its potential as a research tool. Plant Methods 4: doi:10.1186/1746-4811-4-6.
30. Fu Y, Foden JA, Khayter C, Maeder ML, Reyon D, et al. (2013) High-frequency off-target mutagenesis induced by CRISPR-Cas nucleases in human cells. Nat Biotechnol 31: 822–6. doi: 10.1038/nbt.2623. Epub 2013 Jun 23.
31. Xiao A, Wang Z, Hu Y, Wu Y, Luo Z, et al. (2013) chromosomal deletions and inversions mediated by TALENs and CRISPR/Cas in zebrafish. Nucleic Acids Res [Epub ahead of print].
32. Feng Z, Mao Y, Xu N, Zhang B, Wei P, et al. (2013) Multigeneration analysis reveals the inheritance, specificity, and patterns of CRISPR/Cas-induced gene modifications in Arabidopsis. Proc Natl Acad Sci U S A. 2014 Feb 18. [Epub ahead of print]

3

BMPRIA Mediated Signaling Is Essential for Temporomandibular Joint Development in Mice

Shuping Gu[1,9,¤a], **Weijie Wu**[1,2,9], **Chao Liu**[1,¤b], **Ling Yang**[1,3], **Cheng Sun**[1], **Wenduo Ye**[1], **Xihai Li**[1,4], **Jianquan Chen**[5], **Fanxin Long**[5], **YiPing Chen**[1]*

1 Department of Cell and Molecular Biology, Tulane University, New Orleans, Louisiana, United States of America, 2 Department of Dentistry, ZhongShan Hospital, FuDan University, Shanghai, P.R. China, 3 Guanghua School of Stomatology, Sun Yat-sen University, Guangzhou, Guangdong, P.R. China, 4 Academy of Integrative Medicine, Fujian University of Traditional Chinese Medicine, Fuzhou, Fujian, P.R. China, 5 Department of Internal Medicine, Washington University School of Medicine, St. Louis, Missouri, United States of America

Abstract

The central importance of BMP signaling in the development and homeostasis of synovial joint of appendicular skeleton has been well documented, but its role in the development of temporomandibular joint (TMJ), also classified as a synovial joint, remains completely unknown. In this study, we investigated the function of BMPRIA mediated signaling in TMJ development in mice by transgenic loss-of- and gain-of-function approaches. We found that BMPRIA is expressed in the cranial neural crest (CNC)-derived developing condyle and glenoid fossa, major components of TMJ, as well as the interzone mesenchymal cells. *Wnt1-Cre* mediated tissue specific inactivation of *Bmpr1a* in CNC lineage led to defective TMJ development, including failure of articular disc separation from a hypoplastic condyle, persistence of interzone cells, and failed formation of a functional fibrocartilage layer on the articular surface of the glenoid fossa and condyle, which could be at least partially attributed to the down-regulation of *Ihh* in the developing condyle and inhibition of apoptosis in the interzone. On the other hand, augmented BMPRIA signaling by *Wnt1-Cre* driven expression of a constitutively active form of *Bmpr1a* (*caBmpr1a*) inhibited osteogenesis of the glenoid fossa and converted the condylar primordium from secondary cartilage to primary cartilage associated with ectopic activation of Smad-dependent pathway but inhibition of JNK pathway, leading to TMJ agenesis. Our results present unambiguous evidence for an essential role of finely tuned BMPRIA mediated signaling in TMJ development.

Editor: Songtao Shi, University of Southern California, United States of America

Funding: WW was supported by a fellowship from ZhongShan Hospital, FuDan University. LY was supported by a scholarship from the China Scholarship Council (No. 201208440191). This work was supported by the National Institutes of Health grants (R01 DE14044, DE17792) to YC. The funders had no role in this design, data collection and analysis, decision to publish, or preparation of the manuscript.

Competing Interests: The authors have declared that no competing interests exist.

* Email: ychen@tulane.edu

¤a Current address: Shanghai Research Center for Model Organisms, Pudong, Shanghai, P.R. China
¤b Current address: Department of Biomedical Sciences, Baylor College of Dentistry, Texas A&M University, Dallas, Texas, United States of America

9 These authors contributed equally to this work.

Introduction

As an evolutionary creature, the temporomandibular joint (TMJ) is a unique synovial joint generated only in mammals and is involved in food capture and intake, speech, as well as maturation of the facial contour [1]. It is made of specific components originated from the skull base and the low jaw including the glenoid fossa, condyle, articular disc, ligaments, and joint capsule. Although defined as a synovial joint, the developmental process of TMJ differs significantly from the joints of appendicular skeletons that are generated by cleavage or segmentation within a single skeletal condensation [2]. The TMJ develops from two distinct mesenchymal condensations, the glenoid fossa blastema that ossifies primarily through intramembranous bone formation, and the condylar blastema that undergoes endochondral ossification. These two primordia are initially separated widely by intervening mesenchyme that was thought to later contribute to the articular disc and capsule, as well as the synovial lining of joint cavity [3,4] Subsequently, the condylar primordium, arising from the periosteum of the mandibular bone and therefore classified as secondary cartilage [5,6], grows rapidly towards the glenoid fossa, and meanwhile, the articular disc forming from a condensed stripe flanking the apex of the condyle and subsequently separating from the latter, divides the interzone into the upper and lower joint cavities [7]. In mice, the mesenchymal condensation of condyle appears at embryonic day 13.5 (E13.5) and the glenoid fossa at E14.5 [8]. At E15.5, the shape of glenoid fossa and condyle has been established, and at E16.5, the upper synovial cavity becomes discernible with a disc beginning to form. Subsequently at E17.5, the lower joint cavity appears as a definite articular disc separates from the apex of the condyle. This intricate multi-step developmental process is regulated by intrinsic and extrinsic factors.

As for intrinsic constituents, the significance of genetic factors has attracted the attention of the field. Gene targeting studies have revealed essential roles for a number of transcription factors and growth factors in TMJ development, as evidenced by the absence of condylar cartilage in mice carrying mutations in *Sox9*, *Runx2*, or *Tgfbr2*, and by the abnormal development of mice carrying mutations in *Shox2* or *Spry1* and *Spry2* [8–14]. Ihh, which plays a pivotal role in long bone development and digit joint formation [15], has been implicated in TMJ development by initiating the formation of articular disc and instructing the disc to undergo proper morphogenesis and to separate from the condyle, as well as in maintaining proper structure and function of the TMJ after it forms [16–18]. Lack of Ihh or its downstream effector Gli2 results in missing of a distinct disc in the TMJ [16,17]. In addition, extrinsic factors such as biomechanical force also contribute to TMJ development [2].

Bone morphogenetic proteins (BMPs) exert diverse biological functions during development and postnatal homeostasis. BMP signals are transduced into cells through the type I and type II transmembrane serine/threonine kinase complexes by activating Smad-dependent (canonical) pathway, as well as Smad-independent (non-canonical) pathway via activation of the mitogen-activated protein kinase (MAPK) signaling [19]. Extensive studies have established critical roles for BMP signaling in skeletal development and joint morphogenesis, particularly in joint formation of long bones. Joint formation in the appendage skeletons begins with the formation of a condensed cell stripe known as interzone in the developing cartilage template [20]. Cells in the edges of the interzone give rise to the articular cartilage that covers the ends of the adjacent skeletal elements, while cells in the middle of the interzone undergo programmed cell death, leading to physical separation of the contiguous cartilage element and formation of joint cavity [21]. Several members of BMP family, including *Bmp2*, *Bmp4*, *Gdf5*, *Gdf6*, *Gdf7*, are expressed in the interzone along with BMP antagonists *Chondin* and *Noggin* [22–26]. Mice carrying mutations in *Gdf5* or *Gdf6* exhibit lack of joint formation at specific locations [25,27], demonstrating a direct action of BMP signaling in joint morphogenesis. On the other hand, elevated BMP signaling also blocks joint formation, as manifested by failure in joint formation in the limbs of *Noggin* mutant mice [28]. These loss-of- and gain-of-function studies indicate an essential role for tightly regulated BMP activity in synovial joint formation. Furthermore, BMP signaling is also involved in postnatal joint homeostasis and tissue remodeling [26,29].

Being one of the two primary BMP type I receptors (BMPRIA and BMPRIB), BMPRIA plays crucial roles in skeleton patterning and development. In developing limb skeletons, *BmprIa* is expressed in the joint interzone, perichondrium, periarticular cartilage, and hypertrophic chondrocytes [26,30–32]. Tissue specific deletion of *BmprIa* in cartilage lineage leads to chondrodysplasia attributed at least partially to the defective cell proliferation as well as the premature hypertrophy of chondrocytes associated with down-regulation of *Ihh* [33,34]. While joint defect was not identified in mice carrying cartilage specific inactivation of *BmprIa*, mice carrying tissue specific inactivation of *BmprIa* in the interzone indeed exhibited missed joints in the ankles [26], indicating a requirement of BMPRIA mediated signaling in joint formation.

Despite a wealth of documents on BMP signaling in bone and joint formation of appendicular skeletons, little is known about its role in TMJ development. Thus far, the only line of evidence implicating a possible involvement of BMP signaling in TMJ formation is that *Bmp2* and *Bmp7* were found to be expressed in the developing condyle [17,35]. To gain an insight into BMP signaling in TMJ development, in this study, we used transgenic loss-of- and gain-of-function approaches to investigate the function of BMPRIA mediated signaling in TMJ development.

Materials and Methods

Ethics statement

Experiments that involved use of animals in this study was approved by the Institutional Animal Care and Use Committee (IACUC) of Tulane University (protocol number: 0367R) and was in strict accordance with the recommendations in the Guide for Care and Use of Laboratory Animals of the National Institutes of Health.

Animal and sample collection

The generation and identification of transgenic and gene-targeted animals, including *Wnt1-Cre*, *BmprIa^f/f*, and *pMes-caBmprIa* that carries a conditional constitutively active form (with Gln203 to Asp change) of *BmprIa* transgenic allele, have been described previously [15,31,36,37]. *Wnt1-Cre;BmprI^f/f* embryos were obtained by crossing *Wnt1-Cre;BmprIa^f/+* mice with *BmprIa^f/f* line. *Wnt1-Cre;pMes-caBmprIa* embryos were generated by mating *Wnt1-Cre* mice with *pMes-caBmprIa* transgenic line. *Ihh* null mutants were harvested from intercross of *Ihh* heterozygous mice. Embryos with *BmprIa* deficiency in their neural crest cells (*Wnt1-Cre;BmprIa^f/f*) die around E12.5 due to norepinephrine depletion [38,39]. To prevent early embryonic lethality, pregnant females were administrated with the β-adrenergic receptor agonist from 7.5 postcoitum (dpc) on by supplementing drinking water of dams with 200 µg/ml isoproterenol, which would allow *Wnt1-Cre;BmprIa^f/f* embryos to survive to term [40,41]. Embryos were collected from the timed pregnant females, and head samples were dissected in ice cold PBS, fixed individually in 4% paraformaldehyde (PFA) or z-fix (ANATECH Ltd; #170) overnight at 4°C, and tail clip from each embryo was used for PCR-based genotyping, respectively. Mutant and control heads were positioned for serial coronal sections through the TMJ. Comparable sections through the apex of the condyle were picked up for histological, in situ hybridization, immunostaining, and cell apoptosis analyses.

Histology, in situ hybridization, and immunohistochemistry, and Tunel assay

For histological study, paraffin sections were made at 6 µm and subjected for standard Hematoxylin/Eosin staining or Azoncarmine G/Aniline blue staining, as described [42]. Five mutant samples at each stage examined were used to ensure consistency of the phenotype. For in situ hybridization analyses, sections were cut at 10 µm and pretreated with proteinase K and hybridized with appropriate probes. Transcripts were detected by color reaction using BM purple (Roche) as described previously [43]. For immunohistochemical staining, frozen sections, made at 8 µm, were blocked with 4% goat serum and then incubated with primary antibodies against BMPRIA (Abcam; ab38560), Lubricin (Santa Cruz; sc-9854), pSmad1/5, pJNK, pERK, and p-p38 (from Cell Signaling; #9516, #9255, #4370, and #9211), respectively, at 4°C overnight. After washing, samples were incubated with secondary antibodies (Alexa Fluor488 goat anti-rabbit IgG from Invitrogen; #A-11034), counterstained with DAPI, and visualized under fluorescent microscope. Negative controls without primary antibodies were included in parallel. At least three samples of each genotype were used for histology, in situ hybridization, and immunohistochemistry analyses. Terminal deoxynucleotidyl transferase dUTP nick end labeling (Tunel) assay was applied to detect apoptotic cells using In Situ Cell Death Detection Kit (Roche), as described previously [44,45]. Three samples of each genotype were subjected to Tunel assays. Tunel-positive cells within the interzone were counted and presented as percentage of total cells within arbitrarily defined areas. Student's t-test was used to determine the significance of difference between wild type controls and mutants, and the results were presented as P value.

Results

Inactivation of *Bmprla* in CNC lineage leads to defective TMJ formation

To investigate the role of *BmprIa* in TMJ development, we began with examination of BMPRIA expression by immunohistochemistry. At E14.5 when both the primordial condyle and glenoid fossa become discernible, BMPRIA was found present in the developing condyle and glenoid fossa as well as the interzone, with a relatively low level in the condylar cartilage (Fig. 1A). At E15.5, BMPRIA expression retained in the condyle, glenoid fossa, and interzone, with an increased level in chondrocytes undergoing hypertrophy (Fig. 1C). This expression pattern is similar to that in the developing limb skeleton including the joints [26,30–32], suggesting a role for BMPRIA in TMJ morphogenesis. Since the condyle, glenoid fossa, and interzone cells are all derived from CNCs [8], we inactivated *BmprIa* in CNC lineage using the *Wnt1-Cre* transgenic allele. Immunohistochemistry confirmed the absence of BMPRIA in the developing TMJ of *Wnt1-Cre;BmprIa^{f/f}* mice (Fig. 1B).

Histological analyses showed that the initial condensation of the condylar anlage in mutant mice appeared comparable to littermate controls at E13.5 (Fig. 2A, 2B). At E15.5, the morphology of the glenoid fossa did not exhibit an obvious difference between mutants and controls, but the size of the mutant condyle was reduced as compared to controls (Fig. 2C,

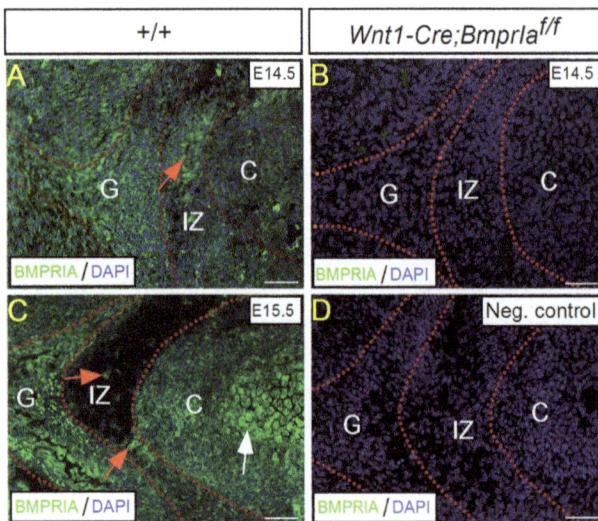

Figure 1. Expression of BMPRIA in the developing TMJ. (A–D) Immunohistochemistry shows expression of BMPRIA in the condylar cartilage, interzone, and glenoid fossa of E14.5 (A) and E15.5 (C) wild type animals, but a lack of staining on the *Wnt1-Cre;BmprIa^{f/f}* TMJ (B) and on the negative control (D). Red arrows point to positive staining in the interzone, and white arrow points to the hypertrophic region where strong expression is detected. Abbreviation: C, condyle; G, glenoid fossa; IZ, interzone. Scale bar = 100 μm.

2D). At E18.5, the control TMJ displayed distinct structures, including a definite articular disc, the upper and lower synovial cavities, and the fibrocartilage/synovial membrane on the articular surface of the glenoid fossa and condyle (Fig. 2E, 2G). However, at this stage, the mutant TMJ exhibited a number of severe defects, including a hypoplastic condyle, lack of a definite disc, failed formation of an upper joint cavity evidenced by the existence of loose connective tissue in the interzone, as well as the absence of the fibrocartilage/synovial membrane of the glenoid fossa (Fig. 2F). Close examination of the mutant condyle at E18.5 revealed the formation of a disc-like structure that failed to separate from the apex of the condyle, leading to an absence of the lower joint cavity (Fig. 2F, 2H). The lack of a synovial joint cavity in the *Wnt1-Cre;BmprIa^{f/f}* TMJ was further confirmed by the absent expression of Lubricin, a key component of joint fluids [46,47], as compared to its abundant expression in controls (Fig. 2I, 2J).

Delayed chondrocyte maturation in the *Wnt1-Cre;BmprIa^{f/f}* condyle

Despite being a secondary cartilage, the growth of condylar cartilage takes the similar chondrogenesis and endochondral ossification process as that in long bone formation. Since BMPRIA mediated signaling is known to regulate primary cartilage differentiation [33,34], we set to examine chondrogenic differentiation process in the *Wnt1-Cre;BmprIa^{f/f}* condyle. In the developing condyle, mesenchymal condensation appears at E13.5, and chondrogenic differentiation occurs at E14.5, and hypertrophy initiates at E15.5 [8]. We found that the timing of initial condensation of the condylar anlagen, as indicated by the expression of *Sox9*, and chondrogenic differentiation, determined by *Col II* expression, was comparable between wild type controls and *Wnt1-Cre;BmprIa^{f/f}* mice (Fig. 3A–D). However, the mutant condyle exhibited a delayed terminal hypertrophy of chondrocytes, as assessed by the delayed *Col X* expression (Fig. 3E–H), and the longer distance between the apex and the beginning of hypertrophic zone in the mutant condyle at E18.5 (Fig. 3G, 3H). This phenotype differs from that in long bones where inactivation of *BmprIa* in condrocytes causes premature chondrocyte differentiation [34], likely due to different properties of primary v.s secondary cartilage.

Down-regulation of *Ihh* and inhibition of apoptosis in the *Wnt1-Cre;BmprIa^{f/f}* TMJ

The similar TMJ phenotype between *Wnt1-Cre;BmprIa^{f/f}* mice and the mice carrying mutations in *Ihh* or its downstream effectors [16,17], particularly the failure of disc separation, persistent interzone cells, and lack of fibrocartilaginous articular surface layer of the glenoid fossa in both mutants (Fig. 2F, 2H; Fig. 4B, 4D), and the overlapped expression pattern of *BmprIa* with *Ihh* in the developing condyle (Fig. 1) [8,17], prompted us to examine *Ihh* expression in the developing condyle of *Wnt1-Cre;BmprIa^{f/f}* mice. In situ hybridization assay revealed a dramatic down-regulation of *Ihh* expression in the developing condyle of the mutants at E14.5 and E15.5, as compared to controls (Fig. 4G–J), consistent with the role of BMPRIA as a positive regulator of *Ihh* expression [34,48]. *BmprIa* thus likely acts through *Ihh* to regulate TMJ formation.

Because *BmprIa* is also expressed in the interzone mesenchymal cells that contribute to the articular disc and synovial membrane of the TMJ, the lack of fibrocartilage layer of the glenoid fossa and the persistence of interzone cells in the *Wnt1-Cre;BmprIa^{f/f}* TMJ could be attributed either directly to the lack of *BmprIa* in these

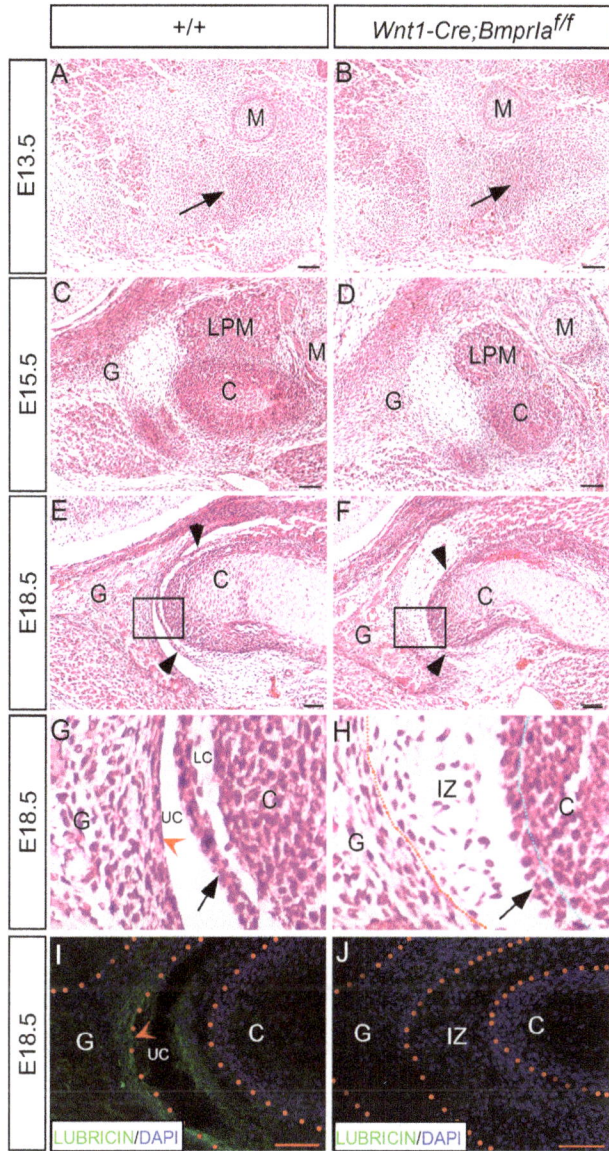

Figure 2. *Wnt1-Cre;Bmprla*$^{f/f}$ **mice display TMJ defects**. (A–H) H&E staining shows histology of the developing TMJ of wild type controls (A, C, E, G) and mutants (B, D, F, H). Note that the initial condensation of the condylar anlagen at E13.5 (A, B) and the morphology of the glenoid fossa at E15.5 (C, D) appear comparable between the controls and mutants. However, the size of the mutant condyle is reduced at E15.5 (D). At E18.5, distinct structures including a definite disc, the upper and lower joint cavities, and the articular surface of the glenod fossa are well present in the control TMJ (E, G). However, in mutants, while a disc-like compact layer could be identified closely associated with the apex of the condyle, it fails to separate to form a distinct disc. In addition, the interzone cells persist, and a fibrocartilage layer fails to form on the articular surface of the glenoid fossa (F, H). (I, J) Immunnohistochemistry reveals expression of Lubricin in the synovial membrane of the control TMJ (I), and the complete absence of Lubricin in the mutant TMJ (J). Arrows in (A, B) point to the condylar condensation, and in (G, H) point to the disc. Arrowheads in (E, F) point to the disc. Red arrowhead points to the articular surface in (G) and the synovial membrane in (I). Abbreviation: C, condyle; G, glenoid fossa; M, Meckel's cartilage; IZ, interzone; LC, lower cavity; UC, upper cavity; LPM, lateral pterygoid muscle. Scale bar = 50 μm.

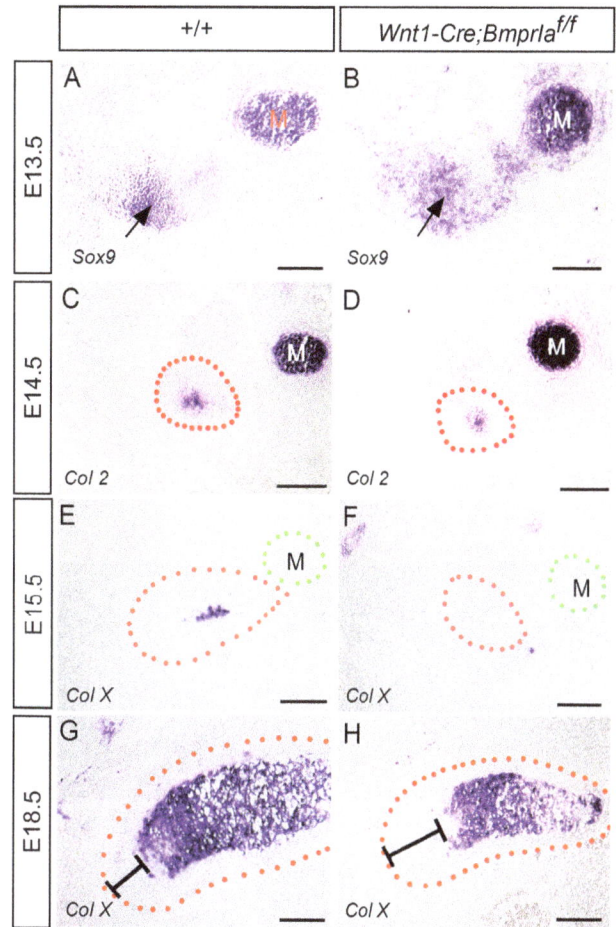

Figure 3. Delayed hypertrophic differentiation in the *Wnt1-Cre;Bmprla*$^{f/f}$ condylar cartilage. (A, B) In situ hybridization shows *Sox9* expression in the condylar condensation (arrow) of wild type (A) and mutant (B) at E13.5. (C, D) *Col II* expression exhibits comparable pattern in the condyle of wild type (C) and mutant (D) at E14.5. (E–H) In situ hybridization reveals *Col X* expression in the wild type condyle at E15.5 (E) and E18.5 (G). However, *Col X* expression is not detected in the mutant condyle at E15.5 (F), but is seen at E18.5 (H). The distance between the apex of the condyle and the beginning of hypertrophic zone is longer than that in the control condyle (G, H). Abbreviation: M, Meckel's cartilage. Scale bar = 100 μm.

cells or indirectly to the significantly reduced level of *Ihh* in the condyle. To distinguish these alternatives, we conducted immunohistochemistry to examine BMPRIA expression in the *Ihh*$^{-/-}$ TMJ at E14.5. We found that although BMPRIA expression appeared comparable in the glenoid fossa of controls and mutants, its expression level was significantly reduced in the interzone and the condyle of the *Ihh* mutant TMJ (Fig. 4E–4F').

While the mechanism of cavitation in TMJ development remains to be addressed, programmed cell death in the interzone is regarded as a critical cellular mechanism for joint cavity formation in long bones [21]. Since BMPRIA mediated signaling is required for programmed cell death in the limb, particularly in the interdigital region [26,49], we wondered if the persistence of the loose connective tissue in the interzone of the *Wnt1-Cre;BmprIa*$^{f/f}$ TMJ is a consequence of reduced level of apoptosis. Tunel assay indeed revealed abundant apoptotic cells specifically in the interzone of the control TMJ at E15.5 (Fig. 4K, 4K'). In contrast, Tunel assay detected a significantly reduced level of

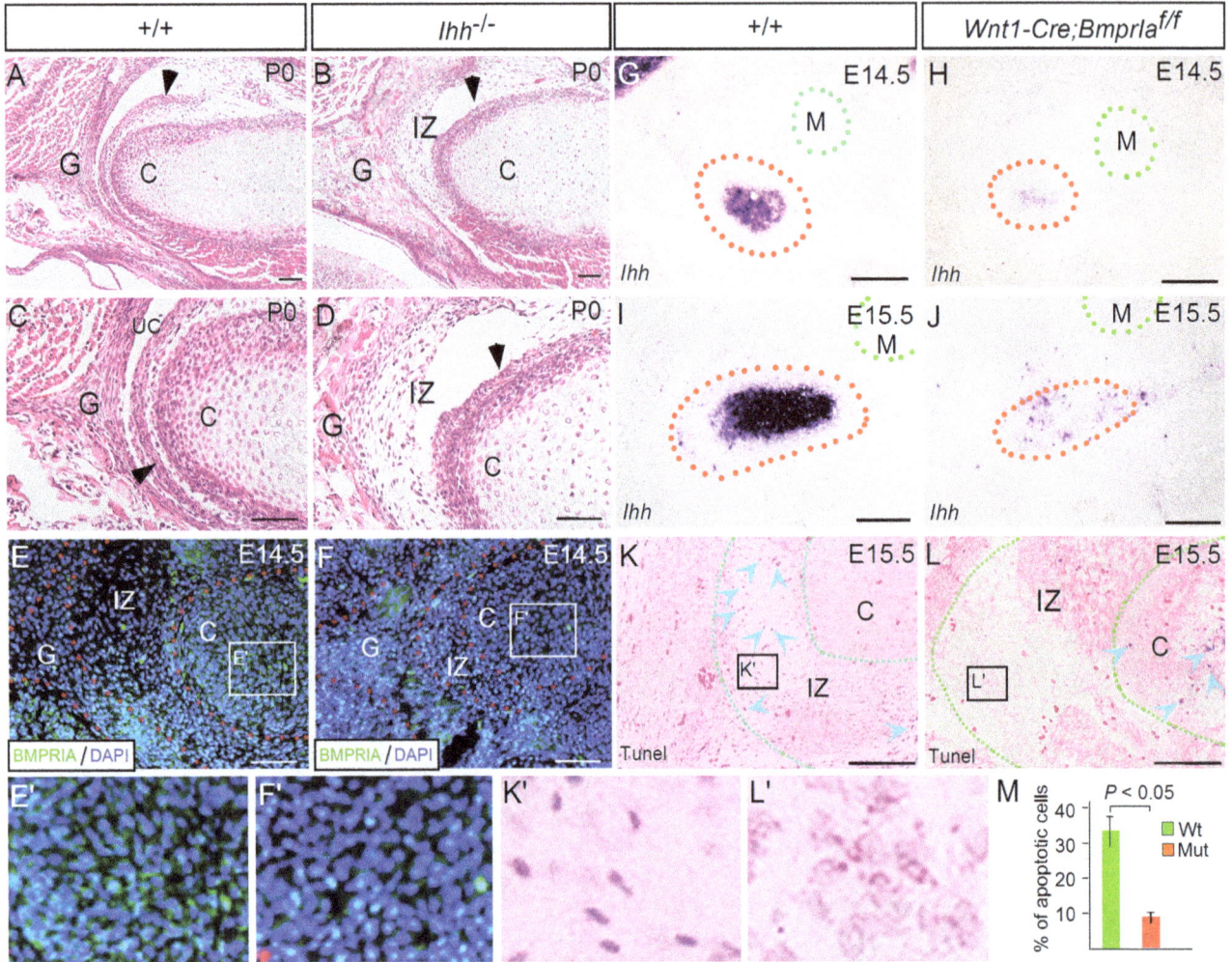

Figure 4. TMJ defects in *Ihh* mutants and reduced *Ihh* expression in the *Wnt1-Cre;Bmprla^{f/f}* condyle. (A–D) H&E staining reveals TMJ defects in the *Ihh* mutant TMJ (B, D), as compared to control (A, C) at P0. In mutant, a disc-like structure (arrowhead) forms but fails to separate from the apex of the condyle, the fibrocartilage layer fails to form on the articular surface of the glenoid fossa, and the interzone cells persist (B, D), as compared to the formation of distinct TMJ structures, including the disc (arrowhead), the fibrocartilginous articular surface, and the clear upper joint cavity, in controls (A, C). (E, E', F, F') Immunohistochemistry reveals significantly down-regulated BMPRIA expression in the condylar cartilage and interzone of *Ihh* mutant at E14.5 (F, F'), as compared to littermate control (E, E'). (G–J) In situ hybridization shows a dramatic down-regulation of *Ihh* in the condylar cartilage of *Wnt1-Cre;Bmprla^{f/f}* mice at both E14.5 (H) and E15.5 (J), as compared to controls (G, I). Tunel assay reveals numerous apoptotic cells (arrowheads) in the interzone of wild type control at E15.5 (K, K'), but very few apoptotic cells in the interzone of the *Wnt1-Cre;Bmprla^{f/f}* TMJ at the same age (L, L'). In contrast, some apoptotic cells (arrowheads) were observed in the mutant condyle (L). (M) Comparison of the percentage of apoptotic cells in the interzone of controls and mutants. Standard deviation values were indicated as the error bars, and the Student's *t*-test was used to determine the significance of difference between control and mutant, as presented as *P* value. Abbreviation: C, condyle; G, glenoid fossa; M, Meckel's cartilage; IZ, interzone; UC, upper cavity. Scale bar = 100 µm.

apoptotic cells in the interzone as well as some apoptotic cells in the condyle of the mutant TMJ (Fig. 4L, 4L', 4M). These observations suggest that similar to joint cavity formation in long bones, programmed cell death in the interzone also represents a critical cellular mechanism for joint cavitation during TMJ formation.

Augmented BMPRIA signaling in CNC lineage leads to TMJ agenesis

To further investigate the role of BMPRIA signaling in TMJ morphogenesis, we took a gain-of-function approach by transgenic expression of a constitutively active form of *BmprIa* (*pMes-caBmprIa*) [37] in CNC cells using the *Wnt1-Cre* allele. In situ hybridization revealed expression of *BmprIa* in the condensing

condylar blastema, the forming site of glenoid fossa, and cells between them, but not in Meckel's cartilage of wild type embryo at E13.5 (Fig. 5A). In *Wnt1-Cre;pMes-caBmprIa* mice at the same stage, strong and wide spread expression of *BmprIa* was found in the TMJ forming region and its surrounding tissues including Meckel's cartilage, indicating successful transgenic expression of *BmprIa* in CNC lineage (Fig. 5B). We and others have shown previously that elevated BMPRIA mediated signaling in CNC cells leads to a spectrum of craniofacial bone defects, including cleft secondary palate, ectopic cartilage formation, and craniosynostosis [50,51]. Histological analysis of the developing TMJ of *Wnt1-Cre;pMes-caBmprIa* mice identified unique TMJ developmental defects. Although the condensation of the condylar blastema occurred similarly to controls at E13.5 (Fig. 5C, 5D), the size of

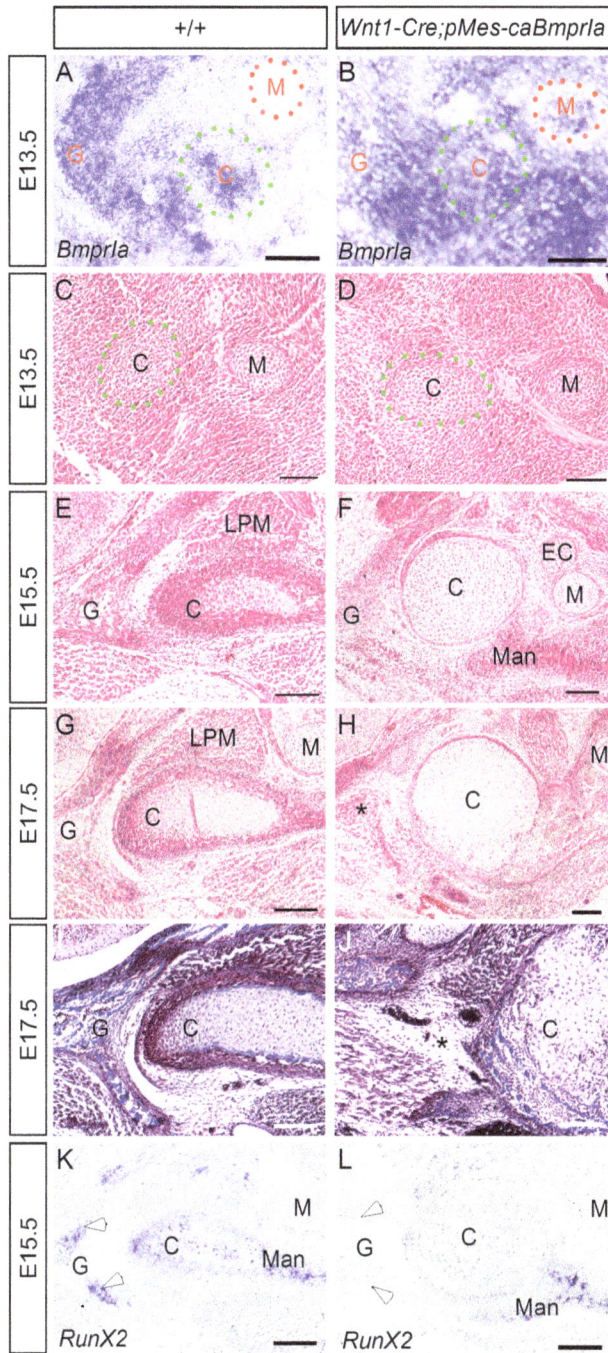

Figure 5. Augmented BMPRIA signaling in CNC cells leads to TMJ agenesis. (A, B) In situ hybridization shows expression of *Bmprla* in the condylar condensation, the future glenoid fossa forming site, and the interzone region, but not in Meckel's cartilage of an E13.5 wild type embryo (A), and an enhanced *Bmprla* expression in the TMJ forming site as well as the surrounding tissues including Meckel's cartilage in an E13.5 *Wnt1-Cre;pMes-cBmprla* embryo (B). (C–H) H&E staining reveals initial condensation of condylar anlagen in control and transgenic animals at E13.5 (C, D), growth and differentiation into primary cartilage of the condylar cartilage and lack of osteogenesis in the glenoid fossa in the transgenic TMJ at E15.5 (F) and E17.5 (H). (I, J) Azocarmine G/Aniline blue staining reveals glenoid fossa degeneration, evidenced by lack of bone formation, in transgenic mouse (J), as compared to control (I). (K, L) In situ hybridization assay shows expression of *Runx2* in the forming genloid fossa, perichondral region of the condyle, and mandibular bone of an E15.5 wild type control (K), but an absent expression of *Runx2* in

the glenoid fossa and a reduced expression in the condylar cartilage of an E15.5 transgenic animal (L). Note retention of *Runx2* expression in the mandibular bone of transgenic mouse (L). Asterisk in (H, J) indicates the site of glenoid fossa degeneration. Open arrowheads in (K, L) point to *Runx2* expression sites in the glenoid fossa. Abbreviation: C, condylar cartilage; G, glenoid fossa; M, Meckel's cartilage; EC, ectopic cartilage; LMP, lateral pterygoid muscle; Man, mandibular bone. Scale bar = 100 μm.

the condylar cartilage in transgenic animals became noticeably enlarged and the entire condylar cartilage appeared to become hypertrophic at E15.5, as compared to controls (Fig. 5E, 5F). Furthermore, unlike in controls that osteogenesis has begun in the glenoid fossa at this stage, the transgenic glenoid fossa failed to take osteogenic differentiation. At E17.5, the glenoid fossa became degenerated in transgenic animal (Fig. 5H). The failure of osteogenic differentiation in the glenoid fossa was further confirmed by the lack of bone formation in the glenoid fossa, assessed by Azocarmine G/Aniline blue staining, and by the absent expression of *Runx2*, a molecular marker for osteoblasts (Fig. 5I–L). Additionally, despite its expression in the mandibular bone, *Runx2* expression is also down-regulated in the condylar cartilage of *Wnt1-Cre;pMes-caBmprIa* mice (Fig. 5L). By E18.5, both the condyle and glenoid fossa degenerated and became unrecognizable (data not shown).

Enhanced BMPRIA signaling converts the condylar primordium from secondary cartilage to primary cartilage by ectopic activation of canonical signaling and inhibition of JNK signaling

As a secondary cartilage, the condylar cartilage expresses type I collagen (Col I), making it distinct from the primary cartilage [52]. Because of its aberrant differentiation, we wondered if the condylar cartilaginous element of *Wnt1-Cre;pMes-caBmprIa* mice retained its secondary cartilage characteristics. In situ hybridization assay revealed *Col I* expression in the control condylar cartilage and mandibular bone, but the absence of *Col I* in the transgenic condylar cartilaginous element despite its expression in the mandibular bone at E14.5 (Fig. 6A, 6B). However, the expression of *Col II* and *Col X* in the transgenic condylar cartilage confirmed its cartilage fate (Fig. 6C–F). We thus conclude that the *Wnt1-Cre;pMes-caBmprIa* condylar primordium adopts a fate of primary cartilage in response to an augmented BMPRIA-mediated signaling. Moreover, Tunel assay revealed an extensive apoptotic event in the transgenic condylar cartilage, beginning at E15.5, as compared to the lack of apoptosis in the control condyle at the same stage (Fig. 6G, 6H), which apparently contributes to the degeneration and disappearance of the condylar cartilage in transgenic animals.

We have shown previously that the expression of *caBmprIa* in CNC lineage induces ectopic activation of Smad1/5/8 signaling as well as p38 signaling in the developing palatal shelves [51]. We therefore set to examine alterations in BMP canonical and non-canonical signaling pathways in the condylar cartilage of *Wnt1-Cre;pMes-caBmprIa* mice by immunohistochemistry. Interestingly, we detected no activation of Smad-dependent as well as p38 and Erk1/2 pathways in the control condyle, as assessed by the lack of pSmad1/5, p-p38, and p-Erk1/2, but observed activity of p-JNK signaling (Fig. 7A, 7C,7E, 7G). In contrast, the condylar cartilage of *Wnt1-Cre;pMes-caBmprIa* mice exhibited an ectopic activation of pSmad1/5, but an absence of pJNK signaling, along with unaltered p38 and pEek1/2 pathways (Fig. 7B, 7D; 7F, 7H). These observations indicate that the switch between BMP canonical and non-canonical signaling pathways likely underlies the fate conversion from the secondary to primary cartilage, with the Smad-dependent signaling favoring the primary cartilage fate.

Discussion

Compared to synovial joint formation in the appendicular skeletons, TMJ development and the underlying molecular mechanisms are relatively under-studied. While the critical roles of BMP signaling in long bone joint development and homeostasis have been well documented [21,26,29], its role in TMJ formation remained completely unknown. In this study, we present evidence that BMPRIA mediated signaling is essential for TMJ morphogenesis, and overly activated BMPRIA signaling is detrimental to TMJ formation. Our results also reveal apoptosis in the interzone as a potential cellular mechanism for cavitation of the TMJ, similar to that in long bone joint formation.

A BMPRIA-Ihh positive regulatory pathway regulates TMJ development

It has been well established that BMP and Ihh signaling interact to regulate chondrocyte proliferation and hypertrophic differentiation [34,53–55]. In the developing limb, BMP signaling, particularly the BMPRIA mediated pathway, positively regulates *Ihh* expression that could also activate in the perichondrium the expression of several BMP ligands, forming a BMP-Ihh positive feedback loop [34,53,54]. Although there is no evidence for an interaction of BMP and Ihh signaling in joint development, the fact that several BMP ligands and receptor are expressed in the interzone and that mutations in either *Noggin* or *Ihh* lead to joint defects including joint ablation in limbs implies the existence of such interaction [15,21,28]. Indeed, in the current study, we found that the ablation of *BmprIa* in CNC lineage produces TMJ defects resembling that in *Ihh* mutant. In both mutants, a functional TMJ failed to form, evidenced by the absent Lubricin expression (Fig. 2) [16]. In addition, both mutants displayed a lack of a distinct articular disc due to failed disc separation from the condyle, persistence of the interzone cells, as well as absent synovial membrane on the articular surface of the glenoid fossa. Consistent with Ihh function in disc formation and separation during TMJ morphogenesis [16,17], we found a dramatic down-regulation of *Ihh* expression in the developing condyle of *Wnt1-Cre;BmprIa^{f/f}* mice. This result also indicates that similar to its role in developing appendicular skeletons, BMPRIA mediated signaling also acts as a positive regulator of *Ihh* expression in the condylar cartilage. On the other hand, *BmprIa* expression was significantly reduced in the condyle and interzone of the *Ihh^{-/-}* TMJ, suggesting the existence of a BMPRIA-Ihh positive feedback loop in the developing TMJ. However, the delayed hypertrophic differentiation observed in the condylar cartilage of *Wnt1-Cre;BmprIa^{f/f}* mice appears to be opposite to the premature hypertrophic differentiation defect seen in the *Ihh^{-/-}* condyle as well as in the long bones of mice carrying *BmprIa* deletion in chondrocyte lineage [16,34]. Although the underlying mechanism is currently unknown, the discrepancy would likely be attributed to the residual *Ihh* expression in the *Wnt1-Cre;BmprIa^{f/f}* condyle as well as the condyle's property as secondary cartilage.

Nevertheless, based on above mentioned observations and the established roles for BMP and Ihh signaling in limb development, we propose a model to summarize the function of the BMPRIA-Ihh regulatory pathway in regulating distinct steps during TMJ morphogenesis (Fig. 8). In this model, BMPRIA and Ihh regulate the expression of each other to coordinate chondrocyte proliferation and differentiation in the developing condyle. Meanwhile,

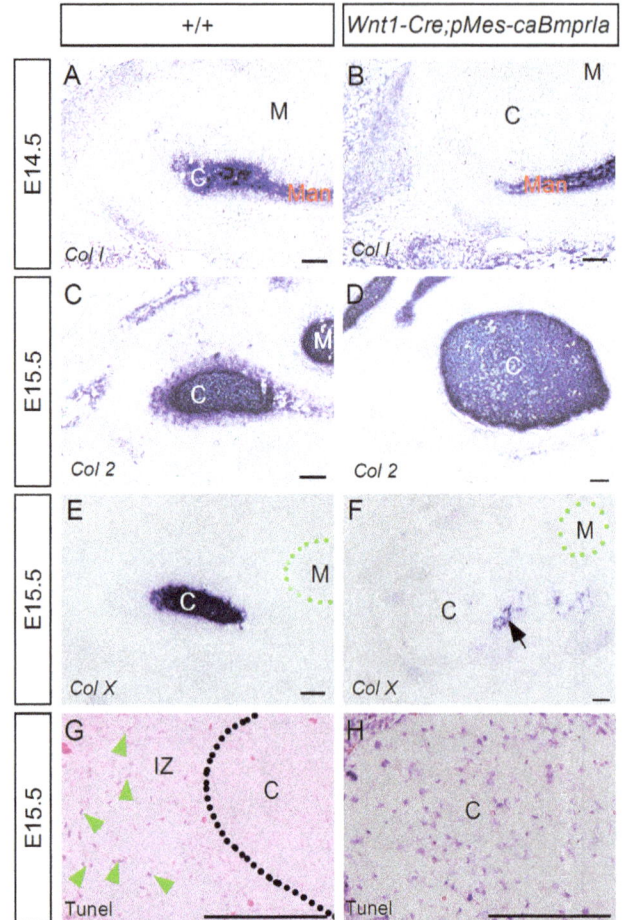

Figure 6. Elevated BMPRIA signaling converts secondary cartilage of the condylar primordium to primary cartilage and induces extensive cell death. (A–F) In situ hybridization detects *Col I* expression in the condylar cartilage and mandibular bone of an E14.5 control embryo (A), and in the mandibular bone of an E14.5 *Wnt1-Cre,pMes-caBmprIa* mice (B). However, *Col I* expression is not detected in the condylar cartilage of transgenic animal (B). *Col II* (C, D) and *Col X* (E, F) expression is observed in the condylar cartilage of control (C, E) and transgenic embryo (D, F) at E15.5. (G, H) Tunel assay reveals numerous apoptotic cells in the interzone but not in the condyle of the E15.5 wild type TMJ, but extensive cell death in the condylar cartilage of E15.5 transgenic embryo (H). Arrow in (F) points to *Col X* expression domain, and arrowheads in (G) point to apoptotic cells. Abbreviation: C, condylar cartilage; M, Meckel's cartilage; IZ, interzone. Scale bar = 100 μm.

Ihh, produced in the condyle, diffuses into the interzone to regulate disc separation and to maintain the expression of BMPRIA that in turn acts in a cell autonomous manner to regulate synovial membrane formation and to trigger apoptosis.

Augmented BMPRIA signaling in CNCs converts the secondary cartilage of the condylar primordium to primary cartilage

Despite being a secondary cartilage, the condyle shares many similarities with primary cartilage in development, including the expression of genes known to be important for cartilage growth and differentiation such as *Bmp2*, *Bmp7*, *Sox9*, *Runx2*, *Osterix*, *Ihh*, *Pthrp*, *Vegf*, *Col II* and *Col X* [8,16,17,35,56,57]. However, the condyle also differs from primary cartilage by its expression of Col I and Col II simultaneously and its capability of differentiating

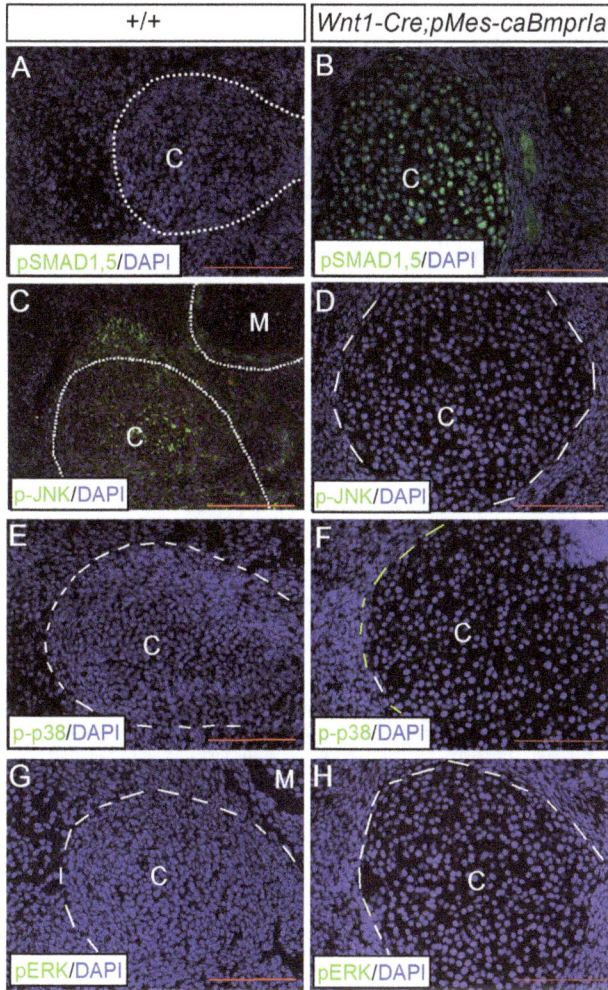

Figure 7. Enhanced BMPRIA signaling activates Smad-dependent pathway but inhibits JNK signaling pathway in the condylar cartilage. Immunohistochemistry shows absent pSmad1/5, p-p38, and pERK, but the presence of pJNK in the E14.5 control condyle (A, C, E, G), and the presence of pSmad1/5, but absent pJNK as well as p-p38 and pERK in the transgenic condylar cartilage (B, D, F, H). Abbreviation: C, condylar cartilage; G, glenoid fossa; M, Meckel's cartilage. Scale bar = 100 µm.

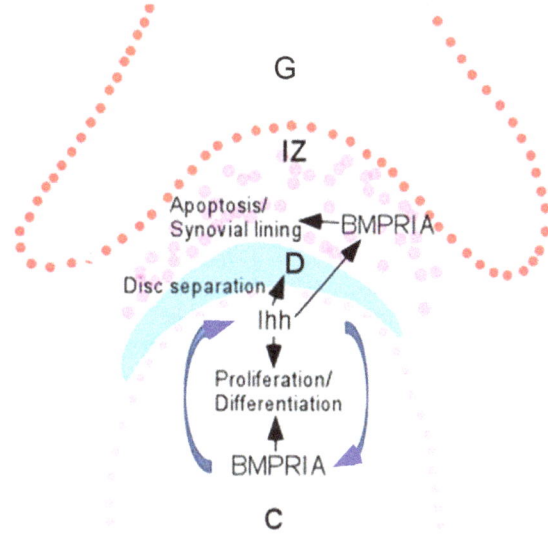

Figure 8. A model illustrating the interaction between BMPRIA and Ihh and their functions in regulating distinct steps during TMJ morphogenesis. Abbreviation: C, condyle; D, disc; G, glenoid fossa; IZ, interzone.

into either chondrocytes or osteoblasts [52]. Although the underlying mechanism for fate determination of primary v.s. secondary cartilage remains elusive, our results that augmented BMPRIA signaling is able to convert the condylar cartilage to a primary cartilage, evidenced by the lack of *Col I* expression, implicate BMP signaling in such fate decision. *BmprIa* is expressed in the cartilage condensations of both long bone and condyle (this study) [33], suggesting its role in the fate decision of both primary and secondary cartilages. It appears that a tightly tuned BMPRIA signaling is essential for fate determination of secondary cartilage. Accompanied with this fate conversion is the switch of the downstream BMP signaling pathway from the non-canonical JNK signaling to the Smad-dependent pathway in the condylar cartilage, suggesting that higher activity of BMPRIA signaling preferentially activates the Smad-dependent pathway, which favors primary cartilage formation. Indeed, the lack of Smad-dependent signaling in the condyle (this study) and the strong expression of pSmad1/5/8 in the limb cartilage condensation [33]

further support this notion. Furthermore, consistent with the role of BMPRIA signaling in apoptosis in the developing limb [26,32], overly activated BMPRIA signaling causes extensive apoptosis in the condylar cartilage and leads to condylar cartilage degeneration, indicating a detrimental effect on chondrocyte survival.

Despite an essential role for *BmprIa* in chondrogenesis and endochondral bone formation, in our current study, we found that inactivation of *BmprIa* did not affect glenoid fossa osteogenesis, suggesting that *BmprIa* may not be essential for intramembranous bone formation. However, elevated BMPRIA signaling instead inhibits osteogenesis in the glenoid fossa. Thus, although BMP signaling is generally accepted as a positive regulator of osteogenesis, elevated BMP signaling could have an opposite effect, depending on the tissue and cell types. Since a normal developing condyle is required to sustain the development of the glenoid fossa [9], the degeneration of the glenoid fossa in *Wnt1Cre;pMes-caBmprIa* mice could be the consequence of failed osteogenesis, or result from an abnormal condylar cartilage with altered property, or both.

Apoptosis as a cellular mechanism of TMJ cavitation

Cell death in the middle of the interzone is considered the cellular mechanism for physical separation of the contiguous cartilage elements during joint formation in long bones [21,58–61]. In the developing TMJ, although the primordial condyle and glenoid fossa form independently and become approximately through condylar growth, disappearance of the interzone cells is necessary for the formation of a joint cavity. The interzone mesenchymal cells are believed to contribute to the articular disc, capsule, and the synovial membrane of the joint cavity [3,4]. However, if apoptosis occurs in the interzone of the TMJ remains arguably. It was reported previously that in the rat TMJ at late developmental stage, apoptotic cells were found only at the subsurface of the condyle and in the region at which the lateral pterygoid muscle attached to the condyle, suggesting that apoptosis may be associated with the lower joint cavity formation of the TMJ [62]. However, in our studies, we found extensive cell death in the interzone of the control TMJ at E15.5, right before the upper cavity becomes discernibly at E16.5. However, in the *Wnt1-Cre;BmprIa^{f/f}* TMJ, such extensive apoptosis was not observed, consistent with the pro-apoptotic role of BMPRIA mediated signaling. The discrepancy between our results and that by Matsuda and colleagues [62] could be attributed to the stage difference. Nevertheless, the lack of apoptosis appears to contribute to the persistence of interzone cells in the mutant TMJ. In addition, BMPRIA signaling is also required for organization of some interzone cells to become synovial lining layer, which could also contribute to the cavitation of the TMJ. We thus propose that the upper joint cavity of the TMJ is formed by organization of the interzone cells into synovial lining layer and capsule, and by removal of excessive cells via apoptosis.

In conclusion, our studies using transgenic loss-of- and gain-of-function approaches reveal the importance of BMPRIA mediated

signaling in TMJ morphogenesis and establish a BMPRIA-Ihh positive regulatory pathway in controlling disc separation, synovial membrane formation, as well as joint cavity formation.

Acknowledgments

The authors thank Dr. Yuji Mishina of the University of Michigan for sharing *BmprIa^{f/f}* mice.

Author Contributions

Conceived and designed the experiments: YC SG WW. Performed the experiments: SG WW CL LY CS WY XL JC. Analyzed the data: SG WW YC. Contributed reagents/materials/analysis tools: JC FL. Contributed to the writing of the manuscript: SG WW FL YC.

References

1. Kermack KA (1972) The origin of mammals and the evolution of the temporomandibular joint. Proc R Soc Med 65: 389–392.
2. Gu S, Chen Y (2013) Temporomandibular joint development. In: Huang, G.T.-J., Thesleff, I. (Eds.), Stem Cells in Craniofacial Development and Regeneration. Wiley-Blackwell; John Wily & Sons, Inc. p71–85.
3. Dixon AD (1997) Formation of the cranial base and craniofacial joints. In: Dixon, A.D., Hoyete, D.A.N., Rönning, O. (Eds.), Fundamentals of Craniofacial Growth. CRC Press LLC. p99–136.
4. Sperber GH (2001) Craniofacial Development. Ontario, BC Decker Inc. Hamilton.
5. Miyake T, Cameron AM, Hall BK (1997) Stage-specific expression patterns of alkaline phosphatase during development of the first arch skeleton in inbred *C57BL/6* mouse embryos. J Anat 190: 239–260.
6. Shibata S, Suda N, Suzuki S, Fukuoka H, Yamashita Y (2006) An in situ hybridization study of *Runx2*, *Osterix*, and *Sox9* at the onset of condylar cartilage formation in fetal mouse mandible. J Anat 208: 169–177.
7. Frommer J (1964) Prenatal development of the mandibular joint in mice. Anat Rec 150: 449–461.
8. Gu S, Wei N, Yu L, Fei J, Chen Y (2008) *Shox2*-deficiency leads to dysplasia and ankylosis of the temporomandibular joint in mice. Mech Dev 125: 729–742.
9. Wang Y, Liu C, Rohr J, Liu H, He F, et al. (2011) Tissue interaction is required for glenoid fossa development during temporomandibular joint formation. Dev Dyn 240: 2466–2473.
10. Purcell P, Jheon A, Vivero MP, Rahimi H, Joo A, et al. (2012) Spry1 and Spry2 are essential for development of the temporomandibular joint. J Dent Res 91: 387–393.
11. Mori-Akiyama Y, Akiyama H, Rowitch DH, de Crombrugghe B (2003) Sox9 is required for determination of the chondrogenic cell lineage in the cranial neural crest. Proc Natl Acad Sci USA 100: 9360–9365.
12. Shibata S, Suda N, Yoda S, Fukuoka H, Ohyama K, et al. (2004) *Runx2*-deficient mice lack mandibular condylar cartilage and have deformed Meckel's cartilate. Anat Embryol 208: 273–280.
13. Oka K, Oka S, Sasaki T, Ito Y, Bringas P Jr, et al. (2007) The role of TGF-beta signaling in regulating chondrogenesis and osteogenesis during mandibular development. Dev Biol 303: 391–404.
14. Oka K, Oka S, Hosokawa R, Bringas P Jr, Brockhoff HC 2nd, et al. (2008) TGF-beta mediated Dlx5 signaling plays a crucial role in osteo-chondroprogenitor cell lineage determination during mandible development. Dev Biol 321: 303–309.
15. St-Jacques B, Hammerschmidt M, McMahon AP (1999) Indian hedgehog signaling regulates proliferation and differentiation of chondrocytes and is essential for bone formation. Genes Dev 13: 2072–2086.
16. Shibukawa Y, Young B, Wu C, Yamada S, Long F, et al. (2007) Temporomandibular joint formation and condyle growth require Indian hedgehog signaling. Dev Dyn 236: 426–434.
17. Purcell P, Joo BW, Hu JK, Tran PV, Calicchio ML, et al. (2009) Temporomandibular joint formation requires two distinct hedgehog-dependent steps. Proc Natl Acad Sci USA 106: 18297–18302.
18. Ochiai T, Shibukawa Y, Nagayama M, Mundy C, Yasuda T, et al. (2010) Indian hedgehog roles in post-natal TMJ development and organization. J Dent Res 89: 349–354.
19. Massagué J (2012) TGFβ signaling in context. Nat Rev Mol Cell Biol 13: 616–630.
20. Mitrovic DR (1977) Development of the metatarsophalangeal joint of the chick embryo: morphological, ultrastructural and histochemical studies. Am J Anat 150: 333–347.
21. Pacifici M, Koyama E, Iwamoto M (2005) Mechanisms of synovial joint and articular cartilage formation: recent advances, but many lingering mysteries. Birth Defects Res C 75: 237–248.
22. Storm EE, Kingsley DM (1996) Joint patterning defects caused by single and double mutations in members of bone morphogenetic protein (BMP) family. Development 122: 3969–3979.
23. Wolfman NM, Hattersley G, Cox K, Celeste AJ, Nelson R, et al. (1997) Ectopic induction of tendon and ligament in rats by growth and differentiation factors 5, 6, and 7, members of the TGF-beta gene family. J Clin Invest 100: 321–330.
24. Francis-West PH, Parish J, Lee K, Archer CW (1999) BMP/DGF-signaling interactions during synovial joint development. Cell Tissue Res 296: 111–119.
25. Settle SH Jr, Rountree RB, Sinha A, Thacker A, Higgins K, et al. (2003) Multiple joint and skeletal patterning defects caused by single and double mutations in the mouse Gdf6 and Gdf5 genes. Dev Biol 254: 116–130.
26. Rountree R, Schoor M, Chen H, Marks M, Harley V, et al. (2004) BMP receptor signaling is required for postnatal maintenance of articular cartilage. PloS Biol 11: e255.
27. Storm EE, Huynh TV, Copeland NG, Jenkins NA, Kingsley DM, et al. (1999) Limb alterations in brachypodism mice due to mutations in a new member of the TGF beta-superfamily. Nature 368: 639–643.
28. Brunet LJ, McMahon JA, McMahon AP, Harland RM (1998) Noggin, cartilage morphogenesis, and joint formation in the mammalian skeleton. Science 280: 1455–1457.
29. Lories RJU, Luyten FP (2005) Bone morphogenetic protein signaling in joint hoemostasis and disease. Cytok Growth Fact Rev 16: 287–298.
30. Dewulf N, Verschueren K, Lonnoy O, Moren A, Grimsby S, et al. (1995) Distinct spatial and temporal expression patterns of two type I receptors for bone morphogenetic proteins during mouse embryogenesis. Endocrinology 136: 2652–2663.
31. Mishina Y, Suzuki A, Ueno N, Behringer RR (1995) *Bmpr* encodes a type I bone morphogenetic protein receptor that is essential for gastrulation during mouse embryogenesis. Genes Dev 9: 3027–3037.
32. Zou H, Wieser R, Massagué J, Niswander L (1997) Distinct roles of type I bone morphogenetic protein receptors in the formation and differentiation of cartilage. Genes Dev 11: 2191–2203.
33. Yoon BS, Ovchinnikov DA, Yoshii I, Mishina Y, Behringer RR, et al. (2005) *Bmpr1a* and *Bmpr1b* have overlapping functions and are essential for chondrogenesis in vivo. Proc Natl Acad Sci USA 102: 5062–5067.
34. Yoon BS, Pogue R, Ovchinikov DA, Yoshii I, Mishina Y, et al. (2006) BMPs regulate multiple aspects of growth-plate chondrogenesis through opposing actions on FGF pathways. Development 133: 4667–4678.
35. Fukuoka H, Shibata S, Suda N, Yamashita Y, Komori T (2007) Bone morphogenetic protein rescues the lack of secondary cartilage in *Runx2*-deficient mice. J Anat 211: 8–15.
36. Danielian PS, Puccino D, Rowitch DH, Michael SK, McMahon AP (1998) Modification of gene activity in mouse embryos in uterus by a tamoxifen-inducible form of Cre recombinase. Curr Biol 8: 1323–1326.
37. He F, Xiong W, Wang Y, Matsui M, Yu X, et al. (2010) Modulation of BMP signaling by Noggin is required for the maintenance of palatal epithelial integrity during palatogenesis. Dev Biol 347: 109–121.
38. Stottmann RW, Choi M, Mishina Y, Meyers EN, Klingensmith J (2004) BMP receptor IA is required in mammalian neural crest cells for development of the cardiac outflow tract and ventricular myocardium. Development 131: 2205–2218.
39. Morikawa Y, Zehir A, Maska E, Deng C, Schneider M, et al. (2009) BMP signaling regulates sympathetic nervous system development through Smad-dependent and -independent pathways. Development 136: 3575–3584.
40. Morikawa Y, Cserjesi P (2008) Cardiac neural crest expression of Hand2 regulates outflow and second heart field development. Circ Res 103: 1422–1429.
41. Li L, Lin M, Wang Y, Cserjesi P, Chen Z, et al. (2011) *BmprIa* is required in mesenchymal tissue and has limited redundant function with *BmprIb* in tooth and palate development. Dev Biol 349: 451–461.
42. Presnell JK, Schreibman MP (1997) Humason's Animal Tissue Techniques. Fifth edition. Baltimore, MD, The Johns Hopkins University Press.

43. St Amand TR, Zhang Y, Semina EV, Zhao X, Hu YP, et al. (2000) Antagonistic signals between BMP4 and FGF8 define the expression of *Pitx1* and *Pitx2* in mouse tooth-forming anlage. Dev Biol 217: 323–332.

44. Zhang Z, Song Y, Zhao X, Zhang X, Fermin C, et al. (2002) Rescue of cleft palate in *Msx1*-deficient mice by transgenic *Bmp4* reveals a network of BMP and Shh signaling in the regulation of mammalian palatogenesis. Development 129: 4135–4146.

45. Alappat SR, Zhang Z, Suzuki K, Zhang X, Liu H, et al. (2005) The cellular and molecular etiology of the cleft secondary palate in *Fgf10* mutant mice. Dev Biol 277: 102–113.

46. Swann DA, Silver FH, Slayter HS, Stafford W, Shore E (1985) The molecular structure and lubricating activity of lubricin isolated from bovine and human synovial fluids. Biochem J 225: 195–201.

47. Marcelino J, Carpten JD, Suwairi WM, Gutierrez OM, Schwartz S, et al. (1999) CACP, encoding a secreted proteoglycan, is mutated in camptodactyly-arthropathy-coxa vara-pericarditis syndrome. Nat Genet 23: 319–322.

48. Seki K, Hata A (2004) Indian hedgehog gene is a target of the bone morphogenetic protein signaling pathway. J Biol Chem 279: 18544–18549.

49. Zou H, Niswander L (1996) Requirement for BMP signaling in interdigital apoptosis and scale formation. Science 272: 738–741.

50. Komatsu Y, Yu PB, Kamiya N, Pan H, Fukuda T, et al. (2012) Augmentation of Smad-dependent BMP signaling in neural crest cells causes craniosynostosis in mice. J Bone Miner Res 28: 1422–1433.

51. Li L, Wang Y, Lin M, Yuan G, Yang G, et al. (2013) Augmented BMPRIA-mediated BMP signaling in cranial neural crest lineage leads to cleft palate formation and delayed tooth differentiation. PLoS ONE 8(6): e66107.

52. Hall BK (2005) Bones and Cartilage: Developmental and Evolutionary Skeletal Biology. San Diego, CA, Elsevier Academic Press.

53. Pathi S, Rutenberg JB, Johnson RL, Vortkamp A (1999) Interaction of Ihh and BMP/Noggin signaling during cartilage differentiation. Dev Biol 209: 239–253.

54. Minina E, Wenzel HM, Kreschel C, Karp S, Gaffield W, et al. (2001) BMP and Ihh/PTHrP signaling interact to coordinate chondrocyte proliferation and differentiation. Development 128: 4523–4534.

55. Minina E, Kreschel C, Naski MC, Ornitz DM, Vortkamp A (2002) Interaction of FGF, Ihh/PThIh, and BMP signaling integrates chondrocyte proliferation and hypertrophic differentiation. Dev Cell 3: 1–20.

56. Fukada K, Shibata S, Suzuki S, Ohya K, Kuroda T (1999) In situ hybridization study of type I, II, X collagens and aggrecan mRNAs in the developing condylar cartilage of fetal mouse mandible. J Anat 95: 321–329.

57. Kuboki T, Kanyama M, Nakanishi T, Akiyama K, Nawachi K, et al. (2003) *Cbfa1/Runx2* gene expression in articular chondrocytes of the mice temporo-mandibular and knee joints in vivo. Arch Oral Biol 48: 519–525.

58. Mitrovic DR (1978) Development of the diathrodial joints in the rat embryo. Am J Anat 151: 475–485.

59. Nalin AM, Greenlee TK, Sandell IJ (1995) Collagen gene expression during development of avian synovial joints: transient expression of type II and X collagen genes in the joint capsule. Dev Dyn 203: 352–362.

60. Kimura S, Shiota K (1996) Sequential changes of programmed cell death in developing fetal mouse limbs and its possible role in limb morphogenesis. J Morph 229: 337–346.

61. Abu-Hijleh G, Reid O, Scothorne RJ (1997) Cell death in the developing chick knee joint. I. Spatial and temporal patterns. Clin Anat 10: 183–200.

62. Matsuda S, Mishima K, Yoshimura Y, Hatta T, Otani H (1997) Apoptosis in the development of the temporomandibular joint. Anat Embryo 196: 383–391.

4

Generation and Characterization of a JAK2V617F-Containing Erythroleukemia Cell Line

Wanke Zhao[1], Kang Zou[2], Taleah Farasyn[1], Wanting Tina Ho[1], Zhizhuang Joe Zhao[1]*

1 Department of Pathology, University of Oklahoma Health Sciences Center, Oklahoma City, Oklahoma, United States of America, 2 Oklahoma School of Science and Mathematics, Oklahoma City, Oklahoma, United States of America

Abstract

The JAK2V617F mutation is found in the majority of patients with myeloproliferative neoplasms (MPNs). Transgenic expression of the mutant gene causes MPN-like phenotypes in mice. We have produced JAK2V617F mice with p53 null background. Some of these mice developed acute erythroleukemia. From one of these mice, we derived a cell line designated J53Z1. J53Z1 cells were stained positive for surface markers CD71 and CD117 but negative for Sca-1, TER-119, CD11b, Gr-1, F4/80, CD11c, CD317, CD4, CD8a, CD3e, B220, CD19, CD41, CD42d, NK-1.1, and FceR1. Real time PCR analyses demonstrated expressions of erythropoietin receptor EpoR, GATA1, and GATA2 in these cells. J53Z1 cells grew rapidly in suspension culture containing fetal bovine serum with a doubling time of ~18 hours. When transplanted into C57Bl/6 mice, J53Z1 cells induced acute erythroleukemia with massive infiltration of tumor cells in the spleen and liver. J53Z1 cells were responsive to stimulation with erythropoietin and stem cell factor and were selectively inhibited by JAK2 inhibitors which induced apoptosis of the cells. Together, J53Z1 cells belong to the erythroid lineage, and they may be useful for studying the role of JAK2V617F in proliferation and differentiation of erythroid cells and for identifying potential therapeutic drugs targeting JAK2.

Editor: Andrew C. Wilber, Southern Illinois University School of Medicine, United States of America

Funding: This work was supported by a pilot grant from the Stephenson Cancer Center of University of Oklahoma. The funder had no role in study design, data collection and analysis, decision to publish, or preparation of the manuscript.

Competing Interests: The authors have declared that no competing interests exist.

* Email: joe-zhao@ouhsc.edu

Introduction

Ph- myeloproliferative neoplasms (MPNs) are clonal hematopoietic malignancies in which one or more myeloid lineages are abnormally amplified. These diseases represent a group of chronic conditions including polycythemia vera (PV), essential thrombocythemia (ET), and primary myelofibrosis (PMF) [1,2]. MPNs mainly affect older people and have an average onset age of 55 years. Complications associated with MPNs include the development of acute leukemia as well as thrombosis, hemorrhage, and myeloid metaplasia. JAK2V617F, a mutant form of tyrosine kinase JAK2, represents a major molecular defect in these diseases and is found in over 95% of PV and over 50% of ET and PMF cases [3–8]. Studies demonstrated that JAK2V617F has enhanced tyrosine kinase activity, causes constitutive activation of down-stream signal transducers when expressed in cells [7], and produces MPN-like phenotypes in transgenic and knock-in mice [9–15]. In earlier studies, we generated JAK2V617F transgenic mice by using the *vav* gene promoter which drives the transgene expression in the hematopoietic system. The transgenic mice display MPN-like phenotypes with much increased numbers of red blood cell and platelets [9]. The constitutive activation nature of JAK2V617F makes it a potential oncoprotein. In searching for other gene mutations that collaborate with JAK2V617F to drive leukemia cell transformation, we recently found that JAK2V617F and loss-function mutation of tumor suppressor p53 co-exist in two well-studied leukemia cell lines, namely, HEL and SET2 [16]. This suggests that JAK2V617F is able to drive leukemic transformation when the function of tumor suppressor p53 is lost. We then crossed JAK2V617F transgenic mice with p53 knockout mice and generated JAK2V617F mice with p53 null background. Interestingly, these mice developed acute leukemia. From one of these mice we derived an erythroleukemia cell line which we designated J53Z1. This study reports some basic feature of this cell line.

Materials and Methods

Materials

Antibodies for flow cytometric analysis of cell surface markers were from BD Biosciences and eBioscience. Antibodies against signaling proteins, including phospho-ERK1/2, phospho-Akt, and phospho-STAT5, were from Cell Signaling Technology. JAK2 inhibitors AZD1480 and ruxolitinib were purchased from Chemietek. All other protein kinase inhibitors were from the Approved Oncology Drugs Set IV of NCI Chemotherapeutic Agents Repository.

Mice

Line A JAK2V617F transgenic mice which carry 13 copies of the JAK2V617F transgene were used in this study as previously described [9]. These mice have been crossed with wild type C57BL/6 mice for over 10 generations [17]. Wild type C57BL/6 and p53 knockout mice (strain name B6.129S2-$Trp53^{tm1Tyj}$/J) were purchased from The Jackson Laboratory [18]. JAK2V617F mice with p53 null background were generated by crossing JAK2V617F transgenic mice with p53 knockout mice. Implantation of cultured cells into wild type C57BL/6 mice was carried out through retro-orbital injections under isoflurane anesthesia. Animals were housed in ventilated cages under standard conditions. This study was carried out in strict accordance with the recommendations in the Guide for the Care and Use of Laboratory Animals of the National Institutes of Health. The protocol was approved by the Institutional Animal Care and Use Committee of the University of Oklahoma Health Sciences Center.

Cell culture

Bone marrow cells (1×10^{-6}) were collected from a JAK2V617F/p53$^{-/-}$ mouse and cultured in Iscove's modified Dulbecco's medium (IMDM) supplemented with 20% fetal bovine serum, and 20 μM 2-mercaptoethanol at 37°C with 5% CO_2. The culture was maintained for three weeks with equal volumes of fresh medium added every 3–4 days. Cells were then subjected to colony culture in semisolid medium containing 1% methylcellulose. Colonies were picked after 7 days of culturing and further expanded in the liquid medium. A total of 12 clones were analyzed, and each showed essentially similar morphology and cell surface markers (see below) typical for erythroleukemia cells. Clone no. 1 was selected for detailed characterization as described below. HCD-57 erythroleukemia cells were obtained from Dr. Maurice Bondurant, Vanderbilt University, and cultured in the above liquid medium plus 1 unit/ml of erythropoietin (EPO) [19]. MV-4-11 leukemia cells (ATCC CRL-9591) were obtained from ATTC and maintained in IMDM containing 10% fetal bovine serum. Cell concentrations were determined by flow cytometric analyses. Dead cells were excluded by staining with 7-amino-actinomycin D, and CountBright absolute counting beads (Invitrogen) were used as a standard for cell counting.

DNA extraction and PCR

Genomic DNAs were purified from mouse tails and cells by using the phenol/chloroform method after proteinase K digestion and PCR-amplified as previously described [9,16,20]. Genotyping of p53 knockout mice and cells was performed with a primer set containing 5′-ACAGCGTGGTGGTACCTTAT, 5′-TATACT-CAGAGCCGGCCT, and 5′-CTATCAGGACATAGCGTT-GG. The expected PCR products were 650bp for the targeted allele and 450 bp for the wild type allele. The JAK2V617F transgene was detected by using 5′-TACAACCTCAGTGGGA-CAAAGAAGAAC and 5′-CCATGCCAACTGTTTAGCAAC-TTCA with an expected PCR product of 594bp. Endogenous mouse Jak2 was detected by using 5′-AGACTTCCAGAACCA-GAACAAAG and 5′-TCACAGTTTCTTCTGCCTAGCTA which gave rise to an 84bp PCR product. PCR products were analyzed on 1.5% agarose gels and visualized by ethidium bromide staining.

Total RNA isolation and real time PCR analysis

Total RNAs were isolated from cultured cells and mouse tissues by using the RNeasy Mini kit (Qiagen), and single strand cDNAs were synthesized with equal amounts of total RNAs by using the QuantiTect reverse transcription kit from Qiagen. Real time PCR was performed with iQ SYBR Green Supermix (Bio-Rad) and primers specific for transgenic human JAK2V617F, mouse Jak2, glyceraldehyde-3-phosphate dehydrogenase (GAPDH), GATA1, GATA2, and erythropoietin receptor EpoR. Melting curves were analyzed to confirm specific amplification of desired PCR, and the identities of final PCR products were verified by separation on agarose gels. For quantification, standard curves were obtained by performing PCR with serial dilutions (covering 5 orders of magnitudes) of purified PCR products in salmon sperm DNA [21]. Levels of transcripts were normalized against that of GAPDH.

Cell and tissue staining

For Wright-Giemsa staining, cells were spun onto glass slides by cytocentrifugation. For histological analysis, tissues were fixed in formaldehyde and embedded in paraffin. Tissue sections (5 μm) were deparaffinized and then stained with Hematoxylin and eosin (H&E). Images were captured by using a DP71 digital camera attached to an Olympus BX51 microscope.

Flow cytometric analyses

Cells were stained with fluorescein isothiocyanate (FITC)-, phycoerythrin (PE)-, or allophycocyanin (APC)-conjugated mono-clonal antibodies specific for mouse CD71 (clone C2), CD117 (clone 2B8), Sca-1 (clone D7), TER-119 (clone Ter-119), CD11b (clone M1/70), Gr-1 (clone RB6-8C5), F4/80 (clone BM8), CD11c (clone N418), CD317 (clone eBio927), CD4 (clone RM4-5), CD8a (clone 53-6.7), CD3e (clone 145-2C11), B220 (clone RA3-6B2), CD19 (clone 1D3), CD41 (clone MWReg30), CD42d (clone 1C2), NK-1.1 (clone PK136), and FceR1 (clone MAR-1) (from BD Biosciences and eBioscience). Stained cells were washed and analyzed by 4-color flow cytometry on a FACSCalibur flow cytometer (BD Biosciences) at the Flow and Image Cytometry Laboratory of University of Oklahoma Health Sciences Center. Data were collected by using the Cell Quest software (BD Biosciences) and analyzed by using the Summit software (Dako Colorado, Inc.). At least 15,000 total events were analyzed. Dead cells were excluded according to staining with 7-amino-actinomy-cin D. For apoptosis analysis, the cells were stained with FITC-Annexin V and propidium iodide.

Cell stimulation and western blot analyses

J53Z1 cells were incubated in plain IMDM for 4 hours and then stimulated with EPO (10 U/ml) or mouse stem cell factor SCF (50 ng/ml) for 10 min. Cells were collected in and washed with ice-cold phosphate-buffered saline (PBS). Protein samples were prepared by adding SDS gel sample buffer directly into the cell pellets. Western blotting analyses were performed with antibodies against phospho-ERK1/2, phospho-Akt, and phospho-STAT5 followed by horseradish peroxidase-conjugated secondary anti-bodies. Captures of enhanced chemiluminescence signals were done by using FluorChem SP imaging system from Alpha Innotech.

Inhibitor screening and cell survival assays

J53Z1 cells were cultured in the presence of various tyrosine kinase inhibitors. After 72 hours of culture, cell viability was assessed by performing XTT assays with XTT and phenazine methosulfate following standard protocols [22].

Statistical analysis

Statistical analyses were performed using the Excel program. Differences between 2 groups of samples were assessed using t tests. P values less than 0.05 (2-tailed) are considered significant.

Results

Generation of the J53Z1 cell line

By crossing JAK2V617F transgenic and p53 knockout mice, we generated JAK2V617F/53$^{-/-}$ mice (Figure 1). It has been reported that JAK2V617F mice developed MPN-like phenotype in two months while homozygous p53$^{-/-}$ develop tumors (principally lymphomas and sarcomas) at three to six months of age [9,18]. Interestingly, we found that JAK2V617F/53$^{-/-}$ mice died of apparent acute leukemia in 4 months with huge spleen and liver. Detailed characterization of these mice will be reported elsewhere. Bone marrow cells from one of these mice were collected and cultured in IMDM medium supplemented with 20% fetal bovine serum. Cell proliferation took off in 3 weeks as a single cell suspension. We then performed colony culture in semisolid medium containing 1% methylcellulose. A total of 12 clones were analyzed, and each showed essentially similar morphology and cell surface markers typical for erythroleukemia cells (see below). Clone no. 1 was selected for detailed characterization as described below. We designated the new cell line J53Z1. JAK2V617F and p53 genotyping verified the presence of JAK2V617F and the absence of wild type p53 in the cells as seen in the parental JAK2V617F/p53$^{-/-}$ mouse (Figure 1). For comparison, we also analyzed a JAK2V617F-negative p53$^{+/-}$ mouse and a previously established erythroleukemia cell line, namely, HCD-57 [23]. As expected, HCD-57 cells do not contain JAK2V617F. Interestingly, however, they did not give rise to any p53 PCR product either. The primers used for amplification of p53 cover intron 6 and exon 7 of the mouse p53 gene. This likely indicates the p53 gene is lost in HCD-57 cells. Indeed, reverse-transcription PCR failed to amplify any p53 transcripts, and no p53 protein was detected by using anti-p53 antibodies (not shown). Finally, as a control, mouse Jak2 was amplified to verify the quality of DNA used for genotyping of JAK2V617F and p53 (Figure 1, bottom panel).

Proliferation and morphology of J53Z1 cells

J53Z1 cells proliferated rapidly in suspension culture with a doubling time of ~18 hr (Figure 2). At present, the cells have continued to grow in liquid medium for 14 months under the same culture condition, showing no changes in morphology and proliferation rate. Theoretically, they have grown over 560 generations. We have made frozen stocks of cells with 10% dimethyl sulfoxide (DMSO) and cryopreserved them in liquid nitrogen. After 12 months of cryopreservation, we were able to recover the cells readily. The growth of J53Z1 cells were significantly retarded in serum-free media, and the cells eventually

died in two weeks. Treatment with erythropoietin (EPO), stem cell factor (SCF), DMSO, and phorbol-12-myristate-13-acetate (PMA) failed to induce differentiation of the cells. We have thus generated a new JAK2V617F-positive and p53$^{-/-}$ cell line. Figure 3 illustrates the morphology of J53Z1 cells together with HCD-57 erythroleukemia cells after Wright-Giemsa staining. Like HCD-57 cells, J53Z1 cells were mostly round to ovoid with a modal diameter of 13 to 20 μm. They displayed large nuclei with a few prominent nucleoli and basophilic cytoplasms. Cytoplasmic vacuoles and protrusions are also present in some cells. These cells showed morphologic characteristics of erythroblasts.

Expressions of erythroid proteins in J53Z1 cells

Flow cytometric analyses of cell surface markers further verified the erythroid lineage of J53Z1 cells. Like HCD-57 cells, J53Z1 cells are strongly positive for immature erythroid cell marker CD71 but negative for TER-119, and the level of CD71 is comparable to that seen in HCD-57 cells and higher than the level observed in normal mouse bone marrow erythroid cells (Figure 4). J53Z1 cells are also positive for expressions of CD117 (cKit) but are negative for hematopoietic stem cell marker (Sca-1), myeloid cell markers (CD11b and Gr-1), monocyte marker (F4/80), dendritic cell markers (CD11c and CD317), T-cell markers (CD4, CD8a, and CD3e), B-cell marker (B220 and CD19), megakaryocyte markers (CD41and CD42d), NK cell marker (NK-1.1), and mast cell marker (FceR1) (see Table 1). Erythroblasts are divided into various stages based on CD71 and TER-119 expression [24]. Since J53Z1 cells were stained negative for TER-119, we believe that they are arrested at a very earlier stage before proerythroblasts during erythroid development. We further employed real time PCR to determine the expressions of GATA1, GATA2, EpoR, mouse Jak2, and transgenic human JAK2V617F in J53Z1 cells (Figure 5). Normal mouse bone marrow and HCD-57 cells were analyzed for comparison. As expected, expressions of EpoR and mouse Jak2 were found in both J53Z1 and HCD-57 cells, at levels above those seen in normal bone marrow cells. In addition, the expression of transgenic human JAK2V617F was detected in J53Z1 cells only, at a level below that of mouse Jak2, which is consistent with its relative expression level in hematopoietic tissues from JAK2V617F transgenic mice [9,17]. Interestingly, substantial expressions of both GATA1 and GATA2 were observed in J53Z1 and HCD-57 cells, at levels much higher than those seen in total mouse bone marrow cells. The GATA family transcription factors have distinct and essential roles in hematopoiesis [25]. As a master regulator of hematopoietic differentiation, GATA1 is expressed in erythroid cells and megakaryocytes. In contrast, GATA2 is essential for maintenance of the hematopoietic stem cell compartment and is

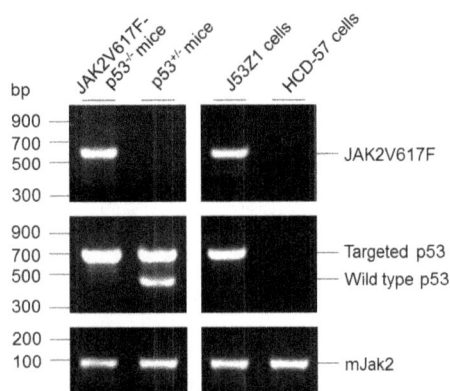

Figure 1. Genotyping of JAK2V617F/p53$^{-/-}$ mice and J53Z1 cells. Genomic DNAs isolated from mouse tails and cultured cells. JAK2V617F transgene, wild type 53, mutant p53, and endogenous mouse Jak2 were PCR-amplified with specific primers described in Materials and Methods. The PCR products were resolved on 3% agarose gels and visualized by ethidium bromide staining. HCD-57, a previously established mouse erythroleukmia cell line, was analyzed for comparison.

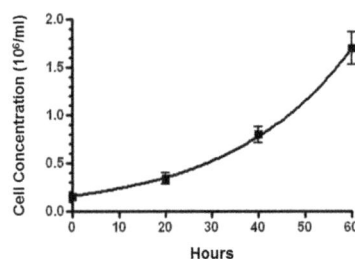

Figure 2. Growth curves of J53Z1 cells in suspension culture. J53Z1 cells were grown in IMDM medium supplemented with 20% fetal bovine serum. Cell numbers were counted by using flow cytometry as described in Materials and Methods. Error bars denote standard deviation (n = 3).

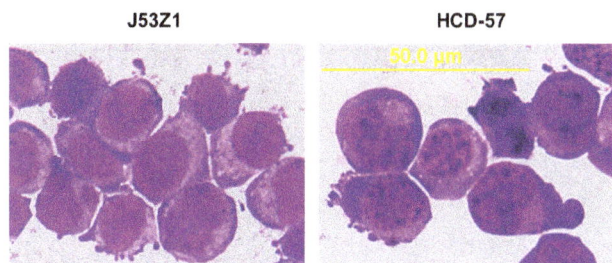

Figure 3. Morphology of J53Z1 cells. Normal growing J53Z1 cells were attached to glass slides by cytospin and then subjected to Wright-Giemsa staining. HCD-57 erythroid leukemia cells were stained for comparison. Photos were taken with a 100x objective lens.

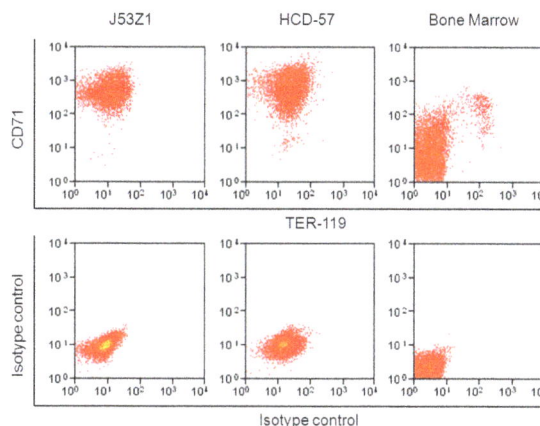

Figure 4. Flow cytometric analysis of CD71 and TER-119 expression on J53Z1 cells. J53Z1, HCD-57, and normal mouse bone marrow cells were labeled with anti-CD71 and TER-119 monoclonal antibodies or with a nonspecific isotype control mouse IgG before flow cytometric analysis.

also involved in the initial activation of GATA1 expression at the first steps of erythroid/megakaryocytic differentiation. During erythroid differentiation, expressions of GATA1 and GATA2 are reciprocal thereby forming the so-called GATA switch [25]. Expression of both GATA1 and GATA2 in J53Z1 and HCD-57 cells suggest that these cells are arrested at a stage that GATA1 and GATA2 expressions overlap. These cells thus represent a unique stage of erythroid development.

Response of J53Z1 cells to growth factors

Primary erythroid progenitor cells rely on two key growth factors, namely, EPO and SCF, for survival and expansion [26]. Although J53Z1 cells are immortal and become independent of specific growth factors, they may still be responsive to these factors. Indeed, when serum-starved J53Z1 cells were treated with EPO and SCF, strong phosphorylation of ERK1/2, Akt, and STAT5 were observed (Figure 6). The EPO- and SCF-induced signal transduction is consistent with the expression of the EPO receptor EpoR and the SCF receptor CD117 (cKit) revealed by real time PCR and flow cytometry, respectively (Figure 5 and Table 1). The data provided further evidence that J53Z1 cells belong to the erythroid lineage. J53Z1 cells should serve as a good cell system for studying cell signaling involved in erythroid development.

Induction of acute leukemia in mice implanted with J53Z1 cells

J53Z1 cells were derived from a mouse with acute erythroleukemia. To find out if they still possess the ability to cause leukemia, we implanted 1×10^6 cultured J53Z1 cells into wild type C57Bl/6 mice through retro-orbital injections. The recipient mice developed anemia and died within 4 weeks with enlarged spleen and liver (Figure 7, upper panel). H&E staining revealed massive accumulation of erythroblast cells in the liver and spleen, characteristic of erythroleukemia (Figure 7, lower panel). This indicates that J53Z1 cells contain cancer stem cells that can initiate cancer *in vivo*.

Inhibition of J53Z1 cells by selective tyrosine kinase inhibitors

As a constitutively active tyrosine kinase, JAK2V617F likely plays a crucial role in supporting the proliferation of J53Z1 cells. Naturally, inhibition of its kinase activity should block cell growth. We analyzed a total of 14 protein kinase inhibitors for their inhibitory effects on J53Z1 cells as shown Figure 8. Except for AZD1480 which is a JAK2 inhibitor currently in clinical trials [27], the rest are FDA-approved anti-cancer drugs. Interestingly,

at 4 µM, several inhibitors including AZD1480, axitinib, crizotinib, dasatinib, erlotinib, ruxolitinib, and sunitinib showed inhibitory effects on the growth of J53Z1 cells, while gefitinib, imatinib, lapatinib, nilotinib, sorafenib, vandetanib, and vemurafenib showed no effects. However, when the concentration was reduced to 0.4 µM, only AZD1480 and ruxolitinib remained strongly inhibitory. More detailed analyses revealed that AZD1480 and ruxolitinib inhibited J53Z1 cells with IC50 values of 0.10 and 0.14 µM, respectively. In contrast, they displayed no inhibitory effects on MV-4-11 leukemia cells which carry a FLT3-ITD mutation [28,29]. The results are quite expected since AZD1480 is known to be a potent, selective JAK2 inhibitor [27], while ruxolitinib is a potent JAK1 and JAK2 inhibitor approved by FDA for treatment of myelofibrosis [30]. It should be noted the inhibitory effects of other kinase inhibitors on J53Z1 cells at the higher concentration may also be attributed to the inhibition of JAK2V617F. In fact, our early studies have demonstrated that EGFR inhibitor erlotinib but not gefitinib inhibits JAK2 [31]. On the other hand, sunitinib is considered a broad range tyrosine kinase inhibitor, and it should not be surprising if it inhibits JAK2 also. Indeed, our in vitro kinase assays by using isolated JAK2 kinase demonstrated that axitinib, crizotinib, dasatinib, and sunitinib but not imatinib, lapatinib, nilotinib, sorafenib, vandetanib, or vemurafenib had strong inhibitory effects on JAK2 kinase activity at micromolar concentrations (not shown). Finally, AZD1480 and ruxolitinib apparently inhibited J53Z1 cells by inducing apoptosis (Figure 8, bottom panel). Additional data demonstrated that inhibited EPO- and SCF-induced activation of ERK1/2, Akt, and STAT5 (not shown). Together, J53Z1 cells provide an excellent cell-based system to screen for JAK2 inhibitors.

Discussion

In the present study, we generated a JAK2V617F-containing erythroleukemia cell line designated J53Z1. J53Z1 cells belong to the erythroid lineage and are arrested at a stage before proerythroblast during erythroid development. They are positive for surface marker CD71 and negative for TER-119. They express EpoR and c-Kit and are responsive to stimulation by EPO and

Table 1. Surface Markers on J53Z1 Cells.

Markers	Positivity	Markers	Positivity
CD71	high	CD4	negative
CD117 (cKit)	high	CD8a	negative
Sca-1	negative	CD3e	negative
TER-119	negative	B220	negative
Gr-1	negative	CD19	negative
CD11b	negative	CD41	negative
F4/80	negative	CD42d	negative
CD11c	negative	NK1.1	negative
CD317	negative	FceR1	negative

SCF. They express both GATA1 and GATA2 transcription factors expression thereby representing a unique stage of erythroid development. These cells should have wide applications in basic research to define the regulation of erythroid development and in translational research to identify JAK2 inhibitors for therapeutic drug development.

Identification of JAK2V617F in MPNs has provided an excellent target for drug development. Many potent JAK2 inhibitors have been developed, and some have shown clinical benefits in treatment of myelofibrosis [30,32,33]. However, the general outcome of clinical trials of JAK2 inhibitors has been disappointing. Obviously, more effective JAK2 inhibitors are needed. In this regards, J53Z1 cells provide an excellent cell-based system for inhibitor screening as illustrated in Figure 8. In addition, since J53Z1 cells are derived from mice, they can induce leukemia in normal mice. In contrast, existing JAK2V617F-positive human cells (e.g., HEL cells) require immunodeficient mice for implantation. Therefore, J53Z1 cells serve as a better system for testing the efficacy of JAK2 inhibitors *in vivo*.

By establishing a JAK2V617F-positive, p53-null, immortal cell line, our study provides further evidence that JAK2V617F collaborates with loss-of-function of p53 to cause leukemic transformation. Malignant transformation usually involves a gain-of-function mutation of oncogenes and a loss-of-function mutation of tumor suppressor genes. Among various tumor suppressors, p53 is the most frequently mutated [34-36]. Interestingly, mutations of p53 are the most common in solid tumors but relatively rare in leukemia [37]. However, many erythroleukemia cell lines identified so far appear to contain p53 mutations. Aside from generating a p53$^{-/-}$ erythroid cell line, our current study also demonstrated the absence of p53 in HCD-57 cells which were derived from a mouse infected at birth with Friend murine leukemia virus [23]. This is consistent with the finding that loss of p53 tumor suppressor function is required for *in vivo* progression of Friend erythroleukemia [38]. Furthermore, our earlier studies identified a M133K p53 mutation in JAK2V617F-positive human erythroleukemia HEL cells [16]. Yet another example of a p53 mutation was found in BCR-Abl-positive human erythroid leukemic K562 cells [39]. These findings support the notion that p53 mutations play a role in erythroblast transformation. Normal p53 suppresses malignant transformation by

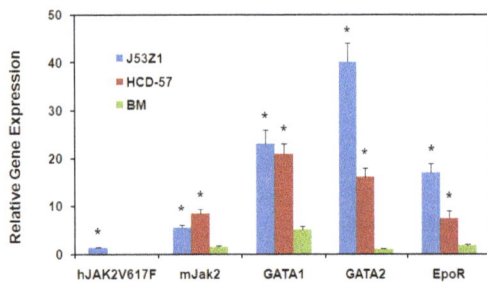

Figure 5. Real time PCR assays of gene expression in J53Z1 cells. Expressions of indicated genes were analyzed by real time PCR using specific PCR primers shown in the top panel. Data represent relative mRNA levels (mean±SD, n=3) normalized to mouse GAPDH which was defined as 1000. HCD-57 and normal mouse bone cells were analyzed for comparison. *P<0.001 in reference to normal bone marrow cells.

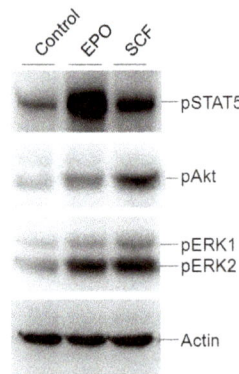

Figure 6. Response of J53Z1 cells to growth factors. J53Z1 cells were serum-starved for 5 hr and then stimulated with 10 Units/ml EPO or 50ng/ml SCF for 10 min. Cell extracts were analyzed for activation of indicated signaling proteins by using phospho-specific antibodies. Equal protein loading was demonstrated by blotting with anti-actin.

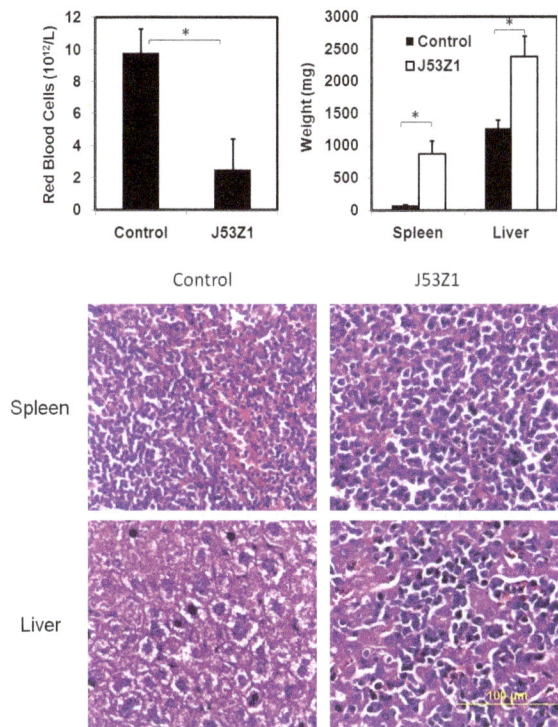

Figure 7. Development of erythroleukemia phenotypes in mice receiving implantation of J53Z1 cells. Cultured J53Z1 cells (1×10^6) were implanted into 12-week-old wild type C57Bl/6 mice through retro-orbital injections. **Upper panel.** Red blood cells, spleen, and liver were analyzed 2 to 4 weeks after implantation. Error bars denote standard deviation ($n \geq 4$), *$P < 0.001$. **Lower panel.** Paraffin sections of spleen and liver from representative control and J53Z1-implanted mice were subjected to H&E staining. Note the loss of normal tissue architecture and infiltration of densely stained erythroleukemia cells in tissues from J53Z1-transplanted mice. Photos were taken with a 40x objective lens.

controlling cell cycle progression, ensuring the fidelity of DNA replication and chromosomal segregation, and inducing apoptosis in response to potentially deleterious events [34-37]. It may also have important role in regulating normal proliferation and differentiation of erythroid cells. In this regard, it will be interesting to see if restoration of p53 function in the p53-deficient erythroleukemia can reinstall erythroid differentiation.

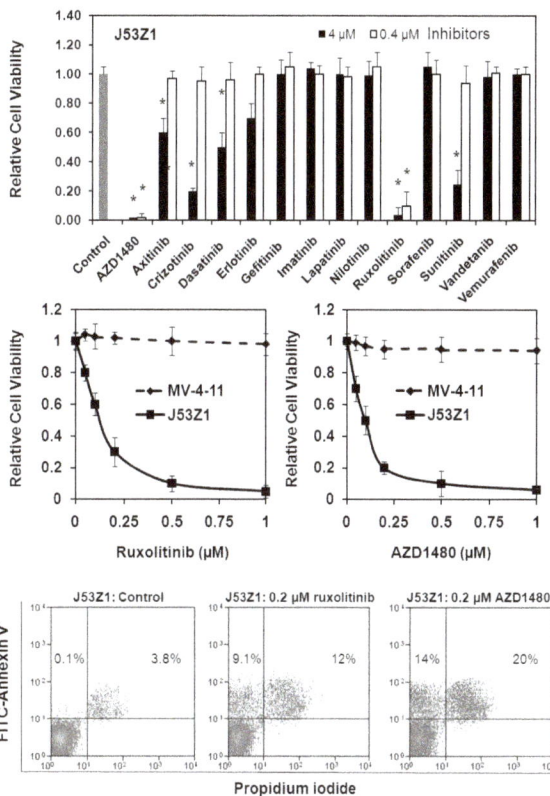

Figure 8. Inhibition of J53Z1 cells by selective protein kinase inhibitors. J53Z1 cells were cultured in the presence of various concentrations of indicated protein kinase inhibitors. **Top and middle panels.** Cell viability was assessed by XTT assays after 72 hr of incubation. Control experiments were performed in the presence of 0.1% DMSO. Error bars denote standard deviation ($n = 3$). *$P < 0.001$ in reference to control. Note that MV-4-11 cells were analyzed for comparison (middle panel). **Bottom panel.** Apoptosis assays were performed with J53Z1 cells after 24 hr of incubation with 0.2 µM of ruxolitinib or AZD1480. Cells were stained with FITC-annexin V and propidium iodide. Percentages of annexin V-positive cells are indicated.

Author Contributions

Conceived and designed the experiments: ZJZ. Performed the experiments: WZ KZ TF WTH. Analyzed the data: WZ KZ TF WTH ZJZ. Contributed reagents/materials/analysis tools: WZ KZ TF WTH ZJZ. Wrote the paper: WZ KZ TF WTH ZJZ.

References

1. Levine RL, Gilliland DG (2008) Myeloproliferative disorders. Blood 112: 2190–2198.
2. Tefferi A (2008) The history of myeloproliferative disorders: before and after Dameshek. Leukemia 22: 3–13.
3. Baxter EJ, Scott LM, Campbell PJ, East C, Fourouclas N, et al. (2005) Acquired mutation of the tyrosine kinase JAK2 in human myeloproliferative disorders. Lancet 365: 1054–1061.
4. Levine RL, Wadleigh M, Cools J, Ebert BL, Wernig G, et al. (2005) Activating mutation in the tyrosine kinase JAK2 in polycythemia vera, essential thrombocythemia, and myeloid metaplasia with myelofibrosis. Cancer Cell 7: 387–397.
5. James C, Ugo V, Le Couedic JP, Staerk J, Delhommeau F, et al. (2005) A unique clonal JAK2 mutation leading to constitutive signalling causes polycythaemia vera. Nature 434: 1144–1148.
6. Kralovics R, Passamonti F, Buser AS, Teo SS, Tiedt R, et al. (2005) A gain-of-function mutation of JAK2 in myeloproliferative disorders. N Engl J Med 352: 1779–1790.
7. Zhao R, Xing S, Li Z, Fu X, Li Q, et al. (2005) Identification of an acquired JAK2 mutation in polycythemia vera. J Biol Chem 280: 22788–22792.
8. Zhao W, Gao R, Lee J, Xing S, Ho WT, et al. (2011) Relevance of JAK2V617F positivity to hematological diseases–survey of samples from a clinical genetics laboratory. J Hematol Oncol 4:4.
9. Xing S, Wanting TH, Zhao W, Ma J, Wang S, et al. (2008) Transgenic expression of JAK2V617F causes myeloproliferative disorders in mice. Blood 111: 5109–5117.
10. Shide K, Shimoda HK, Kumano T, Karube K, Kameda T, et al. (2008) Development of ET, primary myelofibrosis and PV in mice expressing JAK2 V617F. Leukemia 22: 87–95.
11. Tiedt R, Hao-Shen H, Sobas MA, Looser R, Dirnhofer S, et al. (2008) Ratio of mutant JAK2-V617F to wild-type Jak2 determines the MPD phenotypes in transgenic mice. Blood 111: 3931–3940.
12. Mullally A, Lane SW, Ball B, Megerdichian C, Okabe R, et al. (2010) Physiological Jak2V617F expression causes a lethal myeloproliferative neoplasm with differential effects on hematopoietic stem and progenitor cells. Cancer Cell 17: 584–596.
13. Marty C, Lacout C, Martin A, Hasan S, Jacquot S, et al. (2010) Myeloproliferative neoplasm induced by constitutive expression of JAK2V617F in knock-in mice. Blood 116: 783–787.

14. Akada H, Yan D, Zou H, Fiering S, Hutchison RE, et al. (2010) Conditional expression of heterozygous or homozygous Jak2V617F from its endogenous promoter induces a polycythemia vera-like disease. Blood 115: 3589–3597.

15. Li J, Spensberger D, Ahn JS, Anand S, Beer PA, et al. (2010) JAK2 V617F impairs hematopoietic stem cell function in a conditional knock-in mouse model of JAK2 V617F-positive essential thrombocythemia. Blood 116: 1528–1538.

16. Zhao W, Du Y, Ho WT, Fu X, Zhao ZJ (2012) JAK2V617F and p53 mutations coexist in erythroleukemia and megakaryoblastic leukemic cell lines. Exp Hematol Oncol 1:15.

17. Shi K, Zhao W, Chen Y, Ho WT, Yang P, et al. (2014) Cardiac hypertrophy associated with myeloproliferative neoplasms in JAK2V617F transgenic mice. J Hematol Oncol. 7:25.

18. Jacks T, Remington L, Williams BO, Schmitt EM, Halachmi S, et al. (2010) Tumor spectrum analysis in p53-mutant mice. Curr Biol 1994; 4:1–7.

19. Tian C, Gregoli P, Bondurant M (2003) The function of the bcl-x promoter in erythroid progenitor cells. Blood 101:2235–2242.

20. Zhao AH, Gao R, Zhao ZJ (2011) Development of a highly sensitive method for detection of JAK2V617F. J Hematol Oncol 4:40.

21. Jin X, Zhao W, Shi K, Ho WT, Zhao ZJ (2013) Generation of a new congenic mouse strain with enhanced chymase expression in mast cells. PLoS One. 8:e84340.

22. Roehm NW, Rodgers GH, Hatfield SM, Glasebrook AL (1991) An improved colorimetric assay for cell proliferation and viability utilizing the tetrazolium salt XTT. J Immunol Methods. 142:257–265.

23. Ruscetti SK, Janesch NJ, Chakraborti A, Sawyer ST, Hankins WD (1990) Friend spleen focus-forming virus induces factor independence in an erythropoietin-dependent erythroleukemia cell line. J Virol 64:1057–1062.

24. Chen K, Liu J, Heck S, Chasis JA, An X, et al. (2009) Resolving the distinct stages in erythroid differentiation based on dynamic changes in membrane protein expression during erythropoiesis. Proc Natl Acad Sci U S A. 106:17413–17418.

25. Kaneko H, Shimizu R, Yamamoto M (2010) GATA factor switching during erythroid differentiation. Curr Opin Hematol. 17:163–168.

26. Sui X, Krantz SB, You M, Zhao Z (1998) Synergistic activation of MAP kinase (ERK1/2) by erythropoietin and stem cell factor is essential for expanded erythropoiesis. Blood 92:1142–1149.

27. Hedvat M, Huszar D, Herrmann A, Gozgit JM, Schroeder A, et al. (2009) The JAK2 inhibitor AZD1480 potently blocks Stat3 signaling and oncogenesis in solid tumors. Cancer Cell 16:487–497.

28. Quentmeier H, Reinhardt J, Zaborski M, Drexler HG (2003) FLT3 mutations in acute myeloid leukemia cell lines. Leukemia; 17:120–124.

29. Guo Y, Chen Y, Xu X, Fu X, Zhao ZJ (2012) SU11652 Inhibits tyrosine kinase activity of FLT3 and growth of MV-4-11 cells. J Hematol Oncol 5:72.

30. Mesa RA, Cortes J (2013) Optimizing management of ruxolitinib in patients with myelofibrosis: the need for individualized dosing. J Hematol Oncol. 6:79.

31. Li Z, Xu M, Xing S, Ho WT, Ishii T, et al. (2007) Erlotinib effectively inhibits JAK2V617F activity and polycythemia vera cell growth. J Biol Chem 282:3428–3432.

32. Verstovsek S, Kantarjian H, Mesa RA, Pardanani AD, Cortes-Franco J, et al. (2010) Safety and efficacy of INCB018424, a JAK1 and JAK2 inhibitor, in myelofibrosis. N Engl J Med 363: 1117–1127.

33. LaFave LM, Levine RL (2012) JAK2 the future: therapeutic strategies for JAK-dependent malignancies. Trends Pharmacol Sci 33: 574–582.

34. Ferbeyre G, Lowe SW (2002) The price of tumour suppression? Nature 415:26–27.

35. Petitjean A, Mathe E, Kato S, Ishioka C, Tavtigian SV, et al. (2007) Hainaut P, Olivier M. Impact of mutant p53 functional properties on TP53 mutation patterns and tumor phenotype: lessons from recent developments in the IARC TP53 database. Hum Mutat 28:622–629.

36. Donehower LA, Lozano G (2009) 20 years studying p53 functions in genetically engineered mice. Nat Rev Cancer 9:831–841.

37. Liu Y, Elf SE, Asai T, Miyata Y, Liu Y, et al. (2009) The p53 tumor suppressor protein is a critical regulator of hematopoietic stem cell behavior. Cell Cycle 8:3120–3124.

38. Prasher JM, Elenitoba-Johnson KS, Kelley LL (2001) Loss of p53 tumor suppressor function is required for in vivo progression of Friend erythroleukemia. Oncogene 20:2946–2955.

39. Law JC, Ritke MK, Yalowich JC, Leder GH, Ferrell RE (1993) Mutational inactivation of the p53 gene in the human erythroid leukemic K562 cell line. Leuk Res 17:1045–1050.

5

Infection Cycle of *Artichoke Italian Latent Virus* in Tobacco Plants: Meristem Invasion and Recovery from Disease Symptoms

Elisa Santovito[1], Tiziana Mascia[1,2]*, Shahid A. Siddiqui[3], Serena Anna Minutillo[1], Jari P. T. Valkonen[3], Donato Gallitelli[1,2]

1 Dipartimento di Scienze del Suolo, della Pianta e degli Alimenti, Università degli Studi di Bari Aldo Moro, Bari, Italy, 2 Istituto di Virologia vegetale del Consiglio Nazionale della Ricerca, Unità Operativa di Supporto di Bari, Bari, Italy, 3 Department of Agricultural Sciences, University of Helsinki, Helsinki, Finland

Abstract

Nepoviral infections induce recovery in fully expanded leaves but persist in shoot apical meristem (SAM) by a largely unknown mechanism. The dynamics of infection of a grapevine isolate of *Artichoke Italian latent virus* (AILV-V, genus *Nepovirus*) in tobacco plants, including colonization of SAM, symptom induction and subsequent recovery of mature leaves from symptoms, were characterized. AILV-V moved from the inoculated leaves systemically and invaded SAM in 7 days post-inoculation (dpi), remaining detectable in SAM at least up to 40 dpi. The new top leaves recovered from viral symptoms earliest at 21 dpi. Accumulation of viral RNA to a threshold level was required to trigger the overexpression of *RDR6* and *DCL4*. Consequently, accumulation of viral RNA decreased in the systemically infected leaves, reaching the lowest concentration in the 3rd and 4th leaves at 23 dpi, which was concomitant with recovery of the younger, upper leaves from disease symptoms. No evidence of virus replication was found in the recovered leaves, but they contained infectious virus particles and were protected against re-inoculation with AILV-V. In this study we also showed that AILV-V did not suppress initiation or maintenance of RNA silencing in transgenic plants, but was able to interfere with the cell-to-cell movement of the RNA silencing signal. Our results suggest that AILV-V entrance in SAM and activation of RNA silencing may be distinct processes since the latter is triggered in fully expanded leaves by the accumulation of viral RNA above a threshold level rather than by virus entrance in SAM.

Editor: Biao Ding, The Ohio State University, United States of America

Funding: This work was supported, by a grant of University of Bari (http://www.uniba.it), Italy in the framework of Research funds 2010 "Study of plant-pathogen interactions in Artichoke" grant ORBA10Q104. Financial support from the Academy of Finland (http://www.aka.fi/en-GB/A/), grant 1253126 to JPTV, is gratefully acknowledged. The funders had no role in study design, data collection and analysis, decision to publish, or preparation of the manuscript.

Competing Interests: The authors have declared that no competing interests exist.

* E-mail: tiziana.mascia@uniba.it

Introduction

Procedures to obtain virus-free planting material can provide novel cellular and molecular insights in the study of virus-plant interactions [1,2]. Shoot apical meristem (SAM) culture preceded or followed by heat therapy is widely used to produce virus-free plants although its efficiency in virus eradication depends on the virus and the host genotype [1,3,4,5]. During the sanitation of the reflowering globe artichoke variety "Brindisino" with mixed infection of *Artichoke Italian latent virus* (AILV) genus *Nepovirus*, family *Secoviridae* [6] and *Artichoke latent virus* (ArLV), genus *Potyvirus*, family *Potyviridae* [7], ArLV was eliminated by means of SAM culture, while AILV was removed only when two rounds of SAM culture were spaced out with *in vitro* thermotherapy [2]. These preliminary results suggested that AILV was able to enter SAM at some stage of infection and to persist therein. This hypothesis is in line also with the previous finding that AILV is transmitted through seeds of globe artichoke [8], a process that would be dependent on the presence of virus in meristems [9,10].

SAM is a strong photosynthetic sink thus being an ideal destination of viruses although most of them are excluded from there [11,12]. It has been proposed that meristem exclusion is a variation of the RNA silencing (RS)-related "recovery" process that is restricted to the growing point of the infected plant and regulates selectively the entry and persistence of RNA in the shoot apex, including viruses and long-distance post-transcriptional gene silencing (PTGS) signals [13]. On the contrary, the "classical recovery" would be meristem exclusion that operates not only in the meristem but also in the uppermost leaves of the plant that remain free of symptoms [14,15]. In line with this proposal, it was demonstrated that *Cucumber mosaic virus* (CMV) and *Tobacco rattle virus* (TRV) can transiently infect meristem tissues, in contrast to *Potato virus X* (PVX), and for persistence they would need to fully suppress RS implicated in the meristem exclusion process [13,14,16]. The exclusion of PVX would implicate a silencing mechanism that initiates in lower uninfected tissues, moves at or ahead the infection front and involves a long-range RS signaling regulated by host RNA-dependent RNA polymerase RDR6 [14]. Conversely, in plants with SAM transiently infected with TRV and CMV, the priming of RS would require the transient presence of these viruses in SAM and would be independent from the activity of *RDR6* [16,17,18]. The model proposed by Martín-Hernández and Baulcombe [16] suggests also that a "transient accumulation mechanism" would operate in plants infected by TRV and would affect virus accumulation in leaves developed during or after the transient phase of meristem invasion. In the leaves derived from the transiently infected meristem, the virus would persist and continue to accumulate while the leaves derived from virus-free meristem from the post-transient phase exhibit very low virus levels.

In infected plants, recovery usually refers to the condition of new emerging leaves, which develop without symptoms, may contain low concentration of the virus and resist to further virus infection through a sequence-specific RS mechanism, while the infected symptomatic leaves remain virus-infected and continue to show disease symptoms. The first studies linking recovery of plants from viral infection with RS were carried out on nepoviruses [15], but only few of them have addressed viral distribution in SAM in the context of recovery. Dong *et al.* [19] studied dynamics of *Tobacco ringspot virus* (TRSV) distribution in the SAM and in the root apical meristem of tobacco, coming to the conclusion that TRSV persisted in the SAM of tobacco and the asymptomatic leaves of *Nicotiana benthamiana*, but it was transient in the root apical meristem and asymptomatic leaves of tobacco plants. Siddiqui *et al.* [10] reported that transgenic expression of some VSR affected the temperature-dependent infection pattern of the TRSV potato calico strain but no information was provided on the effect of such VSR on the recovery phenotype of TRSV-infected tobacco plants. Jovel *et al.* [20] characterized the interaction of *Tomato ringspot virus* (ToRSV) with host defense responses during symptom induction and subsequent recovery showing the activation of RS, which, however, did not reduce virus titer.

AILV is a member of subgroup B of the genus *Nepovirus* [6], has isometric particles c. 30 nm in diameter, sedimenting as three components with coefficient of 55S (T), 95S (M), and 121S (B). M and B particles, each encapsidate one species of functional single-stranded RNA with estimated Mr of 1.5×10^6 (RNA-2) and 2.4×10^6 (RNA-1), respectively [21,22] but only a partial nucleotide sequence of RNA-2 is available (Acc. No. X87254). The aim of the present study was to better understand the molecular mechanisms behind invasion and persistence of a grapevine isolate of AILV (denoted AILV-V) in SAM, using infected tobacco plants as an amenable experimental system facilitating the study. Time-course experiments were carried out to estimate dynamics of viral RNA accumulation from inoculation up to two months after infection during different stages of tobacco plant growth. Data were related also with virus entrance, distribution and persistence in tobacco SAM, development of disease symptoms and ability of AILV-V to interfere with the RS-based defense response activated in infected plants.

Results

AILV-V Infection in *N. tabacum* Induces Severe Symptoms Followed by Complete Recovery, Undetectable Virus Replication and Resistance to Secondary Infection in Recovered Leaves

Local symptoms of AILV-V infection appeared within 7 days post inoculation (dpi) in inoculated leaves (i.e., in 1st and 2nd fully expanded true leaves; Figure 1) of tobacco plants and consisted of

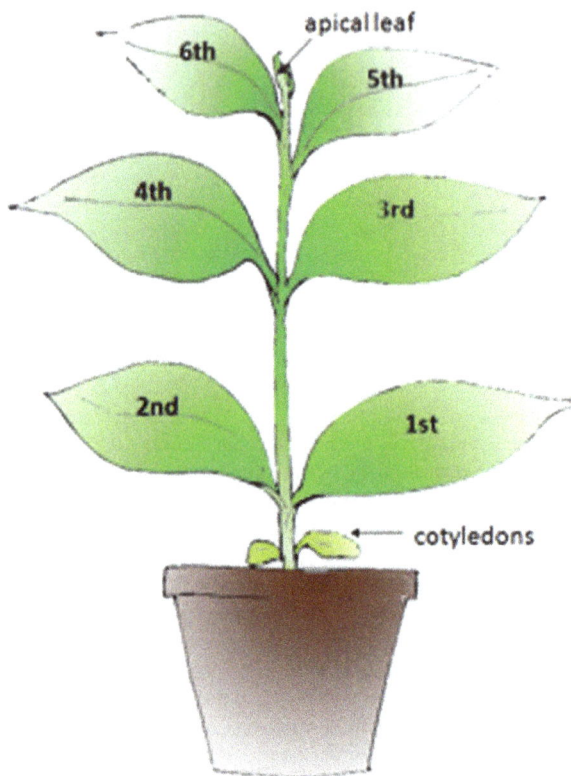

Figure 1. Scheme of a Samsun tobacco plant showing position of leaves used in this study.

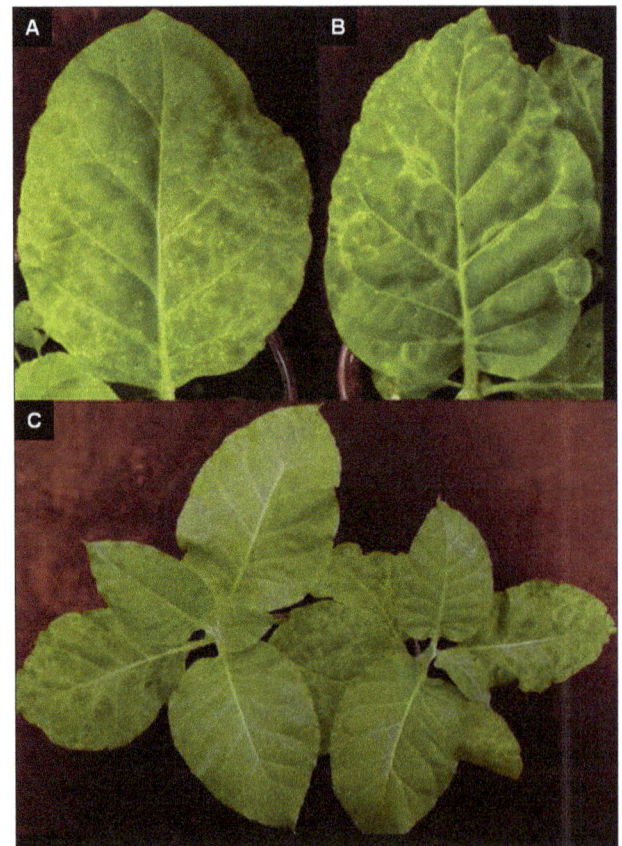

Figure 2. Samsun tobacco plants recover from disease symptoms induced by AILV-V. In **A,** local symptoms in inoculated leaves appeared by 7 dpi and consisted in chlorotic spots, small necrotic rings and line patterns etched in leaf epidermis. In **B,** systemic symptoms are shown in the 3rd and 4th leaf and consist in chlorotic or necrotic ringspots surrounding the veins and peripheral vein clearing. In **C,** young leaves emerged between 21 and 28 dpi showing the recovery phenotype from disease symptoms.

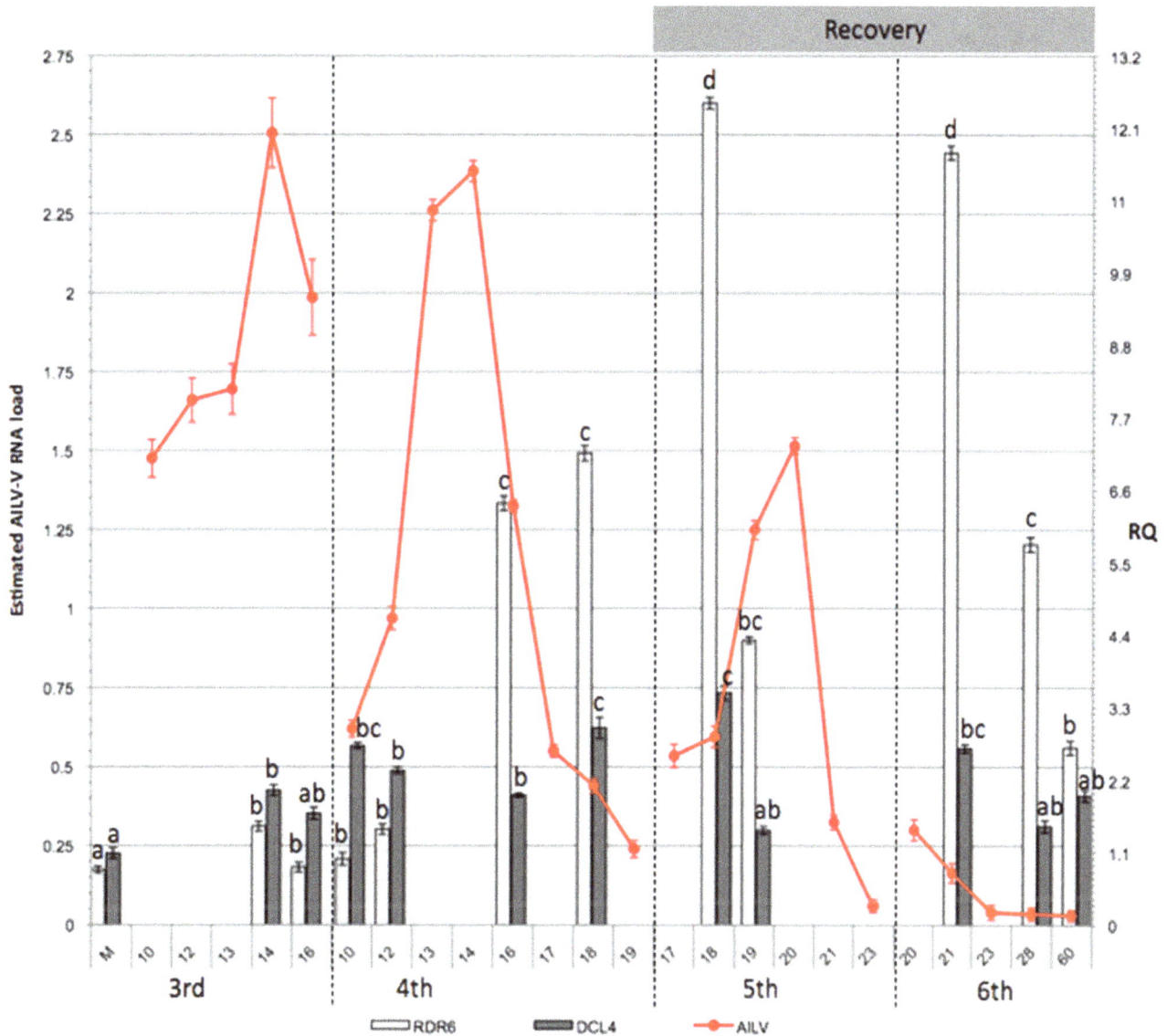

Figure 3. AILV-V RNA-2 accumulation varies in the same leaf, progressively decreasing moving to the successive leaf. Load of viral RNA (lines) was estimated by quantitative dot blot hybridization. RNA data are expressed as means of two independent experiments, were derived from spot intensity values of the target RNAs and were calculated on the basis of a standard curve generated by serial dilutions of a plasmid preparation containing the fragment of the RNA-2 of AILV-V targeted by the probe. Samples were collected from the 3rd, 4th, 5th, and 6th leaves at 24 h intervals from 10 to 23 dpi and then at 28 and 60 dpi. Each point in the line chart represents average of three biological replicates for each of the two experiments and error bars on lines represent the standard error among replicates. Figure shows also the relative quantity (RQ) of *RDR6* and *DCL4* transcripts (columns chart) in samples of tobacco plants collected at selected time points between 10 and 60 dpi with AILV-V. The values were first normalized on the accumulation level of the *GAPDH* mRNA (Δ cycle threshold [Ct] = $Ct_{GAPDH} - Ct_{target}$ RNA) and then used to determine the relative quantification of each target RNA with a calibrator, according to the formula $\Delta\Delta Ct = \Delta Ct_{calibrator} - \Delta Ct_{target}$ RNA. Each target mRNA in an individual mock-inoculated plant served as calibrator (RQ set to 1) for the respective gene. RQ for *RDR6* and *DCL4* transcripts was deduced by the formula expression $2^{-\Delta\Delta Ct}$. Columns represent mean RQ values from three biological replicates for each of the two experiments and different letters represent statistically significant differences values according to separate one-way ANOVA analysis for each target mRNA, using Tukey's test (P<0 05). Vertical bars on columns represent standard deviations among replicates.

small necrotic rings and line patterns etched in leaf epidermis (Figure 2A), followed by the appearance of chlorotic spots that turned necrotic as leaf blade enlarged. Overall local symptoms were severe and caused pronounced leaf blade distortion and precocious senescence. Systemic symptoms appeared in the 3rd and 4th true leaves and reached maximum severity at 14 dpi. They consisted in chlorotic/necrotic ringspots surrounding the veins, peripheral vein clearing and necrotic pinpoints and severe

deformation of leaf blade and margin (Figure 2B). After this severe symptom expression, plants initiated a rapid recovery and by 28 dpi the 5th and 6th unfolded true leaves were free of symptoms (Figure 2C) and plants grew normally up to 60 dpi.

Accumulation of AILV-V RNA was estimated in two experiments (mean values shown in Figure 3 and Table 1), each carried out on 18 plants. Leaf disks (50 mg) were collected from 3rd to 6th true leaves of three plants, representing three biological replicates

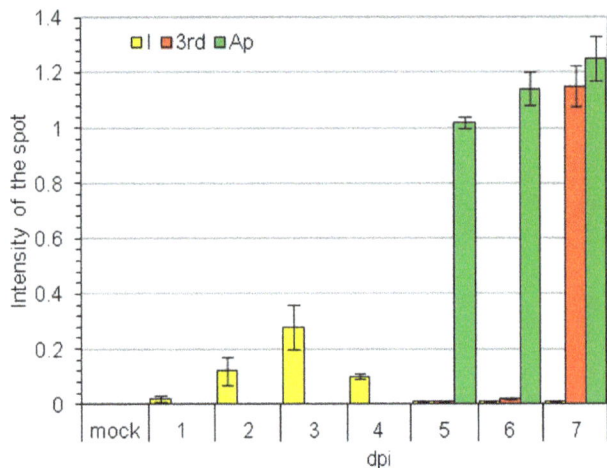

Figure 4. From the inoculated leaf, AILV-V moves first into shoot topmost leaf and then to lower leaves. Dot blot hybridization of samples collected from leaves of tobacco plants from 1 to 7 dpi with AILV-V. Plants at the 1103 growth stage according to the scale for coding growth stages in tobacco – coresta (i.e. with the 3rd leaf unfolded) were inoculated mechanically on the 1st and 2nd leaf. Columns represent mean values of the intensity of dot blot hybridization signal from two technical replicates of samples collected from two plants at each time point. The intensity of hybridization signal with antisense probe to detect viral genome (+)RNA was estimated from serial dilutions of a plasmid preparation, containing the fragment of the RNA-2 of AILV-V targeted by the probe. Vertical bars represent the standard error. I = inoculated leaf; A = apical leaf of the shoot tip. 3rd = third leaf i.e. the first leaf above the two basal leaves used for rub inoculation; Mock = leaf mock-inoculated with buffer.

for each time point. Samples collected between 10 and 23 dpi showed a cyclic variation of AILV-V RNA loads in each leaf, reaching the highest accumulation at 14 dpi in the 3rd leaf (Figure 3 and Table 1) when the maximum symptom severity was observed. Viral RNA concentration was progressively less in the 4th, 5th and 6th leaf to reach a steady-state condition that was maintained up to 60 dpi (Figure 3). Accumulation of AILV-V-RNA was also analyzed daily from 1 to 7 dpi in the samples collected from the inoculated 1st and 2nd unfolded true leaves, the topmost leaf (apical leaf in Figure 1) and the 3rd true leaf (the 3rd leaf in Figure 1). Samples were collected from four plants. To avoid biases in selecting the portion of leaf tissues to be analyzed, due to difference in size between apical and mature leaves, the whole leaf was harvested and crushed with a roller press in six vol (v/w) of alkaline solution and aliquots (5 μl) were used for quantitative dot blot hybridization. Figure 4 shows that viral RNA was detectable in inoculated leaves by 2 dpi and in the apical leaf by 5 dpi, while it was found in the 3rd leaf not earlier than 7 dpi. This pattern of the systemic movement of AILV-V is consistent with phloem transport of photoassimilates from source to sink organs, as shown for TMV in *N. tabacum* [23] and *N. benthamiana* [24].

AILV-V was back-inoculated from the leaves recovered from disease symptoms in three AILV-V-infected plants (40 dpi). Sap from each leaf was inoculated onto two tobacco plants. The inoculated plants both developed severe symptoms of local infection by 7 dpi and systemic infection and disease symptoms similar to the plants tested in the experiments described above (Fig 5A). Total RNA extracted from the recovered leaves was tested by northern blot hybridization using a DIG-labeled positive-sense and negative-sense RNA probes to detect the replication-specific, negative-sense RNA and the genomic positive-sense RNA of the viral RNA-2, respectively. The positive RNA-2 strand was detected (Figure 5B, plot 1), but not the negative strand (Figure 5B, plot 2). In contrast, both the negative and positive strands of viral RNA were detected in the 4th leaf showing symptoms.

Recovered leaves appeared resistant to reinoculation with AILV-V, because at 14 dpi there was no increase in AILV-V RNA-2 amounts in the leaves above the titers detected at the time of inoculation (Figure 5C).

AILV-V Enters and Persists in the Shoot Apical Meristem of Tobacco Plants

AILV-V distribution in shoot apical meristem (SAM) of tobacco plants was determined in two independent experiments by immunolocalization in samples collected from eight plants at 7, 14, 21, 28 and 40 dpi and, as control, from eight mock-inoculated plants. After 7 dpi, the virus was detected in the corpus of tobacco SAM, just beneath the tunica layers (Figure 6), and in leaf primordia. Between 7 and 14 dpi, i.e., concomitantly with maximum severity of disease symptoms, the virus invaded also tunica and persisted in the tissues up to 28 dpi, when full invasion of leaf primordia was observed. By 40 dpi, i.e., when the 5th and 6th leaves recovered from disease symptoms, the virus was detected only in few cells just beneath the corpus.

Transcription profiles of RDR6 and DCL4 during the course of AILV-V infection

To understand how the infected tobacco plants responded to AILV-V infection we analyzed the expression of *RDR6* and *DCL4* genes, which are two of the hallmark enzymes of the RS pathway in the study of plant-virus interactions [25,26]. Variations in the transcription profiles of *RDR6* and *DCL4* (orthologs of *Arabidopsis thaliana*) were monitored in AILV-V-infected tobacco plants from 10 to 60 dpi in two separate experiments (Figure 3 and Table 1) showing that there was a significant correlation between the amounts of *RDR6* mRNA and accumulation of AILV-V RNA-2 (Figure 3 and Table 1). Until 16 dpi, the expression of *RDR6* did not differ significantly in AILV-V-infected and mock-inoculated plants but increased rapidly in the 4th leaf between 16 and 18 dpi and reached the maximum in the 5th leaf and 6th leaf at 18 and 21 dpi, respectively. The upregulation of *RDR6* was almost concomitant with the maximal accumulation of viral RNA in the 3rd and 4th leaf (Figure 3 and Table 1) while in the leaves above the 3rd and 4th leaf, *RDR6* expression decreased progressively towards the top of the plant. It was lowest in the 5th and 6th leaf at 23 dpi and reached a steady-state level equivalent to that of mock-inoculated plants between 28 and 60 dpi.

The transcription levels of *DCL4* showed also a progressive increase, which correlated with accumulation of viral RNA, and reached the maximum at 18 dpi in the 5th leaf (Figure 3 and Table 1). During the recovery phase, i.e., when the viral replication diminished progressively until the time point at which it was almost non-detectable, the transcription level of *DCL4* was similar to that of mock-inoculated plants. These results suggested the upregulation of *RDR6* and *DCL4* was a consequence of AILV-V replication and accumulation of its RNA between 10 and 14 dpi in the 3rd and 4th infected leaves.

The inoculated and systemically infected leaves of tobacco plants were tested for AILV-V specific siRNAs, including plants infected with PVY-SON41 as controls. Despite of two independent experiments carried out to test samples taken at six different time points post-inoculation, it was not possible to clearly detect

Table 1. Comparison of accumulation of viral RNAs and expression level of *RDR6* and *DCL4* in time course experiments with tobacco plants challenged with AILV-V, PVY-SON41 or both viruses.

Leaf position	Sampling time (dpi)	Estimated viral RNA Load[†]				RQ[†]					
		AILV-V*	PVY-SON41**	AILV-V+PVY-SON41		AILV-V*		PVY-SON41**		AILV-V+PVY-SON41	
				AILV-V	PVY-SON41	RDR6	DCL4	RDR6	DCL4	RDR6	DCL4
mock	10	0.00±0.00	0.00±0.00	0.00±0.00	0.00±0.00	0.84±0.05 (a)	1.08±0.08 (a)	1.01±0.14 (a)	0.99±0.21 (a)	1.00±0.06 (a)	1.01±0.08 (a)
3rd	14	2.51±0.05	16.62±0.72	0.13±0.03	3.04±0.08	1.49±0.07 (b)	2.04±0.08 (b)	2.37±0.24 (a)	2.03±0.20 (ab)	1.80±0.07 (a)	0.31±0.06 (a)
	16	1.99±0.05	18.94±0.57	0.16±0.05	4.62±0.08	0.87±0.07 (b)	1.69±0.09 (ab)	3.46±0.29 (a)	2.50±0.18 (ab)	1.33±0.09 (a)	0.58±0.04 (a)
4th	10	0.62±0.03	3.88±0.32	0.20±0.02	0.60±0.09	1.00±0.09 (b)	2.72±0.05 (bc)	20.98±0.33 (bc)	3.49±0.38 (bc)	1.71±0.07 (a)	2.42±0.07 (c)
	12	0.97±0.04	9.85±0.75	0.26±0.03	1.25±0.09	1.45±0.08(b)	1.97±0.05 (b)	34.58±0.44 (d)	3.04±0.28 (bc)	9.57±0.06 (c)	8.18±0.10 (e)
	16	1.33±0.02	36.57±0.78	0.61±0.04	9.07±0.31	6.40±0.11 (c)	1.99±0.03 (b)	29.76±0.29 (cd)	5.20±0.21 (d)	14.94±0.09 (d)	4.30±0.17 (d)
	18	0.44±0.02	46.57±0.93	1.30±0.04	11.59±0.31	7.16±0.12 (c)	2.99±0.16 (c)	67.82±0.47 (e)	4.25±0.22(cd)	1.00±0.08 (a)	1.61±0.12 (bc)
5th	18	0.60±0.03	21.13±0.99	1.09±0.05	12.10±0.35	12.48±0.09 (d)	3.52±0.12 (c)	25.82±0.50 (bcd)	4.15±0.32 (cd)	5.07±0.16 (b)	4.30±0.10 (d)
	19	1.25±0.03	26.32±0.52	1.19±0.04	17.40±0.38	4.32±0.05 (bc)	1.43±0.06 (ab)	13.78±0.45 (b)	5.55±0.33 (d)	3.63±0.12 (bc)	0.39±0.12 (a)
6th	21	0.17±0.03	25.53±0.59	1.85±0.03	12.57±0.27	11.73±0.11 (d)	2.68±0.06 (bc)	19.03±0.36 (b)	7.05±0.25 (d)	16.03±0.27 (d)	0.52±0.13 (a)

[†]Expressed as mean of 3 plants from two experiments;
*values from Figure 3;
**values from Figure 10;
± = standard error. Different letters in brackets represent statistically significant differences of means (column wise) according to analysis of variance (P<0.05) (Tukey test).

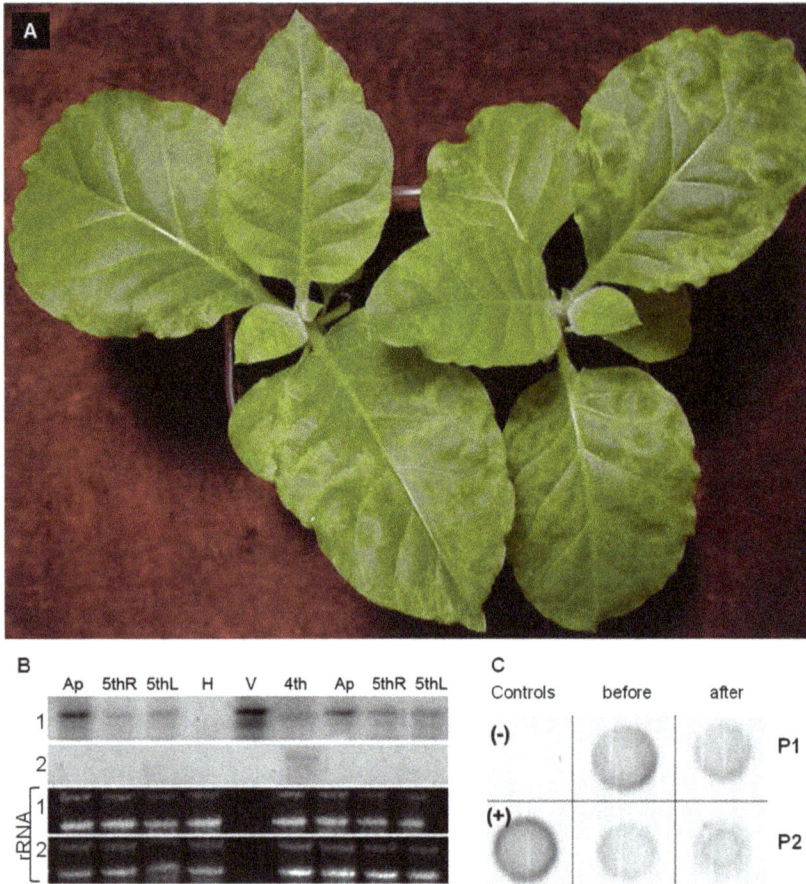

Figure 5. AILV-V would not replicate in leaves recovered from disease symptoms but retains infectivity. In **A.** local and systemic symptoms consisting, respectively, in chlorotic spots and line patterns and, chlorotic/necrotic ringspots surrounding the veins and peripheral vein clearing induced by AILV-V in tobacco at 12 dpi with sap extracted from tobacco leaves, which had recovered from disease symptoms at 40 dpi. In **B.** Northern blot hybridization for detection of (+)RNA (1) and (−)RNA (2) on RNA preparations extracted from the following sources: A = apical leaves at 40 dpi; 5thR and 5thL = samples collected from opposite sites (Right and Left) from the 5th leaf of two tobacco plants at 40 dpi with recovery phenotype; 4th = sample collected from the 4th leaf of a tobacco plant with severe symptoms of systemic infection; V = RNA from a purified preparation of AILV-V particles, used as positive control; H = sample collected from an healthy tobacco plant, used as negative control. The picture shows the presence of a weak signal of hybridization with the plus-sense RNA probe (which detects the replication specific minus-sense RNA) only in correspondence of the sample collected from the 4th leaf with severe symptoms of systemic infection. In **C.** Accumulation of AILV-V RNA2 determined by dot blot hybridization in leaf samples collected from two recovered plants (P1 and P2) at 28 dpi before and 14 days after secondary inoculation. + and − indicate positive (pAILV769 plasmid DNA) and negative (mock-inoculated plant leaf) controls, respectively.

low-molecular-weight RNA specific for AILV-V (only samples from systemically infected leaves at two representative time points at 7 and 14 dpi are shown in Figure 7A). This was not due to the method we used, because the siRNAs specific to PVY-SON41 were readily detected in plants infected by this virus (Figure 7B).

AILV-V Is Unable to Prevent the Establishment or Maintenance of RNA Silencing in Response to Infection or to Revert Pre-Established Silencing

To determine whether AILV-V could interfere with the establishment or maintenance of RNA silencing or revert already silenced genes we used a Samsum tobacco transgenic line expressing the HC-Pro silencing suppressor and induced down-regulation in the expression of the transgene by the infection of PVY-SON41 as shown previously by Savenkov and Valkonen [27]. Twelve transgenic tobacco plants expressing the PVY HC-Pro silencing suppressor (HC plants) and twelve non-transformed (WT plants) tobacco plants were inoculated with AILV-V, PVY-

SON41 or co-inoculated with both viruses and monitored at 14 and 40 dpi for the expression of the HC-Pro protein, the progression of disease symptoms and the accumulation of viral RNAs. Three HC and WT plants were mock-inoculated to serve as negative control.

In WT plants inoculated with PVY-SON41 the amounts of HC-Pro protein increased from 14 to 40 dpi, whereas in the HC-transgenic plants, a peak of HC-Pro protein accumulation was detected at 14 dpi followed by a drastic reduction by 40 dpi (Figure 8A). The initial increase of HC-Pro protein in HC-transgenic plants was higher than in WT plants and mock-inoculated HC plants and was likely due to the accumulation of HC-Pro translated from the transgene transcript and produced by the infecting virus. The drastic reduction of HC-Pro protein exhibited by HC-transgenic plants at 40 dpi indicated that RS targeted both the HC-homologous transgene and viral sequences.

In this experiment, we examined also the progression of disease symptoms and estimated the loads of viral RNA in single and mixed infections in plants inoculated first with PVY-SON41 or

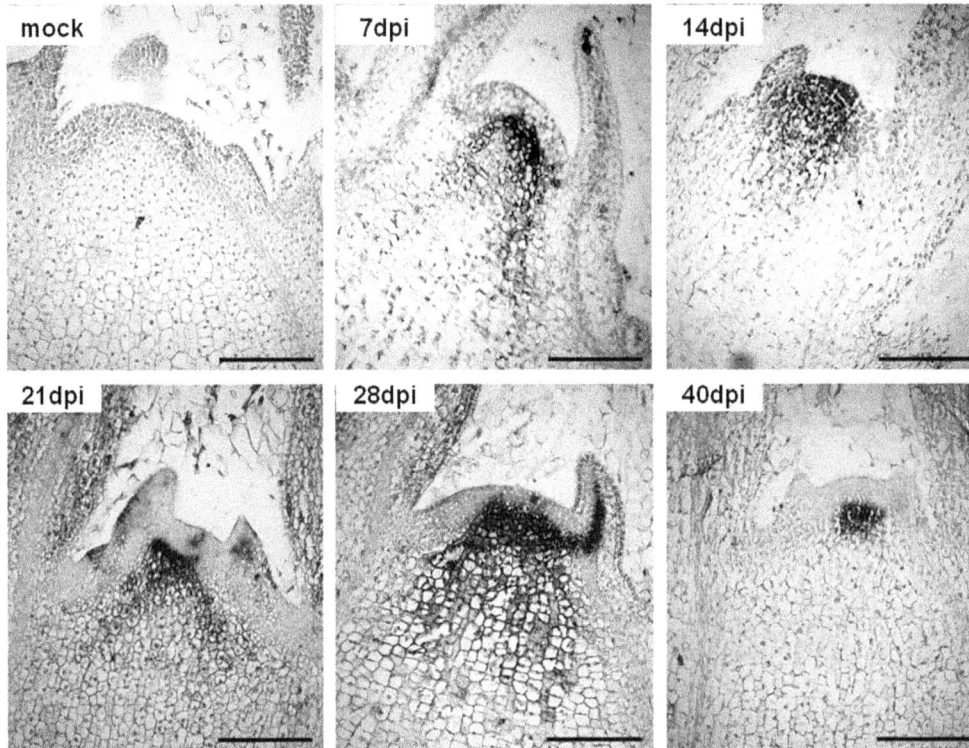

Figure 6. AILV-V enters SAM of Samsun tobacco at a very early stage of infection and persists there up to 40 dpi. One of the two time-course analyses of AILV-V distribution in the SAM of Samsun tobacco showing longitudinal sections of meristem tissues at 7, 14, 21, 28 and 40 dpi. AILV was detected using polyclonal antibodies raised against AILV-V coat protein and signals were developed with alkaline phosphatase diluted 1:500 in PBS/BSA buffer and stained in NBT/BCIP solution. Immunolocalization is demonstrated in cells with dark stain. Pictures show that between 7 and 28 dpi the virus was present in all meristem tissues and in leaf primordia while by 40 dpi, i.e. concomitantly with the recovery phase from disease symptoms, the virus appeared confined between the corpus and the rib meristem. Mock = SAM of healthy tobacco treated with AILV-V antiserum at 21 days after mock- inoculation and used as negative control. Bars = 100 μm.

AILV-V and 14 days later with AILV-V (P+A plants) or PVY-SON41 (A+P plants), respectively. WT plants challenged with AILV-V alone showed severe disease symptoms and high viral RNA accumulation at 14 dpi while by 40 dpi plants displayed a recovery phenotype and approx 6-fold reduction of viral RNA accumulation (Figure 8B). In contrast, the abundance of AILV-V RNA in HC plants was 2 to 6-fold higher than in WT plants at the same time points and none of the transgenic plants exhibited recovery up to 40 dpi (Figure 8B). The disease symptoms in WT plants with mixed virus infection were more severe than in plants infected with AILV-V alone and, unlike HC-transgenic plants, they exhibited a delayed recovery phenotype one week later than observed in the plants that were infected with AILV-V only. In addition, the order of inoculation (A+P or P+A) did not affect the disease progression and phenotypic response.

Accumulation levels of PVY-SON41 RNA were also different in WT and HC plants and in plants with single or double infection. In WT plants infected with PVY-SON41 alone, PVY RNA accumulated at very high levels, whereas in plants co-infected with AILV-V, PVY RNA amounts were 20-fold less and not influenced by the order of inoculation of the two viruses. Accumulation of PVY-SON41 RNA in the HC-transgenic plants was lower in single and mixed infection, regardless of the order of inoculation, than in WT plants infected with PVY-SON41 only (Figure 8B).

Overall these results provide evidence that RS targeted AILV-V, because recovery was delayed in plants expressing PVY HC-Pro while viral RNA concentration was enhanced. In contrast, the

virus did not prevent or revert the down regulation of the HC-Pro protein in HC plants upon infection of PVY-SON41 and did not increase accumulation of the potyviral RNA, suggesting poor or weak ability to interfere with RS.

AILV-V and PVY-SON41 Act Synergistically in Symptom Development but Not in the Accumulation of Viral RNA in Tobacco

To analyze the trilateral interaction between AILV-V, PVY-SON41 and tobacco in more detail, we monitored symptom development, load of viral RNAs and transcription profile of *RDR6* and *DCL4* between 7 and 21 dpi with AILV-V and PVY-SON41 in single and in mixed infection. AILV-V and PVY-SON41 were inoculated alone, each to a group of 18 plants, while in another group of 18 plants AILV-V and PVY-SON41 were inoculated, respectively, to the 1st and 2nd leaf of tobacco to obtain a mixed infection. Plants with single PVY-SON41 infection manifested symptoms of systemic mosaic, which were persistent in the leaves no. 3 to 6 (Figure 9A and Figure 1). Despite of mild disease symptoms, accumulation of potyviral RNA was high, increased progressively and reached the maximum in the 4th leaf at 17 dpi (Figure 10 and Table 1). After this peak, load of PVY-SON41 RNA decreased and reached a steady-state level in the 6th leaf, which was maintained up to 23 dpi when the monitoring was terminated. The transcription level of *RDR6* followed the pattern of PVY-SON41 RNA accumulation and showed the highest upregulation in the 4th leaf at 18 dpi. Upregulation of *RDR6* in

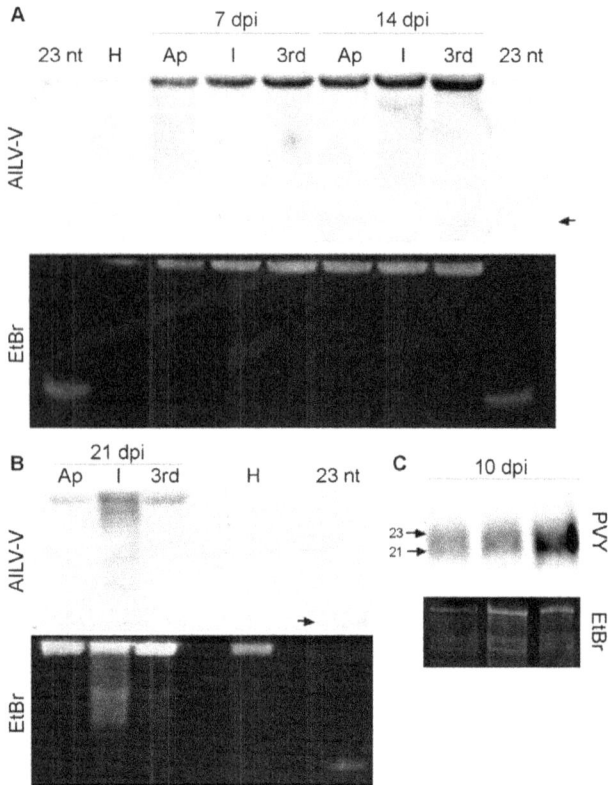

Figure 7. Small interfering RNAs (siRNA) produced in response to AILV-V infection remain below a detectable level. In **A** and **B**, panels represent total RNA preparations enriched in siRNA obtained from samples collected from apical (Ap), rub-inoculated (I) and 3rd (3rd) leaf at 7, 14 and 21 dpi with AILV-V. RNA preparations were first separated in by polyacrylamide gel electrophoresis and stained with ethidium bromide (EtBr)), then transferred to nylon membrane by electroblotting and hybridized with an AILV-V -specific RNA probe (AILV-V) for the last 760 3'-terminal sequences, labeled with digoxigenin and hydrolyzed by alkaline treatment. H = total RNA extracted from an healthy tobacco leaf. P = 23 bp primer. Arrows point the position expected for the 23 bp primer, after hybridization. In **C**, detection of siRNAs in samples collected from leaves of a tobacco plant at 10 dpi with PVY-SON41. Panels show ethidium bromide staining (EtBr) after PAGE analysis and signals produced after hybridization with a PVY-specific RNA probe (PVY) labeled with digoxigenin and hydrolyzed by alkaline treatment. Arrows point position of 21 and 23 bp primers used as size markers.

the 5th and 6th leaf was also substantial (Figure 10 and Table 1) but approx 3- to 4-fold lower than in the 4th leaf. In a similar way the expression profile of *DCL4* showed a progressive increase, paralleling the accumulation of PVY-SON41 RNA (Figure 10 and Table 1).

Progression of disease symptoms, accumulation of viral RNAs and transcriptome analysis of *RDR6* and *DCL4* in plants with single AILV-V infection were in good correlation with data obtained from the previous experiment (Table 1). In plants infected with AILV-V at 16 dpi, the transcription level of *RDR6* was 3.5-fold (in the 3rd leaf) to 5-fold (in the 4th leaf) lower than in plants infected by PVY-SON41 (Table 1). This was consistent with the higher RNA loads of the potyvirus. Similarly, the expression of *DCL4* at 16 dpi in plants infected by AILV-V was 2.5-fold lower (in the 4th leaf) than in plants with single infection by PVY-SON41 (Table 1).

Disease symptoms observed at 21 dpi in mixed infected plants were much more severe than those induced by PVY-SON41 and AILV-V in single infections (Figure 9B and Figure 2). However, in the co-infected plants the maximum load of viral PVY-SON41 RNA was reached not earlier than 18 dpi (in the 5th leaf) and was approx 50% of that at the same time point in plants singly infected by the potyvirus, suggesting an inhibitory effect of AILV-V against PVY-SON41 in co-infected tobacco plants. On the other hand, the load of AILV-V RNA in each leaf did not follow the distribution observed in singly infected plants since it increased progressively from the lower to the upper leaves and the maximum load of RNA was reached not earlier than 21 dpi (in the 6th leaf) in the plants infected with AILV-V only (Table 1). Thus, results suggested that AILV-V and PVY-SON41 interfered with each other at the early stages of concomitant infection.

AILV-V Might Interfere with Cell-to-Cell Movement of the Signal of RNA Silencing

Plant viruses code a wide range of viral silencing suppressor proteins (VSR) targeting different steps of the RNA silencing pathway [26,28,29]. To examine the effect of VSR of a number of RNA and DNA viruses on the dynamic of AILV-V infection, plants of *N. tabacum* cv Xanthi transformed with silencing suppressor genes derived from six different viruses [30] were challenged with AILV-V at the five-leaf stage. Twenty plants per each transgenic line were inoculated and symptoms monitored daily up to 50 dpi. Most of the transgenes did not affect recovery from disease symptoms (Table 2). Recovery was prevented only in plants expressing the P1 VSR of *Rice yellow mottle virus* (RYMV, genus *Sobemovirus*) and it was delayed but not prevented in plants expressing the AC2 VSR of *African cassava mosaic virus* (ACMV, genus *Begomovirus*) or HC-Pro of PVY. Because results from experiments with HC-transgenic plants indicated that AILV-V was not able to inhibit establishment and maintenance of RS, we did additional experiments using agroinfiltrated patch assays in the green fluorescent protein (GFP) transgenic plants of *N. benthamiana* line 16c. Systemic silencing in the 16c plants was induced by inoculation of basal mature leaves with infectious transcripts of the engineered clone of *Tobacco mosaic virus* (genus *Tobamovirus*) carrying the GFP gene (TMV-GFP). In non-transgenic plants the expression of *GFP* from TMV resulted in bright fluorescence both in inoculated and top leaves between 4 and 14 dpi while in 16c *N. benthamiana* line, the vector induced silencing of the GFP in most of the leaves, which appeared red under UV illumination due to autofluorescence of chlorophyll (Figure 11A). At 14 dpi with TMV-GFP the red areas were mechanically inoculated with AILV-V and no suppression of silencing was observed up to 40 dpi while the red fluorescent areas continued to expand (Figure 11B).

In patch assays, the 16c plants were rub-inoculated either with AILV-V, PVY-SON41 or strain B11 of *Potato virus A* (PVA-B11, genus *Potyvirus*) alone, or co-inoculated with AILV-V and PVY-SON41. At 7 dpi, fully enlarged leaves of *N. benthamiana* were agroinfiltrated on the opposite sides of the midrib with *A. tumefaciens* containing the binary vector pBIN-mGFP4 for the expression of GFP under *Cauliflower mosaic virus* 35S promoter (35S). Upon ectopic expression of GFP, a thin border of dark red tissue was visible at 14 dpi in mock-inoculated plants, indicating short-range movement of GFP silencing (Figure 11C). However, the border was not observed in leaves of plants inoculated with AILV-V, suggesting that the virus interfered with cell-to-cell movement of the silencing signal. Green fluorescent areas without red borders visible in leaves inoculated with AILV+PVY, PVY or

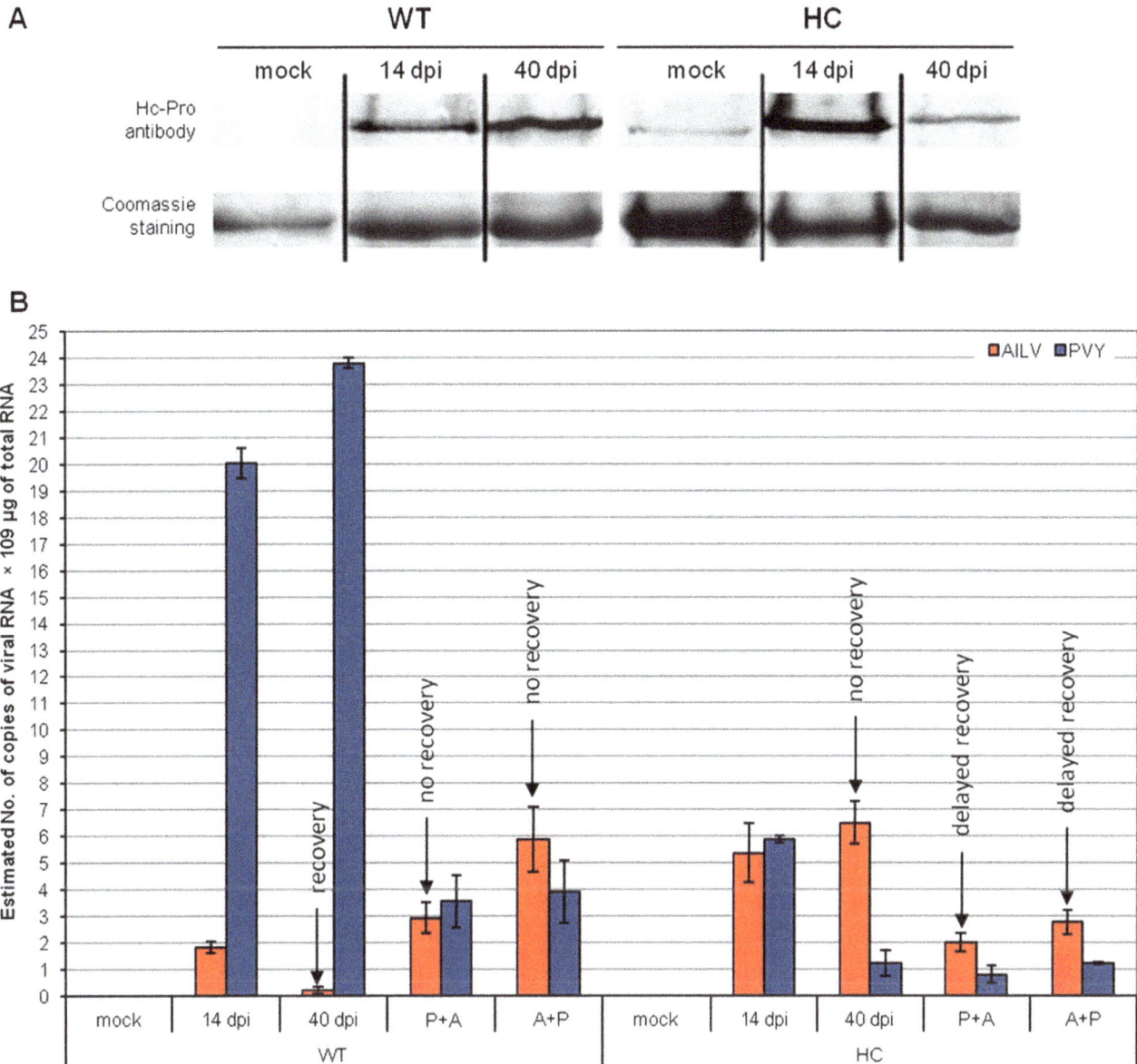

Figure 8. Plants expressing HC-Pro VSR do not enter the recovery phase during AILV-V infection. In **A,** variation in the load of PVY HC-Pro protein detected by western blot in non-transgenic tobacco plants and tobacco plants transformed to express HC-Pro, upon infection with PVY-SON41. WT plants show increasing levels of HC-Pro protein from 14 to 40 dpi while HC-transgenic plants express: i) a steady-state level of the protein after mock-inoculation; ii) an increasing protein load at 14 dpi resulting from the sum of HC-Pro translated from the transgene and from viral transcript and iii) a strong downregulation of the accumulation of the HC-pro protein at 40 dpi caused by the activation of homology-dependent RNA silencing. In **B,** levels of viral RNAs in plants of tobacco wild-type (WT) and transformed with HC-Pro (HC), upon single and mixed infection of AILV-V and PVY-SON41. Columns represent mean values of the number of copies of viral RNAsx10^9 per µg of total RNA estimated from three biological replicates. Vertical bars represent the standard error. Quantitative dot blot was obtained from the intensity of hybridization signal estimated on the basis of a standard curve generated by serial dilutions of a plasmid preparation containing the RNA fragments targeted by the specific probes. Translation of HC-Pro from either transgenic or authentic virus transcripts favors the infection of AILV-V in WT plants so the plants do not show the recovery phenotype. A+P and P+A indicate the order of inoculation (A = AILV-V and P = PVY-SON41) in mixed infection. Samples were collected 14 days after the second inoculation, corresponding to 40 days from the first inoculation, from the newest fully developed leaves.

PVA indicated suppression of silencing driven by VSR coded by PVY-SON41 and PVA-B11 (Figure 11C).

Discussion

This study investigated the dynamics of AILV-V infection in tobacco revealing the following new insights in nepoviral life-cycle: the virus colonized the SAM at 7 dpi and persisted in meristem tissues up to 40 dpi while new leaves showing recovery from severe viral symptoms did not appear before 21 dpi. The asymptomatic leaves, which developed during the recovery phase, contained infectious virus particles but no replication of AILV-V was apparent, and the leaves were protected against reinoculation with AILV-V. We therefore hypothesize that the recovery from symptoms observed in the top leaves might result from a defense response, which, like with PVX [14], was not primed by the entry

Figure 9. Mixed infections of PVY-SON41 and AILV-V in tobacco exacerbate disease symptoms. In **A,** mild mosaic and moderate leaf blade malformation induced in tobacco at 30 dpi with PVY-SON41. In **B,** Chlorotic/necrotic ringspots, severe reduction of leaf lamina and plant growth induced at 30 dpi by a mixed infection of PVY-SON41 and AILV-V.

of AILV in the SAM but was triggered in tissues of systemically infected leaves as soon as accumulation of viral RNA reached a threshold level. The progression of AILV-V infection in tobacco plants was similar to that of the W22 strain of *Tomato black ring virus* (TBRV-W22) [15], another nepovirus of the subgroup B, but it differed from TRSV and ToRSV, which are nepoviruses belonging to subgroup A and C, respectively [6]. Therefore it seems that the dynamics of nepovirus entry and persistence in tobacco SAM, as well as viral RNA accumulation and persistence in asymptomatic leaves recovered from disease symptom, are different among virus species belonging to distinct subgroups of the genus *Nepovirus*.

Quantitative dot blot and immunolocalization analyses provided evidence that AILV-V entered SAM of tobacco plants within 7 dpi and persisted detectable therein during the recovery phase at least up to 40 dpi. Time-course experiments shown in Figure 4 demonstrated that from inoculated leaves the virus moved fastest to the topmost leaf (apical leaf), where it was found at 5 dpi, whereas it was detected in the 3rd leaf (i.e., the first leaf above the two inoculated leaves) not before 7 dpi when the virus was immunolocalized also in the SAM. According to the model proposed by Foster *et al.* [13] and Schwach *et al.* [14] invasion of meristematic tissues should have been prevented by the *RDR*6-mediated amplification of the systemic silencing signal, which was not the case with AILV-V as the virus was detected in SAM one

week before *RDR6* was overexpressed. Therefore, we hypothesize that similarly to TRSV [10], entry of AILV-V in the meristematic tissues was permitted by the very low concentration of viral RNA or virus particles at a time of infection that would be insufficient to trigger an RS-mediated response. Results of immunolocalization between 7 and 28 dpi showed a higher concentration of virus in meristem than in surrounding tissues, as with TRV and TRSV [16,19], suggesting poor or no AILV-V replication and movement in the SAM. The successive reduction of virus titer observed in SAM by 40 dpi might be due to hindered viral transport into the meristem as a consequence of virus clearance in tissues of mature leaves. This model is supported also by the fact that we were unable to demonstrate viral replication in leaves fully recovered from disease symptoms, although these tissues contained viral particles that proved highly infectious in back-inoculation experiments. However, as found in previous studies [31,32,33] we cannot exclude that AILV-V may have replicated in groups of cells.

Roberts et al. [34] proposed that recovery from ringspot symptoms induced by TRSV occurred after viral invasion of the SAM and Ratcliff et al. [15] linked the appearance of the recovery phenotype from TBRV infection in *N. benthamiana* to the activation of RS, which, in turn, was correlated with viral invasion of the SAM. Conversely, the model proposed for TRV in *N. benthamiana* implies that fully expanded leaves showing the recovery phenotype are those deriving from SAM after virus clearance [16]. With AILV-V we provided evidence that the virus was present in tobacco SAM at least two weeks before initiation of the recovery phase in leaves and that reduction of virus titer observed in SAM at 40 dpi was concomitant with the steady-state level of viral RNA loads during the recovery phase in mature leaves. Therefore, we propose that with AILV-V, invasion of the SAM and initiation of recovery are spaced-out processes. Collectively, the model of SAM invasion proposed for AILV-V is consistent with the pattern of meristem invasion in *N. benthamiana* plants proposed for TRV [16]. However, while TRV infection was supported only transiently in meristematic tissues and by the activity of its VSR, with AILV-V we propose that the virus persisted in meristem tissues because of the very low level of accumulation, which was not sufficient to trigger RS. Nonetheless, the activity of a hitherto unknown VSR coded by AILV-V, accounting for its persistence in meristematic tissues, cannot be excluded.

Results of time-course experiments (Figure 3 and Table 1) showed variations in the expression profiles of *RDR6* and *DCL4* during AILV-V infection and suggested activation of an RS-based host response after a threshold of viral RNA concentration was reached between 10 and 14 dpi which coincided with systemic infection of the 3rd leaf (first leaf above the inoculated leaves). We were unable to detect AILV-V- specific siRNAs as evidence for antiviral RS, as reported with plants infected by ToRSV [20] and TRSV [10] as well as in other plant-virus combinations [35,36]. However, indirect evidence that AILV-V was targeted by RS was obtained by the analysis of transgenic lines expressing suppressors of silencing and from the plants in which AILV-V was in mixed infection with PVY. Recovery was delayed in plants expressing PVY HC-Pro and AILV-V RNA concentration was enhanced and was prevented as well in transgenic plants expressing AC2 or RYMV P1. We therefore hypothesize that, similar to TBRV and TRSV [10,15], induction of the recovery phenotype in plants infected by AILV-V might be a consequence of RS activated in lower leaves, which conditioned negatively the accumulation of viral RNA in all the leaves that developed later.

The transcript profile of *DCL4* increased consistently with the accumulation of AILV-V RNA, although with smaller variations

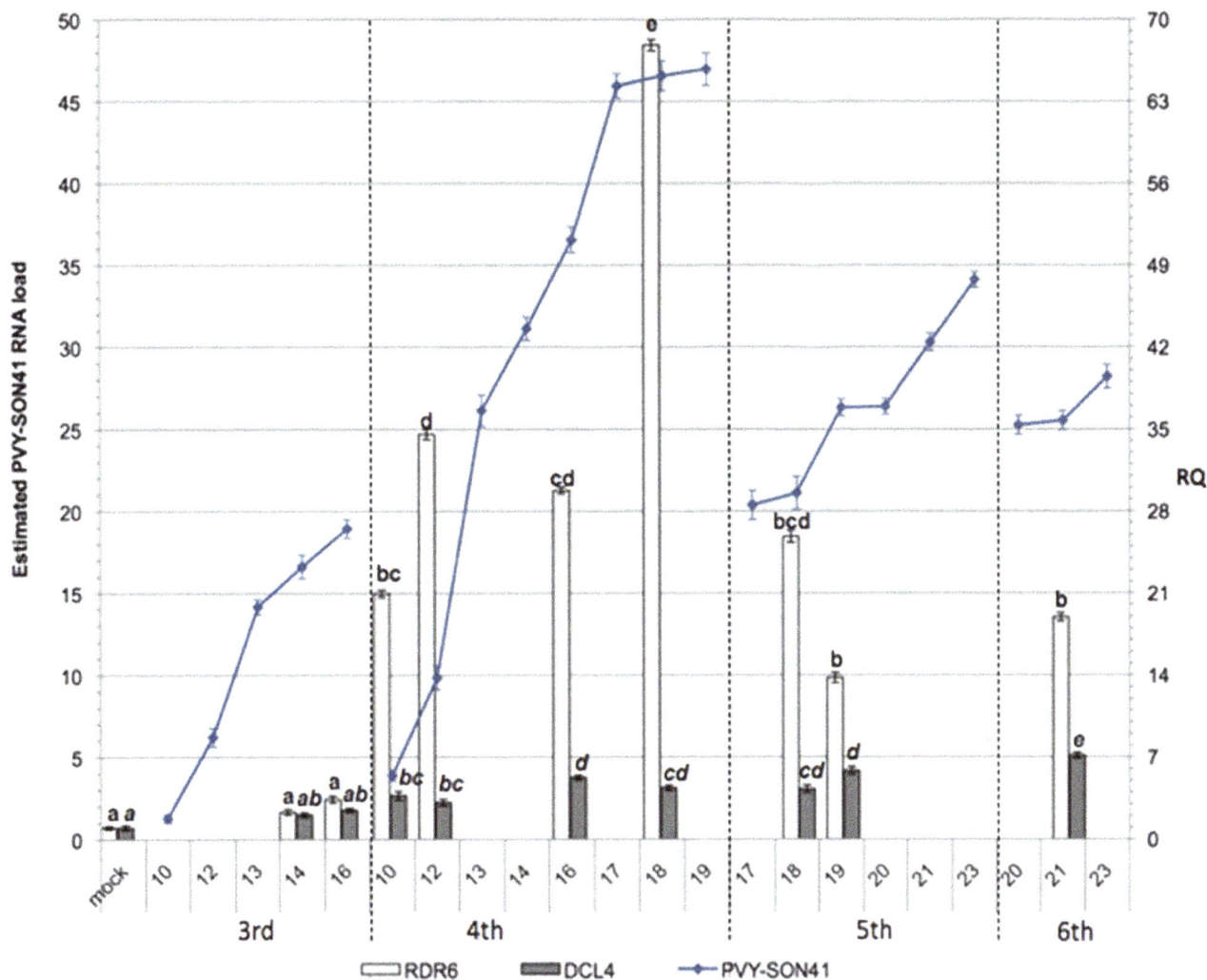

Figure 10. PVY-SON41 infection correlates with viral RNA continuous accumulation and suppression of RNA silencing. Accumulation levels of viral RNA (lines) and transcription profiles of *RDR6* and *DCL4* were estimated by quantitative dot blot hybridization and qPCR, respectively. Symbols and protocols as in Figure 3.

than *RDR6*. In the plant response to viral infections *DCL4* is involved in biogenesis of the bulk of viral siRNAs by dicing perfectly paired double-stranded RNAs generated by either the viral replicase or host-encoded *RDR6* and *RDR1* [26,37]. Upregulation of the transcription levels of *DCL4* was therefore expected at the time points in which maximum accumulation of viral RNA and maximum expression of *RDR6* were recorded, but only modest if any changes were observed. This might be taken as indirect evidence that AILV-V interferes with the activity of *DCL4* in the RS process.

Finally, although the agroinfiltrated patch assay suggested that AILV-V would be able to interfere with cell-to-cell movement of the silencing signal, further experimentation will be required to determine if the AILV-V genome encodes a suppressor of RS and define its mode of action.

Materials and Methods

Plants and Viruses

Plants of *Nicotiana tabacum* cv. Samsun (tobacco) were used in the experiments. Transgenic plants of tobacco cv. Samsun expressing

the helper component proteinase (HC) of *Potato virus Y* (PVY) were obtained from Dr. Peter Palukaitis (Department of Horticultural Sciences, Seoul Women's University, Seoul, South Korea) while transgenic *N. benthamiana* line 16c [38] was kindly provided by Dr. David Baulcombe (Department of Plant Sciences, University of Cambridge, Cambridge, United Kingdom). Lines of tobacco cv. Xanthi transformed with genes for the expression of the following viral RNA silencing suppressor (VRSs) [29] were also used: P1 of *Rice yellow mottle virus* (RYMV), P1 of *Cocksfoot mottle virus* (CoMV), P19 of *Tomato bushy stunt virus*, (TBSV), P25 of *Potato virus X*, (PVX), HC-Pro of PVY, strain N (PVY), 2b of *Cucumber mosaic virus*, strain Kin (CMV), AC2 of *African cassava mosaic virus* (ACMV). Plants were grown in a glasshouse at $24\pm2°C$ with a 16 h light and 8 h dark regime. Inocula were prepared by crushing systemically infected leaves in 100 mM Na_2-K phosphate buffer, pH 7.2, containing 1 mM sodium sulphite and by rubbing the extracts onto celite-dusted 1st and 2nd completely unfolded true leaves (approx. growth stage 1002 of the scale for coding growth stages in tobacco – coresta at http://www.docstoc.com/docs/155125965/A-Scale-for-Coding-Growth-Stages-in-Tobacco---coresta and Figure 1). Plants were screened daily for symptoms appearance

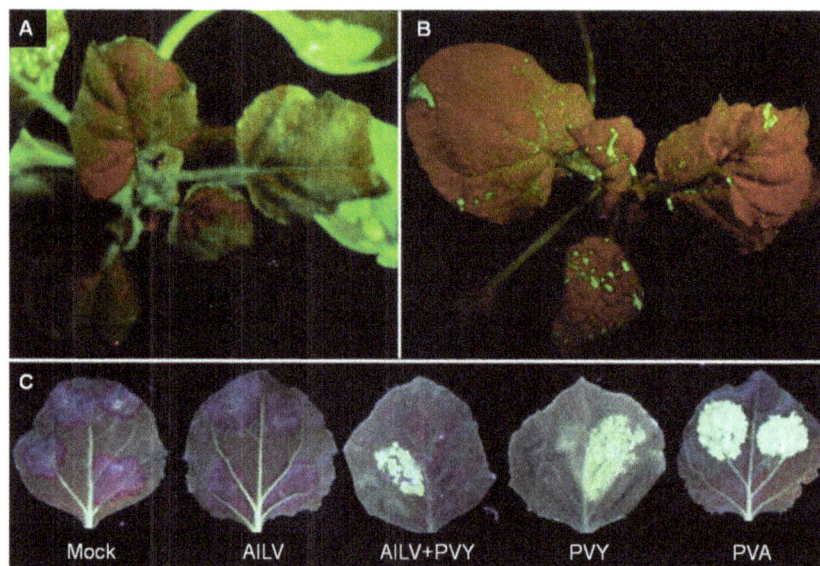

Figure 11. AILV-V is unable to revert GFP silencing while interferes with cell-to-cell movement of silencing signal. In **A,** progression of GFP silencing (indicated by dark red areas along the major veins) in a plant of *N. benthamiana*, line 16c, at 14 dpi with the TMV-GFP vector. Silenced areas were inoculated with AILV-V but no desilencing effects were observed at 30 dpi with AILV-V; rather the silenced areas expanded (in **B**) following the spread of TMV-GFP infection. In **C** Free GFP was expressed transiently in 16c *N. benthamiana* from the binary vector pBIN-mGFP4 carried by *A. tumefaciens*. Prior to agroinfiltration, leaves were mock-inoculated with buffer (Mock) or with AILV-V (AILV), PVY-SON41 (PVY), AILV-V and PVY- SON41 (AILV+PVY) and PVA-B11 (PVA). Upon ectopic expression of GFP, a thin border of dark red tissue was visible at 14 dpi in plants mock-inoculated indicating short-range movement of GFP silencing. This border was not produced in leaves of plants inoculated with AILV-V, suggesting a viral interference with cell-to-cell movement of the silencing signal. Green fluorescent areas visible in AILV+PVY, PVY and PVA infected plants indicate suppression of silencing driven by VSR coded by PVY-SON41 and PVA-B11.

and sampled at 7, 14, 21, 28, 40 and 60 days post-inoculation (dpi) for molecular assays. The following inocula were used: the grapevine isolate of AILV (AILV-V) [39]; the SON41 isolate of PVY (PVY-SON41) kindly provided by B. Moury (INRA, Montfavet, France); an isolate of *Potato virus A* strain B11 (PVA-B11) [40]; and a biologically active transcripts of a recombinant *Tobacco mosaic virus* vector, denoted TMV-GFP, carrying the ORF of the green fluorescent protein (GFP) from the jellyfish (*Aquorea victoria*). Inocula of AILV-V and PVY-SON41 were obtained from systemically infected Samsun tobacco plants while inoculum of PVA-B11 was maintained in Xanthi tobacco plants. Infectious transcripts of TMV-GFP were synthesized from the plasmid pBSG1057 linearized at its *Kpn*I site, using the T7 RNA polymerase and the mMessage mMachine kit (Ambion, Austin, TX, USA), following the protocol of the manufacturer. The plasmid pBSG1057-5 (kindly supplied by Dr. Helene Belanger, Large Scale Biology Corporation, Vacaville CA, USA) contains the TMV genome including the GFP ORF placed under the control of a duplicate of the TMV coat protein subgenomic promoter. Ten microliters of the transcription mixture were rub-inoculated onto each leaf of *N. benthamiana* 16c plants. Plants mock-inoculated with buffer served as negative control.

RNA Extraction and Analysis

Total RNA was extracted from 50 mg plant tissues with the TRIzol reagent (Invitrogen, San Diego, CA, USA) as described by the manufacturer and subjected to RQ1 DNase digestion (Promega, Madison, WI, USA). RNA preparations were suspended in 30 μl RNase-free water and concentration and quality estimated with a Nanodrop ND-1000 spectrophotometer (Nanodrop Technologies, Rockland, DE, USA), electrophoresis through 1.2% agarose gel in TBE buffer (90 mM Tris, 90 mM boric acid,

1 mM EDTA) and Gel-red (Biotium, USA) staining. Final RNA concentrations were adjusted to 1 μg/μl. Plant response to the single and mixed infection of AILV-V and PVY-SON41 was evaluated by the quantification of the expression level of the genes for *RDR6* and *DCL4* in time-course experiments with reverse transcription quantitative real-time PCR (qPCR), using glyceraldehyde 3-phosphate dehydrogenase (*GAPDH*) as housekeeping gene [41]. A group of four plants was inoculated with AILV-V, another group was inoculated with PVY-SON41 and a third group was inoculated with both AILV-V and PVY-SON41 on the 1st and 2nd completely unfolded true leaves of the same plant. Total RNA was extracted from 50 mg leaf tissue collected daily from the 3rd to the 6th leaf between 10 and 60 dpi. In particular, samples from the 3rd and the 4th leaf were collected between 10 and 16 dpi; those from the 4th and 5th leaf between 17 and 19 dpi and those from the 5th and the 6th leaf between 20 and 23 dpi. The 6th leaf was used also to collect samples at 28 and 60 dpi. Total RNA extracted from mock-inoculated plants at each sampling time was used as control. Primers pairs used for the amplification of *RDR6*, *DCL4* and *GAPDH* transcripts were designed with Primer-BLAST (http://www.ncbi.nlm.nih.gov/tools/primer-blast/) and were *RDR6* For (5'-CGACCTCG-CAATGGTCCTA GC-3') and *RDR6* Rev (5'-GCTCGTCCTTCACCCGGAGC-3'), designed on the basis of GenBank Accession No AB361628; *DCL4* For (5'-GTTGA-CAAGTGGCTTCAAAGCAG-3') and *DCL4* Rev (5'-CTTTGCTTGCAGCGCAAATACTC-3'), designed on a Blast alignment between sequence FM986783 from *N. benthamiana* and sequence AM846087 from *N. tabacum* and *GAPDH* For (5'-CGGCCGCCCCGTTGTATC-3') and *GAPDH* Rev (5'-GA-GAGGAGGAGCGAAGTCC-3'), designed on the basis of GenBank accession no. AJ133422). First strand cDNAs were

Table 2. Effect of suppressors of RNA silencing (VSR) derived from different RNA and DNA viruses and expressed as transgenes in lines of *N. tabacum* cv Xanthi on the development of symptoms and induction of the recovery phenotype upon infection of AILV-V.

Plant line*	3	7	10	14	17	21	28	31	35	40	50
WT	LR	RS	RS	M, R	R	R	R	R	R	R	R
pBin61	RS	RS	RS	M, R	R	R	R	R	R	R	R
P1 CfMV	LR	RS, VC, M, B	RS, B	RS	RS	RS	RS	RS	RS	RS	RS
P1 RYMV	M	RS, VC, M	RS, VC, M	RS, VC, M, B	RS, VC, M, B	RS	RS	R	M	R	R
HC-Pro	LR	RS	RS	RS	RS	RS	RS	R	M	R	R
AC2	RS	RS	RS, VC, M	RS, VC, M, B	RS, VC, M, B	RS, B	RS	RS	M	R	R
P25	LR	RS, VC, M, B	RS, B	R	R	R	R	R	R	R	R
2b	RS	RS, VC, M.5	RS, VC, M, B	RS	R	R	R	R	R	R	R
P19	RS	RS, VC, M, B	RS, B	RS	RS	RS	R	R	R	R	R

*Transgenic plant line expressing the following VSR: P1 from *Cocksfoot mottle virus* (P1 CfMV) and from *Rice yellow mottle virus* (P1 RYMV), Hc-Pro from *Potato virus Y* (PVY), AC2 of *African cassava mosaic* (ACMV), P25 from *Potato virus X* (PVX), 2b from *Cucumber mosaic virus* (CMV) and P19 from *Tomato bushy stunt virus* (TBSV). Symptoms were recorded between 3 and 50 dpi with AILV-V. VC = vein clearing; LR = local ringspots; RS = systemic ringspots; M = mosaic; B = leaf blistering R = recovery. WT = untransformed *N. tabacum* cv Xanthi; pBin61 = empty vector.

synthesized from 1 µg of total RNA preparation and 10 pmol random hexamers with High Capacity cDNA Reverse Transcription kit (Applied Biosystems, Foster City, CA, USA) following the protocol of the manufacturer. qPCR was set up in 10 µl of 2X Fast SYBR Green PCR Master Mix (Applied Biosystems), containing 100 ng of first strand cDNA template, and 200 nM each of the forward and reverse primer pairs. Each cDNA sample was amplified in triplicate on a single 48-well optical plate using the StepOne Real-Time PCR system (Applied Biosystems). The cycling profile consisted of 95°C for 2 min followed by 40 cycles of 3 s at 95°C and 10 s at 60°C. Immediately after the final PCR cycle, a melting curve analysis was done to determine the specificity of the reaction. Relative quantification was calculated using the comparative cycle threshold (Ct) method ($RQ = 2^{-\Delta\Delta Ct}$) [42], in which the change in the amount of the target viral RNA was normalized in relation to the endogenous control. Validation experiments were done according to the manufacturer's instructions (Applied Biosystems) to compare the amplification efficiencies of *RDR6* and *DCL4* and the endogenous *GAPDH* mRNA primers. The experiment was repeated twice and statistical significance of the RQ values was assessed by one-way analysis of variance (ANOVA) with Tukey post-hoc test (P<0.05), for each target gene separately, to compare gene expression at each time point.

SiRNA Detection

Preparations of total RNA extracted from leaves of tobacco plants collected daily from 1 to 7 and then at 14 and 21 dpi with either AILV-V or PVY-SON41 were enriched in low-molecular-weight plant RNAs by differential precipitation as described by Bucher et al. [43] with minor modifications. RNAs were separated by denaturing 15% polyacrylamide gel electrophoresis, transferred to positively charged Nylon membranes (Roche) by electroblotting as described by Cillo et al. [44] and cross-linked to the membrane under UV light for 3 min. Low-molecular-weight plant RNAs prepared from tobacco plants infected with PVY-SON41 were used as control. Membranes were hybridized with digoxigenin-labeled RNA probes specific for the last 760 bp and 1800 bp 3′-terminal sequences of AILV-V and PVY-SON41, respectively, and hydrolyzed to small fragments at 60°C for 50 min in 200 mM NaHCO$_3$ and 200 mM Na$_2$CO$_3$. After incubation, the solution was neutralized with acetic acid. Hybridizations and washes were conducted at 42°C. Chemiluminescent signal yielded by hybrids was acquired with 5 min intervals for 90 min of exposure in a ChemiDoc (Bio-Rad Laboratories).

Dot Blot Hybridization Analysis

Dot-blot hybridization was used to verify the presence and estimate the loads of viral RNA in plant tissues. Samples were homogenized with six volumes (v/w) of alkaline solution (50 mM NaOH, 2.5 mM EDTA) and 5 µl from the homogenate were directly applied onto a positively charged Nylon membrane (Roche Diagnostics, Mannheim, Germany) and fixed by UV exposure. Digoxigenin-labeled DNA probes for AILV-V and PVY-SON41 were prepared from plasmids pAILV769 and pPVY-SON41-617, respectively, and used as described previously [45]. Plasmid DNA and unincorporated nucleotides were removed, respectively, using the Whatman FTA kit pK1 (Whatman) and Bio-Rad Bio spin P30 columns (Bio-Rad Laboratories) following the manufacturer's protocol while PCR products were anlayzed for purity by agarose gel electrophoresis. For quantitative dot blot analysis, three biological replicates of samples collected at each sampling time were used and each sample was spotted as two technical replicates. Reproducibility of hybridization signals among biological replicates and samples collected at different time points was assessed by a preliminary dot blot hybridization using *GAPDH* as target and a specific Digoxigenin-labeled DNA probe. A standard curve was obtained by the intensity values of the hybridization signals produced by dilution series of unlabeled PCR products derived from the same insert used to synthesize the probe. Concentration of unlabeled PCR products was estimated with a Nanodrop ND-1000 spectrophotometer. Chemiluminescent signals were acquired after 15 minutes of exposure in a ChemiDoc (Bio-Rad Laboratories), and their intensity estimated by the Quantity One software (Bio-Rad Laboratories, Berkeley, CA, USA).

Immunolocalization

Shoot tips were collected from tobacco plants at weekly intervals, fixed overnight at 4°C with 4% paraformaldehyde in PBS (137 mM NaCl, 2.7 mM, 10 mM Na_2HPO4, 2 mM KH_2PO4, pH 7.4) and dehydrated in an incremental (30, 50, 70, 85 and 99%) ethanol series. Samples were cleared in Histoclear (Natural Diagnostics, Atlanta, GA, USA) and embedded in paraffin (Paraplast; Sigma-Aldrich, St. Louis, MO, USA). Semithin longitudinal sections (10 μm) were collected on polysine slides (Thermo scientific, Braunschweig, Germany) and incubated for 16–18 hours at 37°C. For immunolocalization, sections were treated with Histo-clear for 3 min to remove paraffin, hydrated in ethanol series, pre-incubated with PBS containing 4% bovine serum albumin (BSA) for 30 min and incubated for 3 h at room temperature with a polyclonal serum raised in rabbit against

AILV-V diluted 1:500 in PBS. After washing in PBS three times, the samples were incubated with the secondary mouse-anti rabbit monoclonal antibody conjugated with alkaline phosphatase (Sigma Aldrich, St. Louis, MO, USA) diluted 1:500 in PBS/BSA buffer and stained in NBT/BCIP solution (Sigma-Aldrich) following the protocol of the manufacturer. Slides were observed with a light microscope and pictures taken with a camera integrated to the microscope.

Local Suppression of Silencing in Agroinfiltrated Patch Assays

The *GFP* reporter gene from the pBin-GFPsense pBIN-mGFP4 binary vector [46] was expressed transiently in leaves of *N. benthamiana*, line 16c, by infiltration of a culture of *Agrobacterium tumefaciens* carrying the recombinant plasmid. Prior to agroinfiltration, leaves were either mock-inoculated with buffer or with PVA, AILV-V or PVY-SON41, or co-inoculated with AILV-V and PVY-SON41. To monitor the effect of different viral inocula on the suppression of RNA silencing, leaves were illuminated with a Black Ray long wave UV lamp (UVP, Upland, CA, U.S.A.).

Author Contributions

Conceived and designed the experiments: ES TM SAS DG. Performed the experiments: ES TM SAM. Analyzed the data: ES TM JPTV DG. Contributed reagents/materials/analysis tools: ES TM SAS DG. Wrote the paper: TM JPTV DG.

References

1. Wang Q, Cuellar WJ, Rajamäki ML, Hirata Y, Valkonen JPT (2008) Combined thermotherapy and cryotherapy for efficient virus eradication: relation of virus distribution, subcellular changes, cell survival and viral RNA degradation in shoot tips. Mol Plant Pathol 9: 237–250.
2. Acquadro A, Papanice M, Lanteri S, Bottalico G, Portis E, et al. (2010) Production and fingerprinting of virus-free clones in a reflowering globe artichoke. Plant Cell Tiss Org 100: 329–337.
3. Faccioli G, Marani F (1998) Virus elimination by meristem tip culture and tip micrografting. In: Hadidi A, Khetarpal RK, Koganezawa H editors, Plant Virus Disease Control. St Paul (MN), American Phytopathological Society Press. pp 346–380.
4. Mink G, Wample R, Howell W (1998) Heat treatment of perennial plants to eliminate phytoplasmas, viruses and viroids while maintaining plant survival. In: Hadidi A, Khetarpal RK, Koganezawa H editors, Plant Virus Disease Control. St Paul (MN), American Phytopathological Society Press. pp 332–345.
5. Sharma S, Singh B, Rani G, Zaidi A, Hallan VK, et al. (2008) In vitro production of Indian citrus ringspot virus (ICRSV) free Kinnow plants employing thermotherapy coupled with shoot tip grafting. Plant Cell Tiss Org 92: 85–92.
6. Sanfaçon H, Iwanami T, Karasev AV, Van der Vlugt R, Wellink J, et al. (2011) Family Secoviridae. In King AMQ, Adams MJ, Eric B. Carstens EB, Lefkowitz EJ editors, Virus Taxonomy: Classification and Nomenclature of Viruses. Elsevier Academic Press. pp 881–899.
7. Adams MJ, Zerbini FM, French R, Rabenstein F, Stenger DC, et al. (2011) Family Potyviridae. In King AMQ, Adams MJ, Eric B. Carstens EB, Lefkowitz EJ editors, Virus Taxonomy: Classification and Nomenclature of Viruses. Elsevier Academic Press. pp 1069–1089.
8. Bottalico G, Padula M, Campanale A, Sialer M, Saccomanno F, et al. (2002) Seed transmission of Artichoke Italian latent virus and Artichoke latent virus in globe artichoke. J Plant Pathol 84:167–168.
9. Wang D, Macfarlane SA, Maule AJ (1997) Viral determinants of pea early browning virus seed transmission in pea. Virology 234: 112–117.
10. Siddiqui SA, Sarmiento C, Kiisma M, Koivumäki S, Lemmetty A, et al. (2008) Effects of viral silencing suppressors on tobacco ringspot virus infection in two Nicotiana species. J Gen Virology 89: 1502–1508.
11. Matthews REF (1991) Plant Virology. San Diego, California: Academic Press. 835 p.
12. Hull R (2002) Matthews' Plant Virology. San Diego, California: Academic Press. 1001 p.
13. Foster TM, Lough TJ, Emerson SJ, Lee RH, Bowman JL, et al. (2002) A surveillance system regulates selective entry of RNA into the shoot apex. Plant Cell 14: 1497–1508.
14. Schwach F, Vaistij FE, Jones L, Baulcombe DC (2005) An RNA-dependent RNA polymerase prevents meristem invasion by potato virus X and is required

for the activity but not the production of a systemic silencing signal. Plant Physiol 138: 1842–1852.
15. Ratcliff F, Harrison BD, Baulcombe DC (1997) A similarity between viral defense and gene silencing in plants. Science 276: 1558–1560.
16. Martín-Hernández AM, Baulcombe DC (2008) Tobacco rattle virus 16-kilodalton protein encodes a suppressor of RNA silencing that allows transient viral entry in meristems. J Virol 82: 4064–4071.
17. Mochizuki T, Ohki ST (2004) Shoot meristem tissue of tobacco inoculated with Cucumber mosaic virus is infected with the virus and subsequently recovers from infection by RNA silencing. J Gen Plant Pathol 70: 363–366.
18. Sunpapao A, Nakai T, Dong F, Mochizuki T, Ohki ST (2009) The 2b protein of cucumber mosaic virus is essential for viral infection of the shoot apical meristem and for efficient invasion of leaf primordia in infected tobacco plants. J Gen Virol 90: 3015–3021.
19. Dong F, Mochizuki T, Ohki ST (2010) Tobacco ringspot virus persists in the shoot apical meristem but not in the root apical meristem of infected tobacco. Eur J Plant Pathol 126: 117–122.
20. Jovel J, Walker M, Sanfaçon H (2007) Recovery of Nicotiana benthamiana plants from a necrotic response induced by a nepovirus is associated with RNA silencing but not with reduced virus titer. J Virol 81:12285–12297.
21. Gallitelli D, Mascia T, Martelli GP (2012) Viruses in artichoke. Adv Virus Res 84: 289–324.
22. Martelli GP, Rana GL, Savino V (1977) Artichoke Italian Latent virus. CMI/AAB Descr. Plant viruses No. 176. 4 p.
23. Holmes FO (1930) Local and systemic increase of tobacco mosaic virus. Am J Bot: 789–805.
24. Cheng NH, Su CL, Carter SA, Nelson RS (2000) Vascular invasion routes and systemic accumulation patterns of tobacco mosaic virus in Nicotiana benthamiana. Plant J 23: 349–362.
25. Parent J-S, de Alba AEM, Vaucheret H (2012) The origin and effect of small RNA signaling in plants. Front.Plant Sci. 3:179.doi: 10.3389/fpls.2012.00179.
26. Wang M-B, Masuta C, Smith NA, Shimura H (2012) RNA silencing and plant viral diseases. Mol Plant-Microbe In 25: 1275–1285.
27. Savenkov EI, Valkonen JPT (2002) Silencing of a viral RNA silencing suppressor in transgenic plants. J Gen Virol 83: 2325–2335.
28. Li F, Ding S-W (2006) Virus counterdefense: diverse strategies for evading the RNA-silencing immunity. Ann Rev Microbiol 60: 503–531.
29. Burgyán J, Havelda Z (2011) Viral suppressors of RNA silencing. Trends Plant Sci 16: 265–272.
30. Siddiqui SA, Sarmiento C, Truve E, Lehto H, Lehto K (2008) Phenotypes and functional effects caused by various viral RNA silencing suppressors in transgenic Nicotiana benthamiana and N. tabacum. Mol Plant-Microbe In 21: 178–187.

31. Gammelgård E, Mohan M, Valkonen JPT (2007) Potyvirus-induced gene silencing: the dynamic process of systemic silencing and silencing suppression. J Gen Virol 88: 2337–2346.
32. Hirai K, Kubota K, Mochizuki T, Tesuda S, meshi T (2008) Antiviral RNA silencing is restricted to the marginal region of the dark green tissue in the mosaic leaves of tomato mosaic virus –infected tobacco plants. J Virol 82: 3250–3260.
33. Bedoya LC, Martinez F, Orzáez D, Daròs J-A (2012) Visual tracking of plant virus infection and movement using a reporter MYB transcription factor that activates anthocyanin biosynthesis. Plant Physiol 158: 1130–1138.
34. Roberts D, Christie R, Archer M (1970) Infection of apical initials in tobacco shoot meristems by tobacco ringspot virus. Virology 42: 217–220.
35. Szittya G, Molnár A, Silhavy D, Hornyik C, Burgyán J (2002) Short defective interfering RNAs of tombusviruses are not targeted but trigger post-transcriptional gene silencing against their helper virus. Plant Cell 14: 359–372.
36. Qu F, Morris TJ (2005) Suppressors of RNA silencing encoded by plant viruses and their role in viral infections. FEBS lett 579: 5958–5964.
37. Qu F (2010) Antiviral role of plant-encoded RNA-dependent RNA polymerases revisited with deep sequencing of small interfering RNAs of virus origin. Mol Plant-Microbe In 23: 1248–1252.
38. Voinnet O, Pinto YM, Baulcombe DC (1999) Suppression of gene silencing: a general strategy used by diverse DNA and RNA viruses of plants. P Natl Acad Sci USA 96: 14147–14152.
39. Savino V, Gallitelli D, Jankulova M, Rana G (1977) A comparison of four isolates of artichoke Italian latent virus (AILV). Phytopat Medit 16: 41–50.
40. Puurand Ü, Valkonen JPT, Mäkinen K, Rabenstein F, Saarma M (1996) Infectious in vitro transcripts from cloned cDNA of the potato A potyvirus. Virus Res 40: 135–140.
41. Mascia T, Santovito E, Gallitelli D, Cillo F (2010) Evaluation of reference genes for quantitative reverse-transcription polymerase chain reaction normalization in infected tomato plants. Mol Plant Pathol 11: 805–816.
42. Livak KJ, Schmittgen TD (2001) Analysis of relative gene expression data using real-time quantitative PCR and the $2^{-\Delta\Delta CT}$ Method. Methods 25: 402–408.
43. Bucher E, Hemmes H, de Haan P, Goldbach R, Prins M (2004) The influenza A virus NS1 protein binds small interfering RNAs and suppresses RNA silencing in plants. J Gen Virol 85: 983–991.
44. Cillo F, Finetti-Sialer MM, Papanice MA, Gallitelli D, (2004) Analysis of mechanisms involved in the Cucumber mosaic virus satellite RNA-mediated transgenic resistance in tomato plants. Mol Plant-Microbe In 17: 98–108.
45. Minutillo S, Mascia T, Gallitelli D (2012) Development of a commercial diagnostic kit for the certification of the phytosanitary status in nursery production of globe artichoke. Eur J Plant Pathol 134: 459–465.
46. Haseloff J, Siemering KR, Prasher DC, Hodge S (1997) Removal of a cryptic intron and subcellular localization of green fluorescent protein are required to mark transgenic Arabidopsis plants brightly. P Natl Acad Sci USA 94: 2122–2127.

6

C3HC4-Type RING Finger Protein *Nb*ZFP1 Is Involved in Growth and Fruit Development in *Nicotiana benthamiana*

Wenxian Wu, Zhiwei Cheng, Mengjie Liu, Xiufen Yang, Dewen Qiu*

The State Key Laboratory for Biology of Plant Disease and Insect Pests, Institute of Plant Protection, Chinese Academy of Agricultural Science, Beijing, China

Abstract

C3HC4-type RING finger proteins constitute a large family in the plant kingdom and play important roles in various physiological processes of plant life. In this study, a C3HC4-type zinc finger gene was isolated from *Nicotiana benthamiana*. Sequence analysis indicated that the gene encodes a 24-kDa protein with 191 amino acids containing one typical C3HC4-type zinc finger domain; this gene was named *NbZFP1*. Transient expression of pGDG-*NbZFP1* demonstrated that *Nb*ZFP1 was localized to the chloroplast, especially in the chloroplasts of cells surrounding leaf stomata. Virus-induced gene silencing (VIGS) analysis indicated that silencing of *NbZFP1* hampered fruit development, although the height of the plants was normal. An overexpression construct was then designed and transferred into *Nicotiana benthamiana*, and PCR and Southern blot showed that the *NbZFP1* gene was successfully integrated into the *Nicotiana benthamiana* genome. The transgenic lines showed typical compactness, with a short internode length and sturdy stems. This is the first report describing the function of a C3HC4-type RING finger protein in tobacco.

Editor: Michael Massiah, George Washington University, United States of America

Funding: This research was supported by the National Science Foundation of China (31371984). The website is http://www.nsfc.gov.cn/. The funders had no role in study design, data collection and analysis, decision to publish, or preparation of the manuscript.

Competing Interests: The authors have declared that no competing interests exist.

* E-mail: qiudewen@caas.cn

Introduction

Zinc finger proteins are one of the most abundant types of transcription factors in eukaryotic genomes. Zinc fingers are structurally composed of multiple cysteines and/or histidines, and zinc ions play an important role in the stability of the protein itself. The RING domain of RING finger proteins, which are members of the zinc finger family, was first identified as a DNA-binding motif in the transcription factor TFIIIA from *Xenopus laevis* [1]. In addition to DNA, RING domains also bind RNA, protein, or lipid substrates. The RING motif is relatively small and consists of four pairs of ligands that bind two ions [2]. C3HC4-type RING finger proteins are involved in numerous cellular processes, including transcription, signal transduction, and recombination. Functions attributed to the RING domain itself include protein-protein interactions and ubiquitination. Most RING finger proteins are E3 ubiquitin ligases that mediate the transfer of ubiquitin to target proteins and play important roles in diverse aspects of cellular regulation in plants [3].

C3HC4-type RING finger proteins have been studied on a genomic scale in *Arabidopsis* and rice. *Arabidopsis* RING finger proteins with predicted or known biological functions include *At*COP1 (light) and *At*COP1-interacting protein 8 (AtCIP8; photomorphogenesis) [4], *At*TED3 (light signaling) [5], *At*RMA1 (secretory pathway) [6], *At*PEX10 and *At*PEX12 (peroxisome biogenesis) [7], *At*PRT1 (N-end rule pathway) [8], *At*XB3 (root

development) [9], *At*HUB1 and *At*HUB2 (chromatin modifications) [10], and *At*SDIR1 (stress tolerance) [11]. C3HC4-type RING finger genes have been identified in rice, including *Os*COP1 (*Os*RHC11), *Os*COIN1 (*Os*RHC13), *Os*XB3.1 (*Os*RHC24), and *Os*RHC1. *Os*COP1 is a component of the signal transduction pathway that links light signals to plant development and is the most well-studied C3HC4-type RING finger protein. *Os*COP1 functions as an E3 ubiquitin ligase, which targets photomorphogenesis-promoting transcription factors for ubiquitylation and degradation [4]. Two other C3HC4-type RING finger genes, OsRHC24 and OsRHC1, are related to disease resistance [9].

In our previous study, we used Hrip1, which can result in the formation of necrotic lesions that mimic a typical hypersensitive response and apoptosis-related events including DNA laddering [12], as a bait protein, yeast two hybrid system was performed and five candidate interacting proteins were screened. During the process of further verification about the interaction between Hrip1 and candidate proteins, interestingly, one of the five candidate interacting proteins, which is involved in fruit development in *Nicotiana benthamiana*, in the meanwhile, after blast in NCBI and sol genomics network, there is no relevant report about this gene. In the present study, we emphasized the gene which is involved in fruit development in *Nicotiana benthamiana*, a pair of primers were designed according to the results of yeast two hybrid and sol genomics network information and C3HC4-type RING finger gene was obtained from a tobacco cDNA, this new gene was named *NbZFP1*. Prokaryotic expression of *NbZFP1* in vitro and subcellular localization of *NbZFP1* in tobacco were performed. The function of *NbZFP1* was analyzed by tobacco rattle virus (TRV) based on Virus-induced gene silencing system and over expression in tobacco. This research provides a foundation of the molecular mechanisms of C3HC4-type RING finger proteins in tobacco.

Materials and Methods

Plasmids and Bacterial Strains

The prokaryotic expression vector pGEX-6P-2 was purchased from Amersham Biosciences (Pittsburgh, pennsylvania state, USA). pGEX-6P-2 contains the coding region for glutathione S-transferase (GST). The plant expression vector pBI121, the subcellular localization vector pGDG, and *Agrobacterium GV3101* were derived from lab stocks. *E. coli* strain BL21 (DE3) pLysS was purchased from TransGen Biotech Inc. (China). The VIGS system vectors were a gift from Yule Liu, Tsinghua University.

Plant Cultivation

Tobacco seeds were germinated on 1/2 MS medium in a growth chamber that was maintained at 25°C with 12 h of light and 12 h of darkness. Following germination, the seedlings were transferred to an autoclaved soil mix containing 1:3 (v/v) high-nutrient soil and vermiculite in 8×7.5×7.5-cm pots. One plant per pot was kept in the growth chamber at 25°C with 50% humidity and 16 h of light. The plants were watered on alternate days.

NbZFP1 Cloning and Construction of the Prokaryotic Expression Plasmid

Total RNA of *Nicotiana benthamiana* was extracted using a plant tissue RNA extraction reagent (TransGen Biotech, Beijing, China), and mRNA was used to synthesize first-strand cDNA.

NbZFP1 was amplified from first-strand cDNA of *Nicotiana benthamiana* via polymerase chain reaction (PCR) using the sense primer P1 (5′-CGGGATCCATGTCACTTTCTGGTCGT-3′) and the antisense primer P2 (5′-GCGTCGACTCAGGTTTT-CAGCCCTGT-3′). The PCR thermocycling protocol was as follows: 95°C for 5 min, followed by 30 cycles of 95°C for 30 s, 60°C for 30 s, and 72°C for 1.5 min, with a final extension of 72°C for 10 min. The PCR product was visualized by ethidium bromide-containing agarose gel electrophoresis (1% agarose, 100 V for 20 min) and subsequent UV transillumination. The PCR product was then purified and digested with BamHI and SalI, inserted into the pGEX-6P-2 vector, and transformed into BL21(DE3) pLysS. Positive colonies were selected on Luria-Bertani (LB) agar plates containing ampicillin (100 µg/ml) and screened by direct colony PCR. The extracted plasmid, designated pGEX-6p-2-*NbZFP1*, was subjected to DNA sequencing by Beijing Genomics Institution, Beijing, China.

Expression and Purification of the Recombinant Protein

The cells transformed with pGEX-6p-2-*NbZFP1* were cultured in LB medium containing ampicillin (100 µg/ml) at 37°C with shaking for eight hours. Isopropyl β-D-thiogalactoside (IPTG) was then added to a final concentration of 0.2 mM to induce expression at 16°C for 8 h. The culture transformed with the empty pGEX-6P-2 vector was used as a control. The bacteria were pelleted at 5000 g for 20 min at 4°C. The pellets were resuspended in buffer I (50 mM Tris and 200 mM NaCl, pH 8.0), and the cells were broken by sonication. After adequate sonication, the broken cells were pelleted at 12000 rpm for 1 h at 4°C, and the supernatant was collected. At this point, the samples containing pGEX-6p-2-*NbZFP1* and empty pGEX-6P-2 were ready to be purified. Because pGEX-6P-2 contains the coding region for glutathione S-transferase (GST), we used GST affinity purification technology to purify GST-*NbZFP1* and GST alone in

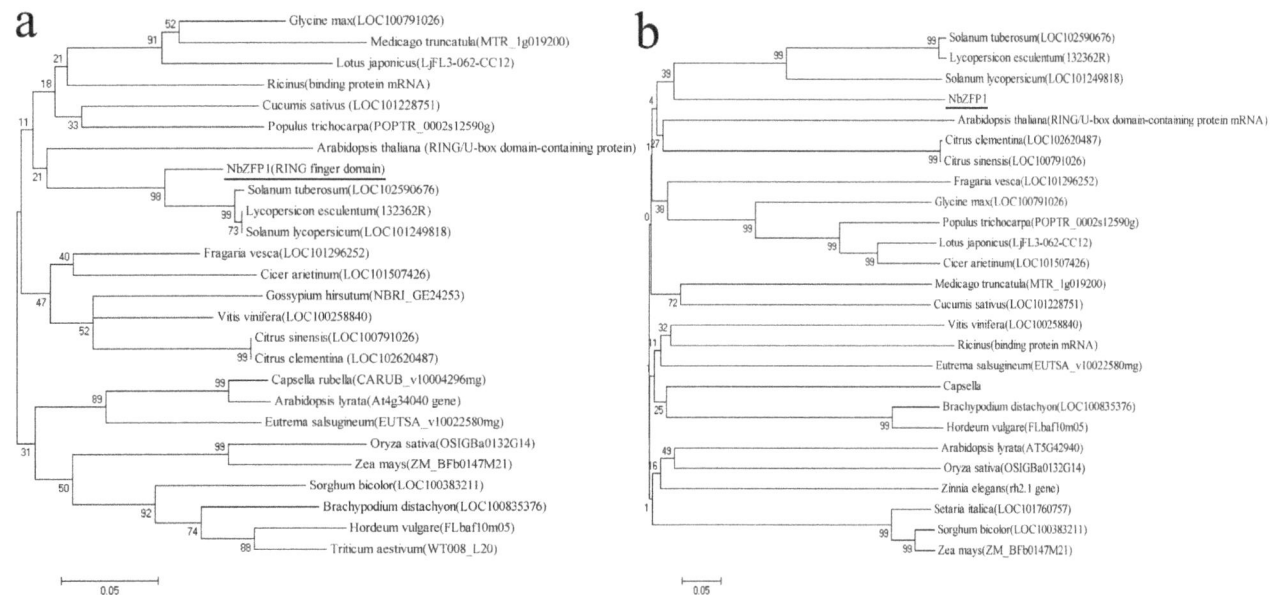

Figure 1. Phylogenetic analysis of the relationships between *NbZFP1* and C3HC4-type RING finger genes in other species. We selected twenty-five typical C3HC4-type RING finger genes with high similarity from different species, and we analyzed the similarity of their RING finger domains and full-length sequences with *NbZFP1*. Phylogenetic trees were generated by the maximum likelihood (ML) method using MEGA4. Bootstrap values from 1000 replicates are indicated at each branch. (a) A maximum likelihood (ML) tree of *NbZFP1* and genes from other species was constructed based on RING finger domain sequences. (b) A maximum likelihood (ML) tree of *NbZFP1* and genes from other species was constructed based on full-length nucleotide sequences.

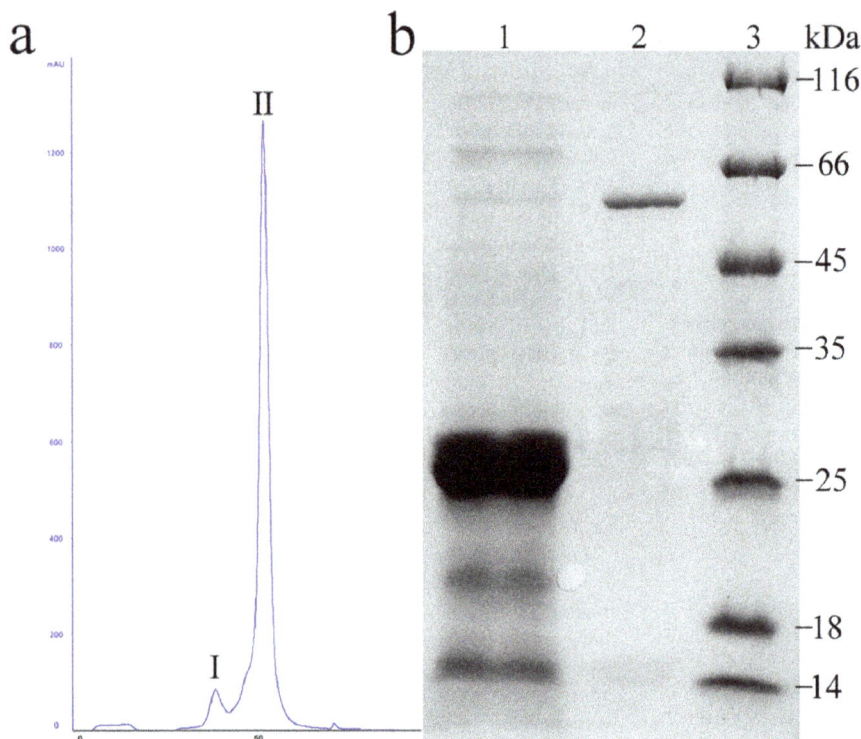

Figure 2. Purification and detection of recombinant protein. The *Nb*ZFP1 protein was expressed in *E. coli* and purified with affinity chromatography (GSTrap™ HP column) and ion exchange chromatography (HiTrap Q HP column). The purified protein showed a single band on SDS-PAGE stained with Coomassie Brilliant Blue R-250. (a) Ion exchange chromatography of a concentrated solution loaded onto a HiTrap Q HP column. (b) SDS-PAGE analysis of the recombinant protein. Lane 1, GST protein after affinity chromatography purification. Lane 2, protein peak II as designated in (a). Lane 3, protein marker.

the first step. After desalination, ion exchange chromatography was applied in the next purification step [13]. Both samples were subjected to sodium dodecyl sulfate-polyacrylamide gel electrophoresis (SDS-PAGE) analysis.

Subcellular Location of the *Nb*ZFP1 Protein

The plasmid pGEX-6P-2-*NbZFP1* was used as a template to amplify *NbZFP1* by PCR with specific primers containing BamHI and SalI restriction enzyme sites. The product was digested with BamHI and SalI and cloned into pGDG that was cut with BamHI and SalI. After successful construction of pGDG-*NbZFP1*, we transformed the plasmid into *Agrobacterium GV3101* using the freeze-thaw method [14]. *Agrobacterium GV3101* harboring pGDG-*NbZFP1* was grown in culture until the optical density of the culture reached 1.0 at 600 nm. The bacteria were pelleted at 5000 g for 15 min at room temperature. The pellets were resuspended in buffer (10 mM MES, 10 mM $MgCl_2$, 200 mM acetosyringone, pH 5.6). Bacterial suspensions were then maintained at room temperature for 2–3 h. Infiltrations were performed by gently inserting a 1-ml disposable syringe into the abaxial surface of fully expanded *Nicotiana benthamiana* leaves that were approximately 2.5 cm wide at the mid-leaf and slowly depressing the plunger [15]. Following agroinfiltration, the plants were maintained in a growth chamber at 25°C with a 16/8 h light/dark photoperiod. The leaves were examined by microscopy between 40 h and 90 h post-infiltration.

VIGS Technique for Silencing *NbZFP1* in *N. benthamiana*

The VIGS system includes the pTRV1 and pTRV2 vectors. Two adaptors with PstI restriction sites were inserted into the pTRV2 vector. This vector allows the insertion of gene silencing fragments by ligation-independent cloning (LIC) [16]. The *NbZFP1* C3HC4-type RING finger domain gene, referred to as *NbZFP1-A1*, was amplified with the primers 5′-CGACGACAAGACCCTTGCTGTGTTTGTCAGGAA-3′ and 5′-GAGGAGAAGAGCCCTTCAGGTTTTCAGCCCTGT-3′. A version of the *NbZFP1* gene lacking the C3HC4-type RING finger domain, referred to as *NbZFP1-A2*, was amplified with the primers 5′-CGACGACAAGACCCTATGTCACTTTCTGGTCGT-3′ and 5′-GAGGAGAAGAGCCCTTGGCTCAACATCCA-CAGG-3′. The PCR products were purified with polyethylene glycol/$MgCl_2$ to remove any nonspecific PCR products and primers. A total of 50 ng of purified PCR product was treated with T4 DNA polymerase (New England Biolabs) in 1× reaction buffer containing 5 mM dATP and dithiothreitol at 22°C for 30 min followed by 20 min of inactivation of T4 DNA polymerase at 70°C. The TRV2-LIC vector was then digested with PstI and similarly treated with T4 DNA polymerase; however, dTTP replaced dATP [17]. A total of 50 ng of treated PCR product and TRV2-LIC vector were mixed and incubated at 65°C for 2 min and subsequently at 22°C for 10 min. Then, 6 µl of the mixture was transformed into *E. coli DH5α* competent cells, and the transformants were tested by colony PCR.

Sequence-validated pTRV2-*NbZFP1-A1* and pTRV2-*NbZFP1-A2* plasmids were each introduced into *Agrobacterium tumefaciens* strain *GV3101* by the freeze-thaw method [14]. Overnight cultures

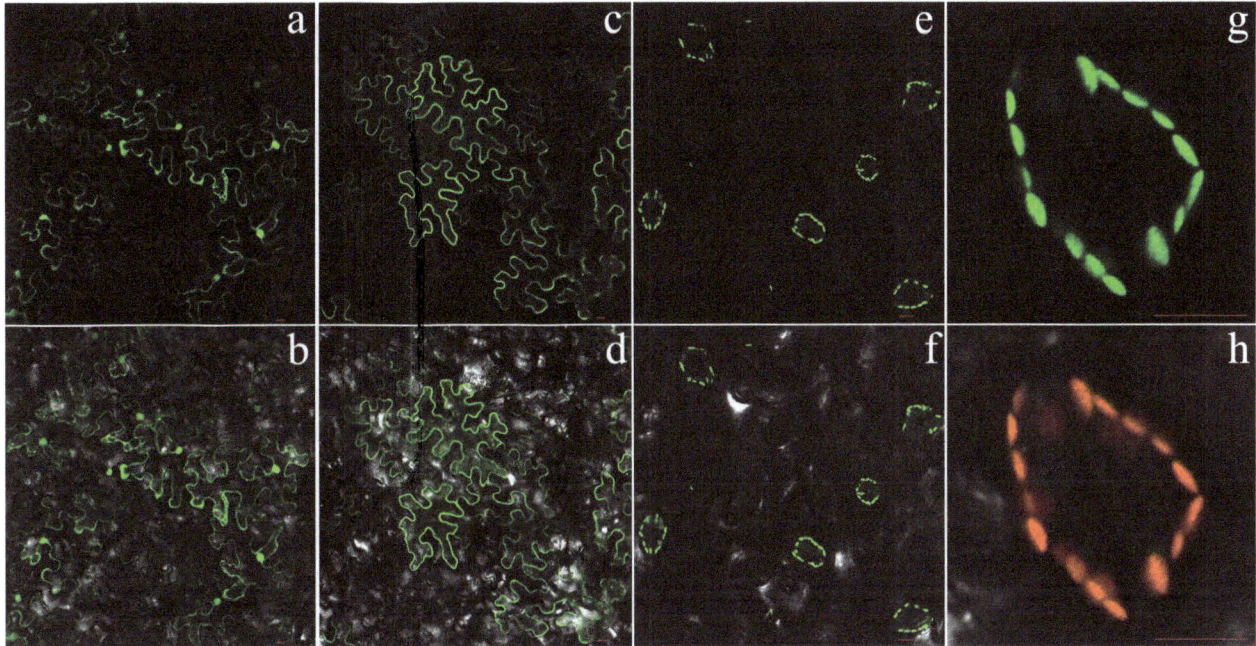

Figure 3. The results about subcellular localization of the NbZFP1 protein. Laser-scanning confocal micrographs showing the fluorescence of leaf cells following infiltration with *Agrobacteria* carrying pGDG, pGDG-*Rac1*, or pGDG-*NbZFP1* plasmids expressing GFP, Rac1-GFP, and *NbZFP1*-GFP proteins, respectively. Scale bar = 10 μm. (a) and (b) GFP expressed from pGDG; (c) and (d) GFP expressed from pGDG-Rac1; (e), (f), (g), and (h) Fluorescence expressed from pGDG-*NbZFP1*. (a), (c), (e), and (g) show the green channel; (b), (d), and (f) show an overlay of the bright-field and green channels; (h) shows an overlay of the bright-field, green, and red channels.

were grown at 28°C in the appropriate antibiotic selection medium. On the following day, the cultures were spun down, and the cells were resuspended in infiltration medium (10 mM MES, 10 mM $MgCl_2$, 200 mM acetosyringone, pH 5.6), adjusted to an OD_{600} of 1, and incubated at room temperature for 3 h. *Agrobacterium* cultures containing pTRV1 and pTRV2 were mixed at a 1:1 ratio and used to infiltrate plants at the 4-leaf stage using a 1-ml needle-less syringe [18]. Total RNA was extracted from the leaves or flowers of wild-type and VIGS *N. benthamiana* plants using the RNeasy plant mini kit (Qiagen). First-strand cDNA was synthesized using 1 mg of total RNA, gene-specific primers, and SuperScript reverse transcriptase (Invitrogen) according to the manufacturer's protocol. The expression level of *NbZFP1* was monitored by real-time PCR, and the *Actin* gene was amplified as a quantitative control. qPCR was conducted with iQ SYBR Green Supermix (Bio-Rad, Hercules, CA, USA) and an iCycler (Bio-Rad) according to the manufacturer's instructions.

Tobacco Transformation

Primer 5.0 was used to design the sense primer, P1 (5′-CGGGATCCATGTCACTTTCTGGTCGT-3′), and the anti-sense primer, P2 (5′-GCGTCGACTCAGGTTTT-CAGCCCTGT-3′). The primers were then used to amplify the ORF of *NbZFP1* by polymerase chain reaction (PCR). Primers P1 and P2 contained BamHI and SalI restriction sites. The PCR products and pBI121 vector were double digested, adapter ligation was performed, and the ligation products were then transformed into *DH5α* cells. Positive colonies were selected on Luria-Bertani (LB) agar plates containing kanamycin (50 μg/ml) and screened by direct colony PCR. The fusion vector pBI121-*NbZFP1* was transformed into *Agrobacterium GV3101*. *Agrobacterium GV3101* harboring pBI121-*NbZFP1* was used to transform *N. benthamiana* according to the method proposed by Horsch et al [19], and the

transformed plants were then transferred to soil for seed setting. T_1 progeny plants from the seeds of independent T_0 transformants were grown on 1/2 MS medium with kanamycin selection, and green plantlets with roots were transferred to soil.

PCR analysis of T_0 plants was performed for the detection of putative transgenic tobacco plants. Genomic DNA from fresh, fully expanded tobacco leaves was used for the PCR analysis. pBI121 harbors a *gus* gene; thus, amplification of a specific fragment of the *gus* gene was performed by adjusting the PCR conditions for 30 cycles. The primers forward-F (5′-ATGGTCCGTCCTGTAGAAACC-3′) and primers reverse-R (5′-GACTGCCTCTTCGCTGTACAG-3′) were used for amplification of the specific *gus* fragment. After positive identification of transgenic plants, we performed Southern blotting to determine the *NbZFP1* copy numbers according to the protocol provided with the DIG High Prime DNA Labeling and Detection Starter Kit II (Amersham Biosciences).

30 Single copy T_1 progeny plants were grown on 1/2 MS medium containing kanamycin, and 30 wild progeny plants were grown without antibiotic as a control. Green plantlets with roots were transferred to pots containing soil. The height of the plants were measured beginning at the four-leaf stage in 10-day intervals. The internode lengths of wild-type and transgenic tobacco were measured at the sixty day after four-leaf stage.

Statistical Analysis

The data were analyzed separately for each experiment with SPSS 16.0 software. The means were compared using Tukey's HSD test. The significance of the different between the means of wild-type and over-expression tobacco in the same growth condition were calculated by chi square test using SPSS 16.0 software. *** indicate significance at the 0.001 of confidence level.

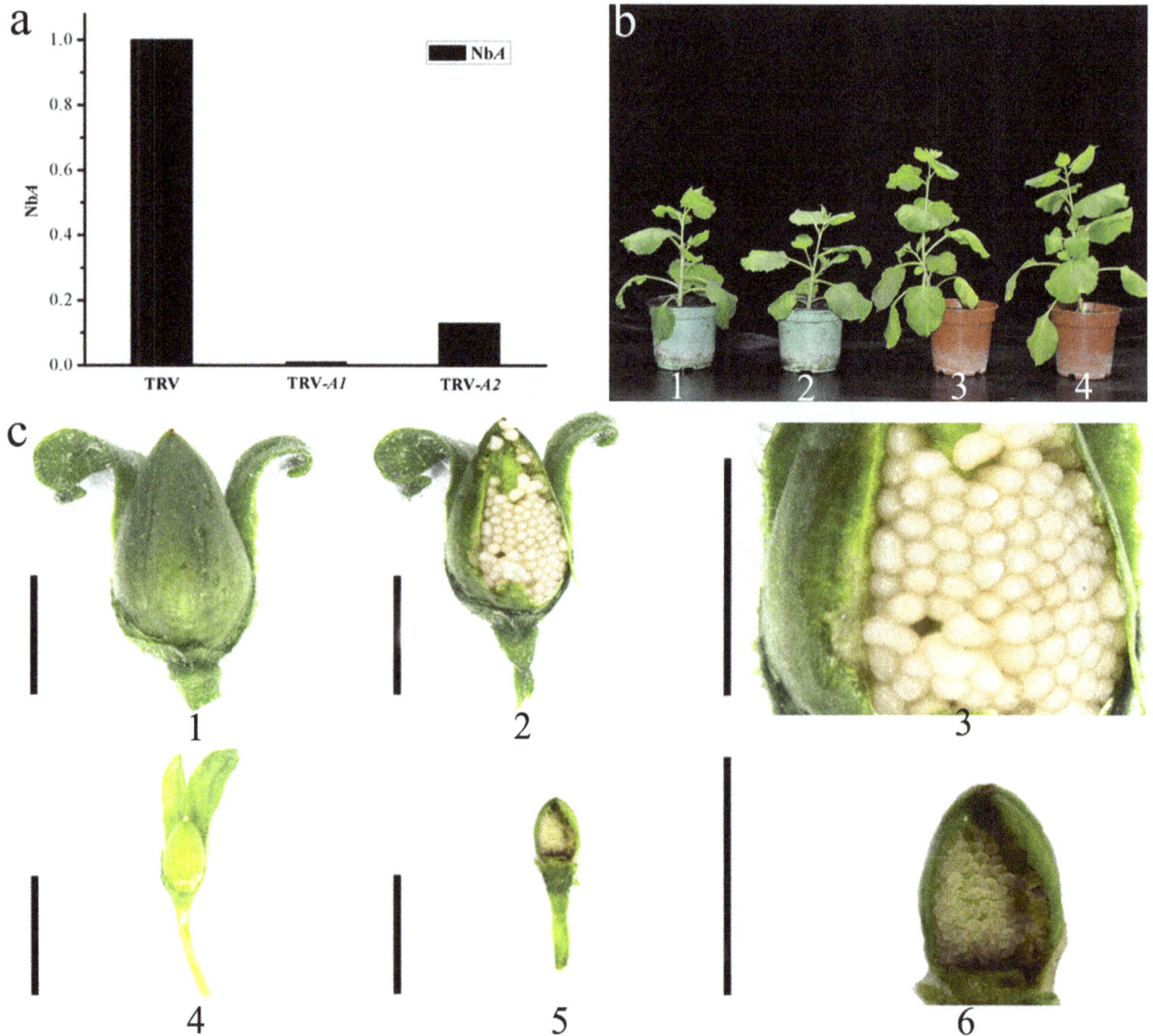

Figure 4. Characterization of VIGS strains by silencing of *NbZFP1* compared with controls. (a) Effect of VIGS on *NbZFP1* (in this graph, *NbA*) transcription in *N. benthamiana*. TRV: The expression level of *NbZFP1* in plants infected with TRV alone; TRV-*A1*: The expression level of *NbZFP1* in plants infected with TRV-*A1* (*A1* is the C3HC4-type RING finger domain gene *NbZFP1*); TRV-*A2*: The expression level of *NbZFP1* in plants infected with TRV-*A2* (A2 is *NbZFP1* lacking the C3HC4-type RING finger domain). (b) The phenotype of VIGS-silenced *NbZFP1* and wild type strains. b1, *NbZFP1*-silenced plant at 20 days after the four-leaf stage; b2, wild type plant at 20 days after the four-leaf stage; b3, *NbZFP1*-silenced plant at 40 days after the four-leaf stage; b4, wild type plant at 40 days after the four-leaf stage. (c) The fruit phenotype of *NbZFP1*-silenced plants compared with controls. c1, c2, and c3: The fruit phenotype of non-silenced control plants. c4, c5, and c6: The fruit phenotype of *NbZFP1*-silenced plants.

Results

NbZFP1 Cloning

Using a pair of specific primers, the gene we have designated as *NbZFP1* was cloned from *N. benthamiana* by RT-PCR. The length of the full open reading frame (ORF: GenBank accession number KJ169550) for this gene is 576bp.

Phylogenic Analysis

To define the relationships between *NbZFP1* and C3HC4-type RING finger genes in other species, phylogenetic analyses were performed with RING finger domain sequences and full-length sequences [20]. The maximum likelihood (ML) tree [21] of the C3HC4-type RING finger domain-containing motif suggested

that C3HC4-type RING finger genes are highly conserved in 26 species. In addition, *NbZFP1* is more closely related to the C3HC4-type RING finger genes of *Solanum tuberosum*, *Lycopersicon esculentum*, and *Solanum lycopersicum* than those of other species (Figure 1a). The tree of full-length sequences of C3HC4-type RING finger genes showed that the similarities between *NbZFP1* and the C3HC4-type RING finger genes in the other 25 species are not high (Figure 1b).

Purification and Detection of Recombinant *Nb*ZFP1 Protein

NbZFP1 was cloned into the prokaryotic expression vector pGEX-6p-2, and the resulting recombinant plasmid, pGEX-6p-2-

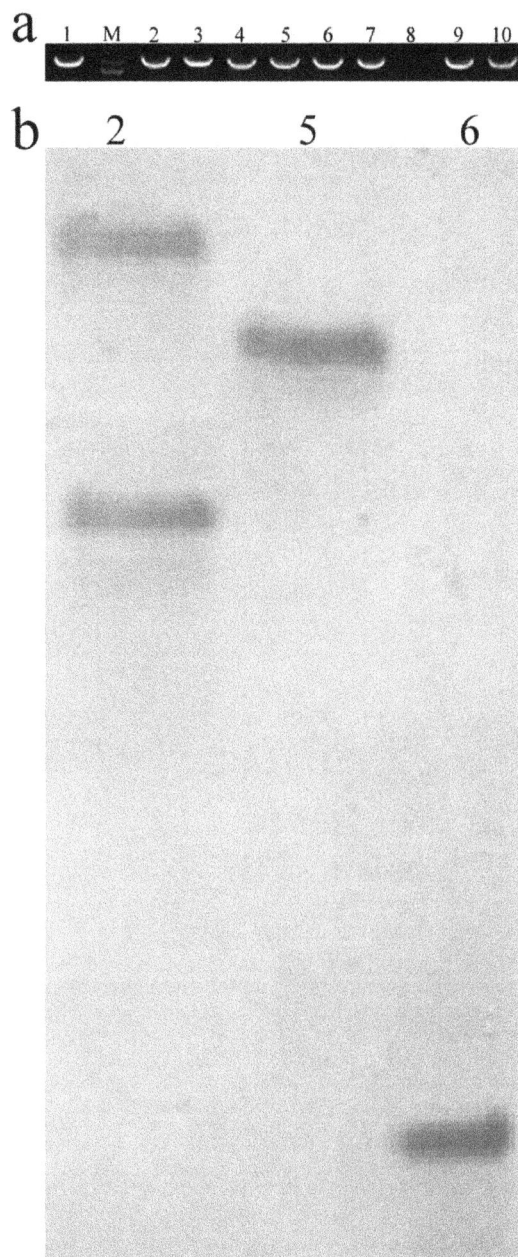

Figure 5. Identification of transgenic tobacco plants. Fresh leaves (100 mg) from T$_1$ plants were collected, and genomic DNA was extracted. A specific fragment of the *gus* gene (approximately 1000 bp in length) was amplified to identify the transgenic lines. (a) PCR identification of transgenic tobacco plants. M, DL8000 marker; 1, positive control; 2–10, transgenic tobacco lines. (b) Southern blot analysis of transgenic tobacco plants. 2, Transgenic line 2; 5, Transgenic line 5; 6, Transgenic line 6.

NbZFP1, was confirmed by colony PCR and DNA sequencing. The recombinant plasmid was then transformed into *E. coli BL21* (DE3) to express the recombinant protein. The expressed protein was soluble in *E. coli* and was purified with a GSTrap HP column followed by a HiTrap desalting column and a HiTrap Q HP column (Figure 2a). The molecular weight of the GST tag is 26 kDa, and SDS-PAGE analysis showed a single band corresponding to the purified recombinant protein with an approximate

molecular mass of 50 kDa (Figure 2b). This result indicated that the *Nb*ZFP1 protein was expressed in *E. coli*.

Subcellular Localization of the *Nb*ZFP1 Protein

To confirm the subcellular localization of the *Nb*ZFP1 protein, we constructed a recombinant vector, pGDG-*NbZFP1*, for transient expression. The empty pGDG vector was used as a negative control. Because the Rac1 protein was shown to be localized to the plasma membrane [22], the pGDG-*Rac1* vector was constructed for expressing Rac1-GFP as a positive control. *Agrobacterium GV3101* containing pGDG, pGDG-*Rac1*, or pGDG-*NbZFP1* was used to infiltrate healthy tobacco plants. Laser scanning confocal micrographs showed that GFP alone localized throughout the whole cell, including the plasma membrane, nucleus, and cytoplasm (Figure 3a and Figure 3b). The fusion protein Rac1-GFP was localized to cell membranes only (Figure 3c and Figure 3d). The protein of interest, *Nb*ZFP1-GFP, was located in the chloroplast, especially in the chloroplasts of cells surrounding leaf stomata (Figure 3e–h).

Developmental Phenotypes Revealed by Silencing *NbZFP1*

TRV-mediated VIGS is an effective tool for assessing the functions of genes in the reproductive tissues of plants [23]. Silencing of phytoene desaturase (PDS) leads to the inhibition of carotenoid synthesis, causing the plants to exhibit a photobleached phenotype [24]. Thus, we assessed the gene silencing efficiency of our TRV-VIGS clones by suppressing the expression of the PDS gene in *N. benthamiana*. The PDS suppression phenotype was visible at 10 days post-infiltration in the upper leaves of the plant and persisted indefinitely. This result indicated that the TRV-VIGS system could be successfully used to induce silencing of other desirable endogenous plant genes [18].

We then cloned two different fragments of *NbZFP1* gene separately into the TRV2-LIC vector for silencing. The first of these fragments was obtained from the *NbZFP1* C3HC4-type RING finger domain gene *NbZFP1-A1*, while the other fragment was obtained from *NbZFP1-A2*, which is a version of *NbZFP1* that lacks the C3HC4-type RING finger domain. We infiltrated plants at the four-leaf stage with a 1:1 mixture of TRV1 and the TRV2-LIC-*NbZFP1* fragment for each of the clones, and we monitored the infiltrated plants throughout their entire lifespan. We then determined the degree of silencing of *NbZFP1* fragments by qPCR. The results revealed a greater than 90% reduction in transcript levels in the silenced plants (Figure 4a). Silencing *NbZFP1* resulted in fruits that were smaller than those of the TRV-infected control (Figure 4c), although no other phenotypic defects were observed (Figure 4b).

Generation of *NbZFP1* Transgenic Tobacco Plants

The plant expression vector pBI121 containing *NbZFP1* was transformed into tobacco via the *Agrobacterium*-mediated method [19]. Nine independent kanamycin-resistant *NbZFP1* transgenic tobacco lines were obtained. These transgenic tobacco plants were verified by PCR, and eight transgenic lines were positive (Figure 5a). We randomly selected 3 transgenic plants and performed a Southern blotting experiment, which showed that *NbZFP1* was successfully integrated into all 3 transgenic tobacco plants and two of the lines had a single copy of *NbZFP1* (Figure 5b).

Phenotypic Effect of *NbZFP1* in Transgenic Lines

We selected T$_0$ plants with a single copy of *NbZFP1* and obtained their seeds. These seeds were then sown on 1/2 MS

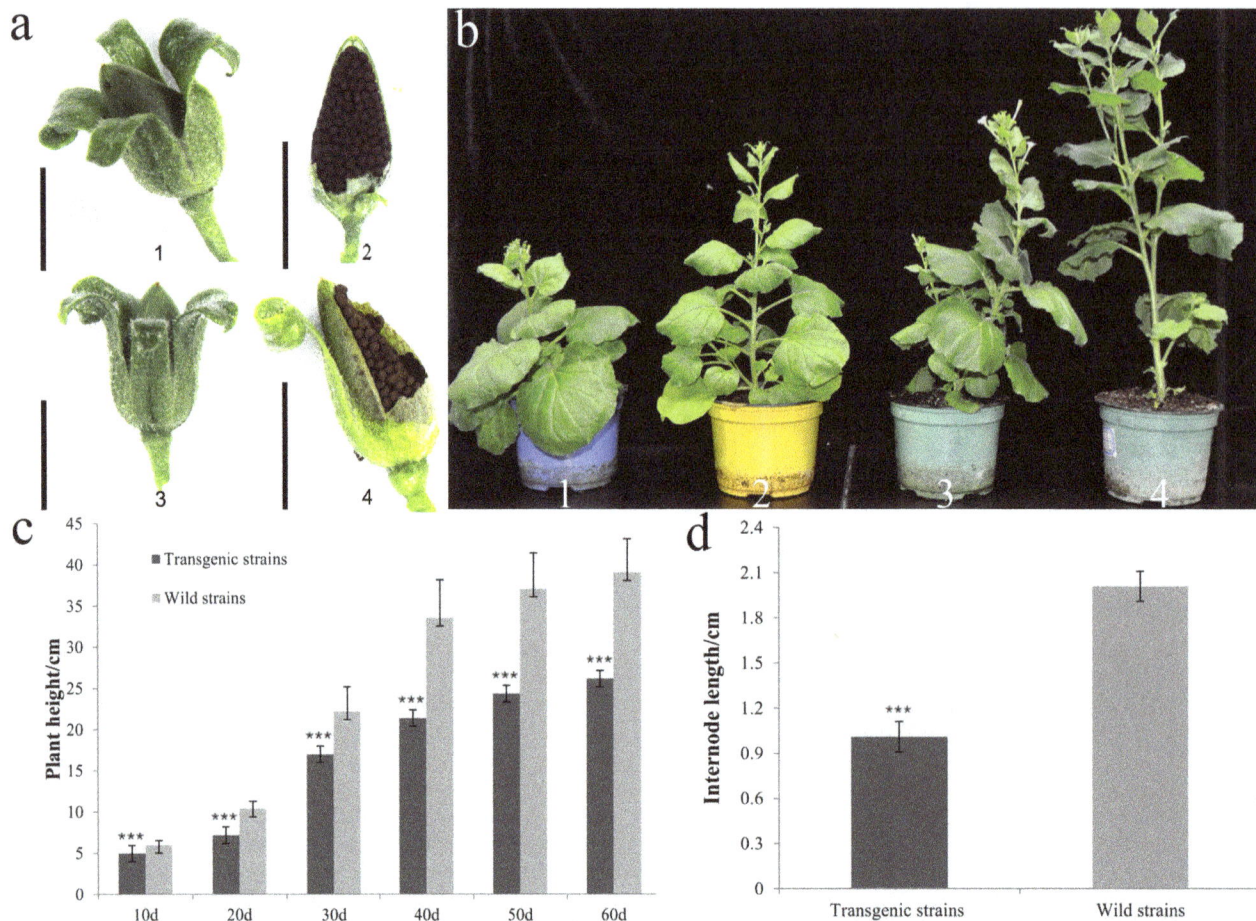

Figure 6. Characterization of T1 single copy lines compared with controls. (a) The fruit phenotype of transgenic lines compared with wild type plants. a1 and a2, the fruit phenotype of transgenic plants. a3 and a4, the fruit phenotype of wild type plants. (b) The phenotype of T_1 single copy lines and wild type lines. b1, T_1 generation single copy transgenic strains at 20 days after the four-leaf stage; b2, wild type strains at 20 days after the four-leaf stage; b3: T_1 single copy transgenic strains at 50 days after the four-leaf stage; b4, wild type strains at 50 days after the four-leaf stage. (c) The height of single copy *NbZFP1* transgenic plants compared with the height of wild type plants at different times. (d) The internode lengths of single copy *NbZFP1* transgenic plants compared with the wild type internode length, 30 plants were measured. Asterisks indicate significant differences from the wild type: [***]$P<0.001$.

medium with kanamycin selection. Wild type plants were grown on 1/2 MS medium without antibiotic. 30 wild-type and transgenic plantlets with roots were transferred to soil, respectively. The height of the plants were measured 10 days after the plants had reached the four-leaf stage, the internode lengths of wild-type and transgenic tobacco were measured at the sixty day after four-leaf stage. We found that transgenic lines were shorter than wild type lines after the four-leaf stage, and the regenerated plants were compact with short internodes and sturdy stems (Figure 6b, Figure 6c, Figure 6d). However, the phenotype of transgenic fruits was normal (Figure 6a).

Discussion

C3HC4-type RING finger proteins play different roles in diverse physiological processes. Many C3HC4-type RING finger genes in plants have been shown to play important roles in the regulation of growth and development [25]. The C3HC4-type RING finger proteins can be divided into three types according to their role. Because the specificity of the ubiquitin proteasome pathway is determined by E3 ligases, the various functions of C3HC4-type RING finger proteins may be explained by the fact

that they act as E3 ligases with different targets in diverse physiological processes. Indeed, C3HC4-type RING finger proteins can regulate many cellular processes, including homeostasis, development, cell division, growth, hormone responses, and stress responses [26]. Some C3HC4-type RING finger genes function as cis-elements and play key roles in the transcriptional regulation of genes controlling various biological processes, including abiotic stress responses [27]. In addition, C3HC4-type RING finger genes also participate hormonal regulation. Phytohormones are the most important signaling molecules in plants, and the concentration of a given phytohormone influences plant growth and development. For example, auxin regulates cell processes and promotes cell elongation. GAs are a large family of plant hormones, some of which are bioactive growth regulators that control seed germination, stem elongation, and flowering [28].

In this study, we provide a look at the C3HC4-type zinc finger protein of tobacco. Sequence analysis indicated that the gene encodes a 24-kDa protein with 203 amino acids, and we refer to this gene as *NbZFP1*. The results of VIGS analysis show that the C3HC4-type zinc finger protein had a clear effect on the process of tobacco seed pod development, although the height of the plants subjected to VIGS treatment was normal. When *NbZFP1* was transformed into tobacco, we found that the fruit phenotype of T_1 transgenic lines containing a single copy of *NbZFP1* was normal, although these lines were characterized by compactness, short internodes, and sturdy stems. We next examined the subcellular localization of *NbZFP1*, and we found that *NbZFP1* was localized in the chloroplast, especially in the chloroplasts of cells surrounding leaf stoma. We therefore hypothesized that *NbZFP1* has a role in stomatal movements. It is known that stomatal opening and closing is mediated by ABA-triggered changes in ion fluxes in guard cells [29,30,31]. ABA is considered to be an important hormone, as it plays a critical role in the response of plants to various stresses. ABA is not only a stress signal but is also required to fine-tune growth and development under non-stress conditions. The physiological processes controlled under these conditions include the regulation of growth, stomatal aperture, hydraulic conductivity, and seed dormancy [32,33,34]. ABA negatively influences the size of guard cells and the internode length, and it also acts together with other phytohormones, such as brassinosteroids, gibberellic acid, and auxin, in the regulation of plant growth and development [35,36,37]. The characteristics of the transgenic lines are consistent with the function of zinc finger proteins in other plants. For example, the EPF zinc finger protein family of *Petunia hybrida* and the SUPERMAN or NNT zinc finger protein of *Arabidopsis thaliana* are both involved in regulating the process of reproductive development [38,39,40] The rice zinc finger protein PROG1 is directly involved in plant growth regulation and domestication [41]. *NbZFP1* might be regulate the concentration of the phytohormone abscisic acid (ABA) through a specific mechanism, as ABA negatively regulates the internode length, or might act together with other phytohormones related to auxin signal

transduction or the auxin/cytokinin signal transduction process, consequently affecting the growth and development of tobacco. However, it is currently unclear how *NbZFP1* influences fruit development. Ma et al. [25] found that among 29 C3HC4-type RING finger genes in rice, 5 genes were preferentially expressed in reproductive tissues or organs. Commonly, a high level of expression or preferential expression in tissues or organs suggests that the highly/preferentially expressed gene may play an important role there. Thus, these 5 rice genes may play an important role at the reproductive stage.

Overall, C3HC4-type RING finger genes have been shown to play important roles in the regulation of growth and development in plants; however, these genes undoubtedly have many functions that have not yet been discovered. The current study provides preliminary data that requires further validation and establishes a basis for further studies. Much work remains to be carried out in the future.

Acknowledgments

We wish to express our gratitude to Dr. Hongmei Zeng (Institute of Plant Protection, Chinese Academy of Agricultural Sciences) and Dr. Lihua Guo (Institute of Plant Protection, Chinese Academy of Agricultural Sciences) for useful discussions. The authors also thank Ms. Cailing Liu (School of Environment, Beijing Normal University) for support in the statistical analysis.

Author Contributions

Conceived and designed the experiments: WW DQ. Performed the experiments: WW ZC ML WW. Analyzed the data: WW ZC ML XY. Contributed reagents/materials/analysis tools: XY ZC. Wrote the paper: WW. Designed the software used in analysis: ZC.

References

1. Berg JM, Shi Y (1996) The galvanization of biology: a growing appreciation for the roles of zinc. Science 271: 1081–1085.
2. Stone SL, Hauksdóttir H, Troy A, Herschleb J, Kraft E, et al. (2005) Functional analysis of the RING-type ubiquitin ligase family of Arabidopsis. Plant physiology 137: 13–30.
3. Ciechanover A (1998) The ubiquitin–proteasome pathway: on protein death and cell life. The EMBO journal 17: 7151–7160.
4. Von Arnim AG, Deng X (1993) Ring finger motif of Arabidopsis thaliana COP1 defines a new class of zinc-binding domain. Journal of Biological Chemistry 268: 19626–19631.
5. Pepper AE, Chory J (1997) Extragenic suppressors of the Arabidopsis det1 mutant identify elements of flowering-time and light-response regulatory pathways. Genetics 145: 1125–1137.
6. Matsuda N, Suzuki T, Tanaka K, Nakano A (2001) Rma1, a novel type of RING finger protein conserved from Arabidopsis to human, is a membrane-bound ubiquitin ligase. Journal of cell science 114: 1949–1957.
7. Schumann U, Wanner G, Veenhuis M, Schmid M, Gietl C (2003) AthPEX10, a nuclear gene essential for peroxisome and storage organelle formation during Arabidopsis embryogenesis. Proceedings of the National Academy of Sciences 100: 9626–9631.
8. Potuschak T, Stary S, Schlögelhofer P, Becker F, Nejinskaia V, et al. (1998) PRT1 of Arabidopsis thaliana encodes a component of the plant N-end rule pathway. Proceedings of the National Academy of Sciences 95: 7904–7908.
9. Wang Y, Pi L, Chen X, Chakrabarty PK, Jiang J, et al. (2006) Rice XA21 binding protein 3 is a ubiquitin ligase required for full Xa21-mediated disease resistance. The Plant Cell Online 18: 3635–3646.
10. Liu K, Wang L, Xu Y, Chen N, Ma Q, et al. (2007) Overexpression of OsCOIN, a putative cold inducible zinc finger protein, increased tolerance to chilling, salt and drought, and enhanced proline level in rice. Planta 226: 1007–1016.
11. Zhang Y, Yang C, Li Y, Zheng N, Chen H, et al. (2007) SDIR1 is a RING finger E3 ligase that positively regulates stress-responsive abscisic acid signaling in Arabidopsis. The Plant Cell Online 19: 1912–1929.
12. Kulye M, Liu H, Zhang Y, Zeng H, Yang X, et al. (2012) Hrip1, a novel protein elicitor from necrotrophic fungus, Alternaria tenuissima, elicits cell death, expression of defence-related genes and systemic acquired resistance in tobacco. Plant, cell & environment 35: 2104–2120.
13. Gräslund S, Nordlund P, Weigelt J, Bray J, Gileadi O, et al. (2008) Protein production and purification. Nature methods 5: 135–146.
14. An G, Ebert PR, Mitra A, Ha SB (1989) Binary vectors.: Springer. 29–47.
15. Goodin MM, Dietzgen RG, Schichnes D, Ruzin S, Jackson AO (2002) pGD vectors: versatile tools for the expression of green and red fluorescent protein fusions in agroinfiltrated plant leaves. The Plant Journal 31: 375–383.
16. Liu Y, Schiff M, Dinesh Kumar SP (2002) Virus-induced gene silencing in tomato. The Plant Journal 31: 777–786.
17. Dong Y, Burch-Smith TM, Liu Y, Mamillapalli P, Dinesh-Kumar SP (2007) A ligation-independent cloning tobacco rattle virus vector for high-throughput virus-induced gene silencing identifies roles for NbMADS4–1 and-2 in floral development. Plant physiology 145: 1161–1170.
18. Liu Y, Schiff M, Marathe R, Dinesh Kumar SP (2002) Tobacco Rar1, EDS1 and NPR1/NIM1 like genes are required for N-mediated resistance to tobacco mosaic virus. The Plant Journal 30: 415–429.
19. Horsch RB, Fry JE, Hoffmann NL, Eichholtz D, Rogers SA, et al. (1985) A simple and general method for transferring genes into plants. Science 227: 1229–1231.
20. Tamura K, Dudley J, Nei M, Kumar S (2007) MEGA4: molecular evolutionary genetics analysis (MEGA) software version 4.0. Molecular biology and evolution 24: 1596–1599.
21. Chiba S, Kondo H, Tani A, Saisho D, Sakamoto W, et al. (2011) Widespread endogenization of genome sequences of non-retroviral RNA viruses into plant genomes. PLoS pathogens 7: e1002146.
22. Simon AR, Vikis HG, Stewart S, Fanburg BL, Cochran BH, et al. (2000) Regulation of STAT3 by direct binding to the Rac1 GTPase. Science 290: 144–147.
23. Chen J, Jiang C, Gookin T, Hunter D, Clark D, et al. (2004) Chalcone synthase as a reporter in virus-induced gene silencing studies of flower senescence. Plant molecular biology 55: 521–530.
24. Kumagai MH, Donson J, Della-Cioppa G, Harvey D, Hanley K, et al. (1995) Cytoplasmic inhibition of carotenoid biosynthesis with virus-derived RNA. Proceedings of the National Academy of Sciences 92: 1679–1683.
25. Ma K, Xiao J, Li X, Zhang Q, Lian X (2009) Sequence and expression analysis of the C3HC4-type RING finger gene family in rice. Gene 444: 33–45.
26. Smalle J, Vierstra RD (2004) The ubiquitin 26S proteasome proteolytic pathway. Annu. Rev. Plant Biol. 55: 555–590.

27. Narusaka Y, Nakashima K, Shinwari ZK, Sakuma Y, Furihata T, et al. (2003) Interaction between two cis-acting elements, ABRE and DRE, in ABA-dependent expression of Arabidopsis rd29A gene in response to dehydration and high-salinity stresses. The Plant Journal 34: 137–148.

28. Yamaguchi S (2008) Gibberellin metabolism and its regulation. Annu. Rev. Plant Biol. 59: 225–251.

29. Levchenko V, Konrad KR, Dietrich P, Roelfsema MRG, Hedrich R (2005) Cytosolic abscisic acid activates guard cell anion channels without preceding Ca2+ signals. Proceedings of the National Academy of Sciences of the United States of America 102: 4203–4208.

30. Vahisalu T, Kollist H, Wang Y, Nishimura N, Chan W, et al. (2008) SLAC1 is required for plant guard cell S-type anion channel function in stomatal signalling. Nature 452: 487–491.

31. Siegel RS, Xue S, Murata Y, Yang Y, Nishimura N, et al. (2009) Calcium elevation-dependent and attenuated resting calcium-dependent abscisic acid induction of stomatal closure and abscisic acid-induced enhancement of calcium sensitivities of S-type anion and inward-rectifying K+ channels in Arabidopsis guard cells. The Plant Journal 59: 207–220.

32. Leung J, Giraudat J (1998) Abscisic acid signal transduction. Annual review of plant biology 49: 199–222.

33. Finkelstein RR, Gampala SS, Rock CD (2002) Abscisic acid signaling in seeds and seedlings. The Plant Cell Online 14: S15–S45.

34. Parent B, Hachez C, Redondo E, Simonneau T, Chaumont F, et al. (2009) Drought and abscisic acid effects on aquaporin content translate into changes in hydraulic conductivity and leaf growth rate: a trans-scale approach. Plant Physiology 149: 2000–2012.

35. De Smet I, Signora L, Beeckman T, Inzé D, Foyer CH, et al. (2003) An abscisic acid-sensitive checkpoint in lateral root development of Arabidopsis. The Plant Journal 33: 543–555.

36. Achard P, Cheng H, De Grauwe L, Decat J, Schoutteten H, et al. (2006) Integration of plant responses to environmentally activated phytohormonal signals. Science 311: 91–94.

37. Zhang S, Cai Z, Wang X (2009) The primary signaling outputs of brassinosteroids are regulated by abscisic acid signaling. Proceedings of the National Academy of Sciences 106: 4543–4548.

38. Kobayashi A, Sakamoto A, Kubo K, Rybka Z, Kanno Y, et al. (1998) Seven zinc-finger transcription factors are expressed sequentially during the development of anthers in petunia. The Plant Journal 13: 571–576.

39. Sakai H, Medrano LJ, Meyerowitz EM (1995) Role of SUPERMAN in maintaining Arabidopsis floral whorl boundaries.

40. Crawford BC, Ditta G, Yanofsky MF (2007) The NTT Gene Is Required for Transmitting-Tract Development in Carpels of Arabidopsis thaliana. Current Biology 17: 1101–1108.

41. Jin J, Huang W, Gao J, Yang J, Shi M, et al. (2008) Genetic control of rice plant architecture under domestication. Nature genetics 40: 1365–1369.

7

Recessive Loci *Pps-1* and *OM* Differentially Regulate *PISTILLATA-1* and *APETALA3-1* Expression for Sepal and Petal Development in *Papaver somniferum*

Sharad K. Singh[1,2], Ashutosh K. Shukla[2], Om P. Dhawan[1]*, Ajit K. Shasany[2]*

1 Genetics and Plant Breeding Division, CSIR-Central Institute of Medicinal and Aromatic Plants, Lucknow, Uttar Pradesh, India, **2** Biotechnology Division, CSIR-Central Institute of Medicinal and Aromatic Plants, Lucknow, Uttar Pradesh, India

Abstract

The involvement of PISTILLATA (PI) and APETALA (AP) transcription factors in the development of floral organs has previously been elucidated but little is known about their upstream regulation. In this investigation, two novel mutants generated in *Papaver somniferum* were analyzed - one with partially petaloid sepals and another having sepaloid petals. Progeny from reciprocal crosses of respective mutant parent genotypes showed a good fit to the monogenic Mendelian inheritance model, indicating that the mutant traits are likely controlled by the single, recessive nuclear genes named "Pps-1" and "OM" in the partially petaloid sepal and sepaloid petal phenotypes, respectively. Both paralogs of *PISTILLATA* (*PapsPI-1* and *PapsPI-3*) were obtained from the sepals and petals of *P. somniferum*. Ectopic expression of *PapsPI-1* in tobacco resulted in a partially petaloid sepal phenotype at a low frequency. Upregulation of *PapsPI-1* and *PapsAP3-1* in the petal and the petal part of partially petaloid sepal mutant and down-regulation of the same in sepaloid petal mutant indicates a differential pattern of regulation for flowering-related genes in various whorls. Similarly, it was found that the recessive mutation *OM* in sepaloid petal mutant downregulates *PapsPI-1* and *PapsAP3-1* transcripts. The recessive nature of the mutations was confirmed by the segregation ratios obtained in this analysis.

Editor: David E. Somers, Ohio State University, United States of America

Funding: This work was funded by the Council of Scientific and Industrial Research (CSIR) through the Twelfth Five Year Plan project (BSC0203). The funders had no role in study design, data collection and analysis, decision to publish, or preparation of the manuscript.

Competing Interests: The authors have declared that no competing interests exist.

* Email: op.dhawan@cimap.res.in (OPD); akshasany@yahoo.com (AKS)

Introduction

The MADS-box gene family encodes a series of transcription factors involved in controlling vegetative development in plants, flowering time and the formation of flowers [1,2,3]. Floral organ identity genes were first described in the model angiosperms *Antirrhinum majus* and *Arabidopsis thaliana*, leading to the proposal of the ABC model of flower development [4]. Most of the genes corresponding to these functions, with the exception of *APETALA2*, are members of the MADS-box family of transcription factors [5]. *PISTILLATA* (*PI*) and its homologs are classified as B-class genes of the MADS-box family and function together with another B-class gene, *APETALA3* (*AP3*), by forming heterodimers for regulating petal and stamen development in eudicots [6,7,8,9,10]. The functions of these genes appear to be conserved across the orthologs analyzed among the core eudicots [8,11,12] and monocots [13,14,15]. A considerable amount of knowledge is available about the molecular mechanisms specifying petal identity in *Arabidopsis* and other core eudicot species however there is little functional evidence regarding homologs with similar roles in petal-identity specification outside of the core eudicots [16] leading to a significant knowledge gap concerning plant organ differentiation, growth and development outside of the most well-studied model systems.

Opium poppy (*Papaver somniferum*) has a long history of practical, medicinal use spanning thousands of years and it continues to be one of the world's most important medicinal plants due to its unique ability to synthesize the drugs morphine, codeine and thebaine and a variety of other biologically active cyclopentano-phenanthrene and benzylisoquinoline alkaloids in its seed pods. Drea et al. [16] described the roles of several MADS-box genes involved in petal specification by demonstrating the duplication and sub-functionalization of *AP3* lineage in *P. somniferum* In poppy, one gene copy influences petal development while the other is responsible for stamen development, contrasting the described role of *AP3* in *Arabidopsis* where *AP3* influences both petal and stamen development. Drea et al. [16] also investigated two paralogs of *PISTILLATA* (*PapsPI-1 and PapsPI-2*) and showed that the *PapsPI-1* gene encodes a product containing the PI-motif as well as a sequence extension at the C-terminus whereas the predicted product of *PapsPI-2* lacks the consensus PI-motif [17] at the C-terminus. This truncation is due to a single nucleotide insertion in the 3' coding region followed by a 2-nucleotide deletion 22 bp downstream that generates a stop codon. Although this domain has been shown to be essential for protein function in *Arabidopsis* PI [18], the *Pisum sativum PI* gene also lacks this conserved domain but has been shown to be capable of rescuing the *Arabidopsis pi*-mutant phenotype [19]. In the present investigation we analyzed different genes involved in flower development by utilizing partially petaloid sepal (Pps-1) and sepaloid petal (OM) mutants that were obtained from the normal sepal and petal phenotypes of I-14 and I-268, respectively. The development of Pps-1 has been described earlier [20,21]. In the Pps-1 mutant, a part of the sepal is converted into petal rather than forming a complete sepal (**Figure 1**) whereas in OM the whole petal is converted into a sepal (**Figure 2**). These analyses indicate the involvement of different recessive mutations for erroneous interconversion of sepals and petals.

Materials and Methods

Plant Material

Plant material consisted of the Pps-1 genotype of *P. somniferum* with partially petaloid sepals, which spontaneously originated from the downy mildew (DM)-resistant genotype I-14. The parent genotype I-14 is characterized by narrow leaves with very deep leaf incisions and white flower petals [20]. In Pps-1, the margins of the sepals are modified into petal-like characters (**Figure 1**). Apart from this, true breeding genotypes I-268 and OM (having mutation 'OM') were selected to test the hypothesis that specific genes are involved in organ conversion. OM was detected in the open pollinated population of the genotype I-268 of opium poppy in which the petals are converted into sepal-like organs (**Figure 2**). All inbred lines (at least 6 selfing cycles) of mutants and their parents were grown and maintained in the research farm of CSIR-Central Institute of Medicinal and Aromatic Plants, Lucknow, India since 2007 and the true-breeding characters were maintained.

Figure 2. Comparison of floral morphology between the parent I-268 (I) and the sepaloid petal mutant (OM; II). A. Flower bud; B. Flower bud without sepal; C. Dissected sepal and petal.

Breeding

Previously, the segregation of *Pps-1* mutation at a ratio of 3:1 has been shown [20], confirming the involvement of single recessive gene in regulating the partially petaloid character. For segregation analysis of the second mutant, the parent genotype I-268 having normal white petals was crossed with OM mutant having sepaloid petals. The crossing was carried out normally and reciprocally taking both as male or female parents. The F_1 and F_2 generation plants were scored for sepaloid petal character. The collected seeds were sown in the field in randomized block design with 3 replications and observations were collected on a single plant basis. Chi-square analysis was applied to test the goodness-of-fit for frequency distributions in the F_2 generations (**Table 1**). A 3:1 segregation of the OM character indicated the involvement of nuclear recessive mutation.

RNA isolation for cDNA preparation and cloning of *PapsPI* gene

Total RNA was isolated from 100 mg of ground tissue samples (from fully developed buds) before anthesis using Trizol reagent (Invitrogen, Cleaveland, OH, USA). RNA was converted into cDNA using the ThermoScript RT-PCR System (Invitrogen, USA) and gene-specific primers [16] were used to amplify the *PapsPI* gene. Amplicons were cloned in pGEM-T easy vector system (Promega) and a total of 20 clones were sequenced for each type of tissue sample (normal sepals of I-14, partially petaloid sepals of Pps-1 and normal petals of both genotypes).

Phylogenetic analysis

Amino acid sequences were aligned (and phylogeny was reconstructed using Bootstrap maximum likelihood method MEGA5 [22]) with MUSCLE multiple sequence alignment [23,24].

Figure 1. Comparison of floral morphology between the partially petaloid sepal mutant (Pps-1; I) and the parent (I-14; II). A. Flower bud; B. Ventral view of sepal; C. Dorsal view of sepal.

Table 1. Segregation pattern of the sepaloid petal (OM) mutant in different generations of the reciprocal crosses involving parent genotype I-268 having wild type phenotype.

Genotype	Characteristics of the genotype	Generation	Number of observed plants		Segregation ratio	Chi-Square	P
			Wild type	Mutant type			
OM	OM, homeotic mutant (with green sepaloid petals)	P_1	0	133		-	-
I-268	Parent genotype from which OM mutant evolved (with white petals)	P_2	106	0	-	-	-
OM×I-268	All plants exhibit white petals with light pink margin	F_1 ($P_1×P_2$)	110	0	-	-	-
I-268×OM	All plants exhibit white petals with light pink margin	r F_1 ($P_2×P_1$)	123	0	-	-	-
OM×I-268	-	F_2 ($P_1×P_2$)	82	23	3:1	0.53	0.50–0.30
I-268×OM	-	r F_2 ($P_2×P_1$)	56	14	3:1	1.18	0.30–0.20

Expression analysis using quantitative and semi-quantitative RT-PCR

Quantitative RT-PCR was carried out using SYBR Green chemistry (Applied Biosystems, USA) as described earlier [25]. Gene-specific primers were designed with Primer Express software (v2.0; Applied Biosystems, USA) and custom-synthesized from Sigma Aldrich, India. The reactions were carried out in 5 biological replicates on the 7900HT Fast Real Time PCR System (Applied Biosystems, USA) and the specificity of the reactions was verified by melting curve analysis with the thermal cycling parameters: initial hold (50°C for 2 min); initial denaturation (95°C for 10 min); and 40 amplification cycles (95°C for 15 s; and 60°C for 1 min) followed by additional steps (60°C for 15 s, 95°C for 15 s and 37°C for 2 min). Relative mRNA levels were quantified with respect to endogeneous control genes (actin [EB740770] in case of *P. somniferum* or ubiquitin [U66264.1] in case of *Nicotiana tabacum*) [26,27]. Sequence Detection System (S.D.S.) software version 2.2.1 was used for relative quantification of gene transcript using the $\Delta\Delta$ C_T method. Threshold cycle (Ct) values obtained after real time PCR were used for calculation of ΔCt value (target-endogenous control). The quantification was carried out by calculating $\Delta\Delta$Ct to determine the fold difference in gene expression [Δ Ct target - Δ Ct calibrator]. Relative quantity (RQ) was determined by $2^{-\Delta\Delta CT}$. Semi-quantitative RT-PCR was performed by following the protocol of Misra et al [26]. Primers were designed on the basis of *P. somniferum* (for *PapsPI-1*, *PapsPI-2*, *PapsAP3-1* and *PapsAP3-2*) gene sequences. Details of the primers used in the semi-quantitative RT- PCR have been provided in **Table S1**.

Tobacco transformation

Specific primers were designed to prepare the overexpression construct for the *PapsPI-1* gene. *Xba*I (forward primer) and *Bam*HI (reverse primer) restriction sites were introduced at either sides of the coding sequence. The amplified PCR-product was cloned in pGEM-T Easy vector and the sequence was confirmed. The plasmid containing the coding region was digested with *Xba*I and *Bam*HI and cloned into pBI121 under the control of the CaMV 35S promoter to yield the final construct *35S::PapsPI-1*. Binary vectors with and without the transgene were separately transformed into GV3101 strain of *Agrobacterium* and used to generate transgenic tobacco plants as described [28,29]. Transformants were observed after 3-4 weeks of selection on kanamycin (200 µg ml^{-1}). Regenerated shoots were excised and rooted. Plantlets with well established root system were hardened for 2 weeks, subsequently transferred to soilrite mix (Keltech Energies Limited, India) and irrigated with diluted MS media. Fully acclimatised plantlets were grown in the greenhouse and genomic DNA samples of transgenic tobacco lines were screened by PCR using *NPTII* and *PapsPI-1* specific primers to verify the transfer of transgene cassettes into the transgenic lines. The non-transformed plants and empty vector-transformed plants did not show any amplification (**Figure S1**).

Results

Expression patterns for genes involved in flowering

Expression level was determined for the four genes of the ABC model in sepals and petals of both the Pps-1 mutant and the wild-type (I-14) through semi-quantitative RT-PCR. Among the genes analyzed, the most significant difference was observed for *PapsPI-1* whose expression was significantly higher in the partially petaloid sepal relative to normal sepal of I-14 (**Figure 3**). Differential expression was not detected in the petals of flowers produced by

MADS Box Genes	Sepal		Petal		Amplification			
					Sepal		Petal	
	I-14	Pps-1	I-14	Pps-1	I-14	Pps-1	I-14	Pps-1
PapsPI-1			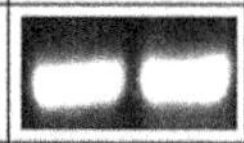		√	√	√	√
PapsPI-2					X	X	√	√
PapsAP3-1			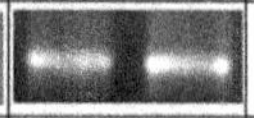		X	√	√	√
PapsAP3-2			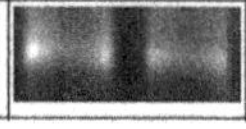		X	X	√	√

Figure 3. Expression of flowering-related genes in the partially petaloid sepal of Pps-1 and the normal sepal of I-14. *PapsPI-1*: *Papaver somniferum PISTILLATA 1*; *PapsPI-2*: *P. somniferum PISTILLATA 2*; *PapsAP3-1*: *P. somniferum APETALA3-1*; *PapsAP3-2*: *P. somniferum APETALA3-2*. Boxes on the right side of the gels indicate amplification (√) or no amplification (**X**), denoting the detection/non detection of homologous genes in *P. somniferum*.

Pps-1 and I-14. A similar expression pattern was also detected for *PapsAP3-1* (**Figure 3**).

Cloning of *PapsPI* gene copies/paralogs and phylogenetic analysis

PapsPI-1 was cloned from cDNA transcribed from the RNA of normal sepals of I-14, partially petaloid sepals of Pps-1 and normal petals of both genotypes. PCR was carried out using gene-specific primers derived from GenBank sequence EF071994 (amino acid ABO13927) [16]. All sequenced amplicons (20 clones each from sepals and petals of both genotypes) were identical, and the *PapsPI-1* sequence (KF550916) from this investigation was 99% similar to the earlier reported *PapsPI-1* (EF071994) [16]. Interestingly, the present study also generated another copy of the *PapsPI* gene (*PapsPI-3*, deposited under Accession No. F550917) that had a stop codon introduced at the 151 amino acid position due to a single base deletion (adenine). This copy was obtained from partially petaloid sepals (of Pps-1), normal sepals (of I-14) and petal tissues of both I-14 and Pps-1. Additionally, seventeen point mutations were also detected in *PapsPI-3* as compared to *PapsPI-1*, of which, nine were before the stop codon in *PapsPI-3*. In phylogenetic analysis the *PapsPI-1* sequence of this investigation (KF550916) and the one reported earlier (ABO13927) [16] clustered together but the sequence of *PapsPI-2* reported earlier (nucleotide EF071995, amino acid ABO13928) [16] was different from that of *PapsPI-3* reported in this investigation (KF550917; **Figure 4**).

PapsPI-1 and *PapsAP3-1* expression in Pps-1 genotype

Higher transcript abundance was observed for *PapsPI-1* in the sepals of Pps-1 as compared to the sepals of I-14. But in the petals of both I-14 and Pps-1, the expression was about 200-fold higher as compared to the sepals of I-14 (**Figure 5**) indicating the causative link between elevated *PapsPI-1* expression and petal tissue specification. When the petaloid part was dissected from the partially petaloid sepal of Pps-1, a 23-fold higher expression for *PapsPI-1* was observed in the petaloid part as compared to the remaining sepal part. Although, a specific trend of expression for *PapsPI-3* was not detected either in the sepal or petal of I-14 and Pps-1, it was interesting to note that higher expression of *PapsPI-3* relative to *PapsPI-1* was observed in the sepal (devoid of petaloid portion) part of Pps-1, as in the case of the true sepal of I-14

(**Figure 5**). Also, in the petaloid part of the Pps-1 sepal and the petals of both I-14 and Pps-1, *PapsPI-3* expression was always found to be lower than that of *PapsPI-1*. *PapsAP3-1* expression was higher in the petaloid part of the partially petaloid sepal of Pps-1 as compared to that in its petaloid-devoid sepal part (**Figure 6**).

Transformation of *PapsPI-1* in tobacco

PapsPI-1 was expressed constitutively in tobacco under the influence of the CaMV 35S promoter. Flowers were obtained in all the twenty transgenic plants screened but only one plant produced a flower having the partially petaloid sepal character (**Figure 7**). Ten plants showed pale green morphology with differential leaf arrangements compared to the control plant. Six-fold increased expression of *PapsPI-1* was observed in the partially petaloid sepal of the transformed flower (**Figure 7D**).

Segregation of *OM* mutation and *PapsPI-1* expression

All the F_1 plants (both normal and reciprocal crosses) showed normal petal phenotype demonstrating the recessive nature of the typical mutant character (*OM*). This also indicated the absence of cytoplasmic control of the mutant trait (sepaloid petal). The segregation pattern of the F_2 populations of both reciprocal crosses also provided a good fit of the monogenic Mendelian ratio (P≥ 0.80-0.70) for the normal wild type (I-268) and the mutant (OM) characters indicating that the mutant trait is controlled by a single recessive nuclear gene "*OM*" (**Table 1**). Interestingly, *PapsPI-3* expression was higher than *PapsPI-1* expression in the normal sepals of I-268 and the sepaloid petal of OM, whereas in the normal petal (of I-268) *PapsPI-1* expression was higher than *PapsPI-3* expression. *PapsPI-1* expression in the normal petal was higher than that in the sepal (**Figure 8**). Relative expression of *PapsPI-1* in the sepal and sepaloid petal (I-268 and OM, respectively) was comparable to the *PapsPI-1* expression in the sepal of I-14 and sepal or the petaloid-devoid sepal part of the partially petaloid sepal of Pps-1.

Discussion

Previously, a spontaneous true breeding homeotic gene mutant Pps-1 with distinct partial petaloid sepals was detected in the population of downy mildew (DM)–resistant elite genotype I-14 during identification of disease resistance sources in opium poppy at CSIR-CIMAP [20]. Analysis of this genotype clearly indicated single, recessive, nuclear gene control of the mutant character and demonstrated that the mutant phenotype is due to mutations at the *Pps-1* locus with a negative control function. In this investigation, a homeotic mutant (OM) was detected in the open-pollinated population of the genotype I-268 of opium poppy in which the petal morphology displayed sepal-like characteristics. This mutant was maintained by several selfing cycles and was observed to be controlled by a recessive mutation.

Expression of genes related to organ identity was measured in the two mutant genotypes, Pps-1 and OM, and compared with their parents, I-14 and I-268, respectively. When sepals of I-14 and Pps-1 genotypes were analyzed, only two genes (*PapsPI-1* and *PapsAP3-1*) showed maximal differential expression. *PapsAP3-1* and *PapsPI-1* have previously been shown to have high expression in petals [16]. Due to significant differential expression in partially petaloid sepals as compared to normal sepals, *PapsPI-1* was taken up for detailed study. As expected, the *PapsPI-1* expression was very high in petals of both I-14 and Pps-1. This is expected as *AP3* and *PI* gene products are believed to form a heterodimer that acts *in vivo* as part of a larger MADS box protein complex, specifying petal as well as stamen identity [8,9,11,30,31,32]. In opium poppy,

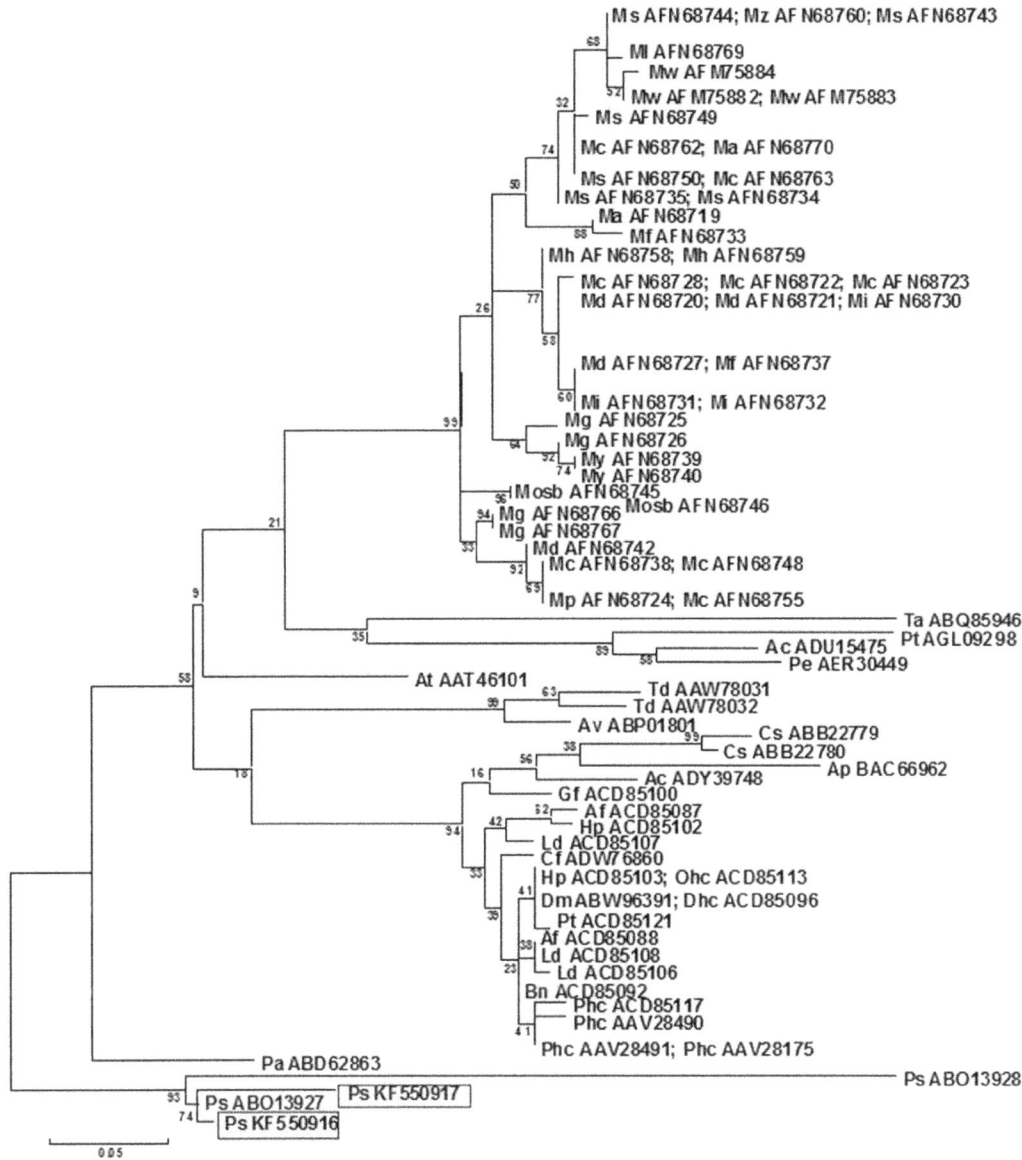

Figure 4. Unrooted maximum likelihood tree comparing the amino acid sequences of *P. somniferum* **PI-1(KF550916), with PI-1 (PISTILLATA) reported from other species.** The symbols for the plants are provided along with the GenBank accession numbers in brackets. *Agapanthus praecox* Ap (BAC66962); *A. praecox*, Ap (ADU15475); *Akebia trifoliate* At (AAT46101); *Ananas comosus* Ac (ADY39748); *Anoectochilus formosanus* Af (ACD85087); *A. formosanus* Af (ACD85088); *Aquilegia vulgaris* Av (ABP01801); *Brassavola nodosa* Bn (ACD85092); *Crocus sativus* Cs (ABB22779); *C. sativus* Cs (ABB22780); *Cymbidium faberi* Cf (ADW76860); *Dendrobium hybrid cultivar Dhc* (ACD85096); *Dendrobium moniliforme* Dm (ABW96391); *Galeola falconeri* Gf (ACD85100); *Habenaria petelotii* Hp (ACD85102); *H. petelotii* Hp (ACD85103); *Liparis distans* Ld (ACD85106); *Ludisia discolor* Ld (ACD85107); *L. discolor* Ld (ACD85108), *Michelia alba* Ma (AFN68719); *Magnolia amoena* Ma (AFN68770); *M. championii* Mc (AFN68738); *M. championii* Mc (AFN68748); *M. championii* Mc (FN68738); *M. coco* Mc (AFN68755); *M. conifera var. chingii* Mc (AFN68728); *M. crassipes* Mc (AFN68722); *M. crassipes* Mc (AFN68723); *M. cylindrica* Mc (AFN68762); *M. cylindrica* Mc (AFN68763); *M. dandyi* Md (AFN68727); *M. delavayi* Md (AFN68745); *M. duclouxii* Md (AFN68720); *M. duclouxii* Md (AFN68721); *M figo* Mf (AFN68733); *M. fordiana* Mf (AFN68737); *M. grandiflora* Mg (AFN68766); *M. grandiflora* Mg (AFN68767); *M. grandis* Mg (AFN68725); *M. hookeri* Mh (AFN68758); *M. hookeri* Mh (AFN68759); *M. insignis* Mi (AFN68730); *M. insignis* Mi (AFN68731); *M. insignis* Mi (AFN68732); *M. liliiflora* Ml (AFN68769); *M. officinalis subsp. biloba* Msob (AFN68745); *M. officinalis subsp. biloba* Mosb (AFN68746); *M. paenetalauma* Mp (AFN68724); *M. paenetalaum* Mp (AFN68726); *M. salicifolia* Ms (AFN68734); *M. salicifolia* Ms (AFN68735); *M. sprengeri* Ms (AFN68743); *M. sprengeri* Ms (AFN68744); *M. stellata* Ms (AFN68749); *M. stellata* Ms (AFN68750); *M. wufengenesis* Mw (AFM75882); *M. wufengenesis* Mw (AFM75883); *M. wufengenesis* Mw (AFM75884); *M. yunnanensis* My (AFN68739); *M yunnanensis* My (AFN68740); *M. zenii* Mz (AFN68760); *Oncidium hybrid cultivar* Ohc (ACD85113); *Papaver somniferum* Ps (KF550916); *P. somniferum* Ps (KF550917); *P. somniferum* Ps (ABO13927); *P. somniferum* Ps (ABO13928); *Paphiopedilum hybrid cultivar* Phc (ACD85117); *Passiflora edulis* Pe (AER30449); *Persea americana* Pa (ABD62863); *Phaius tancarvilleae* Pt (ACD85121); *Phalaenopsis hybrid cultivar* Phc (AAV28175); *P. hybrid cultivar* Phc (AAV28490); *P. hybrid cultivar* Phc (AAV28491); *Populus tomentosa* Pt (AGL09298); *Thalictrum dioicum* Td (AAW78031); *T. dioicum* Td (AAW78032); *Trochodendron aralioids* Ta (ABQ85946).

PapsAP3-1 was able to heterodimerize with *PapsPI-1* like in *Arabidopsis* and *Antirrhinum* [7,16,33]. Although *PapsPI-1* is required to specify petal as well as stamen identity, *PapsAP3-1* functions primarily in the specification of petals and *PapsAP3-2* functions primarily in the specification of stamens [16]. Accordingly, we could observe the differential expression of *PapsAP3-1* in the

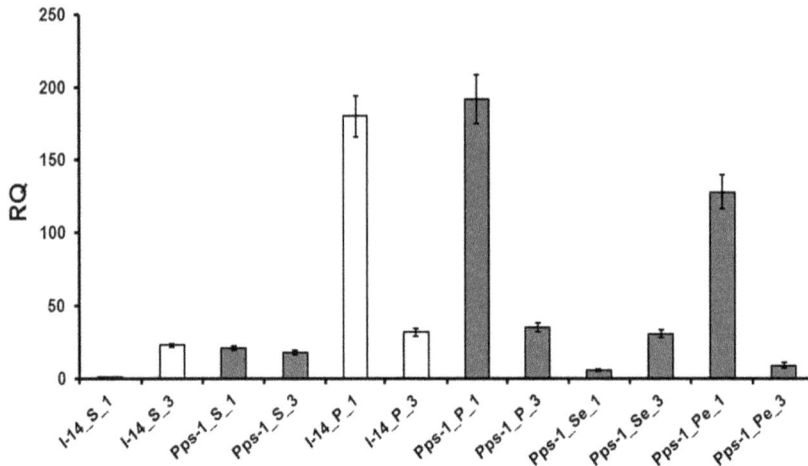

Figure 5. Comparison of quantitative expression levels of *PapsPI-1* **and** *PapsPI-3* **in the petals and sepals of the genotypes Pps-1 and I-14.** The Y-axis represents relative quantities equilibrating the expression of *PapsPI-1* in I-14 sepals as 1RQ value. Data represent mean $\pm$ standard error of 3-5 biological replicates. In X-axis, the names of the genotypes (Pps-1 and I-14) are followed by the organs (S: sepal; P: Petal; Se: Sepaloid part of the sepal and Pe: Petaloid part of the sepal) and *PapsPI* gene expression (1: *PapsPI-1* and 3: *PapsPI-3*). Shaded bar represents Pps-1 genotype.

partially petaloid sepals of Pps-1 genotype compared to the normal sepals of I-14. This confirms the involvement of *PapsPI-1* and *PapsAP3-1* in petal development in *P. somniferum*.

Further, to find out the role of *PISTILLATA* in the petaloid and normal (petaloid-devoid) part of the partially petaloid sepal, the expression of *PapsPI-1* and *PapsPI-3* was compared in petals, sepals, the petaloid portion of the partially petaloid sepals and normal (petaloid-devoid) part of the partially petaloid sepals. Two *PI* gene paralogues (*PapsPI-1* and *PapsPI-3*) were detected in this investigation instead of *PapsPI-1* and *PapsPI-2* as reported earlier [16]. The nucleotide sequence of *PapsPI-3* is different from both *PapsPI-1* and *PapsPI-2*. The expression of *PapsPI-3* was also observed to be always higher when compared to *PapsPI-1* in the normal sepal of I-14 and the normal (petaloid-devoid) part of the partially petaloid sepal of Pps-1, whereas a significantly higher expression of *PapsPI-1* was observed compared to *PapsPI-3* in the

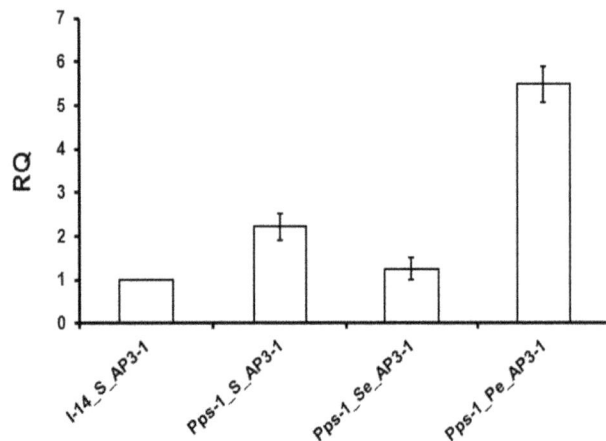

Figure 6. Comparison of quantitative expression levels of *PapsAP3-1* **in the sepals of genotypes Pps-1 and I-14.** The Y-axis represents relative quantities equilibrating the expression of *PapsAP3-1* in I-14 sepals as 1RQ value. Data represent mean $\pm$ standard error of 3-5 biological replicates. On X-axis, S: sepal; Se: Petaloid-devoid part of the sepal and Pe: Petaloid part of the sepal.

petals of both the genotypes and in the petaloid part of the partially petaloid sepal of Pps-1. As described earlier, protein produced by the gene *PapsPI-2* does not dimerize with PapsAP3-1, PapsAP3-2, PapsPI-1 or PapsPI-2. However, this does not obviate the possibility of interaction with other MADS-box gene products that affect its function [16]. Hence the role of *PapsPI-3* cannot be ruled out in partially petaloid sepal character of the Pps-1 flowers.

Ectopic expression of *Antirrhinum Glo* (*GLOBOSA*) in tobacco leads to petaloid sepals, and ectopic expression of both *Def* (*DEFICIENS*) and *Glo* leads to the almost complete conversion of sepals to petals [34]. *Glo* and *Def* are *PISTILLATA* and *APETALA3* orthologs from *Antirrhinum majus*. Ectopic expression of a single homeotic gene, the *Petunia* gene *GREEN PETAL*, has also been described as sufficient to convert sepals to petaloid organs [35]. In this investigation, *PapsPI-1* was expressed ectopically in tobacco under the CaMV 35S promoter and of the twenty transgenic plants (that flowered) screened, only one was observed to be producing flowers with partially petaloid sepal character (**Figure 7**). When some of the transgenic plants were analyzed, all showed *PapsPI-1* gene integration as well as expression in sepals (**Figure S2**). But, the expression of *PapsPI-1* in the sepals of transgenic plant producing flowers with the partially petaloid sepal was highest. Hence, in the case of the mutant Pps-1 of *P. somniferum* the overexpression of *PapsPI-1* in the sepal leads to their conversion to a petal-like phenotype, as corroborated in part by ectopic expression of *PapsPI-1* in tobacco. However, considering the small difference in the relative quantity (RQ) values between the partially petaloid sepal phenotype (II: 5.712) and the transgenic plant having normal flowers (Ta: 5.378), other reasons responsible for the low frequency of *PapsPI-1* transgene phenotype in tobacco cannot be ruled out. It is possible that the very small difference in expression (about 6%) might be the tipping point for initiating a developmental switch. But without other lines showing this phenotype, this is only a speculation and it is also possible that the random insertion of the construct in "Ta" might have caused a gene disruption resulting in the phenotype unrelated to the expression of *PapsPI-1*. The analysis of the transgenic lines Ta, Tb and Tc, not showing the desired phenotype, preclude the possibility of any undesirable effect of the CaMV 35S promoter, which has been described in the past for not yielding desired

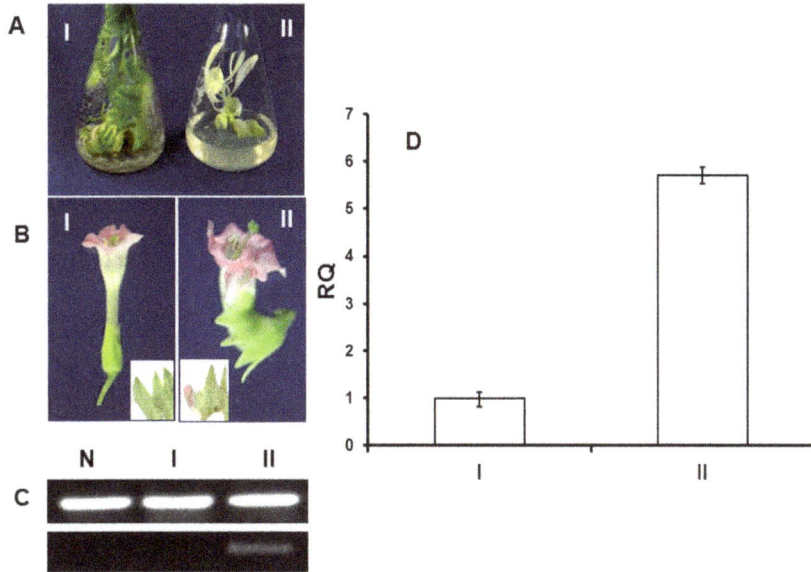

Figure 7. Flower of tobacco plant transformed with *PapsPI-1.* A: Plants in culture (I: Vector transformed; II: Transformed with *PapsPI-1*); B: Flower (I: Vector transformed; II: Transformed with *PapsPI-1*; inset normal sepal and partially petaloid sepal); Semi-quantitative expression of *PapsPI-1* (upper gel shows ubiquitin expression and the lower shows expression of transgene *PapsPI-1*) in: N: Non transformed, I: Vector transformed and plant, II: transformed with *PapsPI-1*); D: Quantitative expression in vector transformed (I) and *PapsPI-1* transformed tobacco sepal (II).

phenotypes, especially for transcription factors expressed under its control [36]. One specific example of misexpression of a component of the flowering regulatory network is ectopic overexpression of the *LFY* gene from *Arabidopsis* [37]. Heterologous expression of transcription factors can also be negatively influenced by the species chosen for overexpression [38]. There may be several reasons for our observing very low frequency of abnormal phenotypes, but the occurrence of a petaloid sepal phenotype while overexpressing the *PapsPI-1* gene cannot be ruled out. The *Pps-1* recessive mutation described earlier [20], was

confirmed in this investigation to be controlling the expression of *PapsPI-1* and *PapsAP3-1*, which is higher in petals of both the plants as well as in the petaloid part of Pps-1 sepals compared to sepals of I-14 and normal (petaloid-devoid) part of the mutant Pps-1 sepals. The functional significance of the heterodimer formed by PapsAP3-1 and PapsPI-1 in determining the petal structure [16] cannot be ignored and in this analysis we observed overexpression of *PapsAP3-1* and *PapsPI-1* in the petaloid part as compared to the normal (petaloid-devoid) part of the sepal in Pps-1. Genes encoding products that function as key regulatory components,

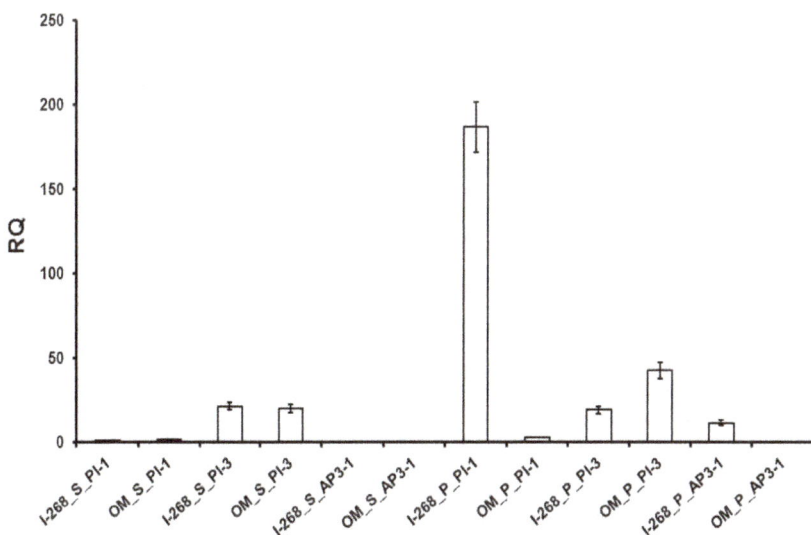

Figure 8. Comparison of quantitative expression levels of *PapsPI-1, PapsPI-3* **and** *PapsAP3-1* **in the petals and sepals of the genotypes, I-268 and OM.** The Y-axis represents relative quantities equilibrating the expression of *PapsPI-1* in 1-268 sepal as 1RQ value. Data represent mean + standard error of 3-5 biological replicates. On the X axis, the names of the genotypes (I-268 and OM) are followed by the organs (S: sepal; P: Petal), the gene expression (PI-1: *PapsPI-1* and PI-3: *PapsPI-3* and AP3-1: *PapsAP3-1*).

such as transcription factors, as well as those participating in large multi-protein complexes (e.g. MADS-domain proteins) [29], appear to be preferentially maintained owing to the requirement for a stoichiometric balance with other components in the pathway [39,40]. Hence, it seems that the recessive *Pps-1* locus might be influencing the expression of *PapsAP3-1* and *PapsPI-1* in sepals during development (**Figures 5, 6**). As the proteins encoded by *PapsAP3-1* and *PapsAP3-2* can heterodimerize with *PapsPI-1*, but *PapsAP3-2* can also homodimerize [16], the role of *PapsPI-3* cannot be ruled out for petaloid conversion of sepal although this type of gene (*PapsPI-3*) has been described to be having limited role in petal morphology as compared to *PapsPI-1*.

Hose in Hose mutants of primrose and cowslip have been found to show dominant homeotic conversion of sepals to petals [41]. The demonstration that in some cases up-regulation of a single B-function MADS box gene can lead to the development of petaloid sepals is consistent with the inheritance of the *Hose in Hose* as a single dominant locus [41]. In contrast, the *CHORIPETALA* and *DESPENTEADO* mutants of *Antirrhinum* are inherited as recessive mutations, which also result in the conversion of sepals to petals [42]. In the present study, the mutations controlling up-regulation of *PapsAP3-1* and *PapsPI-1* in conversion of sepal to petal (*Pps-1*) and down regulation of *PapsAP3-1* and *PapsPI-1* in the conversion of petal to sepal (OM) were confirmed to be recessive in nature. In conclusion, this study indicates a differential pattern of regulation for flowering-related genes in various whorls.

References

1. Jack T (2001) Plant development going MADS. Plant Mol Biol 46: 515–520.
2. Ng M, Yanofsky MF (2001) Function and evolution of the plant MADS-box gene family. Nat Rev Genet 2: 186–195.
3. Zhang XN, Wu Y, Tobias JW, Brunk BP, Deitzer GF, et al. (2008) HFR1 is crucial for transcriptome regulation in the cryptochrome 1-mediated early response to blue light in *Arabidopsis thaliana*. PloS One 3: e3563.
4. Coen ES, Meyerowitz EM (1991) The war of the whorls: genetic interactions controlling flower development. Nature 353: 31–37.
5. Jack T (2004) Molecular and genetic mechanisms of floral control. The Plant Cell 16: Supplement S1–S17.
6. Schwarz-Sommer Z, Hue I, Huijser P, Flor PJ, Hansen R, et al. (1992) Characterization of the *Antirrhinum* floral homeotic MADS-box gene *deficiens*: evidence for DNA binding and autoregulation of its persistent expression throughout flower development. EMBO J 11: 251–263.
7. Trobner W, Ramirez L, Motte P, Hue I, Huijser P, et al. (1992) Globosa – a homeotic gene which interacts with Deficiens in the control of *Antirrhinum* floral organogenesis. EMBO J 11: 4693–4704.
8. Goto K, Meyerowitz EM (1994) Function and regulation of the *Arabidopsis* floral homeotic gene PISTILLATA. Genes and Dev 8: 1548–1560.
9. Riechmann JL, Wang M, Meyerowitz EM (1996) DNA-binding properties of *Arabidopsis* MADS domain homeotic proteins APETALA1, APETALA3, PISTILLATA and AGAMOUS. Nucleic Acids Res 24: 3134–3141.
10. Riechmann JL, Meyerowitz EM (1997) MADS domain proteins in plant development. Biol Chem 378: 1079–1101.
11. Jack T, Brockman LL, Meyerowitz EM (1992) The homeotic gene *APETALA3* of *Arabidopsis thaliana* encodes a MADS box and is expressed in petals and stamens. Cell 68: 683–697.
12. Poupin MJ, Federici F, Medina C, Matus JT, Timmermann T, et al. (2007) Isolation of the three grape sub-lineages of B-class MADS-box *TM6*, *PISTILLATA* and *APETALA3* genes which are differentially expressed during flower and fruit development. Gene 404: 10–24.
13. Ambrose BA, Lerner DR, Ciceri P, Padilla CM, Yanofsky MF, et al. (2000) Molecular and genetic analyses of the Silky1 gene reveal conservation in floral organ specification between eudicots and monocots. Mol Cell 5: 569–579.
14. Whipple CJ, Ciceri P, Padilla CM, Ambrose BA, Bandong SL, et al. (2004) Conservation of B-class floral homeotic gene function between maize and *Arabidopsis*. Development 13:6083–6091.
15. Whipple CJ, Zanis MJ, Kellogg EA, Schmidt RJ (2007) Conservation of B class gene expression in the second whorl of a basal grass and outgroups links the origin of lodicules and petals. Proc Natl Acad Sci USA 104: 1081–1086.
16. Drea S, Hileman LC, de Martino G, Irish VF (2007) Functional analyses of genetic pathways controlling petal specification in poppy. Development 134:4157–4166.
17. Kramer EM, Dorit RL, Irish VF (1998) Molecular evolution of genes controlling petal and stamen development: Duplication and divergence within the *APETALA3* and *PISTILLATA* MADS-box gene lineages. Genetics 149: 765–783.
18. Lamb RS, Irish VF (2003) Functional divergence within the *APETALA3/PISTILLATA* floral homeotic gene lineages. Proc Natl Acad Sci USA 100: 6558–6563.
19. Berbel A, Navarro C, Ferrandiz C, Canas LA, Beltran JP, et al. (2005) Functional conservation of PISTILLATA activity in a pea homolog lacking the PI motif. Plant Physiol 139: 174–185.
20. Dhawan OP, Dubey MK, Khanuja SPS (2007) Detection of a true breeding homeotic gene mutant Pps-1 with partially petaloid sepals in opium poppy (*Papaver somniferum* L.) and its genetic behaviour. J Hered 98: 373–377.
21. Dubey MK, Shasany AK, Dhawan OP, Shukla AK, Khanuja SPS (2009) Genetic variation revealed in the chloroplast-encoded RNA polymerase β # subunit of downy mildew–resistant genotype of opium poppy. J Hered 100:76–85.
22. Hall BG (2013) Bulding phylogenetic trees from molecular data with MEGA. Mov. Bio. Evo. 30: 1229–1235.
23. Edgar RC (2004a) MUSCLE: a multiple sequence alignment method with reduced time and space complexity. BMC Bioinformatics 5:113.
24. Edgar RC (2004b) MUSCLE: multiple sequence alignment with high accuracy and high throughput. Nucleic Acids Res. 32:1792–1797.
25. Maeda H, Fujimoto C, Haruki Y, Maeda T, Kokeguchi S, et al. (2003) Quantitative real-time PCR using TaqMan and SYBR Green for *Actinobacillus actinomycetemcomitans*, *Porphyromonas gingivalis*, *Prevotella intermedia*, *tetQ* gene and total bacteria. FEMS Immunol Med Microbiol 39:81–86.
26. Misra A, Chanotiya CS, Gupta MM, Dwivedi UN, Shasany AK (2012) Characterization of cytochrome P450 monooxygenases isolated from trichome enriched fraction of *Artemisia annua* L. leaf. Gene 510:193–201.
27. Misra M, Pandey A, Tiwari M, Chandrashekar K, Sidhu OP, et al. (2010) Modulation of transcriptome and metabolome of tobacco by *Arabidopsis* transcription factor, AtMYB12, leads to insect resistance. Plant Physiol 152: 2258–2268.
28. Horsch RB, Fry JE, Hoffmann N, Eicholz D, Rogers SG, et al. (1985) A simple and general method for transferring genes into plants. Science 227: 1229–1231.
29. Luo HR, Santamaria M, Benavides J, Zhang DP, Zhang YZ, et al. (2006) Rapid genetic transformation of sweetpotato (*Ipomoea batatas* (L.) Lam) via organogenesis. Afr J Biotechnol 5: 1851–1857.
30. Bowman JL, Smyth DR, Meyerowitz EM (1991) Genetic interactions among floral homeotic genes of *Arabidopsis*. Development 112: 1–20.
31. Honma T, Goto K (2001) Complexes of MADS-box proteins are sufficient to convert leaves into floral organs. Nature 409: 525–529.
32. Pelaz S, Tapia-Lopez R, Alvarez-Buylla ER, Yanofsky MF (2001) Conversion of leaves into petals in *Arabidopsis*. Curr Biol 11: 182–184.
33. Honma T, Goto K (2000) The *Arabidopsis* floral homeotic gene PISTILLATA is regulated by discrete cis-elements responsive to induction and maintenance signals. Development 127: 2021–2030.

Acknowledgments

The authors express their sincere gratitude to the Director, CSIR-CIMAP for keen interest during the study and for providing facilities for the experiments. SKS thanks CSIR for Research Internship and Project Assistantship. The help of Dr Anthony V. Qualley, Manus Biosynthesis, Cambridge, MA, USA, in editing the manuscript is also gratefully acknowledged.

Accession Numbers: Sequence data present in this study can be found in GenBank database (www.ncbi.nlm.nih.gov) of NCBI under the accession numbers *PapsPI-1*: KF550916; *PapsPI-3*: KF550917.

Author Contributions

Conceived and designed the experiments: A. Shasany OPD. Performed the experiments: SKS. Analyzed the data: SKS A. Shukla. Contributed reagents/materials/analysis tools: OPD A. Shasany. Wrote the paper: A. Shasany.

34. Davies B, DiRosa A, EnevaT, Saedler H, Sommer H (1996) Alteration of tobacco floral organ identity by expression of combinations of *Antirrhinum* MADS-box genes. Plant J 10: 663–677.

35. Halfter U, Ali N, Stockhaus J, Ren L, Chua N-H (1994) Ectopic expression of single homeotic gene, *the Petunia* gene *green petal*, is sufficient to convert sepals to petaloid organs. EMBO J 13: 1443–1449.

36. Winfield C, Jordan BR (2006) Biotechnology and floral development, pp 237-266 In: Jordan BR, ed. The Molecular Biology and Biotechnology of Flowering, 2nd edition, CABI International, Wallingford, UK.

37. Weigel D, Alvarez J, Smyth DR, Yanofsky MF, Meyerowitz EM (1992) LEAFY controls floral meristem identity in *Arabidopsis*. Cell 69: 843–859.

38. Brunner AM, Nilsson O (2004) Revisiting tree maturation and floral initiation in the popular functional genomics era. New Phytol 164: 43–51.

39. Evangelisti AM, WagnerA (2004) Molecular evolution in the yeast transcriptional regulation network. J Exp Zool B: Mol Dev Evol 302: 392–411.

40. Birchler JA, Veitia RA (2007) The gene balance hypothesis: From classical genetics to modern genomics. Plant Cell 19: 395–402.

41. Li J, Dudas B, Margaret A, Webster MA, Cook HE, et al. (2010) Hose in Hose, an S locus–linked mutant of *Primula vulgaris*, is caused by an unstable mutation at the Globosa locus. Proc Natl Acad Sci USA 107: 5664–5668.

42. Wilkinson M, Silva ED, Zachgo S, Saedler H, Schwarz-Sommer Z (2000) CHORIPETALA and DESPENTEADO: General regulators during plant development and potential floral targets of FIMBRIATA-mediated degradation. Development 127:3725–3734.

8

Influence of Elastin-Like Polypeptide and Hydrophobin on Recombinant Hemagglutinin Accumulations in Transgenic Tobacco Plants

Hoang Trong Phan[1,6], Bettina Hause[2], Gerd Hause[3], Elsa Arcalis[4], Eva Stoger[4], Daniel Maresch[5], Friedrich Altmann[5], Jussi Joensuu[7], Udo Conrad[1]*

1 Department of Molecular Genetics, Leibniz Institute of Plant Genetics and Crop Plant Research (IPK), Gatersleben, Germany, 2 Cell and Metabolic Biology, Leibniz Institute of Plant Biochemistry (IPB), Halle, Germany, 3 Microscopy Unit, Biocenter, University of Halle-Wittenberg, Halle, Germany, 4 Molecular Plant Physiology, University of Natural Resources and Applied Life Sciences, Vienna, Austria, 5 Department of Chemistry, University of Natural Resources and Applied Life Sciences, Vienna, Austria, 6 Department of Plant Cell Biotechnology, Institute of Biotechnology (IBT), Vietnam Academy of Science and Technology (VAST), Hanoi, Vietnam, 7 VTT Technical Research Centre of Finland, Espoo, Finland

Abstract

Fusion protein strategies are useful tools to enhance expression and to support the development of purification technologies. The capacity of fusion protein strategies to enhance expression was explored in tobacco leaves and seeds. C-terminal fusion of elastin-like polypeptides (ELP) to influenza hemagglutinin under the control of either the constitutive CaMV 35S or the seed-specific USP promoter resulted in increased accumulation in both leaves and seeds compared to the unfused hemagglutinin. The addition of a hydrophobin to the C-terminal end of hemagglutinin did not significantly increase the expression level. We show here that, depending on the target protein, both hydrophobin fusion and ELPylation combined with endoplasmic reticulum (ER) targeting induced protein bodies in leaves as well as in seeds. The N-glycosylation pattern indicated that KDEL sequence-mediated retention of leaf-derived hemagglutinins and hemagglutinin-hydrophobin fusions were not completely retained in the ER. In contrast, hemagglutinin-ELP from leaves contained only the oligomannose form, suggesting complete ER retention. In seeds, ER retention seems to be nearly complete for all three constructs. An easy and scalable purification method for ELPylated proteins using membrane-based inverse transition cycling could be applied to both leaf- and seed-expressed hemagglutinins.

Editor: Shilin Chen, Chinese Academy of Medical Sciences, Peking Union Medical College, China

Funding: This research was supported by the Vietnam Ministry of Education and Training (MoET) and by the BMBF (DLR), Germany. The funders had no role in study design, data collection and analysis, decision to publish, or preparation of the manuscript.

Competing Interests: The authors have declared that no competing interests exist.

* E-mail: conradu@ipk-gatersleben.de

Introduction

The efficient production of recombinant eukaryotic proteins in native conformations is a major goal for the biotechnological production of therapeutic proteins. This goal can only be achieved at high costs in cell-based production systems such as bacteria and animal cells. Such production platforms are either highly capital-intensive, such as mammalian cell cultures based on secreted proteins, or they require complex and efficient refolding of bacterially expressed proteins from insoluble inclusion bodies. Transgenic plants are promising tools to address the limitations of current production systems for recombinant proteins (for reviews see [1–6]). The benefits of plants are the unlimited scale-up potential and relatively low production costs [7]. Notably, the cost of downstream processing steps is generally similar in all recombinant production systems. Downstream costs can account for more than 80% of the overall processing costs [7,8]. The economical production of plant-made recombinant proteins is therefore limited by two major challenges, inadequate accumulation levels and the lack of efficient and scalable purification methods. The design and production of fusion proteins is a general strategy for overcoming these limitations. Recently, three protein fusion systems, ZERA fusions, hydrophobin (HFBI) fusions and fusions with elastin-like polypeptide (ELP), that allow high intracellular accumulation of recombinant proteins in separate and newly formed storage organelles were described. Additionally, these strategies facilitate the development of specific purification processes [9–14]. These three protein fusion systems have very distinct origins; gamma-zein is a plant protein, hydrophobin is a fungal protein, and elastin is of human origin. All three proteins share characteristics that likely cause unique behavior upon plant expression and could cause a significant increase in recombinant protein accumulation. The formation of protein bodies (PBs) induced by both ELP and hydrophobin fusion has recently been investigated using ER-targeted green fluorescent protein (GFP) in transgenic tobacco leaves [12]. An increase in the accumulation and formation of PBs has been shown. An important question is whether this generally holds true for seed expression. The formation of PBs of ELPylated IgG [15] and a slight increase of production levels in seeds have been shown. Another interesting task is to apply these approaches to other therapeutic proteins. ELPylation has been used for the enhancement of expression and purification of several proteins from plants such as antibodies, antigens, cytokines and spider silk proteins (for review see [13]). The ELP fusion strategy is a simple method of purifying recombinant proteins by inverse transition cycling (ITC). The purification process is temperature- and salt-dependent. Increases in salt concentration (NaCl) and temperature result in the formation of insoluble ELP fusion proteins that are separated from the solution by centrifugation at a defined temperature [16]. The insoluble ELP fusion proteins are then re-solubilized in a cool and low-ionic buffer. This purification method was described in 1999 and called centrifugation-based ITC (cITC) [16]. cITC was used to successfully purify antibodies against HIV [15,17], mycobacterial antigens [14], scFv [18], nanobodies [19], spider silk proteins [20], beak and feather disease virus capsid protein [21], erythropoietin [22] and soluble gp130 [23] from plant cells. However, the cITC process had to be optimized to efficiently enrich every individual protein. Another method to enrich ELP fusion proteins from plants called membrane-based ITC (mITC) was developed and optimized by Phan and Conrad (2011) [24]. Hemagglutinin (HA) and neuraminidase from the avian flu virus have been expressed as ELPylated proteins in plants and purified by this simple and scalable method from leaves [24–26].

Influenza viruses continue to be responsible for pandemics [27]. HAs, major antigens of the influenza viruses, are of general interest. Outbreaks of swine flu (H1N1) and avian flu (H5N1) have triggered global concern [28]. Obviously, the proper folding and trimerization of HA antigens requires multiple posttranslational modifications including glycosylation and disulfide bond formation that occur in higher eukaryotic systems. The trimerization could likely result in sufficient antigenicity [29]. Mammalian cell-derived HA trimers induced much higher levels of neutralizing antibodies than similarly produced monomeric HA protein [30]. Plant-generated vaccines focused on HA virus-like particles (VLPs) have been developed [31], and clinical testing is currently underway [32].

To further explore fusion protein strategies in terms of the enhancement of expression and development of scalable purification methods, fusions with hydrophobin and ELP are compared here. Seed-specific expression is included in this evaluation. To assess organ-dependent differences in the formation of induced PBs and the post-translational modification of the recombinant proteins, we analyzed the compartmentalization both in seeds and leaves and took a closer look at the glycosylation of plant-produced HA.

Materials and Methods

Generation of plant vectors

The DNA sequence encoding amino acids 2–564 of hemagglutinin subtype 5 (H5) from the A/Hatay/2004/(H5N1) strain (GenBank accession Q5QQ29) was optimized for expression in tobacco plants and synthesized by GENEART AG (Regensburg, Germany). The previously published constructs were used for expression of H5 and H5-ELP under the control of the CaMV 35S promoter [25] (Figure 1), while the TBAG gene in pRTRA-USP-TBAG and pRTRA-USP-TBAG-ELP (unpublished data) was replaced by the synthesized DNA sequence encoding ectodomain H5 (aa 17–520) to generate expression cassettes (pRTRA-USP-H5 and pRTRA-USP-H5-ELP) under the control of a seed-specific promoter with the introduction of sequence coding for a 6xHis tag (Figure 1).

The DNA fragment coding for the TEV cleavage-recognition site, a flexible (GGGS)$_3$ linker and the hydrophobin gene was amplified by PCR using extension primers and the pTNS9 vector

containing the HFBI coding sequence as a template [33]. The HA1 and H5 coding sequences of the pRTRA-35S-HA1 and pRTRA-USP-H5 vectors were replaced by the PCR product to generate pRTRA-35S-HFBI and pRTRA-USP-HFBI, respectively. The DNA sequence encoding ectodomain H5 (aa 17-520) was cloned into the resulting pRTRA-35S-HFBI and pRTRA-USP-HFBI to form the expression cassettes pRTRA-35S-H5-HFBI and pRTRA-USP-H5-HFBI, respectively.

All expression cassettes in pRTRA vectors were subcloned by *Hin*dIII cleavage into the shuttle vector pCB301-Kan [17,34].

Plant transformation

Tobacco was transformed by agroinfection by the leaf disc method described by Horsch and co-workers [35]. The *Agrobacterium tumefaciens* C58C1 strain transformed with the pCB-kan binary vectors was grown overnight in YEB medium containing 50 µg/ml kanamycin, 50 µg/ml carbenicillin and 50 µg/ml rifampicin. Tobacco leaf discs were submerged for 1 h in the agrobacterium culture and then plated on MS medium at 24°C in the dark for another two days. Infected leaf fragments were transferred to MS medium containing 0.2 mg/L α-naphthalene acetic acid, 1 mg/L 6-benzylaminopurine, 50 mg/L kanamycin and 500 mg/L cefotaxim (NBKC medium). Every 10–14 days, leaf discs were transferred to new NBKC medium until plantlets 2–3 cm in length appeared. These plantlets were transferred to MS medium containing 50 mg/L kanamycin [36] and transferred to soil in the greenhouse. The leaves and seeds of these plants were used for Western blot analysis to screen transgenic plants expressing recombinant proteins under control of the CaMV 35S and seed-specific promoters, respectively.

SDS-PAGE and Western blot

Western blots were performed following the protocol described by Gahrtz and Conrad (2009). In brief, proteins in SDS sample buffer [37] were kept at 95°C for 10 min. The concentration of total soluble protein (TSP) was determined using the Bradford assay. Plant proteins were separated by reducing SDS-PAGE (10% polyacrylamide) and electrotransferred to nitrocellulose membranes. After blocking with 5% (w/v) fat-free milk powder dissolved in TBS (20 mM Tris, 180 mM NaCl, pH 7.8), the membranes were incubated for 2 h at room temperature with a monoclonal anti-c-myc antibody. Antibody binding was detected by the addition of a 1:2,000 dilution of HRP-conjugated sheep anti-mouse IgG. Each membrane was washed three times between each step with TBS containing 0.5% w/v fat-free milk except for the penultimate (TBS only) and final (phosphate-buffered saline, PBS (137 mM NaCl, 2.7 mM KCl, 10 mM Na$_2$HPO$_4$, 1.8 mM KH$_2$PO$_4$, pH 7.4) washes. Antibodies were diluted in TBS with 5% (w/v) fat-free milk powder. The signal was visualized using the enhanced chemiluminescence method (GE Healthcare, UK).

To obtain a rough estimation of the expression level of recombinant proteins, an immunoblotting technique was applied. In general, the samples were serially diluted to achieve band intensities that were similar to the band intensity of a standard protein (^{Nt}anti-hTNFa-VHH-ELP [19]) used at a known amount and containing the c-myc tag for Western blot detection. Band intensities were measured by using totalLab Quant software (Nonlinear Dynamics, USA).

Protein purification by ITC. ELP fusion proteins were purified using mITC optimized to enrich ELPylated proteins from tobacco leaves [24]. Briefly, frozen *N. tabacum* leaves were homogenized in ice-cold 50 mM Tris-HCl (pH 8.0). The plant extract was then cleared by centrifugation (75,600 *g*, 30 min, 4°C). NaCl was added to the extract up to 2 M. The cold extract was

Expression cassettes   **Protein products**   **Molecular weight**

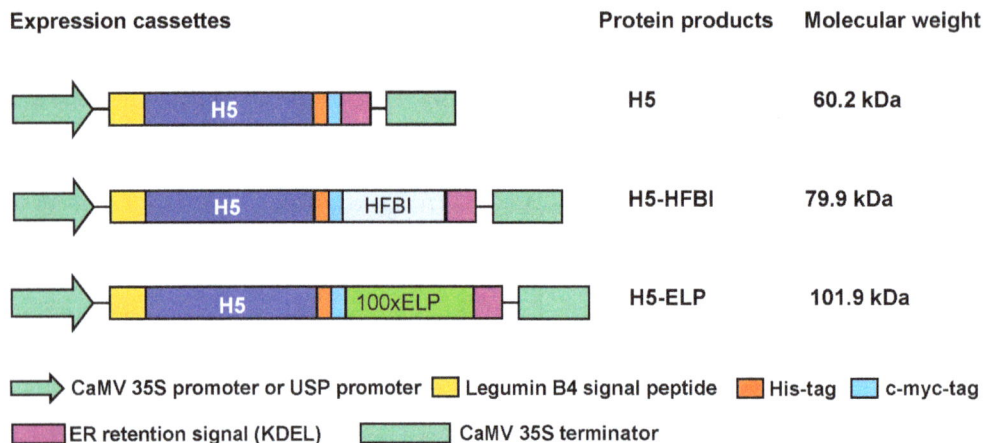

	H5	60.2 kDa
	H5-HFBI	79.9 kDa
	H5-ELP	101.9 kDa

CaMV 35S promoter or USP promoter   Legumin B4 signal peptide   His-tag   c-myc-tag

ER retention signal (KDEL)   CaMV 35S terminator

Figure 1. Expression cassettes for hemagglutinin (HA) in plants. HAs were stably expressed in both leaves and seeds under the control of the CaMV 35S and the seed-specific promoters as the naked form (H5), hydrophobin I fusion protein (H5-HFBI) and ELPylated H5 (H5-ELP). All recombinant HAs contained His and c-myc tags for affinity chromatography purification and Western blotting, respectively. The LeB4 signal peptide and KDEL motif were used to ensure ER retention.

centrifuged again at 75,600 g for 30 min and then passed through a 0.22 μm polyethersulfone membrane (Corning, USA) with the temperature maintained at 4°C. The clear extract was warmed to room temperature and passed through a 0.2 μm cellulose acetate membrane using a vacuum pump (Vacuubrand, Germany). The membrane was washed twice with 2 M NaCl to remove contaminating proteins. Ice-cold MilliQ water was then passed through the filter to elute the ELP fusion proteins.

ELPylated H5 from seeds was initially purified using the mITC procedure described above. mITC purification of ELPylated HA from seeds was improved by introduction of ammonium sulfate precipitation before adding NaCl. The following steps were performed as described above.

Protein Purification by immobilized metal ion chromatography. Leaf/seed samples were ground with a mortar and pestle in liquid nitrogen. Total protein was extracted in 50 mM Tris buffer (pH 8.0). The extract was clarified by centrifugation (18,000 g, 30 min, 4°C) and then filtrated through paper filters. The clear extract was mixed with Ni-NTA agarose resin previously washed twice with water. After mixing for 30 at 4°C, the mixture was applied to a chromatography column. Thereafter, the column was washed with washing buffer (50 mM NaH_2PO_4, 300 mM NaCl, 20 mM imidazole, pH 8.0). Recombinant proteins were then eluted from the column with elution buffer (50 mM NaH_2PO_4, 300 mM NaCl, 125 mM imidazole, pH 8.0). The protein was concentrated with an iCON Concentrator (Thermo Scientific, USA) with a molecular weight cutoff of 9,000 and stored at −20°C.

PNGase F treatment. Purified H5, H5-HFBI and H5-ELP were deglycosylated with the commercial PNGase F enzyme (Catalog No. P0704S; New England Biolabs, Germany). Briefly, 1 μg recombinant purified HA was incubated at 37°C in the presence of PNGase F enzyme for 1 h. The enzyme was inactivated by incubation at 75°C for 10 minutes. After enzyme inactivation, proteins in SDS sample buffer were separated on a 10% reducing SDS-PAGE, blotted and analyzed by Western blot using the anti-c-myc monoclonal antibody as described in detail above.

Immunofluorescence and immunogold labeling. Leaves from transgenic and wild-type tobacco plants were embedded in polyethylene glycol (PEG) [38] and in HM20 after high-pressure freeze fixation and freeze substitution [39] for immunofluorescence and immunogold labeling, respectively. Semithin sections of 3 μm thickness mounted on glass slides for immunofluorescence and sections showing silver interference mounted on copper grids for electron microscopy were blocked in 5% BSA in 0.1 M phosphate buffer saline (PBS, pH 7.4). Sections were then incubated with monoclonal mouse anti-c-myc antibody diluted 1:20 in 5% BSA/PBS at 4°C overnight (fluorescence) or at room temperature for 2 h (electron microscopy). The antibody-antigen reaction was visualized with polyclonal goat anti-mouse serum conjugated to Alexa Fluor 488 or 594 (Invitrogen, UK) for fluorescence microscopy or polyclonal goat anti-mouse serum conjugated to 10 nm gold (British Biocell International, Cardiff, UK) for electron microscopy. Semithin sections of leaves were counterstained by incubation with 0.1 μg/ml 4,6-diamidino-2-phenylindol (DAPI; Sigma-Aldrich, Germany) for 15 min. Micrographs were obtained using a Zeiss Axioimager microscope (Carl Zeiss Microscopy GmbH, Germany) or a LIBRA 120 transmission electron microscope (Carl Zeiss Microscopy GmbH). All micrographs were processed with the Photoshop 12.0.4 program (Adobe, Germany). Seeds expressing H5 alone, H5-ELP and H5-HBFI fusions as well as wild type seeds were fixed, dehydrated and embedded as described [15,40]. The detection of the signal was performed as described for the leaf sections.

N-glycan analysis. The purified HAs were separated by reducing SDS-PAGE in 10% polyacrylamide gels under reducing conditions. Coomassie-stained bands were excised, destained, reduced, carbamidomethylated and digested with trypsin. Tryptic peptides were digested with PNGase A [41]. The released N-linked glycans were then purified using porous graphite carbon spin columns (PGC, Hypercarb; Thermo Scientific) and analyzed by reflectron MALDI-TOF-MS on an Autoflex instrument (Bruker, Germany).

Statistical analysis. Statistical analyses were performed using the SigmaPlot software. The Kruskal-Wallis one way analysis of variance on ranks was applied to test the statistical differences between the median values of the three transgenic plant groups. A P value less than 0.01 was defined as a significant difference.

Results

Expression of influenza hemagglutinin variants in plants

The influenza HA is a transmembrane type I protein containing a signal peptide, an ectodomain, a transmembrane domain and a cytoplasmic tail [42]. To efficiently express H5 in plant cells, the sequence encoding the ectodomain (aa 17–520) of the A/Hatay/2004/(H5N1) influenza virus strain was synthesized with optimized codons and cloned in expression cassettes (Figure 1). The H5 protein was recombinantly expressed alone (H5) or as a fusion with hydrophobin (H5-HFBI) or ELP (H5-ELP) under the control of both the CaMV 35S and USP promoters (Figure 1). Each recombinant protein contained the c-myc tag for downstream detection by Western blot, a His tag for purification by immobilized metal ion chromatography (IMAC) and the KDEL motif at the C-terminal to retain the protein in the endoplasmic reticulum (ER) [43]. The functionality of all binary vectors was validated by assaying the expression of recombinant proteins in transiently transformed *N. benthamiana* leaves and confirmed by Western blot (data not shown). Transgenic tobacco plants were generated by Agrobacterium-mediated leaf disc transformation [35,36]. Transformants were selected on medium containing selecting antibiotics. Regenerated transgenic plants were screened by Western blot using an anti-c-myc monoclonal antibody. The results are shown in Table 1. To monitor the effect of the hydrophobin and ELP fusions on influenza HA accumulation in transgenic plants, the accumulation levels of ELPylated, hydrophobin fused and untagged H5 in leaves and seeds were determined by SDS-PAGE under reducing conditions and Western blot using an anti-c-myc monoclonal antibody and compared to standard proteins with c-myc tag (anti-hTNFα-VHH-ELP [19]). Expression levels of all transgenic plants according the 6 constructs have been analyzed and presented (see Table S1–S6 in File S1). Statistical analyses (P<0.01, Figure 2) show, that ELPylation strongly enhances the expression levels both in leaves and seeds. The hydrophobin fusion had no significant effect on the accumulation of HA in the ER in both leaves and seeds. H5-ELP reached yields 10-fold higher than that of H5-HFBI and untagged H5 in both seeds and leaves (Tables S1–S6 in File S1, Figure 2, 3). H5, H5-HFBI and H5-ELP proteins accumulated to a maximum of 0.04, 0.05 and 0.5% of the total soluble protein in leaves and 0.04, 0.03 and 0.5% in seeds, respectively (Tables S1–S6 in File S1). We also selected among transgenic lines bearing a single locus insertion of transgene (determined by segregation of the *npt II* gene) the two lines with the highest expression for everyone of the six constructs and compared the expression levels directly by Western analyses (Figure 3) to exclude the influence of multiple insertions. The data again indicate, that only ELPylation causes a strong expression enhancement in leaves and seeds as well. This result is in contrast to previous results presented by Joensuu and co-workers [11]. They reported that the hydrophobin fusion significantly enhanced the accumulation of green fluorescence protein and the enzyme glucose oxidase. The apparent molecular weights shown in Figure 3 are higher than the expected sizes predicted from the polypeptide sequences in Figure 1. This could be explained by the fact that glycosylation influences the running behavior during the electrophoretic separation.

Purification of plant-derived recombinant hemagglutinins

In this study, ELPylated HAs were purified from leaves by the procedure described by Phan and co-workers [25] (Figure 4A) and from seeds (Figure 4B) by the mITC protocol described by Phan and Conrad (2011) [24]. Coomassie brilliant blue stained gels show that aggregates of H5-ELP were selectively enriched from leaf extract (Pm, Figure 4A) by mITC, while other plant proteins passed through the 0.2 μm cellulose acetate membrane (Sm, Figure 4A, B). The H5-ELP obtained from mITC was highly pure and concentrated (Pm, Figure 4A). In the case of seed-derived H5-ELP, the target protein was enriched (Pm, Figure 4B), but H5-ELP was still detected by Coomassie brilliant blue staining in the flow through (Sm, Figure 4B), and the purified eluted protein was contaminated with plant proteins (Pm, Figure 4B). This result shows that the mITC procedure optimized for purification of leaf-derived HA was not suitable for the purification of ELPylated HA from seeds. Therefore, the mITC method was adapted to purify seed-derived ELPylated HA. We observed that seed extracts were contaminated with oils. To remove plant oils, proteins were precipitated by ammonium sulfate and then solubilized in Tris buffer (see the Materials and Methods section). The purification was proceeded as in the standard mITC. The Coomassie brilliant blue-stained gel showed that seed-derived ELPylated HA was selectively enriched from seed extract with high purity (Pm, Figure 4C). Furthermore, the target protein was not detected by Coomassie staining and Western blot in the flow through (Sm, Figure 4C). The Western blot results shown in Figure 4 indicate that H5-ELP, with the expected size of approximately 100 kDa, was successfully purified from leaves (Pm, Figure 4A) and seeds (Pm, Figure 4B, C). Only a small portion of the target protein or no target protein was detected in the Sm fraction (Sm, Figure 4A or Sm, Figure 4C, respectively) after performing the optimized procedure.

Hydrophobins are small surface-active fungal proteins that can be purified using a surfactant-based aqueous two-phase system (ATPS; [11,44]). The presence of the hydrophobin tag did not significantly enhance the accumulation of HA in the ER in both leaves and seeds in comparison with pure HA (Figure 2, 3). Therefore, the purification of hydrophobin fusion HAs using ATPS was not included in this study. Instead, H5-HFBI and pure H5 from leaves and seeds were purified by IMAC based on the C-terminal 6x histidine tag (Figure 1) to obtain material for further characterization. Both HAs were analyzed by Coomassie blue stain and Western blot using an anti-c-myc monoclonal antibody (Figure 5).

Subcellular localization of recombinant hemagglutinins in leaves and seeds of transgenic tobacco plants

To monitor the localization of recombinant HAs in transgenic plants, leaves expressing untagged H5, H5-ELP and H5-HFBI were analyzed by fluorescence and electron microscopy *via* c-myc tag detection. Fluorescence microscopy showed that strong fluorescent signals (green color) were detected in leaves expressing recombinant H5 (Figure 6A), H5-HFBI (Figure 6B) and H5-ELP (Figure 6C) compared with the negative control (Figure 6D). Fluorescent signals indicated that recombinant HAs were distributed within the cytoplasm, mainly around nuclei which were stained by DAPI (blue color). Therefore, we hypothesized that recombinant HAs are localized in the ER. Electron microscopy was applied to address this question.

Electron micrographs showed the distribution of untagged H5 that was mainly found in the ER lumen (Figure 7A). In addition, H5 labelling was occasionally detected in PB at a size of 231.68 ± 136.73 nm (n = 10) (Figure 7B). In contrast to H5, the H5-ELP fusion was predominantly detected in the bigger PBs that were typically 835 ± 323.36 nm (n = 35) (Figure 7C). The membrane surrounding the PBs was decorated with ribosomes (arrows, Figure 7C). This result suggested that ELP-induced PBs were

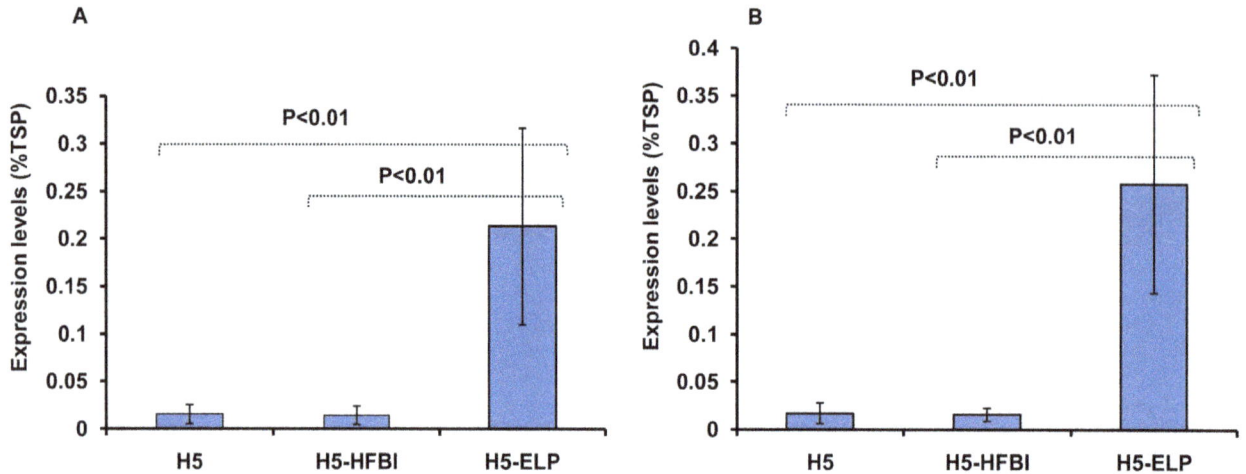

Figure 2. Influence of hydrophobin and elastin-like polypeptide on hemagglutinin accumulation in transgenic plants under the control of either the CaMV 35S promoter (A) or the seed-specific USP promoter (B). Each column shows the mean value of expression level of the transgenic plants, and the error bars indicate the standard deviation. TSP: total soluble protein.

derived from the ER membrane and distributed in the cytoplasm as a novel organelle. When H5 was fused to hydrophobin, PBs were also observed (Figure 7D); however, they were more abundant than those observed in the leaves expressing untagged H5 and less abundant than those observed in H5-ELP leaves

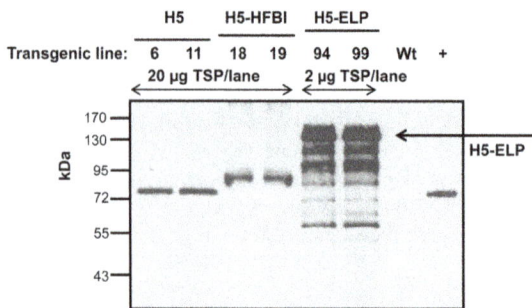

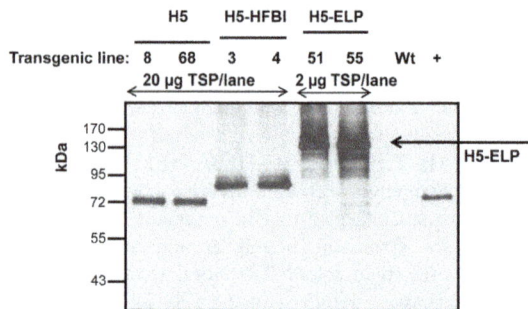

Figure 3. Expression of influenza HAs in transgenic tobacco plants. The extracted proteins from transgenic tobacco leaves (A) or seeds (B) were separated by 10% SDS-PAGE, blotted and detected with anti-c-myc monoclonal antibody followed by horseradish peroxidase-linked sheep anti-mouse IgG as a secondary antibody. "+": anti-hTNFα-VHH-ELP was used as a Western blot standard [19]; Wt: wild type; TSP: total soluble protein. The numbers refer to independent primary transgenic plants.

(Figure 7). H5-HFBI bodies were 426.21 ± 122.44 nm (n = 17) in size.

HA-derived PBs have been observed in transgenic tobacco seeds. When expressed alone, H5 was found within the protein storage vacuoles (PSV) in both the endosperm and the embryo (Figure S1A). H5-ELP fusions formed PBs both in the endosperm and the embryo cells (Figure 8). In the endosperm cells, ELP bodies have the same appearance as observed when fused to antibody chains [15]. Thus, they are irregular in size and shape (Figure 8A) and have a flocular, loosely packed content (Figure 8C). As for the embryo, cells contain a high number of ELP bodies, more abundant than that observed in the antibody fusions [15]. They exhibit the same appearance in endosperm cells and form large masses found within the cytoplasm (Figure 8C, D). No clear labelling was observed in the apoplast or any other compartment of endosperm or embryonic cells apart from a weak nonspecific signal within the crystalloids of the PSVs (Figure 8). PBs observed after ELPylation were not detected in wild-type seeds (Figure S1B).

HA-hydrophobin bodies were found both in the endosperm and the embryo, although in the endosperm they were rather scarce (Figure 9A). In contrast to ELP, hydrophobin bodies are regular in size and on average much smaller (~300 nm). These bodies are tightly packed with material of medium electron density (Figure 9B, C) and are most likely ER derived because their membrane appears to be decorated with ribosomes (Figure 9B, C). Some of the hydrophobin bodies, another phase of a loosely packed material, were also labelled with anti-c-myc antibody (Figure 9C). No significant labelling was found in the PSVs or any other compartment of the cell. Hydrophobin bodies were not present in wild-type seeds (Figure S1B).

Glycosylation profiling of plant-derived hemagglutinins

Influenza HA contains N-linked oligosaccharides [45]. Amino acid sequence analysis of the HA of the A/Hatay/2004/(H5N1) strain has predicted six potential N-linked glycosylation sites (Asn10, 11, 23, 154, 165 and 286). The glycosylation of recombinant HAs was confirmed by the digestion of purified H5, H5-HFBI, and H5-ELP using the commercial PNGase F enzyme that removes N-linked glycans between N-acetylglucosamine (Gn) and asparagine residues from glycoproteins. Deglycosylation of plant-derived HAs was visualized by Western blot

Table 1. Transgenic tobacco plants expressing recombinant hemagglutinins.

Transgene	Number of regenerated plants	Number of transgenic plants
CaMV 35S promoter		
H5*	25	18
H5-HFBI	51	38
H5-ELP*	55	43
USP promoter		
H5	36	25
H5-HFBI	65	39
H5-ELP	91	50

Transgenic plants were screened by Western blot using an anti-c-myc monoclonal antibody. H5: hemagglutinin; H5-HFBI: hemagglutinin fusion with hydrophobin I; H5-ELP: ELPylated hemagglutinin.
*Plants expressing H5 and H5-ELP under the control of the CaMV 35S promoter were described by Phan and co-workers [25].

A. Purification of leaf-derived H5-ELP using mITC

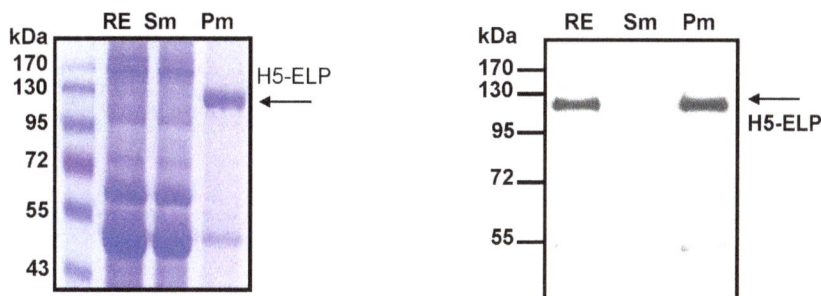

B. Purification of seed-derived H5-ELP using mITC

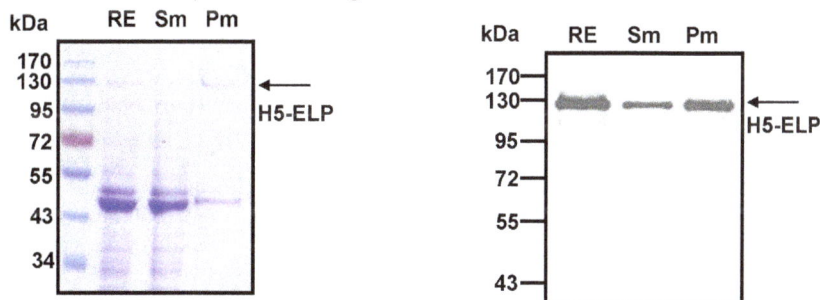

C. Purification of seed-derived H5-ELP using improved mITC

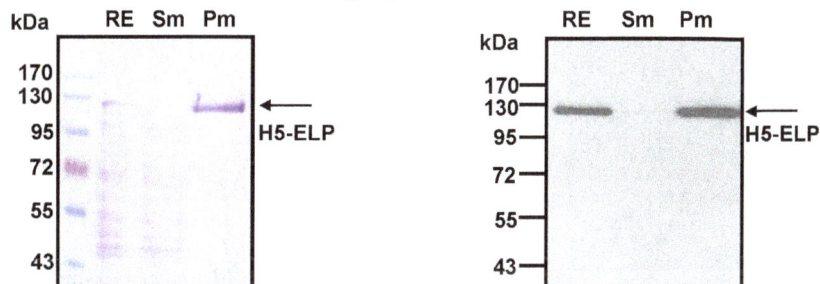

Figure 4. Purification of ELPylated hemagglutinin (H5-ELP) from transgenic leaves and seeds by membrane-based ITC. ELPylated hemagglutinins from leaves (A) and seeds (B) were purified by the standard or improved mITC methods (C) described in the Materials and Methods section. Proteins in the raw plant extract (RE), in the supernatant after passage through a 0.2 μm cellulose acetate membrane (Sm) and in the eluent (Pm) were collected during the mITC purification process and separated by 10% SDS-PAGE. Recombinant proteins were then detected using Coomassie Brilliant Blue staining (left) or an anti-c-myc monoclonal antibody (right).

A. Affinity chromatography purification of leaf-derived H5 and H5-HFBI

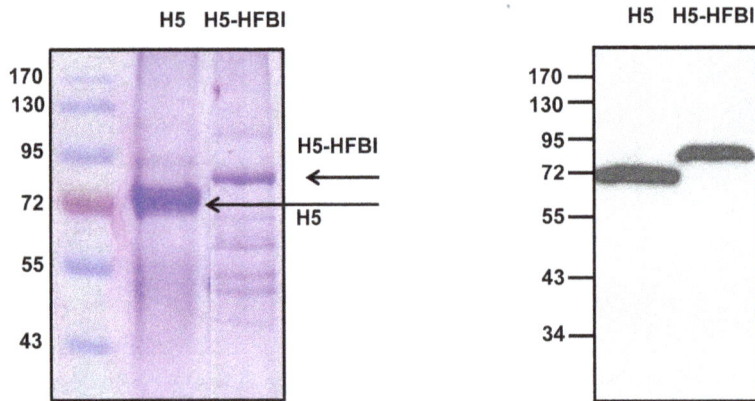

B. Affinity chromatography purification of seed-derived H5 and H5-HFBI

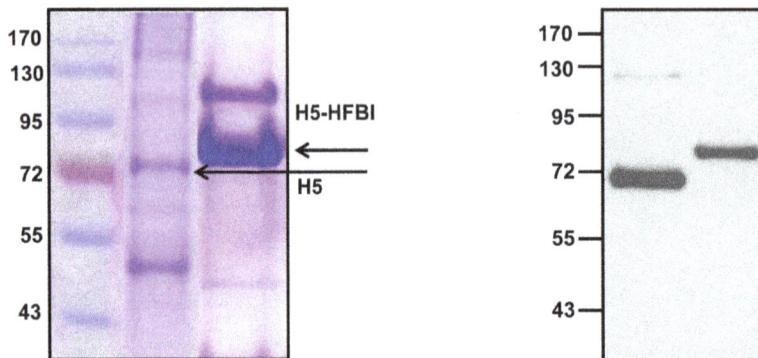

Figure 5. Purification of non ELPylated HA from transgenic leaves and seeds by His tag-based affinity chromatography. Non ELPylated HAs from leaves (A) and seeds (B) were purified by the IMAC method described in the materials and methods section. Purified proteins were separated by 10% SDS-PAGE and then detected using Coomassie Brilliant Blue staining (left) or by an anti-c-myc monoclonal antibody (right).

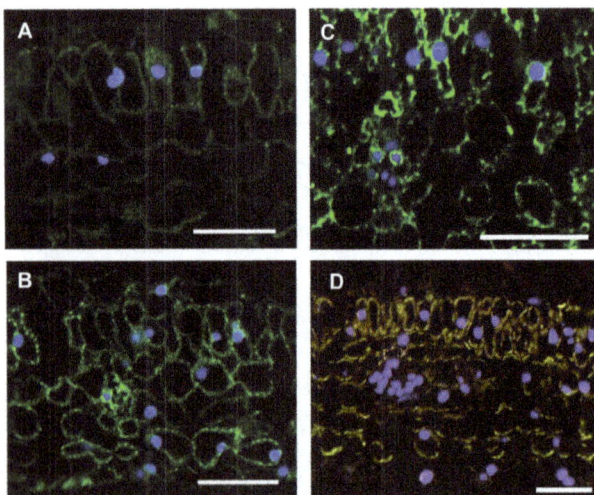

Figure 6. Immunofluorescence analysis of recombinant HAs in plant leaves. Leaves were fixed, embedded in PEG and sectioned. Recombinant HAs were immunodecorated with an anti-c-myc monoclonal antibody followed by incubation with secondary antibody (anti-mouse-IgG conjugated with AlexaFluor488) and counterstaining with DAPI. A. H5; B. H5-HFBI; C. H5-ELP; D. wild type. Bars represent 50 μm.

analysis using an anti-c-myc antibody (Figure 10). Immunodetection revealed that, after removal of the N-glycans, the apparent molecular weights of the recombinant HA proteins shifted to the expected molecular weight calculated using the amino acid sequence (Figure 10). This result confirmed that recombinant HAs were post-translationally modified by the addition of N-linked glycans.

To further investigate the N-linked glycans of the recombinant proteins, HAs purified from leaves and seeds were separated by SDS-PAGE (Figure 4, 5). The bands were excised and digested with trypsin. Mass spectrometry analyses revealed that carbohydrates bound to leaf-derived H5 and H5-HFBI contained 70–72% oligo-mannose type (OMT), with Man_7GnGn being the most dominant form. Twenty-seven to thirty percent of the glycoforms were complex type (CT) N-glycans with typical plant glycans containing fucose and xylose (Table 2). These results indicate that the KDEL sequence-mediated retention of H5 and H5-HFBI in the ER of leaf cells was efficient but not complete for all constructs. Leaf-derived H5 and H5-HFBI were not completely retained in the ER. In contrast, H5-ELP contained only the OMT form; the most abundant glycoforms were Man_7GnGn (72.32%) and Man_8GnGn (27.68%). No CT glycans were detected in this case (Table 2). This result indicates that the ELP fusion tag efficiently retained the recombinant H5-ELP in the ER. Interestingly, H5 and H5-ELP accumulating in seeds contained only the OMT form

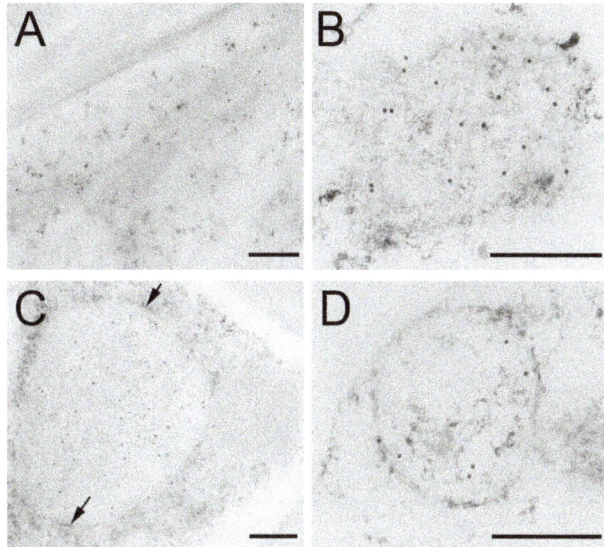

Figure 7. Localization of recombinant hemagglutinins in leaves visualized by electron microscopy. Thin leaf sections mounted on copper grids were probed with monoclonal mouse anti-c-myc antibody followed by the goat anti-mouse-IgG conjugated to 10 nm gold particles. A and B. H5; C. H5-ELP; D. H5-HFBI. The membrane of immunodecorated PB is surrounded by ribosomes (arrows, C). Bars represent 250 nm.

Figure 8. Localization of hemagglutinin-ELP fusions in tobacco seeds. A, B. Fluorescence microscopy.C, D. Electron microscopy.A, C. Endosperm. B, D. Embryo. Note the ELP bodies (arrowheads, A, B) and those that are loosely packed (arrowheads, C, D). Cell wall (cw), oil bodies (ob), protein storage vacuole (arrows), nucleus (n). Bars 50 μm (A, B), 1 μm (C, D).

or a very small proportion of the CT glycan form (6.32%, in the case of H5-HFBI) (Table 2).

Discussion

The expression of foreign proteins using plants is a promising method for the safe and low-cost production of antigens for vaccination purposes. However, its widespread use is curtailed by problems related to expression levels, antigen purity and proper post-translational modification which are important for induction of robust immune responses. Currently, ELP and hydrophobin fusion strategies were among the most attractive ways to enhance target protein accumulation in transgenic plants and to provide a simple means for the purification of recombinant proteins (for reviews see [11,13].

In this study, the major antigen of avian influenza A virus, HA [46], was expressed with C-terminally fused ELP and HFBI. We show here that C-terminal fusion of ELP to HA resulted in increased accumulation of H5-ELP in both leaves and seeds under control of either the constitutive CaMV 35S or the seed-specific USP promoter in comparison to the untagged H5 and H5-HFBI (Figure 2, 3). Our results are consistent with previous studies that reported that expression levels of ELPylated-proteins, such as antibodies [15,17], anti TNFα nanobodies [19], human interleukin and spider silk proteins [20,47] and vaccines [14,25], were higher than those of proteins without ELP fusions. In general, accumulation enhancement by ELPylation was dependent on the specific protein and ranged between two- and 100-fold (for review see [13]). In contrast to ELP, HA fusion with hydrophobin did not significantly enhance the expression level compared to untagged H5, whereas hydrophobin-fused GFP increased two [11] or three-fold in concentration [12] in comparison to untagged GFP. This may be explained by the high accumulation of GFP itself in plant cells (18% and 0.3% of TSP in transiently and stably transformed leaves, respectively [11,12], while HA accumulated in transgenic

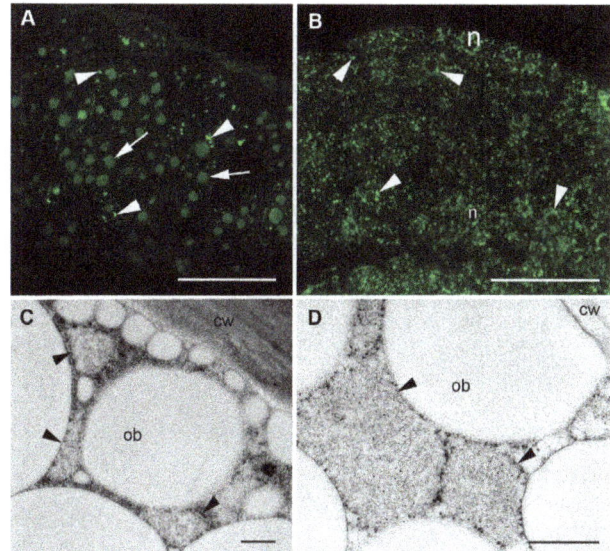

plants at a rather low level (0.04% TSP). Again, the enhancement of the expression levels of recombinant proteins in plants caused by hydrophobin is likely depending on the protein fused to HFBI.

In the actual study, we found that increasing expression levels of ER-targeted recombinant proteins are highly related to the presence of novel PBs in the leaves and seeds of transgenic tobacco. ER-targeted pure HA could induce such protein particles in leaves, but they were very rare and small. Novel protein particles containing H5-HFBI were more frequent and larger than those of the H5 alone, whereas H5-HFBI showed only a slightly enhanced accumulation compared to untagged H5 (Figure 7B, D). In contrast to the cases mentioned above, PBs containing H5-ELP were much bigger and formed as novel organelles located in the cytoplasm. Novel PBs induced by HFBI and ELP in leaves were previously reported by Conley and co-workers [48] and by Floss and co-workers for tobacco seeds [15]. The question of whether PB formation is only a result of increased accumulation remains open.

Enhancement of the accumulation of HAs by ELP fusion results in a higher concentration of the ELPylated target proteins in the initial aqueous extraction. Higher levels of ELP-fused proteins facilitate subsequent purification of ELPylated proteins by ITC [49,50]. ELP fusion proteins could be purified by the ITC method, which is based on precipitation of ELPylated proteins. Micrometer-sized aggregates of ELPylated proteins were collected/harvested by centrifugation or through a 0.2 μm membrane, which are called centrifugation or membrane-based ITC, respectively [24,51]. We previously reported that avian flu antigens were successfully enriched using an adapted membrane-based ITC method from stably [24] or transiently transformed leaf materials with an efficient recovery rate [25]. In this study, ELPylated hemagglutinin was successfully purified from transgenic seeds using the same mITC procedure for leaf material (Figure 4). However, the target protein was still detected in the flow through after passing through a 0.2 μm membrane (Figure 4B). This result indicates that ELPylated HA from seeds was not completely

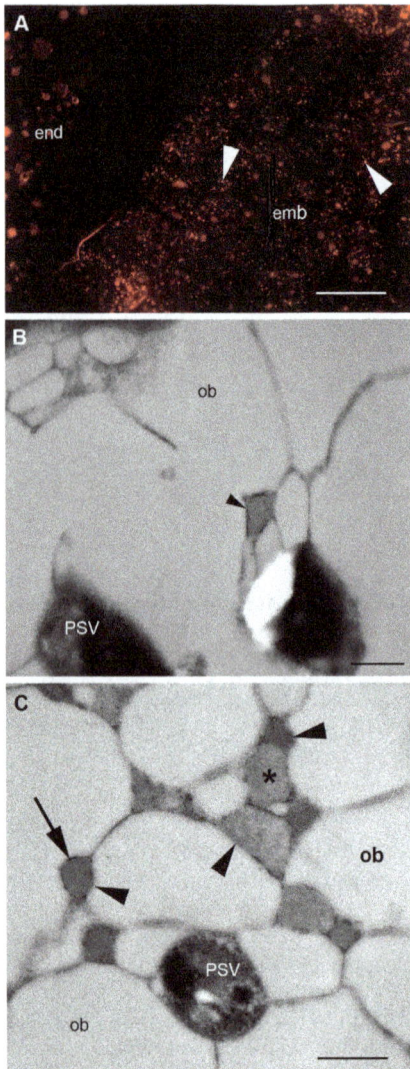

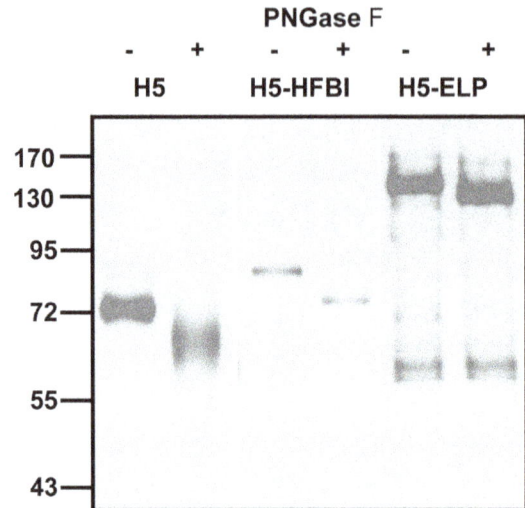

Figure 10. Western blot analysis of purified HAs treated/untreated with PNGase F. Purified HAs from leaves were deglycosylated using the commercial PNGase F enzyme described in the Materials and Methods section. PNGase F-treated and untreated proteins were then separated in 10% SDS-PAGE. Recombinant proteins were detected using an anti-c-myc monoclonal antibody. "−" and "+" indicate PNGase F-untreated and treated samples, respectively.

Figure 9. Localization of hemagglutinin-hydrophobin I fusions in tobacco seeds. A. Fluorescence microscopy. B, C. Electron microscopy. B. Endosperm. C. Embryo. Scarce hydrophobin bodies in the endosperm (arrowheads, A, B). Abundant hydrophobin bodies in the embryo cells (arrowheads, A, C). Hydrophobin bodies show non-uniform electron density (*, C). Endosperm (end), embryo (emb), protein storage vacuole (PSV), ribosomes (arrow). Bars 25 μm (A), 0.5 μm (B, C).

recovered. An improved mITC method using ammonium sulfate to precipitate proteins and simultaneously remove plant oils was applied to completely recover ELPylated HA from transgenic seeds, suggesting that oils in the protein extract may affect the ability to retain precipitated ELPylated HA on the membrane surface. Our results indicate that ELPylated proteins can be purified from different plant tissues by a simple and inexpensive purification method. Purified ELPylated proteins retain their functionality, such as enzymatic activity [24], the antigen-binding activity of ELPylated antibodies [15] or the receptor-binding activity of ELPylated HA [25]. The low immunogenicity [14,25] and biocompatibility [52,53] of ELPs are promising factors in the further application of ELPylation technology for the production and purification of therapeutic proteins for human and veterinary medicine.

In general, glycoproteins are synthesized by ribosomes associated with the ER membrane, translocated into the ER lumen, and trafficked from the Golgi apparatus (GA) toward the plasma membrane and the apoplastic space along the secretory pathway. N-linked glycosylation is co-translationally performed on the glycoprotein consensus sequence N-X-T/S (for a review see [54]). N-linked glycans of glycoproteins undergo several maturation steps including glycose trimming and mannose addition to form high mannose-type glycans in the ER. N-linked glycans are then transported and modified further in the GA from cis to medial and $trans$ cisternae to form typical plant glycans with the addition of xylose and fucose to the core glycans. In this study, the KDEL motif was used to retain recombinant proteins in the ER [43]. The N-linked glycan profiles of these recombinant proteins showed that leaf-derived H5 and H5-HFBI predominantly possess OMT glycans (70–72%) that supposedly matured in the ER compartment and 27–30% CT glycans with typical plant glycans (GnGnX and GnGnF) (Table 2). The abundance of OMT and lower abundance of CT in untagged HA has been reported in previous studies of KDEL-tagged H1 and H5 expressed in transiently transformed plants [55,56]. In contrast to these studies, we found that H5-ELP exclusively harbored OMT glycans (Table 2). These results suggested that cellular trafficking of H5, H5-HFBI and H5-ELP may be different or the fusion partner could have a shielding effect. H5-ELP is preferentially located in the ER-derived PBs, whereas portions (approximately 70%) of leaf-derived H5 and H5-HFBI are preferentially deposited in the ER-derived PBs. Another portion (approximately 30%) may be trafficked to the GA. Munro and Pelman (1987) reported that KDEL-fused proteins are trafficked into the GA and then back into the ER [57]. Otherwise, the presence of CT glycans in seeds is below 10%. It seems that there is less "escape" of recombinant proteins from the ER. The glycosylation pattern plays an important role for the quality and functionality of plant-produced proteins. From this general point it is important to note, that fusion protein strategies do not only enhance expression levels and purification, but also influence

Table 2. Glycosylation profile of recombinant HAs from leaves and seeds.

Recombinant protein	Glycoform (%)				
	Man$_6$GnGn	Man$_7$GnGn	Man$_8$GnGn	GnGnX	GnGnF
Leaf-derived HA					
H5		64.08	5.91	25.50	4.50
H5-HFBI		55.12	17.47	27.41	
H5-ELP		72.32	27.68		
Seed-derived HA					
H5		75.56	24.44		
H5-HFBI	10.22	61.42	22.04	6.32	
H5-ELP	3.93	44.74	51.33		

Man$_{6, 7, 8}$: Mannose $_{6, 7, 8}$; Gn: *N*-acetylglucosamine; X: Xylose; F: Fucose.

posttranslational modifications due to modified intracellular sorting.

The different subcellular localizations of recombinant HAs were visualized by transmission electron microscopy. The distribution of HAs was consistent with the *N*-linked glycan profiles. We found that the frequency of PB formation and their sizes were dependent on the expression level of recombinant proteins in plant cells. Leaf-derived H5-ELP was predominantly stored in large PBs (diameter 800 nm). H5-HFBI was also deposited in PBs; however, H5-HFBI-induced PBs were much smaller than those of H5-ELP and less abundant. In contrast, H5 alone was located in tubular ER structures and occasionally detected in PBs.

PB formation induced by HA alone suggests that PB formation is not only dependent on ELP or HFBI tags. However, the presence of these tags enhanced the expression levels of recombinant proteins and also increased the frequency of the appearance of PBs. There may be a threshold value for PB formation as noted by Gutierrez and co-workers. They estimated a threshold value of 0.2% TSP [12].

The PBs containing H5-ELP and the absence of CT glycans suggested that this recombinant protein did not arrive in the GA, but is stored in the PBs in the cytoplasm without contact with the GA enzymes that transfer xylose and fucose glycans into core glycans, while H5 and H5-HFBI are likely partially trafficked into the GA and then transported back to the ER.

In summary, in this study we showed that fusing H5 to either HFBI or ELP and ER targeting can induce PBs in leaves as well as in seeds. ELPylation can cause a significant enhancement of expression in both leaves and seeds. In addition, an easy and scalable purification method for ELPylated proteins such as mITC

that was previously shown with high recovery efficiency of leaf [24,25] and seed (in this study) -derived recombinant proteins could be scalably applied to hemagglutinins in the future.

Supporting Information

Figure S1 Localization of hemagglutinin in tobacco seeds by immunofluorescence microscopy. A. Cross-section of transgenic seeds expressing H5. Note the clear signal within the PSVs (arrows) in both the endosperm and in the embryo. B. Negative control using wild type seeds. No labelling was found within any cell compartment in either the embryo or the endosperm. end, endosperm; emb, embryo. Bars represent 20 μm.

Acknowledgments

We thank Christine Helmold, Isolde Tillack, Ulrike Gresch and AnnekatrinRother for their excellent technical help, Armin Meister for help with the statistical analyses and the members of the Cost Action FA0804, Molecular Pharming, for helpful discussion.

Author Contributions

Conceived and designed the experiments: UC ES. Performed the experiments: HTP EA BH GH DM. Analyzed the data: HTP DM EA BH GH FA. Contributed reagents/materials/analysis tools: BH GH ES FA JJ UC. Wrote the paper: HTP UC EA ES BH.

References

1. Fischer R, Schillberg S (2004) Plant-made pharmaceuticals and technical proteins. Weinheim: WILEY-VCH Verlag GmbH & Co. KGaA. 315 p.
2. Sharma AK, Sharma MK (2009) Plants as bioreactors: recent developments and emerging opportunities. Biotechnol Adv 27: 811–832.
3. Floss DM, Rose-John S, Scheller J (2013) ELP fusion technology for biopharmaceuticals. In: Schmidt SR, editor. Fusion protein technologies for biopharmaceuticals: applications and challenges. Hoboken, New Jersey, USA.: John Wiley & Sons, Inc.pp. 211–226.
4. Stoger E, Ma J, Fischer R, Christou P (2005) Sowing the seeds of success: pharmaceutical proteins from plants. Curr Opin Biotechnol 16: 167–173.
5. Davoodi-Semiromi A, Samson N, Daniell H (2009) The green vaccine A global strategy to combat infectious and autoimmune diseases. Human Vaccines 5: 488–493.
6. Egelkrout E, Rajan V, Howard J (2012) Overproduction of recombinant proteins in plants. Plant Sci 184: 83–101.

7. Yusibov V, Rabindran S (2008) Recent progress in the development of plant derived vaccines. Expert Rev Vaccines 7: 1173–1183.
8. Evangelista RL, Kusnadi AR, Howard JA, Nikolov ZL (1998) Process and economic evaluation of the extraction and purification of recombinant beta-glucuronidase from transgenic corn. Biotechnol Prog 14: 607–614.
9. Torrent M, Llompart B, Lasserre-Ramassamy S, Llop-Tous I, Bastida M, et al. (2009) Eukaryotic protein production in designed storage organelles. BMC Biol 7: 5.
10. Torrent M, Llop-Tous I, Ludevid MD (2009) Protein body induction: a new tool to produce and recover recombinant proteins in plants. Methods Mol Biol 483:193–208.
11. Joensuu JJ, Conley AJ, Lienemann M, Brandle JE, Linder MB, et al. (2010) Hydrophobin fusions for high-level transient protein expression and purification in *Nicotiana benthamiana*. Plant Physiol 152: 622–633.

12. Gutierrez S, Saberianfar R, Kohalmi S, Menassa R (2013) Protein body formation in stable transgenic tobacco expressing elastin-like polypeptide and hydrophobin fusion proteins. BMC Biotechnol 13: 40.

13. Floss DM, Schallau K, Rose-John S, Conrad U, Scheller J (2010) Elastin-like polypeptides revolutionize recombinant protein expression and their biomedical application. Trends Biotechnol 28: 37–45.

14. Floss DM, Mockey M, Zanello G, Brosson D, Diogon M, et al. (2010) Expression and immunogenicity of the mycobacterial Ag85B/ESAT-6 antigens produced in transgenic plants by elastin-like peptide fusion strategy. J Biomed Biotechnol 2010: 1–15.

15. Floss DM, Sack M, Arcalis E, Stadlmann J, Quendler H, et al. (2009) Influence of elastin-like peptide fusions on the quantity and quality of a tobacco-derived human immunodeficiency virus-neutralizing antibody. Plant Biotechnol J 7: 899–913.

16. Meyer DE, Chilkoti A (1999) Purification of recombinant proteins by fusion with thermally-responsive polypeptides. Nat Biotechnol 17: 1112–1115.

17. Floss DM, Sack M, Stadlmann J, Rademacher T, Scheller J, et al. (2008) Biochemical and functional characterization of anti-HIV antibody-ELP fusion proteins from transgenic plants. Plant Biotechnol J 6: 379–391.

18. Joensuu JJ, Brown KD, Conley AJ, Clavijo A, Menassa R, et al. (2009) Expression and purification of an anti-Foot-and-mouth disease virus single chain variable antibody fragment in tobacco plants. Transgenic Res 18: 685–696.

19. Conrad U, Plagmann I, Malchow S, Sack M, Floss DM, et al. (2011) ELPylated anti-human TNF therapeutic single-domain antibodies for prevention of lethal septic shock. Plant Biotechnol J 9: 22–31.

20. Scheller J, Henggeler D, Viviani A, Conrad U (2004) Purification of spider silk-elastin from transgenic plants and application for human chondrocyte proliferation. Transgenic Res 13: 51–57.

21. Duvenage L, Hitzeroth II, Meyers AE, Rybicki EP (2013) Expression in tobacco and purification of beak and feather disease virus capsid protein fused to elastin-like polypeptides. J Virol Methods 191: 55–62.

22. Conley AJ, Joensuu JJ, Jevnikar AM, Menassa R, Brandle JE (2009) Optimization of elastin-like polypeptide fusions for expression and purification of recombinant proteins in plants. Biotechnol Bioeng 103: 562–573.

23. Lin M, Rose-John S, Grotzinger J, Conrad U, Scheller J (2006) Functional expression of a biologically active fragment of soluble gp130 as an ELP-fusion protein in transgenic plants: purification via inverse transition cycling. Biochem J 398: 577–583.

24. Phan HT, Conrad U (2011) Membrane-based inverse transition cycling: an improved means for purifying plant-derived recombinant protein-elastin-like polypeptide Fusions. Int J Mol Sci 12: 2808–2821.

25. Phan HT, Pohl J, Floss DM, Rabenstein F, Veits J, et al. (2013) ELPylated haemagglutinins produced in tobacco plants induce potentially neutralizing antibodies against H5N1 viruses in mice. Plant Biotechnol J 11: 582–593.

26. Phan HT, Floss DM, Conrad U (2013) Veterinary vaccines from transgenic plants: highlights of two decades of research and a promising example. Curr Pharm Des 19: 5601–5611.

27. Watanabe Y, Ibrahim MS, Suzuki Y, Ikuta K (2012) The changing nature of avian influenza A virus (H5N1). Trends Microbiol 20: 11–20.

28. World Health Organization. Influenza at the human-animal interface (HAI). Available:http://www.who.int/influenza/human_animal_interface/en/. Accessed 25 june 2013.

29. Cornelissen LAHM, de Vries RP, de Boer-Luijtze EA, Rigter A, Rottier PJM, et al. (2010) A single immunization with soluble recombinant trimeric hemagglutinin protects chickens against highly pathogenic avian influenza virus H5N1. PLoS ONE 5: e10645.

30. Wei CJ, Xu L, Kong WP, Shi W, Canis K, et al. (2008) Comparative efficacy of neutralizing antibodies elicited by recombinant hemagglutinin proteins from avian H5N1 influenza virus. J Virol 82: 6200–6208.

31. D'Aoust MA, Lavoie PO, Couture MMJ, Trépanier S, Guay JM, et al. (2008) Influenza virus-like particles produced by transient expression in *Nicotiana benthamiana* induce a protective immune response against a lethal viral challenge in mice. Plant Biotechnol J 6: 930–940.

32. Landry N, Ward BJ, Trépanier S, Montomoli E, Dargis M, et al. (2010) Preclinical and clinical development of plant-made virus-like particle vaccine against avian H5N1 influenza. PLoS ONE 5: e15559.

33. Nakari-Setala T, Aro N, Kalkkinen N, Alatalo E, Penttila M (1996) Genetic and biochemical characterization of the *Trichoderma reesei* hydrophobin HFBI. Eur J Biochem 235: 248–255.

34. Xiang C, Han P, Lutziger I, Wang K, Oliver DJ (1999) A mini binary vector series for plant transformation. Plant Mol Biol 40: 711–717.

35. Horsch RB, Fry JE, Hoffmann NL, Eichholtz D, Rogers SD, et al. (1985) A simple and general method for transferring genes into plants. Science 227: 1229–1231.

36. Floss DM, Conrad U (2010) Expression of complete antibodies in transgenic plants. In: Kontermann R, Dübel S, editors. Antibody Engineering: Springer Berlin Heidelberg. pp. 489–502.

37. Gahrtz M, Conrad U (2009) Immunomodulation of plant function by in vitro selected single-chain Fv intrabodies. Methods Mol Biol 483: 289–312.

38. Isayenkov S, Mrosk C, Stenzel I, Strack D, Hause B (2005) Suppression of allene oxide cyclase in hairy roots of *Medicago truncatula* reduces jasmonate levels and the degree of mycorrhization with Glomus intraradices. Plant Physiol 139: 1401–1410.

39. Thieme F, Szczesny R, Urban A, Kirchner O, Hause G, et al. (2007) New type III effectors from *Xanthomonas campestris pv. vesicatoria* trigger plant reactions dependent on a conserved N-myristoylation motif. Mol Plant Microbe Interact 20: 1250–1261.

40. Arcalis E, Marcel S, Altmann F, Kolarich D, Drakakaki G, et al. (2004) Unexpected deposition patterns of recombinant proteins in post-endoplasmic reticulum compartments of wheat endosperm. Plant Physiol 136: 3457–3466.

41. Kolarich D, Altmann F (2000) N-Glycan analysis by matrix-assisted laser desorption/ionization mass spectrometry of electrophoretically separated nonmammalian proteins: application to peanut allergen Ara h 1 and olive pollen allergen Ole e 1. Anal Biochem 285: 64–75.

42. Veit M, Thaa B (2011) Association of influenza virus proteins with membrane rafts. Adv Virol 2011: 14 pages.

43. Wandelt CI, Khan MR, Craig S, Schroeder HE, Spencer D, et al. (1992) Vicilin with carboxy-terminal KDEL is retained in the endoplasmic reticulum and accumulates to high levels in the leaves of transgenic plants. Plant J 2: 181–192.

44. Linder MB, Qiao M, Laumen F, Selber K, Hyytia T, et al. (2004) Efficient purification of recombinant proteins using hydrophobins as tags in surfactant-based two-phase systems. Biochemistry 43: 11873–11882.

45. Keil W, Geyer R, Dabrowski J, Dabrowski U, Niemann H, et al. (1985) Carbohydrates of influenza virus. Structural elucidation of the individual glycans of the FPV hemagglutinin by two-dimensional 1H n.m.r. and methylation analysis. EMBO J 4: 2711–2720.

46. Gerhard W (2001) The role of the antibody response in influenza virus infection. Curr Top Microbiol Immunol 260: 171–190.

47. Patel J, Zhu H, Menassa R, Gyenis L, Richman A, et al. (2007) Elastin-like polypeptide fusions enhance the accumulation of recombinant proteins in tobacco leaves. Transgenic Res 16: 239–249.

48. Conley AJ, Joensuu JJ, Menassa R, Brandle JE (2009) Induction of protein body formation in plant leaves by elastin-like polypeptide fusions. BMC Biol 7: 48.

49. Floss DM, Conrad U (2012) Plant molecular pharming - veterinary applications. In: MeyersRA, editor. Encyclopedia of Sustainability Science and Technology, Springer. pp. 8073–8080.

50. Christensen T, Trabbic-Carlson K, Liu W, Chilkoti A (2007) Purification of recombinant proteins from *Escherichia coli* at low expression levels by inverse transition cycling. Anal Biochem 360: 166–168.

51. Ge X, Trabbic-Carlson K, Chilkoti A, Filipe CDM (2006) Purification of an elastin-like fusion protein by microfiltration. Biotechnol Bioeng 95: 424–432.

52. Urry DW, Parker TM, Reid MC, Gowda DC (1991) Biocompatibility of the bioelastic materials, poly(GVGVP) and its {gamma}-Irradiation cross-linked matrix: summary of generic biological test results. J Bioact Compat Polym 6: 263–282.

53. Rincon AC, Molina-Martinez IT, de Las Heras B, Alonso M, Bailez C, et al. (2006) Biocompatibility of elastin-like polymer poly(VPAVG) microparticles: in vitro and in vivo studies. J Biomed Mater Res A 78: 343–351.

54. Gomord V, Fitchette AC, Menu-Bouaouiche L, Saint-Jore-Dupas C, Plasson C, et al. (2010) Plant-specific glycosylation patterns in the context of therapeutic protein production. Plant Biotechnol J 8: 564–587.

55. Zhang S, Sherwood RW, Yang Y, Fish T, Chen W, et al. (2012) Comparative characterization of the glycosylation profiles of an influenza hemagglutinin produced in plant and insect hosts. Proteomics 12: 1269–1288.

56. Shoji Y, Farrance CE, Bautista J, Bi H, Musiychuk K, et al. (2011) A plant-based system for rapid production of influenza vaccine antigens. Influenza Other Respir Viruses 6: 204–210.

57. Munro S, Pelham HR (1987) A C-terminal signal prevents secretion of luminal ER proteins. Cell 48: 899–907.

9

A Novel Gene *SbSI-2* Encoding Nuclear Protein from a Halophyte Confers Abiotic Stress Tolerance in *E. coli* and Tobacco

Narendra Singh Yadav[1], Vijay Kumar Singh[1], Dinkar Singh[1], Bhavanath Jha[1,2]*

1 Discipline of Marine Biotechnology and Ecology, CSIR-Central Salt and Marine Chemicals Research Institute, Bhavnagar, Gujarat, India, **2** Academy of Scientific and Innovative Research, CSIR, New Delhi, India

Abstract

Salicornia brachiata is an extreme halophyte that grows luxuriantly in coastal marshes. Previously, we have reported isolation and characterization of ESTs from *Salicornia* with large number of novel/unknown salt-responsive gene sequences. In this study, we have selected a novel salt-inducible gene *SbSI-2* (*Salicornia brachiata salt-inducible-2*) for functional characterization. Bioinformatics analysis revealed that SbSI-2 protein has predicted nuclear localization signals and a strong protein-protein interaction domain. Transient expression of the RFP:SbSI2 fusion protein confirmed that SbSI-2 is a nuclear-localized protein. Genomic organization study showed that *SbSI-2* is intronless and has a single copy in *Salicornia* genome. Quantitative RT-PCR analysis revealed higher *SbSI-2* expression under salt stress and desiccation conditions. The *SbSI-2* gene was transformed in *E. coli* and tobacco for functional characterization. pET28a-SbSI-2 recombinant *E. coli* cells showed higher tolerance to desiccation and salinity compared to vector alone. Transgenic tobacco plants overexpressing *SbSI-2* have improved salt- and osmotic tolerance, accompanied by better growth parameters, higher relative water content, elevated accumulation of compatible osmolytes, lower Na$^+$ and ROS accumulation and lesser electrolyte leakage than the wild-type. Overexpression of the *SbSI-2* also enhanced transcript levels of ROS-scavenging genes and some stress-related transcription factors under salt and osmotic stresses. Taken together, these results demonstrate that *SbSI-2* might play an important positive modulation role in abiotic stress tolerance. This identifies *SbSI-2* as a novel determinant of salt/osmotic tolerance and suggests that it could be a potential bioresource for engineering abiotic stress tolerance in crop plants.

Editor: Girdhar K. Pandey, University of Delhi South Campus, India

Funding: The financial assistance received from Council of Scientific and Industrial Research (CSIR) (www.csir.res.in), New Delhi (BSC0109: SIMPLE) is duly acknowledged. The funders had no role in study design, data collection and analysis, decision to publish, or preparation of the manuscript.

Competing Interests: The authors have declared that no competing interests exist.

* Email: bjha@csmcri.org

Introduction

The world population is increasing rapidly and may reach 6 to 9.3 billion by the year 2050, whereas crop production is decreasing rapidly because of the negative impact of various environmental stresses; therefore, it is now very important to develop stress tolerant varieties to cope with this upcoming problem of food security [1]. Major abiotic stresses includes high salinity, drought, temperature extremes, water logging, high light intensity, and mineral deficiencies. Abiotic stresses reduce plant growth and development, causing poor productivity or plant death in extreme conditions. Abiotic stresses are the primary causes of crop loss worldwide, reducing average yields of major crop plants by more than 50% [2]. Plants adapt to environmental stresses via a plethora of responses, including the activation of molecular networks that regulate stress perception, signal transduction and the expression of both stress-related genes and metabolites. Plants have stress-specific adaptive responses as well as responses which protect the plants from more than one environmental stress [3]. Various genes induced by abiotic stresses are grouped under two categories, namely functional genes and regulatory genes. The first category of genes generally facilitates production of important enzymes and metabolic proteins, which include osmolytes, transporters/channel proteins, antioxidative enzymes, lipid biosynthesis genes, polyamines and sugars. The second category of genes consists of regulatory proteins, such as Transcription factors (TFs) belonging to the bZIP, DREB, MYC/MYB, and NAC families, which control expression of many downstream stress tolerance genes [4,5,6]. A number of abiotic stress-related genes, as well as some transcription factors and regulatory sequences in plant promoters have been characterized [4,7]. Whole genome sequencing and microarray analysis have provided valuable insight towards the understanding of molecular mechanism of abiotic stress tolerance involving a number of functional and regulatory genes [7,8]. Transcription factors modulate expression of specific groups of genes through sequence specific DNA binding and protein-protein interaction. They can act as activators or repressors of gene expression, leading to specific cellular responses. Abiotic stress related TFs follow ABA dependent and independent pathways. Identification of key regulatory TFs and their regulatory activators and repressors, their target genes and protein partners is essential to understand the regulatory complex networks. Studies in *Arabidopsis* and *Oryza sativa* indicated that a number of *cis*-elements and their corresponding binding proteins, i.e. TFs, are involved in plant stress responses [4]. It has been reported that transgenic plants overexpressing genes encoding key transcription factors showed enhanced tolerance to various abiotic stresses [6,9,10,11,12,13].

Halophytes are useful organisms to study salt tolerance mechanisms because they are well adapted to salinity and can overcome this problem more efficiently than glycophytes [14]. Halophytes have a unique genetic makeup allowing them to grow and survive under salt stress conditions [15]. Experimental studies in our laboratory have concentrated on an extreme halophyte, *Salicornia brachiata* Roxb., in an effort to identify and characterize novel/unknown genes that enable salt tolerance. *S. brachiata* (*Amaranthaceae*), a leafless succulent annual halophyte, commonly grows in the salt marshes of Gujarat coast in India. *Salicornia* can grow in a wide range of salt concentrations (0.1–2.0 M) and can accumulate quantities of salt as high as 40% of its dry weight [15]. This unique characteristic provides an advantage for the study of salt tolerance mechanisms. Therefore, this plant may serve as a potential bioresource for salt-responsive genes study.

Expressed sequence tags (ESTs) analysis is a rapid and powerful method for elucidating information regarding gene expression and also provides an opportunity to identify new genes involved in biological functions. EST databases have been developed in many glycophytic plant species in response to different stresses like cold, desiccation, high salinity and ABA [2,16] and also in some halophytes. In addition to known functional genes, unknown and hypothetical genes provide a good candidate pool to find novel stress tolerance genes. A large number of unknown genes, which lack similarity with known genes in the NCBI database, have been reported. It can be envisaged that these unknown genes constitute the unique genetic make-up of the plant helping it to sustain itself under stress condition. EST databases of different halophytic plants show a large percentage of unknown genes like *Sueda salsa* (22%, [17]), *Mesembryanthum crystallinum* [18], *Thellungiella halophilla* (32%, [19]), *Avicennia marina* (30%, [20]), *Limonium sinense* (37%, [21]), *Aleuropus littoralis* (20%, [22]), *Spartina alterniflora* (13%, [23]), *Macrotyloma uniflorum* (30%, [24]), *S. brachiata* (29%, [25]), *Tamarix hispida* (21%, [26]), Alfalfa (22%, [27]) and *Chenopodium album* (42%, [28]). Previously, we have identified approximately 1000 ESTs in response to salt stress from *S. brachiata* [25] and also characterized some important abiotic stress tolerant genes (*SbGST*, [29]; *SbMAPKK*, [15]; *SbDREB2A*, [30]; *SbNHX1*, [31]; *SbASR1*, [32]; *SbSOS1*, [33]; *SbSI-1*, [34]). The *S. brachiata* EST database contains large number of novel/unknown/hypothetical genes [25]. These genes might be playing an important role in providing salinity stress adaptation to *Salicornia*, and therefore can serve as a valuable bioresource for engineering abiotic stress tolerance in crop plants. With this aim we have characterized a novel salt-inducible gene *SbSI-2* in response to salt and osmotic stress, through its heterologous expression in *E. coli* and tobacco.

Materials and Methods

Ethics statement

Plant samples were collected from open coastal areas. Locations are not the part of any national parks or protected areas, thus do not require any specific permits. It is further to confirm that the field studies did not involve endangered or protected species.

Plant growth and stress treatments

Salicornia brachiata seeds were harvested from dried plants collected from the coastal area near Bhavnagar (Latitude 21° 45′N, Longitude 72° 14′E), Gujarat, India. The seeds were germinated in plastic pots containing garden soil and the plants were grown in natural conditions. One-month-old seedlings were carefully uprooted and transferred to hydroponic culture (½ major and minor MS stock, [35]) in a culture room with a dark/light cycle of 8/16 h at 25°C for one month. The nutrient solution was renewed twice in a week. Plants were given different stress treatments like salt stress (250 mM NaCl) and desiccation by wrapping in tissue paper for 0, 6, 12 and 24 h. Upon completion of the treatments, shoot tissues were collected, frozen in liquid nitrogen and stored at −80°C.

Cloning of *SbSI-2* gene

The EST of *SbSI-2* was made full length and characterized for its role in abiotic stress tolerance. Total RNA was extracted from salt stressed plants of *S. brachiata* by GITC method [36]. The 5′-RACE reaction was performed according to manufacturer's protocol (Invitrogen, San Diego, CA, USA). The first strand of cDNA was synthesized with a gene-specific primer GSP R1 (5′-TGATAATACATCCGGGCAGTT-3′) and Superscript RT II. The mRNA was removed with RNase H, and a homopolymeric tail was added to 3′-end of the cDNA. The dC-tailed cDNA was subjected to PCR amplification with gene specific primer GSP R2 (5′-ACCCCTGCATCTATCAACTCTG-3′) and an AAP (Abridged Anchor primer) primer (5′-GGCCACGCGTCGAC-TAGTAC(G)$_{16}$-3′) supplied with kit. Further, nested PCR amplification was performed using a nested, gene-specific primer GSP R3 (5′-AGGGTTAGGGCAAGAAAGAAAG-3′) and AUAP primer (5′-GGCCACGCGTCGA CTAGTAC-3′) supplied with kit. The amplicon was purified from agarose gel and cloned into the pGEM-T Easy vector system II (Promega, Madison, Wisconsin) and sequenced (Macrogen Inc., Seoul, South Korea).

To perform 3′-RACE reactions, the first strand of cDNA was synthesized using PK1(oligo dT primer) adapter primer (5′-CCAGTGAGCAGAGTGACGAGGACTCGAGCTCA AGC(T)$_{17}$-3′). Following synthesis of the first strand of cDNA, PCR was performed with a gene-specific primer, GSP F1 (5′-AACTGCCCG-GATGTATTATCAC-3′), and an adaptor primer, PK2 (5′-CCAGTGAGCAGAGTGACG-3′). Further, a nested PCR was setup by gene specific primer GSP F2 (5′-AAGGAAGCTCTTCTG-GAGTTGA -3′) and an adaptor primer PK3 (5′-GAGGACTC-GAGCTCAAGC-3′). The nested amplified fragments were purified from agarose gel and cloned into the pGEM-T Easy vector system II and sequenced.

The *SbSI-2* EST and the 5′- and 3′-RACE reactions were sequenced, and contiguous sequences were assembled to obtain the full-length *SbSI-2* gene. After determining the open reading frame, the full length *SbSI-2* cDNA was PCR-amplified with AccuPrime Pfx DNA polymerase (Invitrogen) in conjunction with SbSI2F (5′- CGC GGATCCATGGGATTTCATTCCTTTG - 3′) and SbSI2R (5′-CCGGAATTCTCAACAAAT CGAAT-GAAGAA-3′) primers, containing *BamH1* and *EcoRI* sites, respectively. The amplification product was then cloned into a pJET1.2/blunt cloning vector (MBI Fermentas) and sequenced (Macrogen Inc., Seoul, South Korea).

In silico analysis

The NCBI database was used as a search engine for nucleotide and protein sequences. TMpred online software was used for the prediction of transmembrane domains and ClustalW for sequence alignment. Indication of conserved domains of SbSI-2 gene was obtained by BLASTp (http://www.ncbi.nlm.nih.gov). Secondary structure prediction was carried out by Expasy tools (http://www.expasy.ch/tools/). Nuclear localization signals (NLS) of SbSI-2 protein were predicted by the WoLF pSORT [37] and CELLO Prediction server (http://cello.life.nctu.edu.tw/cgi/main.cgi). Discriminate score for being a nuclear protein, calculated from the presence of NLS motif, pat4, pat7, bipartite motif, and the amino acid composition [38]. Leucine-rich nuclear export signals (NES) were predicted by NetNES 1.1 server (http://www.cbs.dtu.dk/services/NetNES/) using a combination of neural networks and hidden Markov models. Protein-protein interaction domains were detected by PROFisis PredictProtein server (https://www.predictprotein.org/). Phosphorylation motifs were predicted by NetPhosK 1.0.

Isolation of SbSI-2 genomic clone

Genomic DNA from *Salicornia* plant was isolated using CTAB-DNA extraction method [39]. PCR was conducted to amplify the *SbSI-2* genomic fragment using SbSI2F and SbSI2R primers, which were used to amplify the complete open reading frame of *SbSI-2* from the cDNA clone. The amplicon was gel purified, cloned in pGEM-T Easy vector and sequenced.

Copy number analysis of SbSI-2 gene by southern blotting

Southern blotting was performed to determine *SbSI-2* gene copy number. Genomic DNA (20 μg) from *S. brachiata* was digested with *EcoRI*, *HindIII*, and *SmaI* separated by electrophoresis (0.8% agarose gel) and transferred onto a Hybond N+ membrane (Amersham Pharmacia, UK) using alkaline transfer buffer (0.4 N NaOH with 1 M NaCl). Blot was hybridized with PCR generated probe for *SbSI-2* gene labeled with DIG-11-dUTP, following the manufacturer's user guide (Roche, Germany). Pre-hybridization and hybridization were carried out at 42°C overnight in DIG EasyHyb buffer solution (Roche, Germany). The hybridized membrane was detected by using CDP-Star chemiluminescent as substrate, following manufacturer user guide (Roche, Germany) and signals were visualized on X-ray film after 30 min.

Subcellular localization of SbSI-2 protein

A translational fusion of SbSI-2 with RFP (red fluorescent protein) was made using Gateway technology [40]. The full length *SbSI-2* cDNA was PCR-amplified with AccuPrime Pfx DNA polymerase in conjunction with SbSI2CAF (5′-CACCATGG-GATTTCATTCCTTTG-3′) and SbSI2CAR (5′-TCAA-CAAATCGAATGAAGAA-3′) primers. The blunt-end PCR product was then cloned into a pENTER/D-TOPO Entry vector (Invitrogen, USA) and sequenced. Thereafter, the LR recombination reaction was performed between an attL-containing Entry clone pENTER/D-TOPO-SbSI2 vector and an attR-containing destination vector pSITE-4CA by Gateway LR Clonase II enzyme mix (Invitrogen, USA). The resulting LR reaction was used to transform *E. coli* DH5α cells. Colonies growing on streptomycin containing media were checked for insertion of *SbSI-2* gene in Destination vector by PCR amplification. The resulting fusion construct (expression clone) was isolated and insertion of gene was confirmed through sequencing. The fusion construct (RFP:SbSI-2) was transferred into onion epidermal cells by particle bombardment with gene gun (PDS-1000/He Biolistic, Biorad, USA). The pSITE-4CA (RFP) vector was used as control. After incubation on MS plate for 12–24 h, the onion epidermal cells were observed for transient expression of RFP with an epifluorescence microscope (Axio Imager, Carl Zeiss AG, Germany). DAPI staining used as standard control for nuclear localization.

Quantitative RT-PCR analysis

Total RNA was isolated from control and treated plant samples from *S.brachiata* using GITC method [36] and quantified using ND-1000 spectrophotometer (Nanodrop technologies, USA). The cDNA was prepared using 5 μg total RNA by Superscript RT III first-strand cDNA synthesis kit (Invitrogen, San Diego, CA). Real-time qPCR was performed on a Bio-Rad IQ5 detection system (Bio-Rad, U.S.A.) with QuantiFast Kit (Qiagen, USA). The PCR reactions was carried out containing 5 pmol of gene specific primers (forward 5′-CCCAGAAAGAAAAAGGCAAGA-3′ and reverse 5′-CTCCAGAA GAGCTTCCTTTGC-3′) and β-tubulin (forward 5′-GGAGTCACCGAGGCAGAG-3′ and reverse 5′-ATCACATATCAGAAACCACAA-3′) at 95°C-5 min followed by 95°C-10 s, 60°C-30 s and 72°C-30 s for 40 cycles, and continued for melting curve analysis to check the specificity of PCR amplification. The amplified product was resolved on a 1% agarose gel to check specificity of PCR product. Experiments were repeated twice independently. Fold changes were calculated using the CT method. CT values for individual variants were compared to CT values for a reference control (β-tubulin) for all treated samples and data was analysed using untreated plants at every time point as baseline control [41].

The expression patterns of reactive oxygen species (ROS) related genes (*NtSOD*, *NtAPX*, *NtCAT*) and some stress-responsive TFs (*NtDREB2* and *AP2*-domains containing TF) were also analyzed by qRT-PCR in both transgenic and WT plants after salt stress. Gene-specific primer pairs of ROS-related genes (*NtSOD*, *NtAPX* and *NtCAT*; primers sequence taken from Huang et al. [42]) and stress-responsive TFs *NtDREB2* (DREB2F 5′-GCCGACGCTAAGGATA TTCA-3′ and DREB2R 5′-TGCAAAACAGAGCTTCCTCA-3′) and *AP2*-domains containing TF (AP2dF 5′-AAGGGCGAGGAAGAACAAAT-3′ and AP2dR 5′-GTGGCTCTGGAA AGTTGA-3′) were utilized for expression studies, whereas QACTF (5′-CGTT TGGATC-TTGCTGGTCGT-3′) and QACTR (5′-CAGCAATG CCAGG-GAACATAG-3′) primers were used for actin. qRT-PCR reactions were carried out as described above and repeated three times to ensure reproducibility.

Cloning of SbSI-2 cDNA in pET28a expression vector and recombinant protein expression

The *SbSI-2* gene was excised from pJET1.2-SbSI-2 vector using *BamH1* and *EcoR1* restriction endonucleases and cloned in pET28a vector. *E. coli* BL21 (DE3) cells were transformed with recombinant plasmid (pET28a-SbSI-2) or pET28a vector alone. The recombinant protein was expressed by adding 1 mM IPTG at 0.6 OD_{600}. Recombinant protein production was induced after 2 h of treatment and reached maximum at 6 h.

Functional validation of SbSI-2 gene in E. coli BL21 (DE3) cells under salt and desiccation stresses

Spot assay. A spot assay was carried out to ascertain the function of *SbSI-2* in *E. coli* cells. BL21 (DE3) cells were transformed with recombinant plasmid (pET28a-SbSI-2) and vector alone. Cells were grown in LB medium to 0.6 OD_{600}. Thereafter, 1 mM IPTG was added and cells were grown for 12 h at 30°C. Next day cultures were diluted to 0.6 OD_{600}, and then diluted to 10^{-3}, 10^{-4} and 10^{-5}. Ten microliters from each dilution was spotted on LB basal plates or supplemented with 500 mM NaCl, 500 mM KCl or 600 mM Mannitol.

Liquid culture assay. Functional analysis was also carried out in liquid culture using LB basal medium, as well as supplemented with NaCl, KCl, PEG and mannitol. *E. coli* BL21 (DE3) cells with recombinant plasmid or vector alone were grown as mentioned above, diluted to 0.6 OD_{600} and 400 μl cells were inoculated in 10 ml LB medium containing 500 mM NaCl,

500 mM KCl, 10% PEG (6000) and 600 mM mannitol, and incubated at 30°C. The bacterial suspension was harvested at every 2 h till 12 h and OD_{600} was measured.

Construction of plant transformation vector and tobacco transformation

To perform plant transformation, *SbSI-2* cDNA was PCR-amplified with AccuPrime Pfx DNA polymerase in conjunction with SbSI2PF (5′-TCCGAGCTCATGGGATTTCATTCCTT-TG-3′) and SbSI2PR (5′-CGCGGATCC TCAACAAATCGA-ATGAAGAA-3′) primers, which contained *SacI* and *BamHI* sites, respectively. The *SbSI-2* gene was cloned into the pRT100 vector [43] to add the *35S* promoter and terminator. The pRT100 plant expression vector contains strong and constitutive *35S* promoter from cauliflower mosaic virus and ampicillin resistance gene for bacterial selection. The *SbSI-2* amplicon was digested with *SacI* and *BamHI* restriction endonucleases. The pRT100 vector was also linearized by using the same set of restriction endonucleases and ligated overnight with *SacI/BamHI* digested *SbSI-2* at 8°C, which places *SbSI-2* under the control of *CaMV 35S* promoter and polyadenylation signal. The recombinant pRT100 vector was further transformed in *E. coli* DH5α cells. Colonies growing on ampicillin containing media were checked for insertion of *SbSI-2* gene by PCR amplification using gene specific primers. The above gene construct (pRT100-SbSI-2) was digested with the double cutter restriction enzyme *PstI* to get the entire expression cassette containing the *CaMV 35S* promoter, the *SbSI-2* gene, and the terminator. Thereafter, the entire expression cassette (35S-SbSI-2-terminator) was cloned into the pCAMBIA2301 binary vector at the *PstI* site. The resulting vector was mobilized into *Agrobacterium tumefaciens* (EHA 105) and used to transform tobacco (*Nicotiana tabacum* cv. Petit Havana) plants according to a standard protocol [44]. Putative transgenic plants regenerated directly from leaf edges in the presence of kanamycin were transferred to culture bottles that contained MS basal medium supplemented with kanamycin (100 mg/l). Transgenic lines were screened via GUS assay and PCR amplification analysis. Seeds of the selfcrossed transgenic lines were harvested for subsequent experiments.

Confirmation of putative transgenic tobacco plants

Confirmation by PCR analysis. Genomic DNA was isolated from different lines via the CTAB (N-cetyl-N,N,N-trimethylammonium bromide) method [39]. To verify the presence of the transgene, PCR was conducted with gene-specific primers and *gus*-specific primers (gusAF 5′-GATCGC-GAAAACTGTGGAAT-3′ and gusAR 5′-TGAGCGTC GCA-GAAC ATTAC-3′). PCR products were analyzed on 0.8% agarose gel with appropriate size DNA marker.

Confirmation by histochemical GUS staining. GUS activity was visualized in leaf tissue with a β-glucuronidase Reporter Gene staining kit (Sigma, USA). Seedlings from transgenic plants were dipped into GUS staining solution, vacuum infiltrated for 2 min and then incubated overnight at 37°C in the dark. The tissues were then rinsed with 80% ethanol for 4 h to overnight to remove chlorophyll.

Confirmation of transgene integration and determination of copy number. Transgene integration and copy number was determined by Southern hybridization; for this, 20 μg of genomic DNA from each transgenic lines and WT plants were digested with *Hind III*. Digested DNA fragments were separated on 0.8% agarose and blotted onto a Hybond (N^+) membrane (Amersham Pharmacia, UK) using alkaline transfer buffer (0.4 N NaOH with 1 M NaCl). A DIG-11-dUTP labeled gene specific DNA probe was synthesized by PCR according to the manufacturer's protocol

(Roche, Germany). Hybridization was carried out at 42°C overnight in DIG EasyHyb buffer solution (Roche, Germany). The hybridized membrane was detected using CDP-Star as substrate (Roche, Germany) and signals were visualized on X-ray film.

***SbSI-2* transgene expression analysis by semi-quantitative RT-PCR.** To check the mRNA levels of overexpressed *SbSI-2* gene in transgenic plants, semi-quantitative RT-PCR was carried out. Total RNA was isolated from WT and transgenic plant samples using GITC buffer and was quantified with a ND1000 spectrophotometer (Nanodrop Technology, USA). The cDNA was prepared using 5 μg of total RNA with a SuperScript RT III first-strand cDNA synthesis kit. The synthesized cDNA (1 μl, diluted 1:5) was used as a template, and actin was used as an internal control for RT-PCR analysis. The *SbSI-2*-specific primer pair, RTF (5′- CCCAGAAAGAAAAAGG-CAAGA-3′) and RTR (5′-CTCCAGAAGAGCTTCCTTTGC-3′), was utilized for expression study of the *SbSI-2*, whereas QACTF (5′-CGTTTGGATCTTGCTGGTCGT-3′) and QACTR (5′-CAGCAATG CCAGGGAACATAG-3′) primers were used for actin. PCR reactions were carried out in 1× PCR buffer supplemented with 200 μM dNTPs, 1.25 U Taq DNA polymerase and 5 pmol of each of the gene-specific primers according to the following conditions: an initial denaturation at 95°C for 3 min, 25 cycles at 94°C for 30 s, 60°C for 30 s and 72°C for 30 s, followed by a final extension at 72°C for 7 min. RT-PCR experiments were repeated three times, and the amplified products were analyzed via agarose gel electrophoresis.

Evaluation of transgenic plants exposed to salt and osmotic stress

T_1 transgenic lines were confirmed by PCR amplification with gene specific primers and *gus*-specific primers. To analyze the stress tolerance of *SbSI-2*-overexpressing tobacco plants, seeds were germinated in MS medium supplemented with 0, 200 mM NaCl (salt stress) and 300 mM mannitol (osmotic stress) in culture room conditions. The percentage of seed germination was scored 18 days after seed inoculation. T_1 seedlings were also analyzed for different growth parameters under salt and osmotic stresses. At eight days, kanamycin-positive seedlings were transferred to MS medium supplemented with 0, 200 mM NaCl or 300 mM mannitol in petri dishes. Shoot length, root length, leaf surface area, fresh weight (FW), dry weight (DW) and relative water content (RWC) of the seedlings were measured after 30 days for salt and osmotic stress. Seedling tissues were collected after 45 days of salt and osmotic stress and subjected to various physiological and biochemical analyses. To study the growth for longer duration, the WT and transgenic lines seeds were first germinated on the MS basal medium, and after one week of germination, the kanamycin-positive seedlings were transferred in jars on the MS basal medium or supplemented with 200 mM NaCl.

Chlorophyll estimation. Seedling chlorophyll content of transgenic and WT plants, grown under different stresses, were estimated according to Arnon [45] and chlorophyll content was calculated per gram of fresh tissue weight [46].

Electrolyte leakage. Electrolyte leakage was measured as described by Lutts et al. [47].

Measurement of proline content. Free proline content in the seedlings was determined using acid ninhydrin as previously described by Bates et al. [48]

In vivo localization of O_{2-} and H_2O_2 content in the transgenic seedlings. *In vivo* detection of O_2^- and H_2O_2 was accomplished by histochemical staining with nitro blue tetrazolium (NBT) and 3,3′- diaminobenzidine (DAB) as described by Shi et al. [49].

Na⁺, K⁺ and Ca²⁺ ion content analysis. Ion content was determined via the method described by Shukla et al. [50].

Statistical analyses

Each experiment was performed three times with three replicates and data from fifteen plants were recorded in each replicates. One-way ANOVA between subject factors was performed by ezANOVA (http://www.cabiatl.com/mricro/ezanova/) for analysis of variance to determine the least significant difference between means. Mean values that were significantly different at $p \leq 0.05$ within treatment from each other are indicated by different letters (a, b and c).

Results

Isolation and sequence analysis of *SbSI-2* cDNA

Previously, we have identified about 1000 ESTs in response to salt stress, from the extreme halophyte *Salicornia brachiata*. Among these, *SbSI-2* EST (Gen-Bank accession number EB485109) was made full length using the 5′ and 3′ RACE. The *SbSI-2* cDNA (Gen-Bank accession number JX872273) was 537 bp long, contained a 26 bp 5′ UTR, a 423 bp open reading frame and an 88 bp 3′ UTR region (Fig. S1A). The cDNA encoded a polypeptide of 140 amino acid residues with a predicted molecular mass of 15.93 kDa and an isoelectric point of 10.34. *SbSI-2* did not reveal homology with known gene by NCBI protein blast analysis and showed matching with unknown/hypothetical genes. Hydropathicity analysis by TMpred program (http://www.ch.embnet.org/cgi-bin/TMPRED_form_parser) showed that SbSI-2 has no transmembrane domains. The secondary structure of SbSI-2 was analyzed by PSIPRED protein structure prediction software showed that peptides contain 2 alpha helixes, 5 extended strands and 8 random coils (Fig. S1B).

ScanProsite (http://au.expasy.org/) has revealed three distinct regions (1–60, 61–120 and 121–140) based on 3 distinct amino acids sequence patterns (Fig. S2). The WoLF PSORT program revealed nuclear localization signals (Fig. S3 A) in SbSI-2 protein. A possible cleavage site is located between amino acids 24 and 25. The SbSI-2 protein contained three pat4 motifs, two pat7, two bipartite motifs showing NLS Score above 2 (Fig. S3 A). CELLO Prediction server also predicted SbSI-2 to be a nuclear protein (Fig. S3B). NetNES 1.1 predicted a leucine-rich nuclear export signal (NES, 131–136 aa) in SbSI-2 (Fig. S3 C, D). PROFisis Predict protein revealed that SbSI-2 has a strong protein-protein interaction domain (97–116, Fig. S4). SbSI-2 shows the possibility of phosphorylation with PKC (Protein kinase C), PKA (Protein kinase A), cdc2 and CKI (Caseine kinase I). Highest Score was 0.92 for PKC at position T-55 (Fig. S5).

Genomic organization study

We have amplified *SbSI-2* ORF from genomic DNA and cDNA. Both resulting PCR products were same size on agarose gel (Fig. 1A). The amplified fragments were purified from agarose gels, cloned into a pGEM-T Easy vector and sequenced. Comparison of the genomic clone sequence with cDNA clone showed that *SbSI-2* gene has single exon structure, which is also called intronless gene.

Copy number of *SbSI-2* gene

Southern analysis was undertaken to detect the copy number of *SbSI-2* in the *S.brachiata* genome. It was observed that *SbSI-2* probe hybridized to only single fragments of genomic DNA, digested

with different restriction enzymes (*EcoRI*, *HindIII*, and *SmaI*). Southern blot revealed the presence of a single copy *SbSI-2* gene in *S. brachiata* genome (Fig. 1B).

The SbSI-2 protein is localized in the nucleus

In silico sequence analysis revealed that the SbSI-2 protein has nuclear localization signals. To corroborate this we tested *in vivo* subcellular localization by transient expression assays using onion epidermal cells with pSITE-4CA constructs expressing RFP alone and the RFP:SbSI-2 fusion construct (Fig. 2A). When onion cells were transformed with RFP alone, red fluorescence signals were distributed evenly in the entire cell region, whereas in RFP:SbSI-2 fusion construct the fluorescence was accumulated in the nucleus only (Fig. 2B). These results indicate that SbSI-2 is localized in the nucleus.

Differential expression of *SbSI-2* transcripts under salt and desiccation stresses

Expression analysis of *SbSI-2* gene was carried out by quantitative real-time PCR using cDNA from salt (NaCl) and desiccation treated plants for different time period (0, 6, 12 and 24 h). In the presence of 250 mM NaCl, *SbSI-2* transcript increased 1.5 to 4-fold (Fig. 2C) and under desiccation condition the transcript was up-regulated 2 to 70-fold (Fig. 2D). *SbSI-2* showed maximum fold change in desiccation conditions.

Expression analysis of SbSI-2 protein in *E. coli* by SDS-PAGE

The recombinant protein was expressed by adding 1 mM IPTG at 0.6 OD_{600}. The recombinant protein was induced after 2 h of treatment and reached maximum at 6 h (Fig. S6A). Presence of recombinant protein was also confirmed during liquid assay experiment after 12 hours of growth (Fig. S6B).

Overexpression of novel *SbSI-2* in *E. coli* enhances growth during salt and osmotic stresses

Spot assay. pET28a-SbSI-2 recombinant cells were spotted on LB basal medium and medium supplemented with NaCl, KCl, and mannitol (Fig. 3). Recombinant (pET28a-SbSI-2) and control cells showed similar growth on LB medium in overnight grown culture (Fig. 3A). In NaCl treatment pET28a-SbSI-2 recombinant *E. coli* cells showed slightly high growth, whereas in KCl similar growth was observed compared to vector alone (Fig. 3B, C). In desiccation treatment pET28a-SbSI-2 recombinant *E. coli* cells showed significantly better growth compared to vector alone (Fig. 3D).

Liquid assay. Growth was also analyzed in LB liquid medium; 400 µl aliquots of pET28a-SbSI-2 recombinant and control *E. coli* BL21 (DE3) cells were inoculated in 10 ml LB liquid medium and medium supplemented with NaCl, KCl, PEG and Mannitol (Fig. 3E–I). In LB liquid medium pET28a-SbSI-2 recombinant cells and vector alone (pET28a) showed similar growth at different time points (Fig. 3E). In NaCl treatment pET28a-SbSI-2 recombinant *E. coli* cells showed higher growth 8 h after inoculation, whereas in KCl similar growth was observed compared to vector alone (Fig. 3F, G). In desiccation treatment pET28a-SbSI-2 recombinant *E. coli* cells showed better tolerance compared to vector alone (Fig. 3H, I). In the presence of 10% PEG, bacterial growth was similar up to 2 h after inoculation, but it was significantly increased in pET28a-SbSI-2 recombinant *E. coli* cells thereafter (Fig. 3H). Mannitol inhibited the growth until 4 h in both control and pET28a-SbSI-2 recombinant cells; however, after 4 h growth was significantly improved in

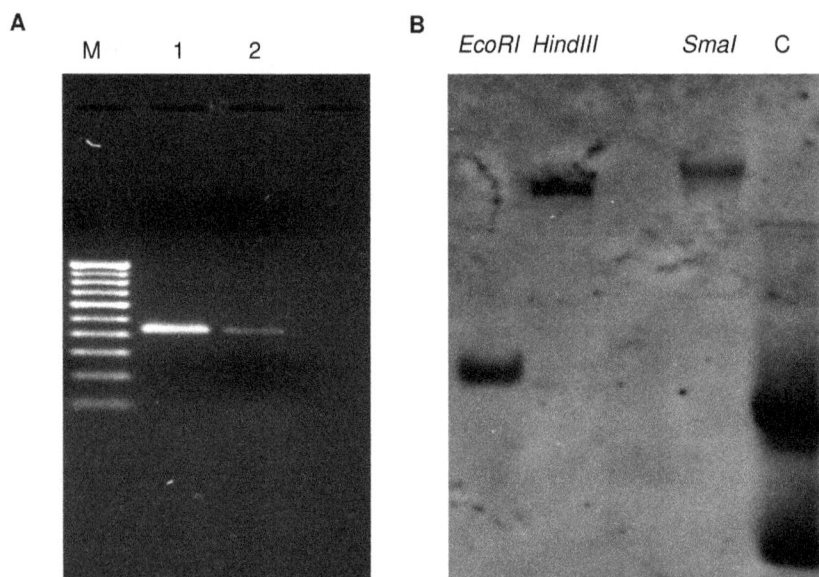

Figure 1. Genomic organization analysis. (A) Genomic organization study, Lane 1: Amplified *SbSI-2* PCR product from cDNA Lane 2: Amplified *SbSI-2* PCR product from genomic DNA M: 100 bp DNA ladder (B) Southern blot of *SbSI-2* gene from *Salicornia brachiata* genomic DNA. C: positive control (*SbSI-2* gene cloned in pGEMT-T Easy vector).

pET28a-SbSI-2 recombinant cells compared to control cells (Fig. 3I). The liquid culture assay data showed a similar pattern of results observed with spot culture assays.

Overexpression of *SbSI-2* enhances salinity tolerance of transgenic tobacco plants

The pCAMBIA2301-35S:SbSI-2 construct (Fig. 4A) was introduced into tobacco plants for *in vivo* functional characterization of *SbSI-2* gene. Putative transgenic lines were selected on kanamycin-containing medium and were subsequently verified by GUS analysis. GUS-positive transgenic lines were further verified by PCR analysis with gene specific primers and *gus*-specific primers. Thirty nine GUS and PCR positive individual transgenic lines derived from independent transgenic events were subsequently transferred to plastic pots containing garden soil and further to earthen pots after 15 days of hardening. There were no morphological differences observed between transgenic lines and WT plants under normal conditions. Seeds of 35S:SbSI-2 transgenic plants exhibited the expected 3:1 ratio of Kan^r/Kan^s during germination in kanamycin-containing medium. Three independent transgenic lines (L11, L17 and L22) were selected on the basis of GUS intensity and were further analyzed for *SbSI-2* transgene expression via semi-quantitative RT-PCR (Fig. 4B, C). The *SbSI-2*-overexpressing transgenic lines showed different levels of *SbSI-2* expression, whereas the expression of *SbSI-2* was not observed in WT plants (Fig. 4C). The L17 transgenic line exhibited maximum expression of *SbSI-2* gene (Fig. 4C). Southern blot analysis of transgenic lines L11, L17 and L22 showed single copy insertion of *SbSI-2* (Fig. 4D).

To study the effect of salt stress on germination, WT and transgenic lines seeds (L11, L17 and L22) were germinated in MS medium supplemented with 0 or 200 mM NaCl. Under normal conditions (0 mM NaCl) there was no difference observed between WT and transgenic seeds (Fig. 4E–G). The efficiency of germination was reduced under NaCl stress for both WT and transgenic seeds. However, transgenic seeds exhibited better germination efficiency than WT seeds under 200 mM NaCl (Fig. 4F, G). In addition to seed germination assays, the growth of

transgenic seedlings exposed to salt stress condition was also examined (Fig. 5A, B). Seeds of WT and transgenic tobacco were allowed to germinate in MS medium for 8 days. Subsequently, seedlings were transferred to medium containing 0 or 200 mM NaCl. Transgenic lines of L11 and L17 exhibited significant enhancements in root and shoot length relative to WT seedlings (Fig. 6A, B). All three transgenic lines exhibited significant difference in leaf area relative to WT seedlings (Fig. 6C). Transgenic lines exhibited significant increases in fresh weight (FW), dry weight (DW) and relative water content relative to WT (Fig. 6D–F). Transgenic lines also exhibited better growth than their WT counterparts when subjected to salt stress in culture jars for long period (Fig. 5C, D).

Overexpression of *SbSI-2* led to higher chlorophyll content, reduction in electrolyte leakage and increase in accumulation of compatible osmolytes under salt stress

The Chlorophyll content of WT and transgenic seedlings was similar under non-stress conditions and decreased upon salt stress in both WT and transgenic lines (Fig. 7A). However, the transgenic lines showed less reduction in chlorophyll content than WT seedlings (Fig. 7A). Transgenic seedlings also exhibited significantly reduced electrolyte leakage relative to WT under salt stress (Fig. 7B). Proline functions as osmoprotectent and can prevent cell dehydration and enhance stress tolerance in plants [51]. At normal condition, proline content was almost equal in WT and transgenic seedlings; however, in the presence of 200 mM NaCl, the transgenic seedlings had higher proline content relative to WT (Fig. 7C).

Overexpression of *SbSI-2* gene reduced accumulation of reactive oxygen species (ROS) under salinity stress

WT seedlings exhibited more NBT and DAB staining than transgenic seedlings after salt stress (Fig. 7D, E). These results demonstrated that WT seedlings accumulated more O_2^- and H_2O_2 relative to transgenic seedlings, confirming that *SbSI-2* helps to minimize oxidative stress in plants.

A

2x35S:RFP

2x35S:RFP-SbSI-2

B

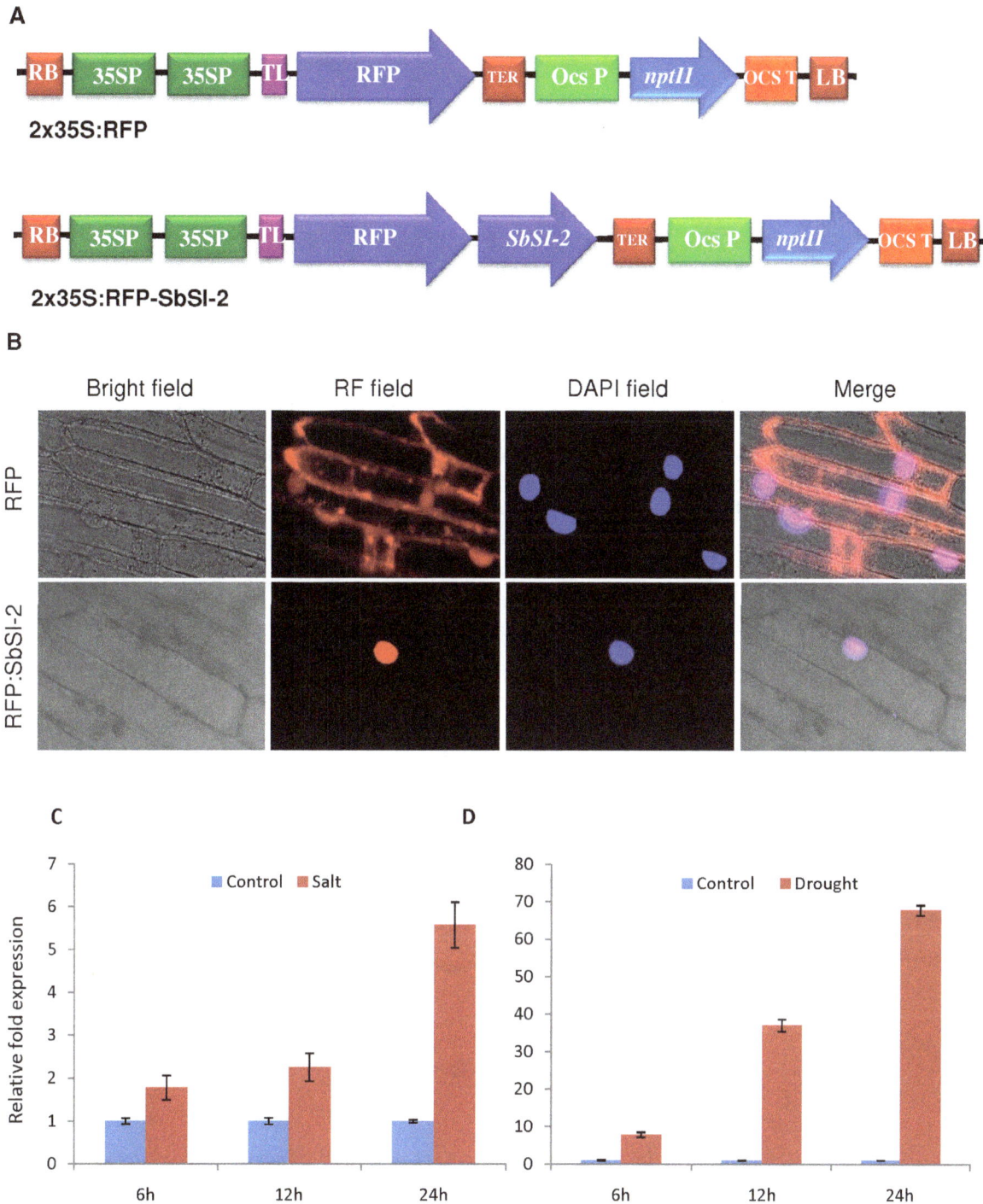

Figure 2. Subcellular localization of RFP:SbSI-2 fusion protein in onion epidermal cells. (A) Schematic representation of the pSITE-3CA-2X35S:RFP:SbSI-2 construct (RFP:SbSI-2) used for transient expression. (B) Cells with constructs expressing red fluorescence protein (RFP) alone and the RFP:SbSI-2 fusion protein were analyzed under bright and red fluorescence field. (C–D) Quantitative real-time PCR analysis of *SbSI-2* under salt and desiccation conditions for different time period in *S.brachiata*. The relative fold expression of *SbSI-2* at different time points under stress was calculated using the Ct value of untreated plants (control plant) at respective time points.

Ion content analysis of *SbSI-2*-overexpressing tobacco plants under salt stress

The Na$^+$, K$^+$ and Ca^{2+} content was measured in transgenic and WT seedlings grown in 0 mM and 200 mM NaCl. Under non-stress conditions transgenic and WT plants exhibited almost equal Na$^+$ content (Fig. 8A). After salt stress, seedling tissues exhibited increased Na$^+$ content in both transgenic lines and WT, however transgenic seedlings accumulated lower Na$^+$ compared to WT seedlings (Fig. 8A). Transgenic as well as WT seedlings showed reduction in K$^+$ content under NaCl stress. However, transgenic lines L11 and L17 exhibited higher K$^+$ ion content relative to WT

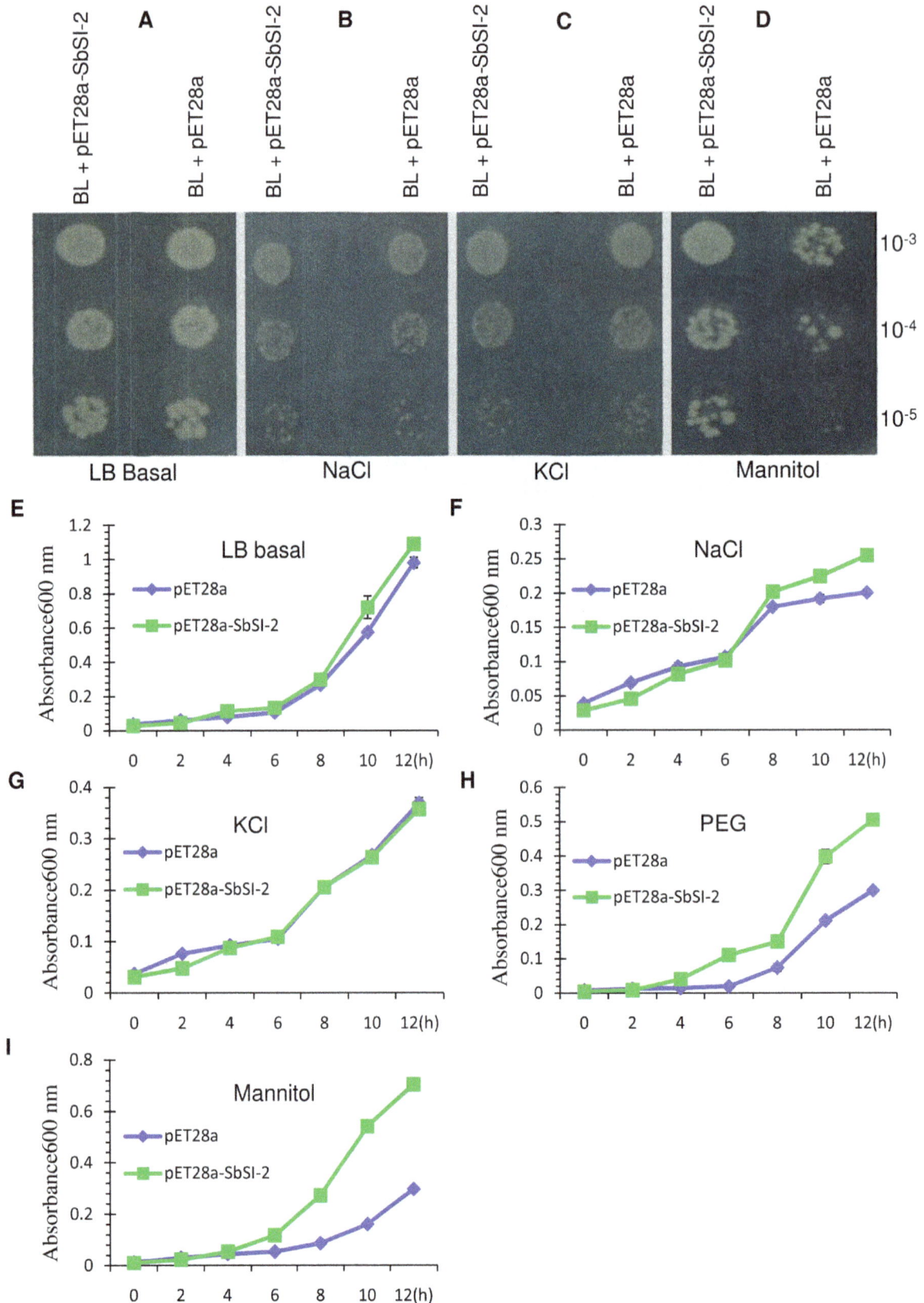

Figure 3. Growth analysis of recombinant *E. coli* cells having *SbSI-2*. (A–D) Spot assay of BL21 (DE3)/pET28a-SbSI-2 and BL21 (DE3)/pET28a on LB basal plates and LB supplemented with NaCl, KCl and Mannitol. Ten microliters from 10^{-3} to 10^{-5} dilutions were spotted on (A) LB basal plates, (B) LB supplemented with 500 mM NaCl, (C) 500 mM KCl, and (D) 600 mM Mannitol. (E–I) Growth analysis of novel gene *SbSI-2* was carried out in LB liquid medium with different supplements. (E) LB medium, (F) 500 mM NaCl, (G) 500 KCl, (H) 10% PEG, and (I) 600 mM Mannitol. O.D$_{600}$ was recorded

at 2 h interval up to 12 h and mean values are represented in graph.

seedlings under salt stress (Fig. 8B). The L11 and L17 transgenic seedlings showed an improved K$^+$/Na$^+$ ratio at NaCl stress conditions relative to WT (Fig. 8C). The transgenic lines showed an increase in Ca^{2+} content as compared to WT under salt stress (Fig. 8D). It is observed that the changes in ion content are caused by NaCl both in WT and transgenic lines. However, the mode and

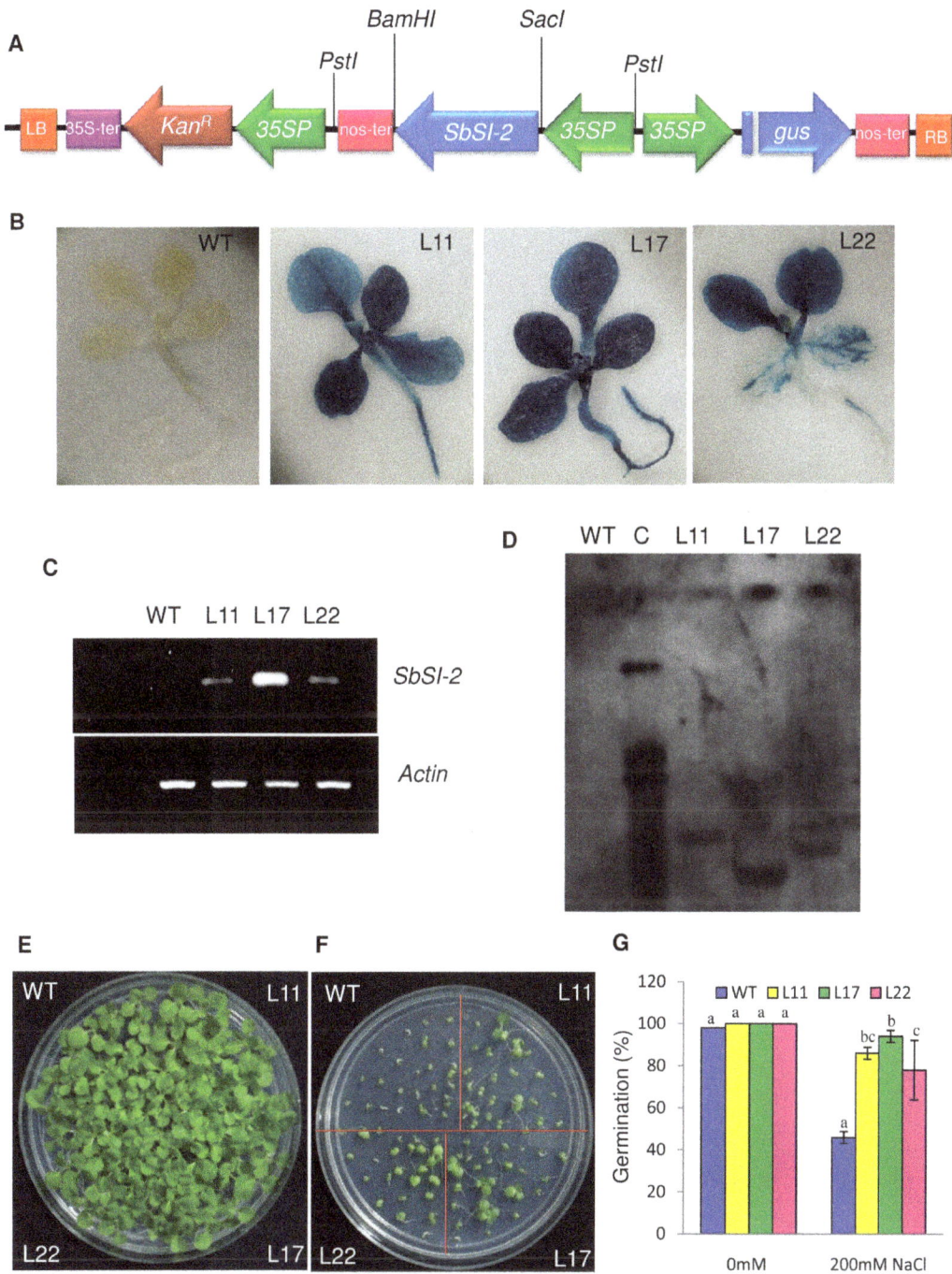

Figure 4. Confirmation of transgenic tobacco plants. (A) Schematic representation of the pCAMBIA2301-35S:SbSI-2 construct used to transform tobacco plants with the SbSI-2 gene, (B) GUS assay of seedlings, showing positive GUS expression in the transgenic lines, (C) Transcript levels of the SbSI-2 gene in transgenic lines and WT plants via semi-quantitative RT-PCR, (D) Southern analysis of transgenic lines, (E–F) Germination of seeds from transgenic lines (L11, L17 and L22) and WT plants in (E) 0 mM, and (F) 200 mM NaCl and (G) Graphs represent the percentage germination of transgenic lines (L11, L17 and L22) and WT plants in salt stress and normal condition. Mean values that were significantly different at p≤0.05 within treatment from each other are indicated by different letters (a, b and c).

Figure 5. Phenotypic comparison of the growth of WT and transgenic lines overexpressing the _SbSI-2_ gene under salt stress. (A–B) Growth comparison of transgenic lines (L11, L17 and L22) and WT seedlings after 30 days in (A) 0 mM, and (B) 200 mM NaCl. (C–D) Growth of whole plants from transgenic lines (L11, L17 and L22) and WT plants in (C) 0 mM and (D) 200 mM NaCl in culture bottles.

magnitude of change are different in wild type and transgenic lines. For example, the decrease in Ca^{++} ion content in WT is 67% and that in L11, L17 and L22 is 59%, 60% and 53%, respectively (Figure 8). Therefore, the observed difference in transgenic lines vis-à-vis WT may be due to _SbSI-2_.

SbSI-2 expression conferred osmotic tolerance in transgenic tobacco

We performed osmotic tolerance tests in both WT and transgenic plants to gain a better understanding of _SbSI-2_ function under dehydration. To study the effect of osmotic stress on germination, seeds of WT and transgenic lines (L11, L17 and L22) were germinated in MS medium supplemented with 0 or 300 mM mannitol. Under non-stress conditions, transgenic and WT seeds

showed almost equal germination efficiency (Fig. 9A, C). At 300 mM Mannitol, transgenic seeds showed higher germination than WT seeds (Fig. 9B, C). In addition to seed germination assays, the growth of transgenic seedlings exposed to osmotic stress condition was also examined (Fig. 9D, E). Seeds of WT and transgenic tobacco were allowed to germinate in MS medium for 8 days. Subsequently, seedlings were transferred to medium containing 0, or 300 mM mannitol.

All transgenic seedlings exhibited significant difference in shoot length, root length and leaf area relative to WT under osmotic stress (Fig. 10A, B, C). Transgenic seedlings exhibited significant increases in both FW and DW relative to WT (Fig. 10D, E). WT seedlings showed signs of dehydration in the presence of mannitol

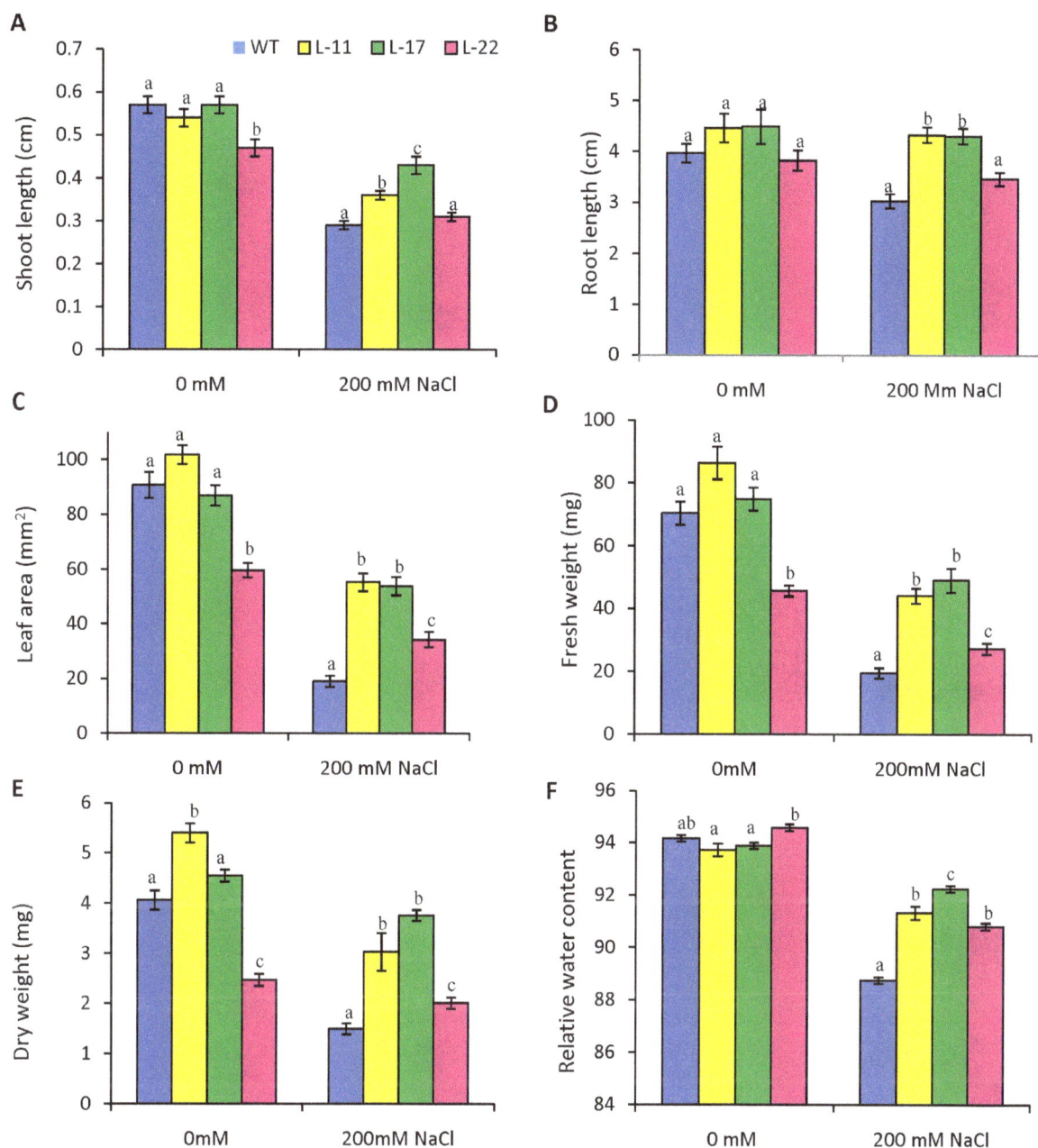

Figure 6. Comparison of growth parameters of seedlings from transgenic lines (L11, L17 and L22) and WT in 0 mM, and 200 mM NaCl. (A) shoot length, (B) root length, (C) leaf area, (D) fresh weight, (E) dry weight and (F) relative water content (RWC). Mean values that were significantly different at p≤0.05 within treatment from each other are indicated by different letters (a, b and c).

stress, whereas the transgenic seedlings had better water status (Fig. 10F).

SbSI-2 transgenic lines showed higher chlorophyll content, lower electrolyte leakage and higher accumulation of proline in response to osmotic stress

After osmotic stress, chlorophyll content reduced in both transgenic and WT seedlings compared to non-stress conditions (Fig. 11A). However, L11 and L22 showed significantly higher chlorophyll content compared to WT plants upon osmotic stress (Fig. 11A). During osmotic stress, transgenic seedlings exhibited significantly reduced electrolyte leakage as compared to WT

seedlings (Fig. 11B). Proline content was also measured in osmotic stress conditions. At non-stress condition, proline content was almost similar in WT and transgenic seedlings; however, in the presence of 300 mM mannitol, L11 and L17 had higher proline content than WT (Fig. 11C).

Overexpression of SbSI-2 gene reduced the accumulation ROS under osmotic stress

In vivo localization study demonstrated that WT seedlings accumulated more O_2^- and H_2O_2 than transgenic seedlings under osmotic stress (Fig. 11D, E).

A

B

C

D

E

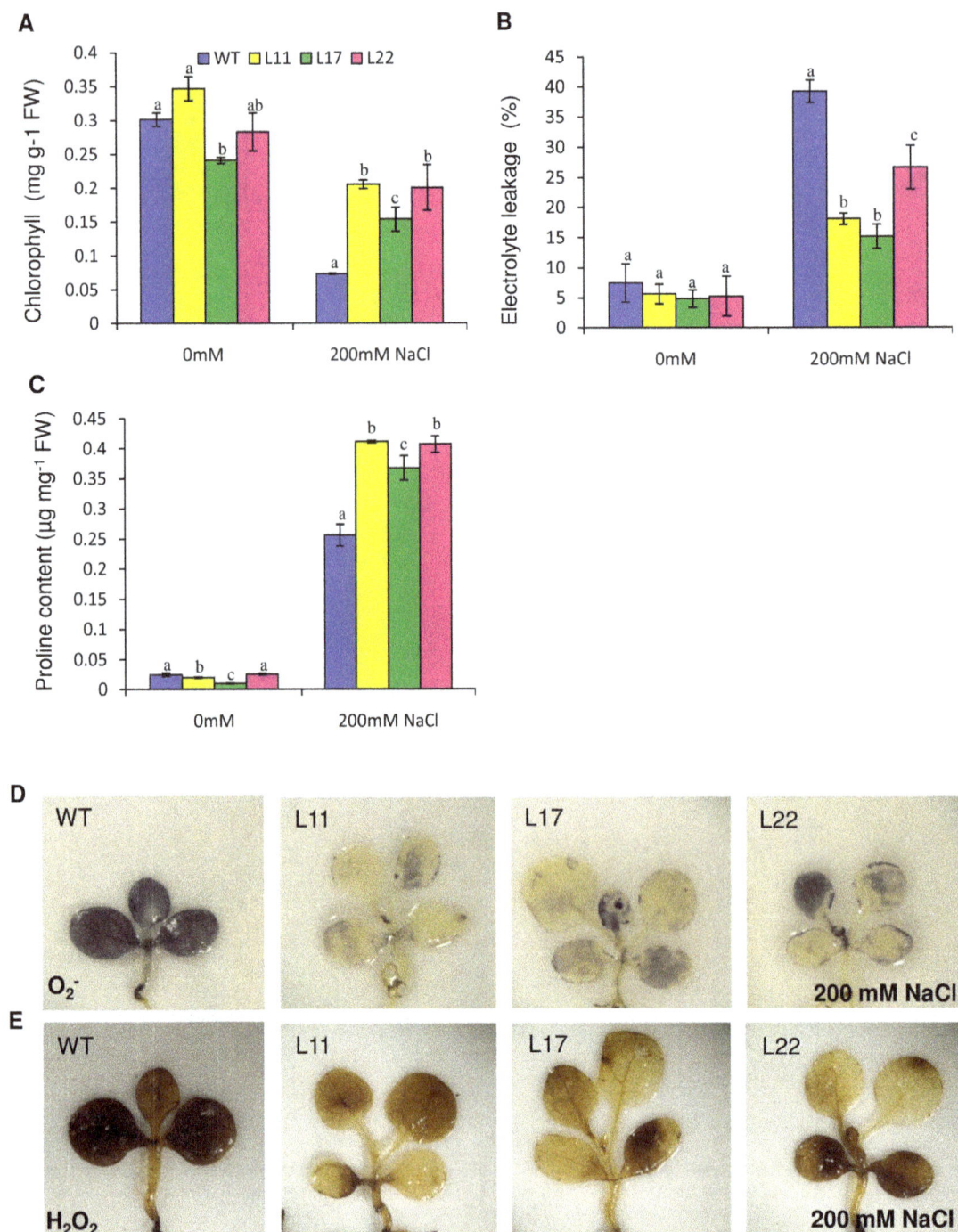

Figure 7. Comparison of various biochemical and physiological parameters of transgenic lines (L11, L17 and L22) and WT under salt stress. Chlorophyll (A), Electrolyte leakage (B), and proline (C) contents of transgenic (L11, L17 and L22) and WT seedlings grown in 0 mM, and 200 mM NaCl. (D–E) *In vivo* localization of O_2^- and H_2O_2 in seedlings of 35S:SbSI-2 transgenic lines and WT under salt stress. (D) Localization of O_2^- by NBT staining, (E) Localization of H_2O_2 by DAB staining. Mean values that were significantly different at p≤0.05 within treatment from each other are indicated by different letters (a, b and c).

Expression analysis of ROS-related genes in *SbSI-2*-overexpressing transgenic tobacco plants under salinity and osmotic stress

To gain further insight into molecular mechanism(s) underlying the enhanced salinity and osmotic tolerance of *SbSI-2*-overexpressing transgenic tobacco plants, we performed qRT-PCR

analysis of ROS-related genes. The *NtSOD*, *NtAPX* and *NtCAT* genes encode enzymes for ROS-scavenging in plants. Transcript levels of these ROS-related genes were higher in transgenic tobacco plants as compared to WT under both salt and osmotic stress conditions (Fig. 12 A–F). Under control condition, transgenic lines also showed higher expression of ROS-related

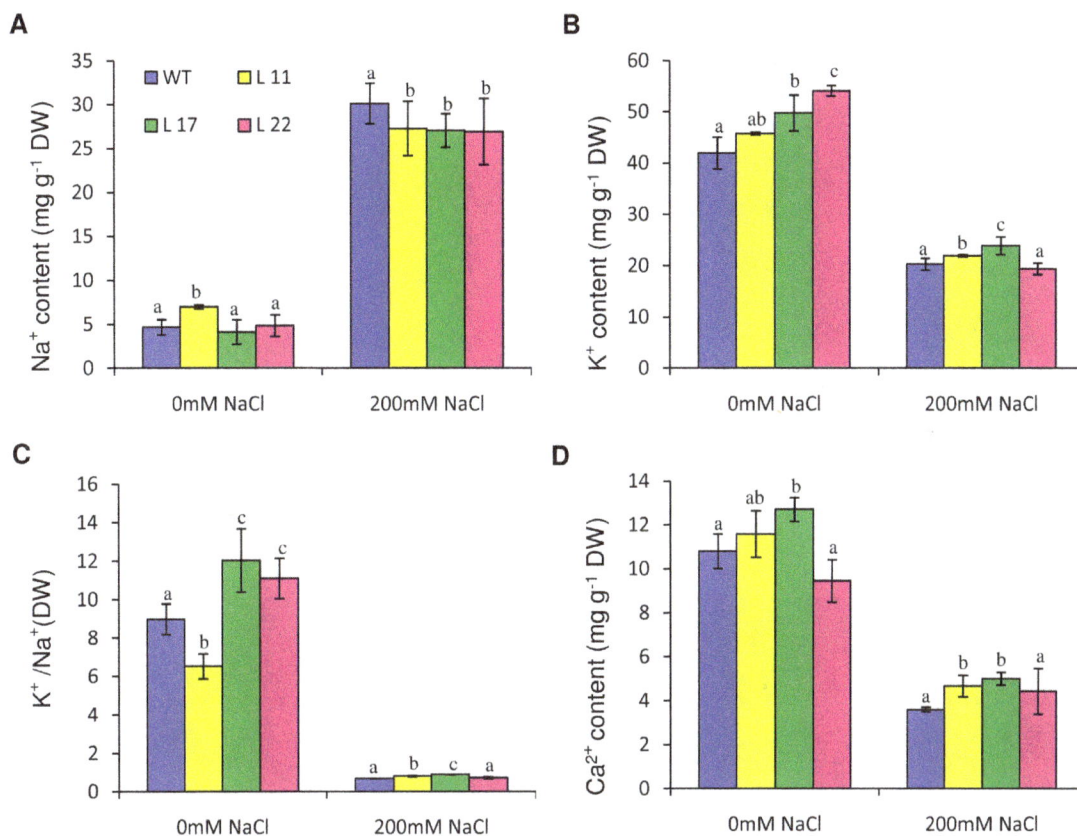

Figure 8. Ion content analysis. Na^+ (A), K^+ (B) and Ca^{2+} (D) contents in seedlings of transgenic lines (L11, L17 and L22) and WT grown in 0 mM, and 200 mM NaCl. Individual K^+/Na^+ ratios are shown in (C). Mean values that were significantly different at $p \leq 0.05$ within treatment from each other are indicated by different letters (a, b and c).

genes. These results indicate that overexpression of the *SbSI-2* gene in tobacco positively modulates expression of ROS-related genes.

Effect of *SbSI-2* overexpression on the expression of stress-responsive TFs

To investigate how *SbSI-2* increases salt and osmotic stress tolerance, the expression levels of some stress-responsive TFs (*NtDREB2* and *AP2*-domains containing TF) were evaluated in transgenic and WT plants. Expression levels of *NtDREB2* and *AP2*-domains containing TF were higher in transgenic tobacco plants as compared to WT under both stress and control conditions (Fig. 13). However relative fold increase of transcript (NtDREB2) is seen only in L11 and L17 under salt stress (Figure 13A), but no relative fold increase is observed under osmotic stress relative to control condition (Figure 13B). The NtAP2 transcription factor gene showed relative fold increase in transcript level only in L17 and L22 line under salt stress relative to control condition (Figure 13C). In case of osmotic stress, all transgenic lines showed increase in transcript level of NtAP2 TF (Figure 13D). On the basis of these results, we speculate that over-expression of *SbSI-2* positively modulates expression of stress responsive TFs.

Discussion

Abiotic stress reduces plant growth and survival. Plants survive various stresses by controlling responses at both cellular and molecular level. Plant adaptation to abiotic stress involves a plethora of genes related to ion homeostasis, compatible osmolytes

synthesis, ROS-scavenging and antioxidant defence mechanism. Characterization of unknown genes is an important and challenging task in deciphering their role in stress tolerance. In our previous study we have identified 270 unknown/hypothetical genes and 12 miRNAs [25,52]. Out of 270 unknown/hypothetical genes 90 unknown genes confirmed their role in salt stress by reverse Northern analysis [25]. The *SbSI-2* cDNA fragment spanning the entire open reading frame (ORF) was cloned and sequenced (Gen-Bank accession number JX872273). Amino acid sequence analysis of SbSI-2 by different *in silico* tools revealed important features like three distinct regions, nuclear localization signals (NLS), leucine-rich nuclear export signals (NES), strong protein-protein interaction domain and phosphorylation sites (Figs. S2, S3, S4, S5). Transient expression of the RFP:SbSI2 fusion protein also showed that SbSI-2 is a nuclear-localized protein (Fig. 2B), suggesting that SbSI-2 may function in the nucleus. Nuclear localization is an integral part of abiotic stress response. Many stress-associated proteins, like some TFs (bZIP, DREB, MYC/MYB, NAC, C_2H_2 zinc finger protein Msn2P), some kinase gene and a number of functional proteins belong to a subset of cellular proteins that localize to the nucleus [11,12,53,54,55,56,57,58,59]. Genomic organization study showed that *SbSI-2* is an intronless gene (Fig. 1A). Genes without introns are a characteristic feature of prokaryotes, but there are still a number of intronless genes in eukaryotes. Study these eukaryotic genes that have prokaryotic architecture could help to understand the evolutionary patterns of related genes and genomes [60]. Intronless genes are good candidate pool for

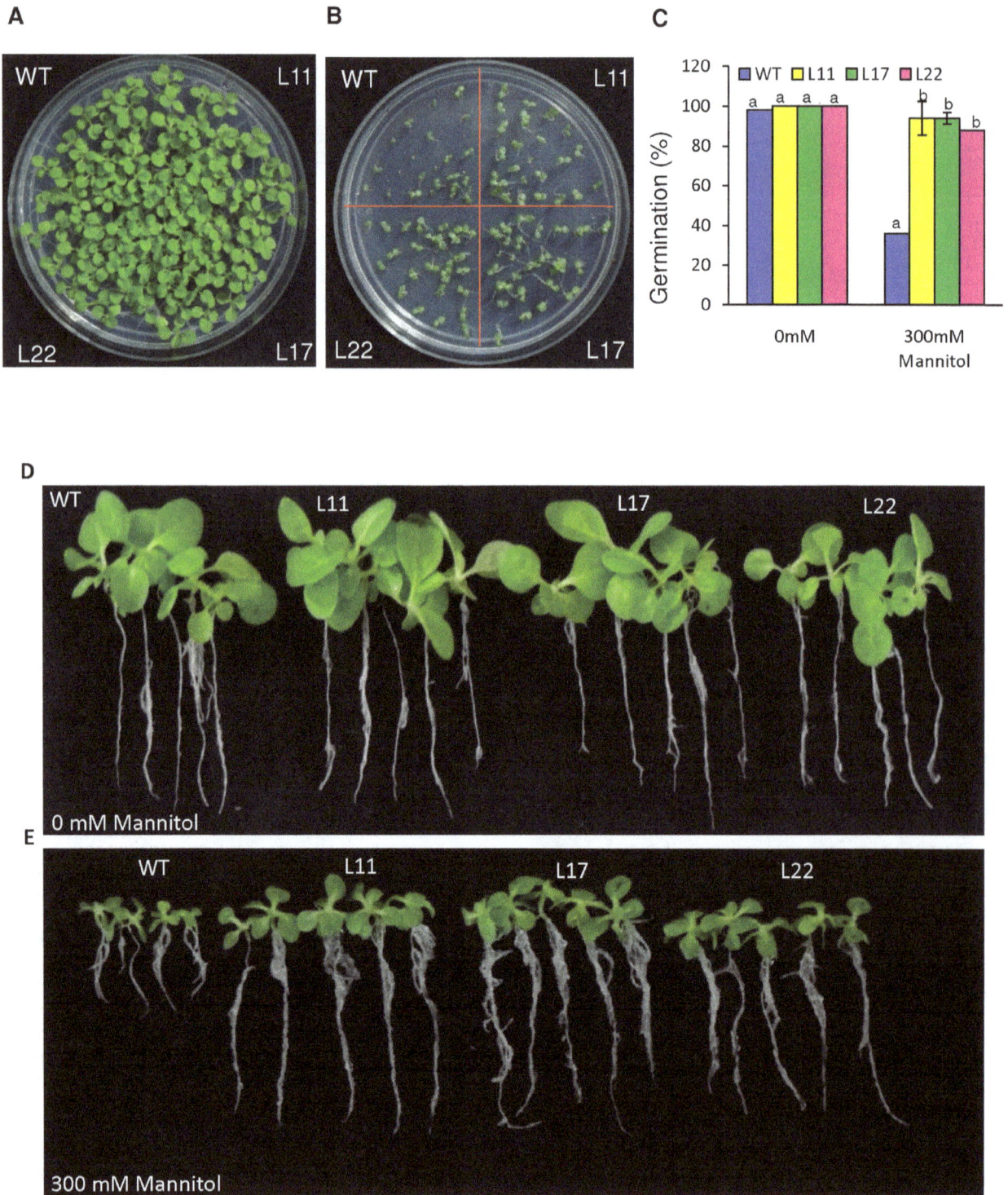

Figure 9. Phenotypic comparison of the growth of WT and transgenic lines overexpressing the *SbSI-2* gene under osmotic stress. (A–B) Germination of seeds from transgenic lines (L11, L17 and L22) and WT plants in (A) 0 mM, and (B) 300 mM mannitol. (C) Graphs represent the percentage germination of transgenic lines (L11, L17 and L22) and WT plants in osmotic stress (300 mM mannitol) and normal condition. (D–E) Growth comparison of transgenic lines (L11, L17 and L22) and WT seedlings in (d) 0 mM, and (e) 300 mM mannitol. Mean values that were significantly different at p≤0.05 within treatment from each other are indicated by different letters (a, b and c).

subsequent functional and evolutionary analysis. It has been also reported that many intronless genes remained conserved in archaea, bacteria, fungi, plants, metazoans, and other eukaryotes during evolutions [60]. Southern blot analysis revealed the presence of a single copy of *SbSI-2* in the *S. brachiata* genome (Fig. 1B).

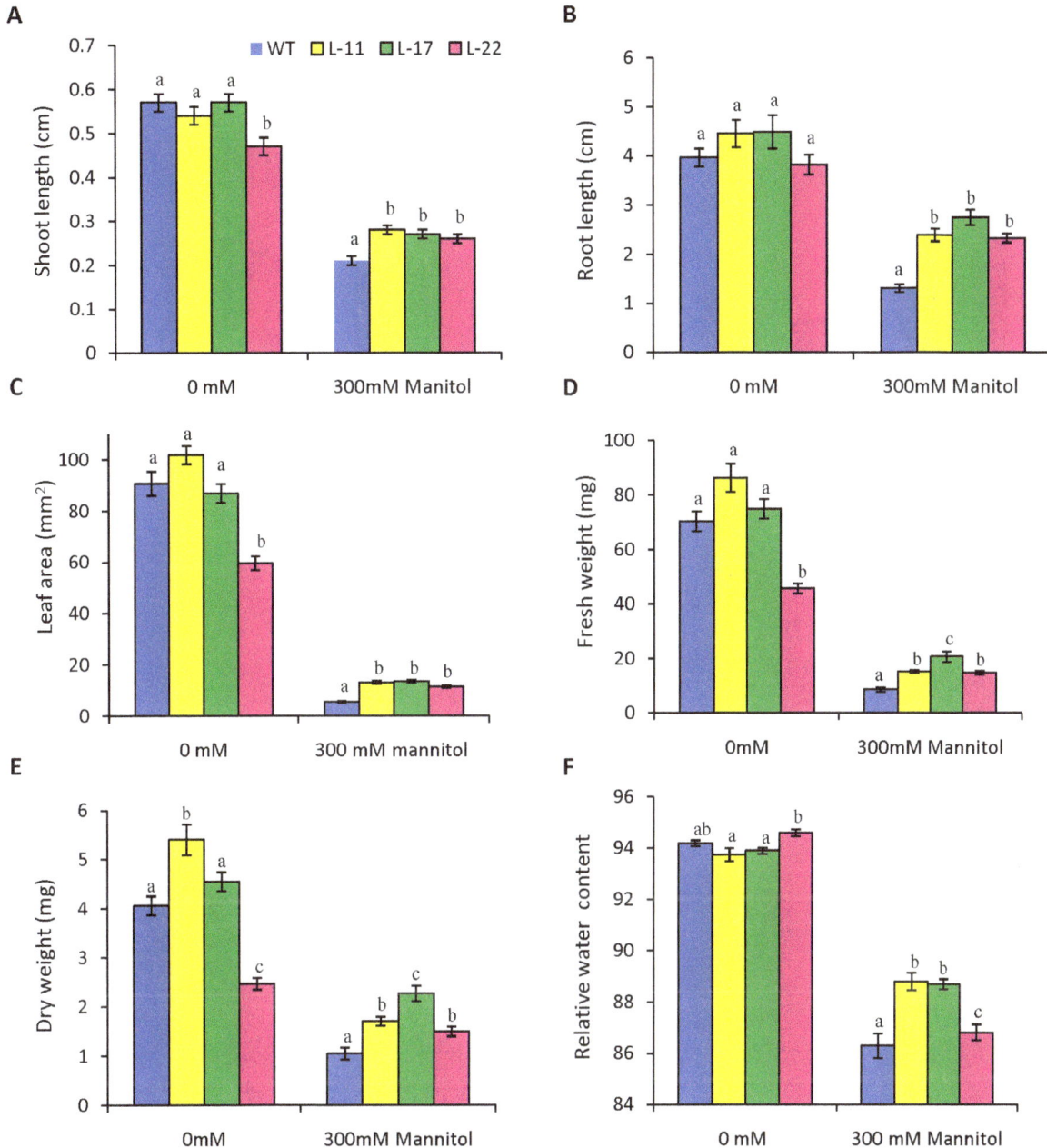

Figure 10. Comparison of growth parameters of seedlings from transgenic lines and WT in 0 mM, and 300 mM mannitol (osmotic stress). (A) shoot length, (B) root length, (C) leaf area, (D) fresh weight, (E) dry weight and (F) relative water content (RWC). Mean values that were significantly different at p≤0.05 within treatment from each other are indicated by different letters (a, b and c).

SbSI-2 showed the increased mRNA expression by salt stress and desiccation, signifying that *SbSI-2* plays an important role in abiotic stress tolerance. Heterologous expression of *SbSI-2* in *E. coli* cells demonstrated that pET28a-SbSI-2 recombinant *E. coli* cells showed better tolerance to desiccation and salinity compared to vector alone (Fig. 3). Similar to our study, a few earlier publications have also reported better growth of *E. coli* cells by overexpression of other plant stress-associated genes [30,34,61,62,63,64,65]. PM2, a group 3 LEA protein from soybean, conferred salt stress tolerance in *E. coli* [61]. Expression of phytochelatin synthase in *E. coli* resulted in better protection to heat, salt, carbofuran (pesticide), cadmium, copper and UV stress

[61]. It has been reported that the SbDREB2A transcription factor conferred tolerance to different stress conditions in *E. coli*. Yadav et al. [34] reported that a novel salt-inducible gene *SbSI-1* confers salt and desiccation tolerance in *E. coli*. Recently, Jin-long et al. [65] showed that the expressed novel dirigent protein ScDir from sugarcane had increased *E. coli* tolerance to PEG and NaCl. To further understand the function of *SbSI-2* under abiotic stress, we developed *SbSI-2*-overexpressing transgenic tobacco plants. Testing for a range of physiological parameters it was found that tobacco plants overexpressing *SbSI-2* have improved salt and osmotic tolerance, accompanied by better growth, higher seed germination, better water status, and higher photosynthetic rate as

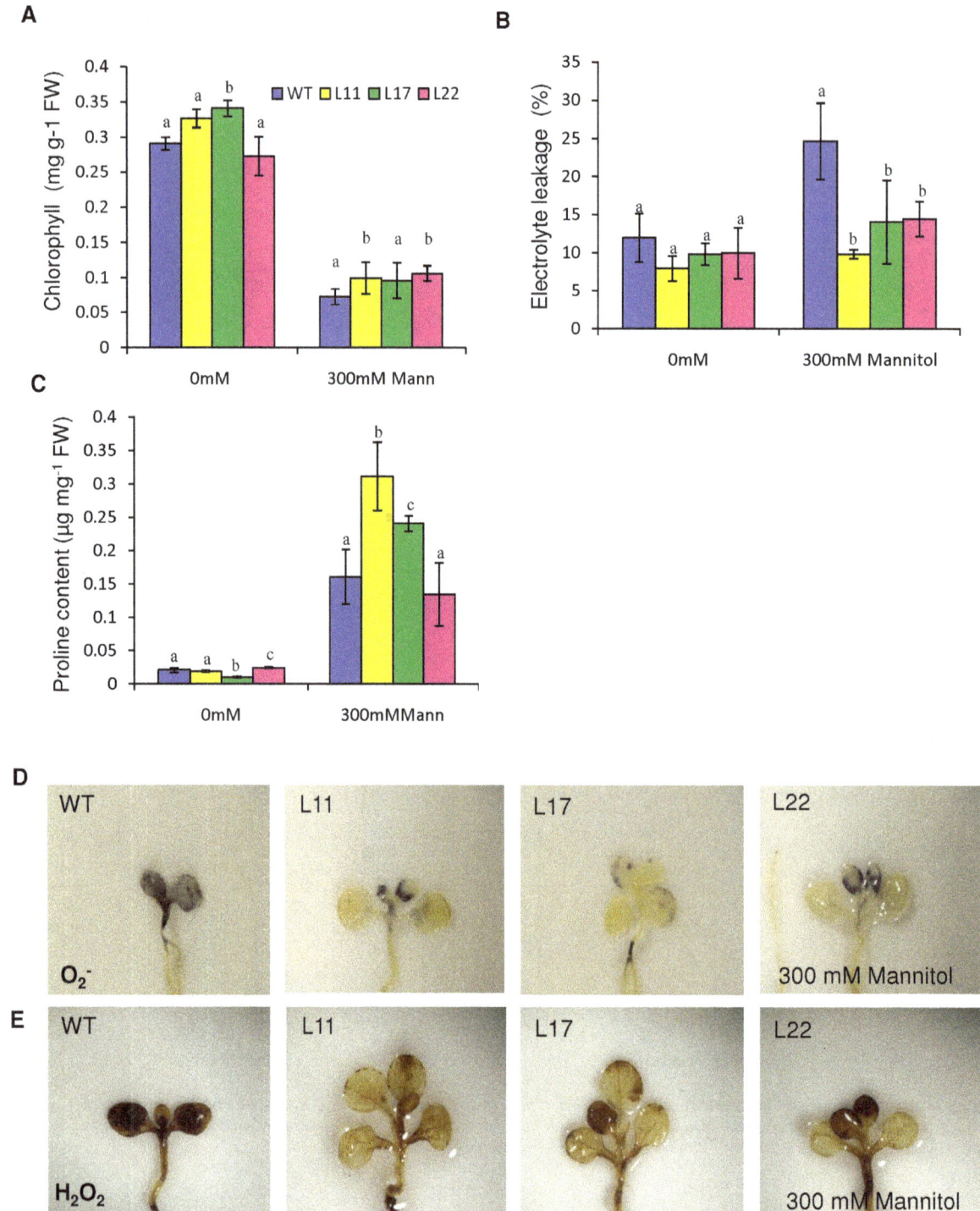

Figure 11. Comparison of various biochemical and physiological parameters of transgenic lines (L11, L17 and L22) and WT under osmotic stress. Chlorophyll content (A), Electrolyte leakage (B), and proline contents (C) of transgenic lines (L11, L17 and L22) and WT plants grown in the presence of 0 mM, and 300 mM mannitol. (D,E) *In vivo* localization of O_2^- and H_2O_2 in seedlings of 35S:SbSI-2 transgenic lines and WT under osmotic stress. (D) Localization of O_2^- by NBT staining, (E) Localization of H_2O_2 by DAB staining. Mean values that were significantly different at p≤ 0.05 within treatment from each other are indicated by different letters (a, b and c).

compared to WT plants (Figs. 4; 5; 6; 7A; 9; 10; 11A). The better water status in transgenic plants, indicate that SbSI-2 helps in water retention during NaCl and osmotic stress.

The relative abundance of proline and total soluble sugar are important biochemical indicators of salinity and drought stress in plants [66]. It has been reported that increased proline content under various environmental stresses significantly improved plant

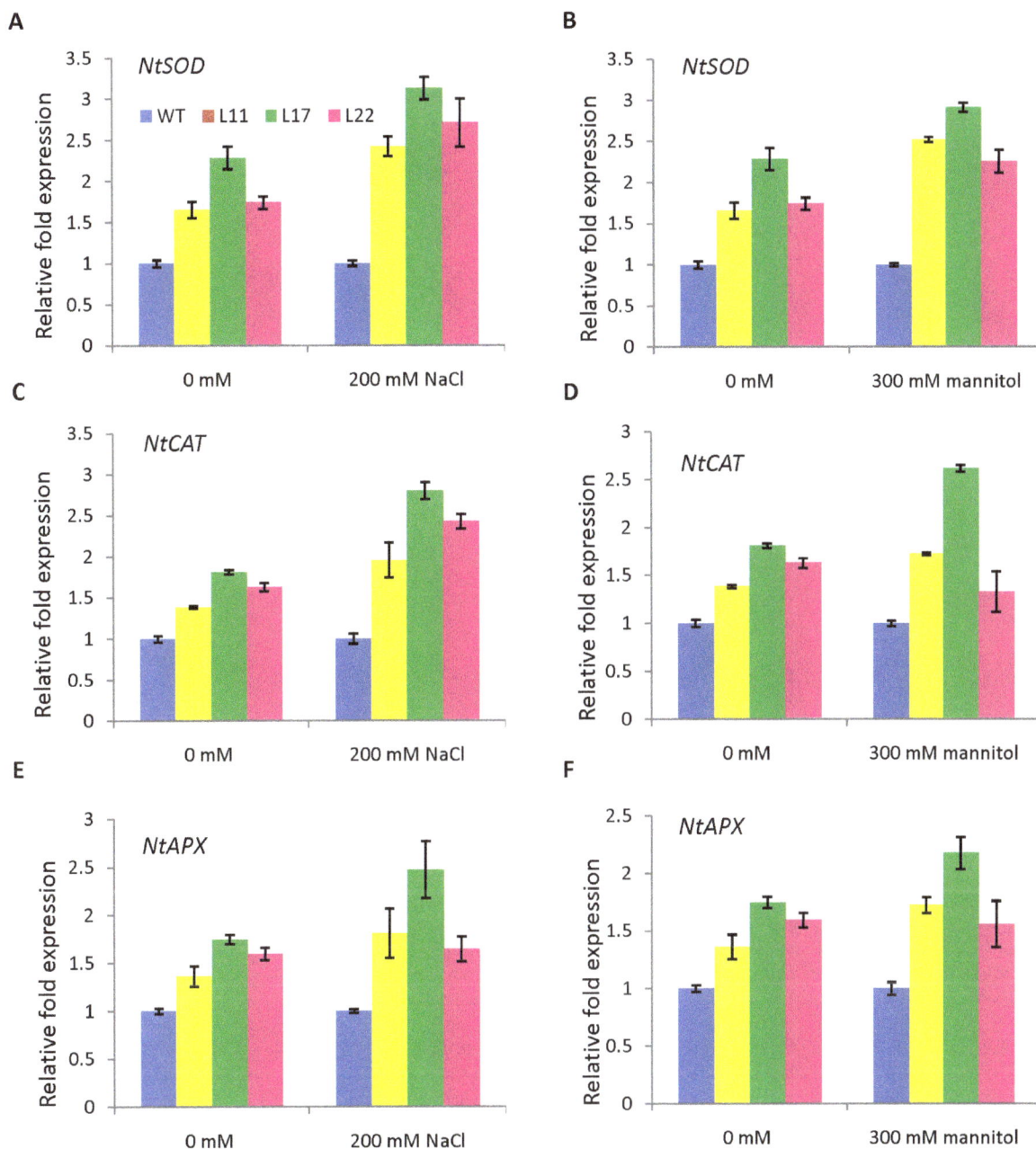

Figure 12. Expression analysis of ROS-related genes (*NtSOD, NtCAT, NtAPX*) in WT and *SbSI-2*-overexpressing plants by qRT-PCR. (A, C and E) Expression analysis under salt stress (200 mM NaCl) and (B, D and F) Expression analysis under osmotic stress (300 mM mannitol).

stress tolerance [67,68,69]. Proline protects the plants in response to salt and drought stresses by osmoprotection and ROS-scavenging, which contributes to membrane stability and mitigates the disruptive effect of stress [70]. Transgenic plants overexpressing *SbSI-2* accumulate higher proline relative to WT plants during salt and osmotic stress (Figs. 7C; 11C).

In the present study, transgenic plants facing salt and osmotic stress exhibited low electrolyte leakage (Figs. 7B; 11B). Plants experiencing abiotic stress often exhibit symptoms of oxidative stress as evidenced by elevated accumulation of ROS and MDA contents [71]. It has been reported that plants maintain their ROS pools at low levels in order to minimize cellular damage caused by oxidative stress [72,73]. ROS accumulation depends on the

balance between production and contemporaneous scavenging [74]. Plant cells have a complex antioxidant defence system for ROS-detoxification [75]. The overexpression of *SbSI-2* reduced accumulation of ROS in response to salt and osmotic stresses, which indicates a reduced oxidative damage resulting from stress (Figs. 7D, E; 11D, E).

Salinity stress leads to increase in cellular Na^+ and decrease in K^+, which causes an ion-toxic effect in cells, physiological drought, and lack of nutrients [76]. Therefore, maintenance of a low intracellular Na^+ concentration or a high cytosolic K^+/Na^+ ratio is crucial for salt tolerance in plants [77,78]. Studies have shown that, among glycophytes such as wheat, Na^+ efflux and high K^+/Na^+ ratio are the key mechanisms involved in salt tolerance

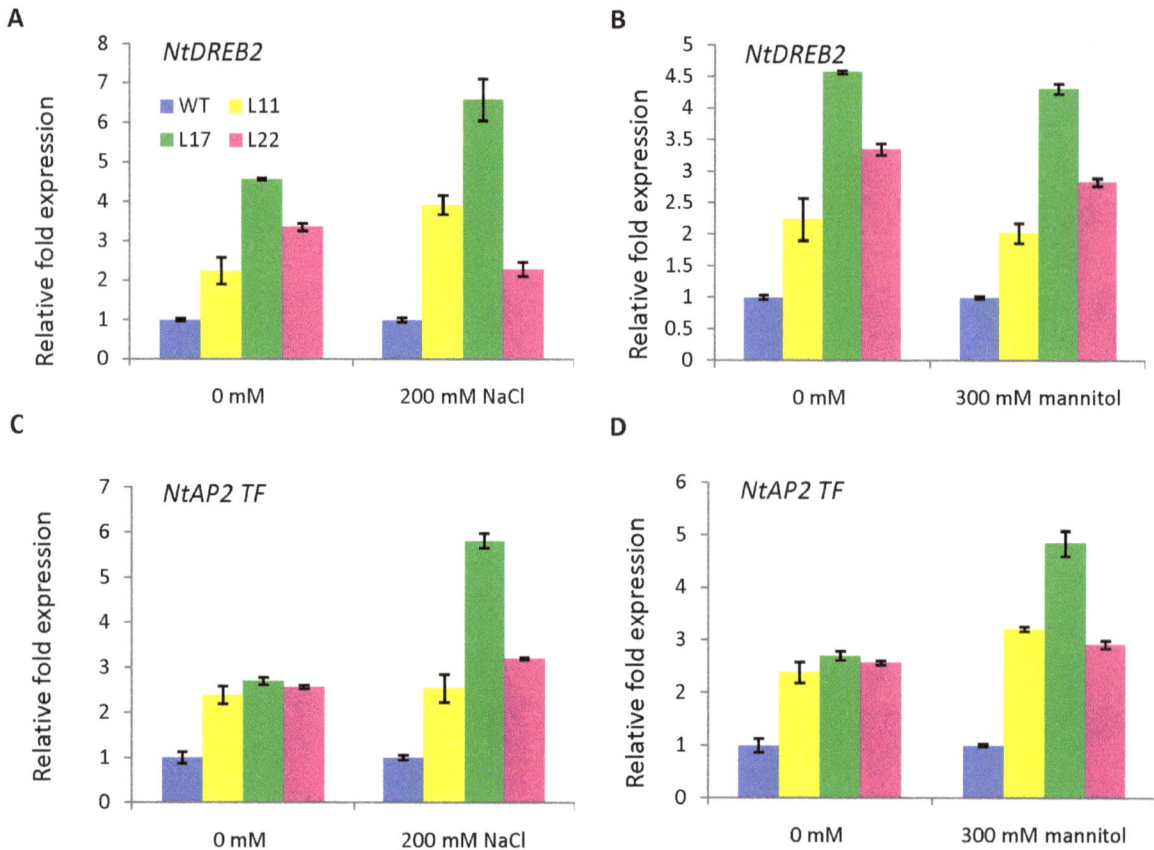

Figure 13. Expression analysis of stress-related transcription factors (*NtDREB2* and *AP2*-domain containing TF) in WT and *SbSI-2*-overexpressing plants by qRT-PCR. (A and C) Expression analysis under salt stress (200 mM NaCl), and (B and D) Expression analysis under osmotic stress (300 mM mannitol).

[79,80]. To decipher the mechanism by which *SbSI-2* overexpression improves salt tolerance, ion contents analysis were carried out in transgenic and WT plants under non-stress and salt-stress conditions. *SbSI-2*-overexpressing transgenic tobacco seedlings accumulated lower Na^+ and higher K^+ content after salt stress, with improved K^+/Na^+ ratio (Fig. 8), suggesting that *SbSI-2* overexpression ensures better physiological activities and imparts salt tolerance by increasing the K^+/Na^+ ratio. On the basis of these physiological and biochemical analysis, we can speculate that improved salt and osmotic tolerance in the *SbSI-2*-overexpressing transgenic lines is correlated with high water retention capacity, higher accumulation of osmolytes, high K^+/Na^+ ratio, reduced electrolyte leakage and less accumulation of ROS.

To gain further insight into enhanced abiotic stress tolerance in *SbSI-2*-overexpressing transgenic tobacco at the molecular level, the expression levels of ROS-scavenging genes (*NtSOD, NtAPX* and *NtCAT*) and some stress-associated TFs (*NtDREB2* and *AP2*-domains containing TF) were evaluated in transgenic and WT plants. *SbSI-2*-overexpressing lines showed higher expression of genes encoding ROS-scavenging enzymes (SOD, APX and CAT) under salt or osmotic stress, which was consistent with the lower levels of ROS in transgenic seedlings relative to WT seedlings (Fig. 12). In present study, the bioinformatics analysis and transient expression assays of the RFP:SbSI2 fusion protein showed that SbSI-2 is a nuclear-localized protein and has a strong protein-protein interaction domain which possibly interact with transcription factors that regulate the expression of the abiotic stress-responsive genes. To further investigate, we have carried out

expression analysis of two stress-associated TFs *DREB2* and *AP2*-domains containing TF in transgenic lines and WT plants. *SbSI-2*-overexpressing plants showed up-regulated expression of *DREB2* and *AP2*-domains containing TF, which in turn enhanced the expression of abiotic stress-responsive genes. These results demonstrate that *SbSI-2* might play vital positive regulatory role in abiotic stress tolerance. Taken together, we propose that the improved salt and osmotic tolerance in *SbSI-2*-overexpressing transgenic plants might be achieved by elevated expression of stress-associated TFs, which in turn up-regulate the expression of abiotic stress-responsive genes. However, further study is needed to confirm this.

In conclusion, we have cloned and characterized a novel salt-inducible gene *SbSI-2* from the extreme halophyte *Salicornia brachiata*. The *SbSI-2* gene showed up-regulation by different abiotic stresses. Subcellular localization study indicated that the SbSI-2 protein is nuclear-localized. The *SbSI-2* gene was transformed in *E. coli* and tobacco for functional characterization. pET28a-SbSI-2 recombinant *E. coli* cells showed higher tolerance to desiccation and salinity compared to vector alone. Further, overexpression of the *SbSI-2* gene in tobacco conferred salt- and osmotic tolerance by promoted seed germination, improved growth parameters, higher relative water content, higher K^+/Na^+ ratio, higher chlorophyll and elevated accumulation of of compatible osmolytes as compared to control plants. Transgenic plants exhibited reduction in electrolyte leakage and ROS in response to salt and osmotic stresses. Overexpression of *SbSI-2* also enhanced the transcript levels of ROS-scavenging genes and some stress-responsive TFs under salt and osmotic stresses. The present study demonstrates that *SbSI-2* might play an important positive modulation role in abiotic stress tolerance and suggests that it could be a potential bioresource for bioengineering abiotic stress tolerance in crop plants.

Supporting Information

Figure S1 Nucleotide sequence with deduced amino acid sequence and predicted secondary structure of SbSI-2 protein. (A) Full-length cDNA and deduced amino acid sequence of *SbSI-2*. Start codon (ATG) and stop codon (TGA) are indicated by green and red colour, respectively. The 5′ and 3′-UTR regions are indicated by blue color. (B) Predicted secondary structure viz. helix, strands and coils as indicated by pink rods, arrow and solid lines, respectively.

Figure S2 ScanProsite results together with ProRule-based predicted intra-domain features.

Figure S3 *In silico* localization analysis by various bioinformatics softwares. (A) Nuclear localization signals and their positions by the WoLF PSORT. (B) Cello prediction result for localization. (C) and (D) Leucine-rich nuclear export signals (NES) prediction by NetNES 1.1 server using a combination of neural networks and hidden Markov models. The prediction server calculates NES score from Hidden Markov Models (HMMs) and Artificial Neural Network (ANN) scores (all three values are given for each residue).

Figure S4 Protein-protein binding domain detected by PROFisis PredictProtein server. Blue underlined text shows the strong protein-protein interaction domain.

Figure S5 Predicted Phosphorylation sites in SbSI-2 protein by NetPhosK 1.0 software.

Figure S6 SDS-PAGE analysis of expression of recombinant protein (shown by arrow) in *E. coli* BL21 (DE3) cells expressing *SbSI-2* gene. (A) M marker (kDa), Lane 1: uninduced protein, Lane 2: induced protein 2 h, Lane 3: induced protein 4 h, Lane 4: induced protein 6 h. (B) M marker (kDa), Lane 1 and 2: induced protein 6 h, Lane 3 and 4: uninduced protein, Lane 5: induced protein in liquid assay after 12 h of growth.

Acknowledgments

CSIR-CSMCRI Communication No. PRIS-149/2013. The authors are grateful to Arabidopsis Biological Resource Centre, the Ohio State University, USA for providing the pSITE-4CA vector. Dr. P. K. Agarwal is thankfully acknowledged for his help in the initial stage of the study.

Author Contributions

Conceived and designed the experiments: BJ. Performed the experiments: NSY VKS DS. Analyzed the data: NSY DS VKS. Contributed reagents/materials/analysis tools: BJ. Wrote the paper: NSY BJ.

References

1. Mahajan S, Pandey GK, Tuteja N (2008) Calcium- and salt-stress signaling in plants: Shedding light on SOS pathway. Arch Biochem Biophys 471: 146–158.
2. Vinocur B, Altman A (2005) Recent advances in engineering plant tolerance to abiotic stress: achievements and limitations. Curr Opin Biotechnol 16: 123–132.
3. Huang GT, Ma SL, Bai LP, Zhang L, Ma H, et al. (2012) Signal transduction during cold, salt, and drought stresses in plants. Mol Biol Rep 39(2): 969–87.
4. Yamaguchi-Shinozaki K, Shinozaki K (2006) Transcriptional regulatory networks in cellular responses and tolerance to dehydration and cold stresses. Annu Rev Plant Biol 57: 781–803.
5. Hadiarto T, Tran LS (2011) Progress studies of drought-responsive genes in rice. Plant Cell Rep 30: 297–310.
6. Nakashima K, Tran LS, Van Nguyen D, Fujita M, Maruyama K, et al. (2007) Functional analysis of a NAC-type transcription factor OsNAC6 involved in abiotic and biotic stress-responsive gene expression in rice. Plant J 51: 617–630.
7. Agarwal PK, Jha B (2010) Transcription factors in plants and ABA dependent and independent abiotic stress signalling. Biol plant 54: 201–212.
8. Seki M, Narusaka M, Ishida J, Nanjo T, Fujita M, et al. (2002) Monitoring the expression profiles of 7000 Arabidopsis genes under drought, cold and high-salinity stresses using a full-length cDNA microarray. Plant J 31: 279–292.
9. Dubouzet JG, Sakuma Y, Ito Y, Kasuga M, Dubouzet EG, et al. (2003) OsDREB genes in rice, Oryza sativa L., encode transcription activators that function in drought-, high-salt- and cold responsive gene expression. Plant J 33: 751–763.
10. Xiang Y, Tang N, Du H, Ye HY, Xiong LZ (2008) Characterization of OsbZIP23 as a key player of the basic leucine zipper transcription factor family for conferring abscisic acid sensitivity and salinity and drought tolerance in rice. Plant Physiol 148: 1938–1952.
11. Li C, Lv J, Zhao X, Ai X, Zhu X, et al. (2010) TaCHP: a wheat zinc finger protein gene down-regulated by abscisic acid and salinity stress plays a positive role in stress tolerance. Plant Physiol 154: 211–221.
12. Hao YJ, Wei W, Song QX, Zhang FW, Zou HF, et al. (2011) Soybean NAC transcription factors promote abiotic stress tolerance and lateral root formation in transgenic plants. Plant J 68: 302–313.
13. He Y, Li W, Lv J, Jia Y, Wang M, et al. (2012) Ectopic expression of a wheat MYB transcription factor gene, TaMYB73, improves salinity stress tolerance in *Arabidopsis thaliana*. J Exp Bot 63: 1511–1522.
14. Gong Q, Li P, Ma S, Indu Rupassara S, Bohnert HJ (2005) Salinity stress adaptation competence in the extremophile *Thellungiella halophila* in comparison with its relative *Arabidopsis thaliana*. Plant J 44: 826–839.
15. Agarwal PK, Gupta K, Jha B (2010) Molecular characterization of the *Salicornia brachiata SbMAPKK* gene and its expression by abiotic stress. Mol Biol Rep 37: 981–986.
16. Amtmann A, Bohnert HJ, Bressan RA (2005) Abiotic stress and plant genome evolution: search for new models. Plant Physiol 138: 127–130.
17. Zhang L, Xiu-Ling Ma, Zhang Q, Chang-Le Ma, Wang PP, et al. (2001) Expressed sequence tags from a NaCl-treated *Suaeda salsa* cDNA library. Gene 267: 193–200.
18. Kore-eda S, Cushman MA, Akselrod I, Bufford D, Fredrickson M, et al. (2004) Transcript profiling of salinity stress responses by large-scale expressed sequence tag analysis in *Mesembryanthemum crystallinum*. Gene 341: 83–92.
19. Wang ZL, Li PH, Fredricksen M, Gong ZZ, Kim CS, et al. (2004) Expressed sequence tags from *Thellungiella halophila*, a new model to study plant salt-tolerance. Plant Sci 166: 609–616.
20. Mehta PA, Sivaprakash K, Parani M, Venkataraman G, Parida AK (2005) Generation and analysis of expressed sequence tags from the salt-tolerant mangrove species *Avicennia marina* (Forsk) Vierh. Theor Appl Genet 110: 416–424.
21. Chen SH, Guo SL, Wang ZL, Zhao JQ, Zhao YX, et al. (2007) Expressed sequence tags from the halophyte *Limonium sinense*. J Seq Map 18: 61–67.
22. Zouri N, Ben Saad R, Legavre T, Azaza J, Sabau X, et al. (2007) Identification and sequencing of ESTs from the halophyte grass *Aeluropus littoralis*. Gene 404: 61–69.
23. Baisakh N, Subudhi PK, Varadwaj P (2008) Primary responses to salt stress in halophyte, smooth cordgrass (*Spartina alterniflora* Loisel.). Funct Int Gene 8: 287–300.
24. Reddy PCO, Sairanganayakulu G, Thippeswamy M, Reddy PS, Reddy MK, et al. (2008) Identification of stress induced genes from the drought tolerant semi-arid legume crop horsegram (*Macrotyloma uniflorum* (Lam.) Verdc.) through analysis of subtracted expressed sequence tags. Plant Sci 175: 372–384.
25. Jha B, Agarwal PK, Reddy PS, Lal S, Sopory SK, et al. (2009) Identification of salt-induced genes from *Salicornia brachiata*, an extreme halophyte through expressed sequence tags analysis. Genes Genet Syst 84: 111–120.
26. Li HY, Wang YC, Jiang J, Liu GF, Gao CQ, et al. (2009) Identification of genes responsive to salt stress on *Tamarix hispida* roots. Gene 433(1–2): 65–71.
27. Jin H, Sun Y, Yang Q, Chao Y, Kang J, et al. (2010) Screening of genes induced by salt stress from Alfalfa. Mol Biol Rep 37: 745–753.
28. Gu L, Xu D, You T, Li X, Yao S, et al. (2011) Analysis of gene expression by ESTs from suppression subtractive hybridization library in *Chenopodium album* L. under salt stress. Mol Biol Rep 38(8): 5285–95.

29. Jha B, Sharma A, Mishra A (2011) Expression of *SbGSTU* (tau class glutathione S-transferase) gene isolated from *Salicornia brachiata* in tobacco for salt tolerance. Mol Biol Rep 38: 4823–4832.

30. Gupta K, Agarwal PK, Reddy MK, Jha B (2010) SbDREB2A, an A-2 type DREB transcription factor from extreme halophyte *Salicornia brachiata* confers abiotic stress tolerance in *Escherichia coli*. Plant Cell Rep 29: 1131–1137.

31. Jha A, Joshi M, Yadav NS, Agarwal PK, Jha B (2011) Cloning and characterization of the *Salicornia brachiata* Na⁺/H⁺ antiporter gene *SbNHX1* and its expression by abiotic stress. Mol Biol Rep 38: 1965–1973.

32. Jha B, Lal S, Tiwari V, Yadav SK, Agarwal PK (2012) The *SbASR -1* Gene Cloned from an Extreme Halophyte *Salicornia brachiata* Enhances Salt Tolerance in Transgenic Tobacco. Mar. Biotechnol 14: 782–792.

33. Yadav NS, Shukla PS, Jha A, Agarwal PK, Jha B (2012) The *SbSOS1* gene from the extreme halophyte *Salicornia brachiata* enhances Na⁺ loading in xylem and confers salt tolerance in transgenic tobacco. BMC Plant Biol 12: 188.

34. Yadav NS, Rashmi D, Dinkar Singh, Agarwal PK, Jha B (2012) A novel salt-inducible gene *SbSI-1* from *Salicornia brachiata* confers salt and desiccation tolerance in *E. Coli*. Mol Biol Rep 39: 1943–1948.

35. Murashige T, Skoog F (1962) A revised medium for rapid growth and bioassay with tobacco tissue cultures. Physiol Plantarum 15: 473–497.

36. Chomczynski P, Sacchi N (1987) Single-step method of RNA isolation by acid guanidinium thiocyanate–phenol–chloroform extraction. Anal Biochem 162: 156–159.

37. Horton P, Park KJ, Obayashi T, Fujita N, Harada H, et al. (2007) "WoLF PSORT: Protein localization predictor". Nucleic Acids Res 35: 585–587.

38. Nakai K, Kanehisa M (1992) A knowledge base for predicting protein localization sites in eukaryotic cells. Genomics 14: 897–911.

39. Saghai Maroof MA, Solima KM, Jorgenson RA, Allard RW (1984) Ribosomal DNA spacer-length polymorphisms in barley: Mendelian inheritance, chromosomal location, and population dynamics. Proc Nat Acad Sci USA 81: 8014–8018.

40. Walhout A, Temple G, Brasch M, Hartley J, Lorson M, et al. (2000) GATEWAY recombinational cloning: application to the cloning of large numbers of open reading frames or ORFeomes. Methods Enzymol 328: 575–592.

41. Livak KJ, Schmittgen TD (2001) Analysis of relative gene expression data using real- time quantitative PCR and the 2(-Delta Delta C(T)) method. Methods 25: 402–408.

42. Huang XS, Luo T, Fu XZ, Fan QJ, Liu JH (2011) Cloning and molecular characterization of a mitogen-activated protein kinase gene from *Poncirus trifoliata* whose ectopic expression confers dehydration/drought tolerance in transgenic tobacco. J Exp Bot 62: 5191–5206.

43. Topfer R, Matzeit V, Gronenborn B, Schell J, Steinbiss HH (1987) A set of plant expression vectors for transcriptional and translational fusions. Nucleic Acids Res 15: 5890.

44. Horsch RB, Fry JE, Hoffmann NL, Eichholtz D, Rogers SG, et al. (1985) A simple and general method for transferring genes into plants. Science 227: 1229–1231.

45. Arnon DI (1949) Copper enzymes in isolated chloroplasts. Polyphenoloxidase in *Beta vulgaris*. Plant Physiol 24: 1.

46. Lichtenthaler HK (1987) Chlorophylls and carotenoids: pigments of photosynthetic Biomembranes. Method Enzymol 148: 350–382.

47. Lutts S, Kinet JM, Bouharmont J (1996) NaCl-induced senescence in leaves of rice (*Oryza sativa* L.) cultivars differing in salinity resistance. Ann Bot 78: 389–398.

48. Bates LS, Waldern R, Teare ID (1973) Rapid determination of free proline for water stress studies. Plant Soil 39: 205–207.

49. Shi J, Fu XZ, Peng T, Huang XS, Fan QJ, et al. (2010) Spermine pretreatment confers dehydration tolerance of citrus in vitro plants via modulation of antioxidative capacity and stomatal response. Tree Physiol 30: 914–922.

50. Shukla PS, Agarwal PK, Jha B (2012) Improved salinity tolerance of *Arachis hypogaea* (L.) by the interaction of halotolerant plant-growth-promoting rhizobacteria. J Plant Growth Regul 31: 195–206.

51. Niu CF, Wei W, Zhou QY, Tian AG, Hao YJ, et al. (2012) Wheat WRKY genes *TaWRKY2* and *TaWRKY19* regulate abiotic stress tolerance in transgenic Arabidopsis plants. Plant Cell Environ 35: 1156–1170.

52. Singh D, Jha B (2014) The isolation and identification of salt –responsive novel micro RNAs from *salicornia brachiata*, an extreme halophyte. Plant Biotechnol rep doi:10.1007/s11816-014-0324-5

53. Gorner W, Durchschlag E, Martinez-Pastor MT, Estruch F, Ammerer G, et al. (1998) Nuclear localization of the C2H2 zinc finger protein Msn2p is regulated by stress and protein kinase A activity. Genes Dev 12: 586–597.

54. Knowlton AA, Salfity M (1996) Nuclear localization and the heat shock proteins. J Biosci 21: 123–132.

55. Boudsocq M, Lauriere C (2005) Osmotic signaling in plants: multiple pathways mediated by emerging kinase families. Plant Physiol 138: 1185–11194.

56. Li S, Xu C, Yang Y, Xia G (2010) Functional analysis of *TaDi19A*, a salt-responsive gene in wheat. Plant Cell Environ 33: 117–129.

57. Kong X, Pan J, Zhang M, Xing X, Zhou Y, et al. (2011) *ZmMKK4*, a novel group C mitogen-activated protein kinase kinase in maize (Zea mays), confers salt and cold tolerance in transgenic Arabidopsis. Plant Cell Environ 34: 1291–1303.

58. Yang A, Dai X, Zhang WH (2012) A R2R3-type MYB gene, *OsMYB2*, is involved in salt, cold, and dehydration tolerance in rice. J Exp Bot 63: 2541–2556.

59. Zhang L, Li Y, Lu W, Meng F, Wu C, et al. (2012) Cotton *GhMKK5* affects disease resistance, induces HR-like cell death, and reduces the tolerance to salt and drought stress in transgenic *Nicotiana benthamiana*. J Exp Bot 63: 3935–3952.

60. Zou M, Guo B, Shunping H (2011) The Roles and Evolutionary Patterns of Intronless Genes in Deuterostomes. Comparative Funct Genom. doi:10.1155/2011/680673.

61. Liu Y, Zheng Y (2005) PM2, a group 3 LEA protein from soybean, and its 22-mer repeating region confer salt tolerance in *Escherichia coli*. Biochem Biophys Res Commun 31: 325–332.

62. Chaurasia N, Mishra Y, Rai LC (2008) Cloning expression, analysis of phytochelatin synthase (pcs) gene from *Anabaena* sp. PCC 7120 offering multiple stress tolerance in *Escherichia coli*. Biochem Biophys Res Commun 376: 225–230.

63. Hong GX, Jiang J, Wang BC, Li HY, Wang YC, et al. (2010) ThPOD3, a truncated polypeptide from *Tamarix hispida*, conferred drought tolerance in *Escherichia coli*. Mol Biol Rep 37: 1183–1190.

64. Reddy PS, Mallikarjuna G, Kaul T, Chakradhar T, Mishra RN, et al. (2010) Molecular cloning and characterization of gene encoding for cytoplasmic Hsc70 from *Pennisetum glaucum* may play a protective role against abiotic stresses. Mol Genet Genomics 283: 243–254.

65. Jin-long G, Li-ping X, Jing-ping F, Ya-chun S, Hua-ying F, et al. (2012) A novel dirigent protein gene with highly stem-specific expression from sugarcane, response to drought, salt and oxidative stresses. Plant Cell Rep 31: 1801–1812.

66. Ashraf M, Harris PJ (2004) Potential biochemical indicators of salinity tolerance in plants. Plant Sci 166: 3–16.

67. Abraham E, Rigo G, Szekely G, Nagy R, Koncz C, et al. (2003) Light-dependent induction of proline biosynthesis by abscisic acid and salt stress is inhibited by brassinosteroid in Arabidopsis. Plant Mol Biol 51: 363–372.

68. Verbruggen N, Hermans C (2008) Proline accumulation in plants: a review, Amino Acids 35: 753–759.

69. Cvikrová M, Gemperlová L, Dobrá J, Martincová O, Prášil IT, et al. (2012) Effect of heat stress on polyamine metabolism in proline-overproducing tobacco plants, Plant Sci 182: 49–58.

70. Szekely G, Abraham E, Cseplo A, Rigo G, Zsigmond L, et al. (2008) Duplicated P5CS genes of Arabidopsis play distinct roles in stress regulation and developmental control of proline biosynthesis. Plant J 53: 11–28.

71. Verslues PE, Kim YS, Zhu JK (2007) Altered ABA, proline and hydrogen peroxide in an Arabidopsis glutamate:glyoxylate aminotransferase mutant. Plant Mol Biol 64: 205–217.

72. Harb A, Krishnan A, Ambavaram MMR, Pereira A (2010) Molecular and physiological analysis of drought stress in Arabidopsis reveals early responses leading to acclimation in plant growth. Plant Physiol 154: 1254–1271.

73. Foyer CH, Shigeoka S (2011) Understanding oxidative stress and antioxidant functions to enhance photosynthesis. Plant Physiol 155: 93–100.

74. Pitzschke A, Djamei A, Bitton F, Hirt H (2009) A major role of the MEKK1–MKK1/2–MPK4 pathway in ROS signalling. Mol Plant 2: 120–137.

75. Miller G, Suzuki N, Ciftci-Yilmaz S, Mittler R (2010) Reactive oxygen species homeostasis and signaling during drought and salinity stresses. Plant Cell Environ 33: 453–457.

76. Zhu N, Jiang Y, Wang M, Ho CT (2001) Cycloartane triterpene saponins from the roots of Cimicifuga foetida. J Nat Prod 64: 627–629.

77. Lynch J, Läuchli A (1984) Potassium transport in salt-stressed barley roots. Planta 161: 295–301.

78. Maathuis FJM, Amtmann A (1999) K⁺ nutrition and Na⁺ toxicity: the basis of cellular K⁺/Na⁺ ratios. Ann Bot 84: 123–133.

79. Munns R, James RA, Läuchli A (2006) Approaches to increasing the salt tolerance of wheat and other cereals. J Exp Bot 57: 1025–1043.

80. Rodríguez-Navarro A, Rubio F (2006) High-affinity potassium and sodium transport systems in plants. J Exp Bot 57: 1149–1160.

10

Auto-Regulation of the *Sohlh1* Gene by the SOHLH2/SOHLH1/SP1 Complex: Implications for Early Spermatogenesis and Oogenesis

Shuichi Toyoda[1☯]**, Takuji Yoshimura**[1,2☯]**, Junya Mizuta**[1]**, Jun-ichi Miyazaki**[1]*

1 Division of Stem Cell Regulation Research, Osaka University Graduate School of Medicine, Osaka, Japan, **2** Laboratory of Reproductive Engineering, the Institute of Experimental Animal Sciences, Osaka University Medical School, Osaka, Japan

Abstract

Tissue-specific basic helix-loop-helix (bHLH) transcription factor proteins often play essential roles in cellular differentiation. The bHLH proteins SOHLH2 and SOHLH1 are expressed specifically in spermatogonia and oocytes and are required for early spermatogonial and oocyte differentiation. We previously reported that knocking out *Sohlh2* causes defects in spermatogenesis and oogenesis similar to those in *Sohlh1*-null mice, and that *Sohlh1* is downregulated in the gonads of *Sohlh2*-null mice. We also demonstrated that SOHLH2 and SOHLH1 can form a heterodimer. These observations led us to hypothesize that the SOHLH2/SOHLH1 heterodimer regulates the *Sohlh1* promoter. Here, we show that SOHLH2 and SOHLH1 synergistically upregulate the *Sohlh1* gene through E-boxes upstream of the *Sohlh1* promoter. Interestingly, we identified an SP1-binding sequence, called a GC-box, adjacent to these E-boxes, and found that SOHLH1 could bind to SP1. Furthermore, chromatin-immunoprecipitation analysis using testes from mice on postnatal day 8 showed that SOHLH1 and SP1 bind to the *Sohlh1* promoter region *in vivo*. Our findings suggest that an SOHLH2/SOHLH1/SP1 ternary complex autonomously and cooperatively regulates *Sohlh1* gene transcription through juxtaposed E- and GC-boxes during early spermatogenesis and oogenesis.

Editor: Klaus Roemer, University of Saarland Medical School, Germany

Funding: This work was supported by a Grant-in-Aid (No. 22590190) from the Japanese Society for the Promotion of Science (http://www.jsps.go.jp/english/index.html). The funders had no role in study design, data collection and analysis, decision to publish, or preparation of the manuscript.

Competing Interests: The authors have declared that no competing interests exist.

* Email: jimiyaza@nutri.med.osaka-u.ac.jp

☯ These authors contributed equally to this work.

Introduction

Transcriptional regulation is essential for cellular differentiation. Previous studies have demonstrated that a number of transcriptional factors play important roles in early spermatogenesis and oogenesis [1,2]. Recently, several gene-knockout studies revealed that the germ cell-specific basic helix-loop-helix (bHLH) proteins SOHLH2 and SOHLH1 are expressed in spermatogonia and early oocytes [3–6] and are required for their differentiation [4–8]. The *Sohlh2* transcript is upregulated shortly after birth, and the SOHLH2 protein is expressed in the adult testis by a portion of A_s spermatogonia throughout differentiation [6]. In mouse oogenesis, the *Sohlh2* transcript is upregulated before birth [6], and its protein is expressed in primordial through primary oocytes in the ovary [3,6]. Although the *Sohlh1* and *Sohlh2* expression patterns are similar, the *Sohlh1* transcript is upregulated following *Sohlh2* expression in both early spermatogenesis and oogenesis [6]. The SOHLH1 protein is expressed by A_{al} spermatogonia throughout differentiation [3]. Since both male and female *Sohlh2*- and *Sohlh1*-null mice are infertile, and these mice have similar abnormalities in gonad histology and gene expression patterns, *Sohlh2* may be upstream to *Sohlh1* in the gene regulatory hierarchy.

The bHLH proteins are known to form heterodimers or homodimers to bind to the consensus E-box DNA sequence CANNTG. Some bHLH proteins, such as ARNT, can transactivate target genes through homodimerization [9], while others, such as MAX-MYC, transactivate their target genes through heterodimerization [10,11]. It has been reported that SOHLH2 and SOHLH1 can form a heterodimer [6], and that the *Sohlh1* mRNA levels are significantly reduced in the *Sohlh2*-null testis and ovary compared to the levels in wild-type gonads [6–8]. Since *Sohlh1* contains several E-box (CACGTG) motifs in its promoter region (see below), it is possible that the SOHLH proteins regulate the *Sohlh1* gene.

Transcription factors often function by forming complexes with other proteins. The bHLH proteins sometimes form ternary complexes with SP1, which is a zinc finger-type transcription factor [12] that binds to the consensus DNA sequence GGGGCGGGGC, called a GC-box [13]. The ternary complex binds to juxtaposed E- and GC-boxes and synergistically transactivates the adjacent promoter, as seen in Myogenin/SP1 and NeuroD1/SP1 complexes [14,15]. SP1 is widely expressed in various cell types, including spermatogonia and oocytes [16,17]. Interestingly, here we identified juxtaposed E- and GC-box sequences in the upstream region of the *Sohlh1* gene of various mammalian species, and found evidence that the SOHLH proteins form a ternary complex with SP1 to regulate the *Sohlh1* gene. We also identified the motifs in the *Sohlh1* promoter involved in this regulation. These findings improve our understanding of the molecular mechanisms that regulate *Sohlh1* in male and female germ-cell differentiation.

Materials and Methods

Ethics statement

Experiments involving animals were carried out in accordance with institutional guidelines under protocols (No. 21–089) approved by the Animal Care and Use Committee of the Osaka University Graduate School of Medicine.

Cell culture

HEK293 cells (BioWhittaker, Walkersville, MD) were cultured in Minimum Essential Medium (Sigma-Aldrich, St. Louis, MO; Cat#M0643) supplemented with 10% heat-inactivated fetal calf serum, at 37°C.

Western blotting and immunoprecipitation assay

Samples were homogenized in RIPA buffer (10 mM Tris-HCl, pH 7.4, 1 mM EDTA, 150 mM NaCl, 1% NP-40, 0.1% SDS, and 0.1% sodium deoxycholate). The extracted protein was mixed 2:1 v/v with 3x sample buffer (New England BioLabs, Beverly, MA; Cat#B7703S) and a 1/30 volume of 1.25 M dithiothreitol (DTT), after which it was heated at 99°C for 5 min, separated in SDS-polyacrylamide gel, and analyzed by western blotting as described previously [18]. The following primary antibodies were used in this study: mouse anti-FLAG antibody (Sigma-Aldrich; Cat#F3165, 2 µg/ml at final concentration), rabbit anti-Myc-tag antibody (MBL, Nagoya, Japan; Cat#562, 1:1000 dilution), rabbit anti-SOHLH2 antibody [6], rabbit anti-SOHLH1 antibody (Abcam, Cambridge, MA; Cat#ab49272, 1:5000 dilution), and rabbit anti-SP1 antibody (Bethyl Laboratories, Montgomery, TX; Cat#IHC-00208, 1:200 dilution). Secondary antibodies used in this study were as described previously [6]. An immunoprecipitation assay was performed as described previously, using agarose beads conjugated with an anti-FLAG antibody (Sigma-Aldrich; ANTI-FLAG M2 Affinity Gel, Cat#F2426) [6].

Vector construction

The promoter region of the mouse *Sohlh1* gene (−1036 to −1 bp upstream of the *Sohlh1* translational start site) was obtained by PCR from the genomic DNA of E14, a mouse embryonic stem cell line derived from a 129/Ola mouse strain. For E- and GC-box mutagenesis, we used a PCR-based method using primers with mutated sequences. Promoters containing a deletion were prepared by PCR or with the appropriate restriction enzymes. These promoters were inserted into the multi-cloning sites of the pGL3-Basic vector (Promega, Madison, WI; Cat#E1751) and used for reporter assays. The pCMV-FLAG-Sohlh1, pCMV-FLAG-Sohlh2, pCAG-Sohlh1, and pCAG-Sohlh2 plasmid vectors were constructed as described previously [6]. The pcDNA3-Sohlh2-Myc and pcDNA3-Sohlh1-Myc vectors were obtained by inserting mouse *Sohlh2* and *Sohlh1* cDNA, respectively, into the pcDNA3-Myc-His vector (Invitrogen, Carlsbad, CA; Cat#V855-20). Mouse *Sp1* cDNA obtained from testis RNA by reverse transcription followed by PCR was inserted into a pCAG-IP plasmid vector [19]. All the PCR-amplified fragments were confirmed by sequencing. The primers used in this study are available upon request.

Reporter assays

HEK293 cells were plated on 24-well plates at a density of 2×10^4 cells per well, 24 hours prior to transfection. The cells were then co-transfected with 200 ng of a reporter vector, 0.32 to 200 ng of expression vectors, and 0.1 ng of pRL-CMV normalization vector per well using HilyMax (Dojindo Molecular Technologies, Kumamoto, Japan; Cat#H357–10). After 48 hours, the total cell extracts were obtained and subjected to luciferase assays using the Dual-Luciferase Reporter Assay System (Promega).

Chromatin-immunoprecipitation (ChIP) assay

Testes were isolated from three wild-type mice on postnatal day (P) 8 and were fixed in 500 µl of fixation buffer (1% formaldehyde, 4.5 mM HEPES, 9 mM NaCl, 0.09 mM EDTA) for 10 min at room temperature followed by adding 55 µl of 1.5 M glycine to stop the crosslinking reaction. ChIP experiments were performed using the EZ ChIP kit (Millipore, Billerica, MA; Cat#17–371) according to the manufacture's instruction. After washing three times with 1 ml of ice-cold phosphate-buffered saline (PBS), testicular cells were lyzed in 400 µl of SDS lysis buffer and sonicated with a sonicator (Branson, Danbury, CT). After centrifugation at 18,000 *g* for 5 min, 50 µl of the supernatant was diluted with 450 µl ChIP dilution buffer containing 0.5% protease inhibitor cocktail. Magna beads and rabbit anti-SOHLH1 antibody (Abcam), rabbit anti-SP1 (ChIPAb+ Sp1, Millipore; Cat#17–601) antibody, or normal rabbit IgG were added to the samples, and incubated overnight at 4°C. Then, the samples were washed once with Low Salt Immune Complex Wash Buffer, once with High Salt Immune Complex Wash Buffer, once with LiCl Immune Complex Wash Buffer, and twice with TE buffer. The precipitated DNA was liberated from the immune complex by adding 100 µl of ChIP Elution Buffer and 1 µl of Protenase K followed by heating at 62°C for 2.5 hours. DNA was recovered using Spin filter column, eluted in 100 µl of TE buffer, and applied to qPCR. Genomic regions upstream of the *Sohlh1* gene were amplified using specific primer pairs: 5′-TGCCCCTA-GAAATCCACTAGAGACG-3′ and 5′-GATAGCTTG-CAGCTCTGTTTCTGAC-3′ for the *Sohlh1* promoter region (−371 to −284); 5′-TGACACTGTCCACAACAGGAAGGAC-3′ and 5′-ATCCAGGCTGCCTTTCACTTTCTGC-3′ for a control region far upstream of the *Sohlh1* promoter (−8946 to −8834). Accumulation of fluorescent products was monitored using the StepOnePlus Real-Time PCR System (Applied Biosystems, Foster, CA).

Results

SOHLH2 and SOHLH1 form homodimers

Previously, we demonstrated that SOHLH2 forms a heterodimer with SOHLH1 [6]. To determine whether SOHLH2 and SOHLH1 can also form homodimers, we transiently co-expressed FLAG-SOHLH2 and FLAG-SOHLH1 with SOHLH2-Myc and SOHLH1-Myc, respectively, in HEK293 cells, which endogenously express neither SOHLH2 nor SOHLH1 (**Figure 1A and 1B**). Control western blot experiments shown in **Figure 1A and 1B** confirmed that the anti-FLAG antibody did not cross-react with SOHLH2-Myc or SOHLH1-Myc, and that the anti-Myc antibody did not cross-react with FLAG-SOHLH2 or FLAG-SOHLH1. Immunoprecipitation-assay bands indicating homodimerization were detected for both SOHLH2 (**Figure 1C, lane 1**) and SOHLH1 (**Figure 1D, lane 1**). These observations suggested that SOHLH2 and SOHLH1 could form homodimers *in vivo*, in agreement with another recent report that SOHLH2 and SOHLH1 form both heterodimers and homodimers [20].

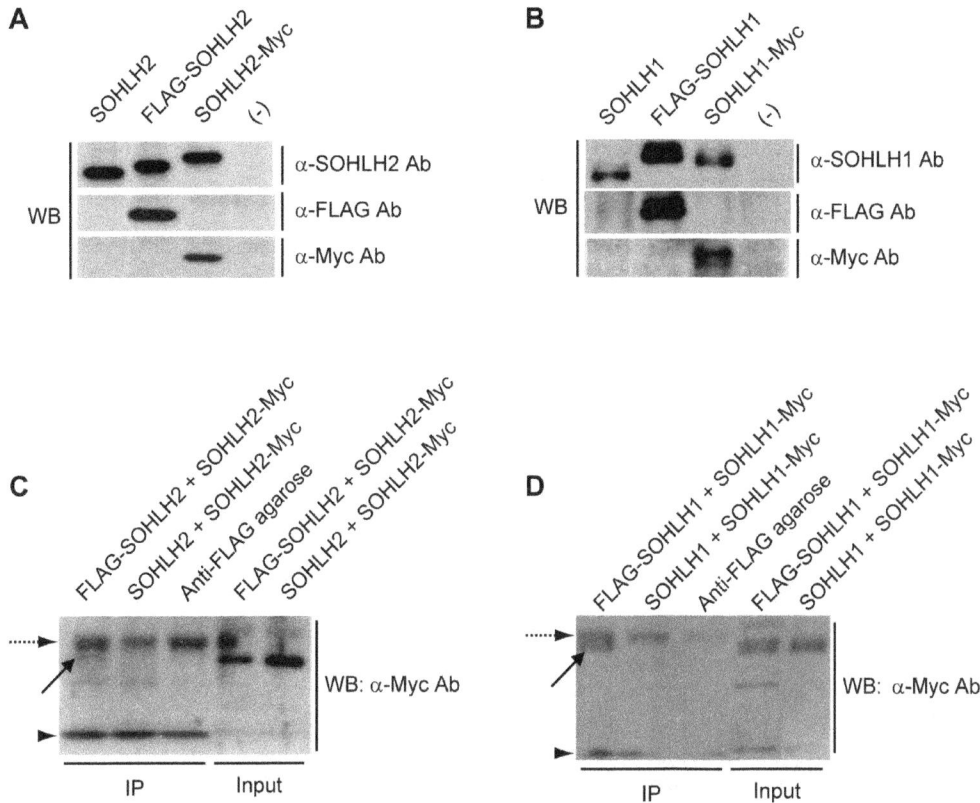

Figure 1. SOHLH2 and SOHLH1 form homodimers. (A) Western blots of lysates of HEK293 cells overexpressing SOHLH2, FLAG-SOHLH2, and SOHLH2-Myc, using anti-SOHLH2, anti-FLAG, and anti-Myc antibodies. (B) Western blots of lysates of HEK293 cells overexpressing SOHLH1, FLAG-SOHLH1, and SOHLH1-Myc, using anti-SOHLH1, anti-FLAG, and anti-Myc antibodies. (C) Lysates of HEK293 cells overexpressing FLAG-SOHLH2, SOHLH2-Myc, or SOHLH2 were immunoprecipitated with anti-FLAG agarose beads and subjected to western blotting using an anti-Myc antibody. Arrow: an SOHLH2-Myc band. Pre-immunoprecipitation lysates were used as input. (D) Lysates of HEK293 cells overexpressing FLAG-SOHLH1, SOHLH1-Myc, or SOHLH1 were immunoprecipitated with anti-FLAG agarose beads and subjected to western blotting using an anti-Myc antibody. Arrow: an SOHLH1-Myc band. Pre-immunoprecipitation lysates were used as input. Anti-FLAG agarose was loaded to indicate the IgG heavy chain (dashed arrow) and IgG light chain (arrowhead) bands.

The SOHLH2/SOHLH1 heterodimer upregulates the Sohlh1 promoter

During testicular and ovarian development, *Sohlh2* mRNA is upregulated prior to *Sohlh1* expression. The *Sohlh1* expression in *Sohlh2*-null mice remains low in the testis or ovary [6]. These observations indicated that the SOHLH2 protein might regulate *Sohlh1* gene activity by forming a homodimer or a heterodimer with SOHLH1. To evaluate the roles of SOHLH2 and SOHLH1 in regulating the *Sohlh1* promoter activity, we introduced a luciferase reporter plasmid vector containing the 1036-bp promoter region of the mouse *Sohlh1* gene, along with various amounts of plasmid vectors expressing SOHLH2 or SOHLH1, into HEK293 cells. As shown in **Figure 2A**, SOHLH2 or SOHLH1 alone did not markedly transactivate the *Sohlh1* promoter more strongly than the reporter alone. However, introducing the reporter plasmid with 0.32 ng, 1.6 ng, 8 ng, or 40 ng each of SOHLH2 and SOHLH1 expression plasmids significantly increased the reporter gene expression 2.8-, 3.4-, 5.1-, and 7.8-fold, respectively, relative to that of the reporter plasmid alone (**Figure 2A**). Thus, co-expressing SOHLH2 and SOHLH1 caused a dose-dependent increase in *Sohlh1* promoter activity.

Species-conserved E-boxes in the Sohlh1 promoter are important for transactivation

To determine which sequences in the *Sohlh1* gene promoter are required for its transcriptional activation by SOHLH2 and SOHLH1, HEK293 cells were transfected with reporter-gene plasmids containing various lengths of the 5′ upstream sequence of the *Sohlh1* promoter, along with 40 ng each of SOHLH2- and SOHLH1-expression plasmid vectors. Reporter plasmids containing the 1036-bp and 321-bp upstream sequences of the *Sohlh1* gene produced comparable luciferase activity (**Figure 2B**). However, a reporter plasmid containing the 154-bp upstream region of the *Sohlh1* gene showed 70% less luciferase activity, indicating that the region from −321 to −154 of the *Sohlh1* promoter contains important sequences for *Sohlh1*'s regulation by SOHLH2 and SOHLH1.

Since conservation between species often highlights important functional sequences, we analyzed sequences in the publicly available NCBI genomic database (http://www.ncbi.nlm.nih.gov/). We found that the mouse *Sohlh1* gene contains three E-boxes (CACGTG) from −240 bp to −284 bp upstream of its coding region, and that these sequences are well conserved in the rat. This conservation suggested that the SOHLH proteins might regulate the *Sohlh1* gene through these E-boxes, which were designated E1, E2, and E3 (proximal, middle, and distal, respectively) (**Figure 3**).

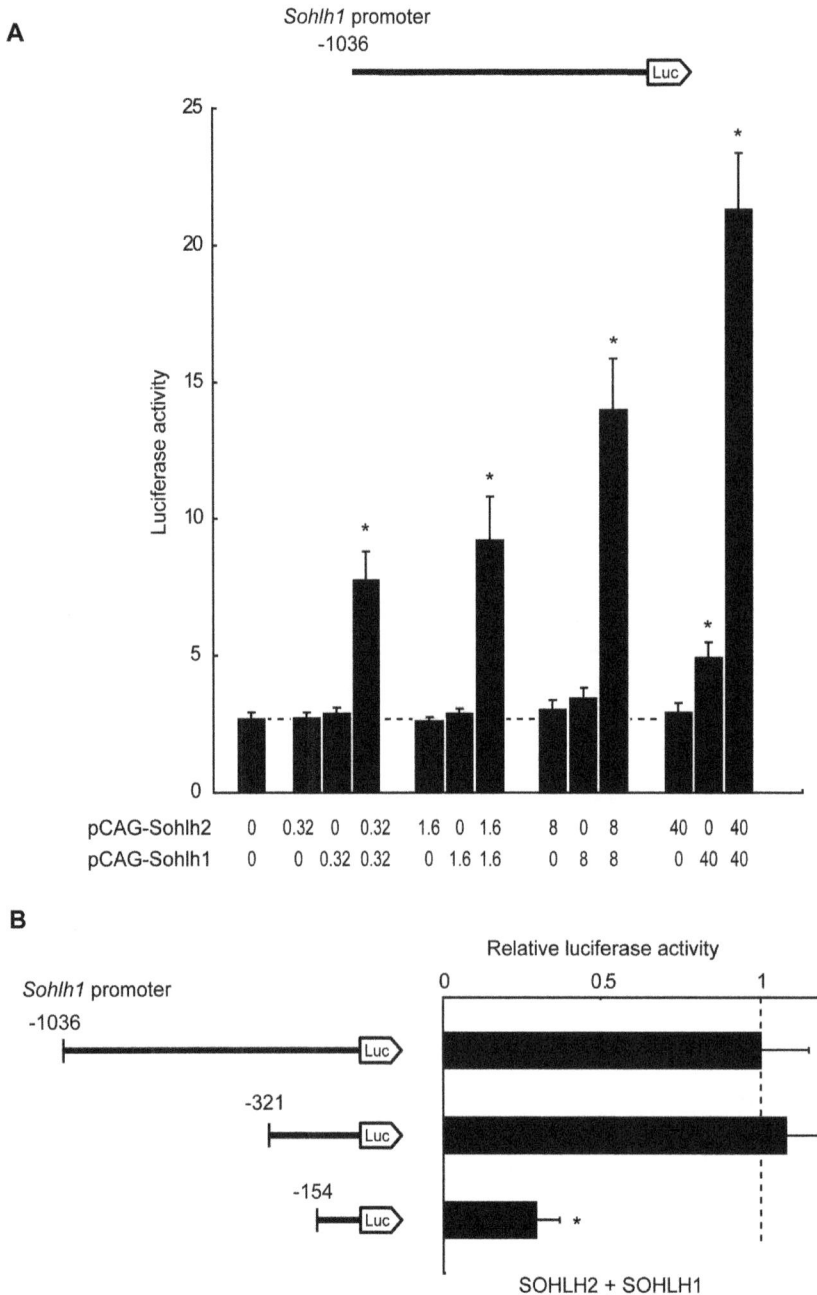

Figure 2. SOHLH2/SOHLH1 heterodimer regulates the *Sohlh1* promoter through a restricted upstream region. (A) Reporter assay using the pGL3-Basic vector containing the −1036 bp mouse *Sohlh1* promoter (200 ng), a pCAG-Sohlh2 expression vector (0 to 40 ng), a pCAG-Sohlh1 expression vector (0 to 40 ng), and a pRL-CMV normalization vector (0.1 ng). Results show the *Sohlh1* promoter-driven firefly luciferase activity relative to CMV promoter-driven Renilla luciferase activity. (B) Reporter assays using pGL3-Basic vectors containing various lengths of the mouse *Sohlh1* promoter (200 ng), a pCAG-Sohlh2 expression vector (200 ng), a pCAG-Sohlh1 expression vector (200 ng), and a pRL-CMV normalization vector (0.1 ng). Results show the firefly/Renilla luciferase activity relative to that obtained with the −1036 bp promoter fragment, which was arbitrarily set at 1. Error bars represent the S.E.M. of the means of 3–5 separate experiments done in triplicate. *P* values were calculated by Student's *t*-test. *$P < 0.05$. SOHLH2 and SOHLH1 regulate the activity of the mouse *Sohlh1* promoter through a region −154 to −321 bp upstream from its translational start site.

The E-boxes are not equal in regulating the *Sohlh1* gene

To determine whether the SOHLH2/SOHLH1 heterodimer regulates the *Sohlh1* gene through these E-boxes, we constructed *Sohlh1* promoters containing mutations in specific E-boxes (CACGTG to GGATCC) and expressed them with SOHLH2 and SOHLH1 in reporter assays. While the promoter with an E1 or E2 mutation showed approximately 25% and 50% less luciferase activity, respectively, compared with the intact promoter, the E3-mutant promoter showed approximately 75% less activity (**Figure 4**). A comparable reduction was seen with an E1/E2/E3 triple-mutant promoter. These observations suggested that

```
        -355                                          -321
          |                                             |
Mouse  acgggttacttggaactgtgttgctactttctgacaaaggggtcagaaacagagctgcaa
  Rat  acctgttgaacagaaccgtgtagctactttctggcaaagcggccaaaagcagagctgcaa
       **    ***     **** **** ********** ***** ** ** ** **********
          |
        -344

        -284                    -264                   -245
          |                       |                      |
Mouse  gctatCACGTGggctgtgatcaggtCACGTGgtctgaaggtctgCACGTGaagcagggag
  Rat  gcaatCACGTGggctgtgatcaggtCACGTGgtctgaagggctgCACGTGaagcagtgag
       ** ************************************** *************** ***
           |___E3            |___E2               |___E1
        -273               -253                  -234

            -216
              |
Mouse  tcaaacacgagaaGGGGCGGGGGCaagcccagcacgaagtgggcagg--tggggcggggag
  Rat  tcaaacacgaggaGGGGCGGGGGCaagctcagcacgaggtgggcaggcttgggagggggag
       ***********  **************** ******* ********  ****  ******
                   |___GC___
                -205

                      -154          -141      -1
                        |             |         |
Mouse  gcggggggggggggcg-tggagtgagacgcttgcataa....ccaATGgcg
  Rat  g-gggtggcgagacgcttgcatgaga-gtgggca-gc....ccaATGgcg
       * **** ** * ** *  ***** *   ***   ********
                                      |         |
                                    -130       -1
```

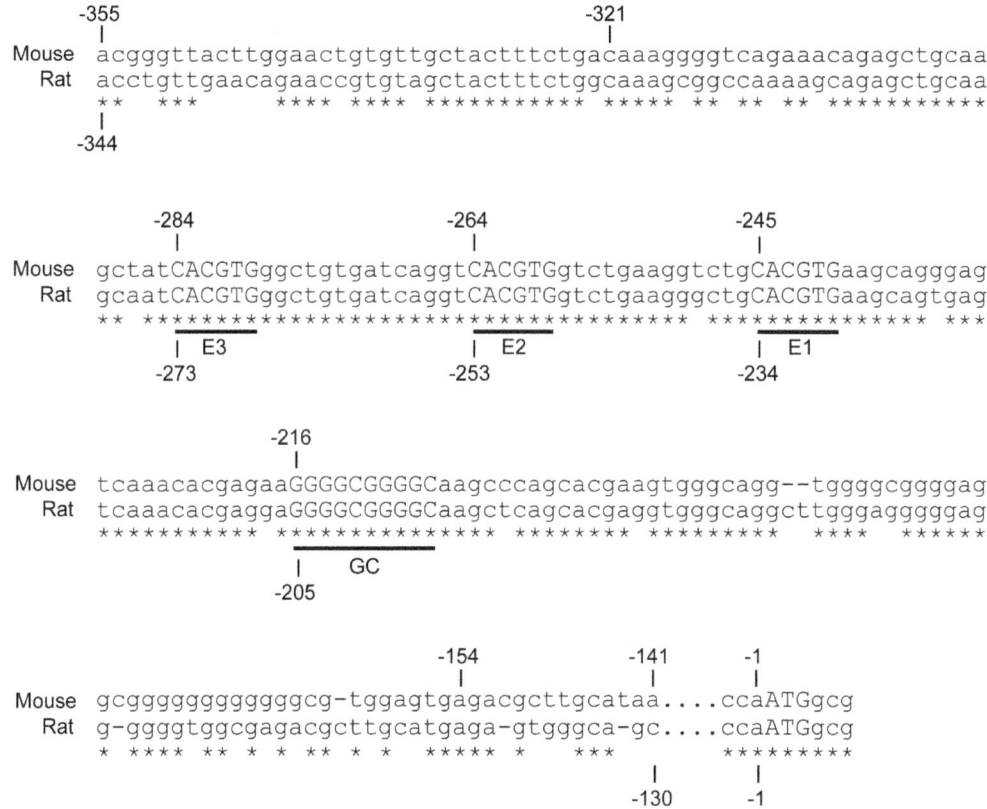

Figure 3. Conserved regulatory regions of the mouse and rat *Sohlh1* gene. The underlined E- and GC-box sequences are conserved between the mouse and rat. These E- and GC-box sequences are also found in the *Sohlh1* promoter region of the chimpanzee and human (not shown).

the E3 box was the most important site of SOHLH2/SOHLH1 heterodimer binding.

SOHLH2, SOHLH1, and SP1 are functionally associated in *Sohlh1* promoter activation

While searching for conserved sequences in the region from −321 to −154 bp upstream of the *Sohlh1* gene, we also found a species-conserved GC-box (GGGGCGGGGC), which contains the binding sequence of the widely expressed transcription factor SP1, neighboring the E-boxes (**Figure 3**). Some bHLH proteins interact with SP1 to cooperatively activate target genes by binding to juxtaposed E- and GC-boxes [14,15]. Therefore, it was possible that the SOHLH proteins cooperate with SP1 to activate *Sohlh1* through E- and GC-boxes in its promoter region.

To investigate the importance of the GC-box, we introduced the reporter vector containing the *Sohlh1* promoter and expression

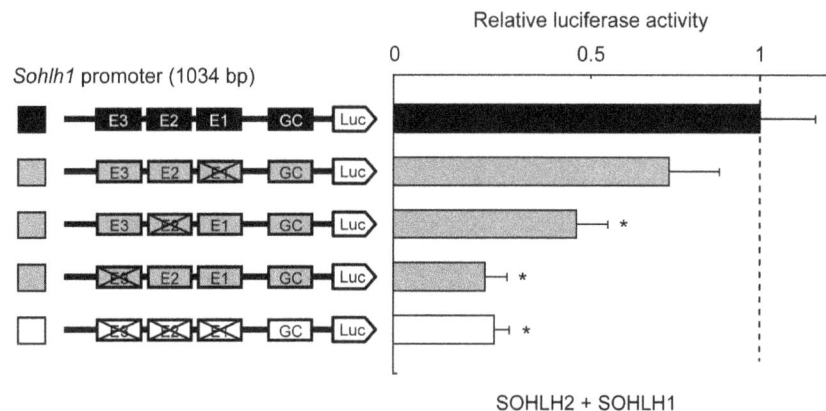

Figure 4. The three *Sohlh1*-promoter E-boxes are not equal in regulating *Sohlh1*. Reporter assays using the pGL3-Basic vector with various mutations of the E-boxes in the *Sohlh1* promoter, along with pCAG-Sohlh1 and pCAG-Sohlh2 expression vectors (200 ng each). Results show the firefly/Renilla luciferase activity relative to that of the −1036 bp intact promoter, which was arbitrarily set at 1. Error bars represent the S.E.M. of the means of 3–6 separate experiments done in triplicate. P values were calculated by Student's *t*-test. *P<0.05.

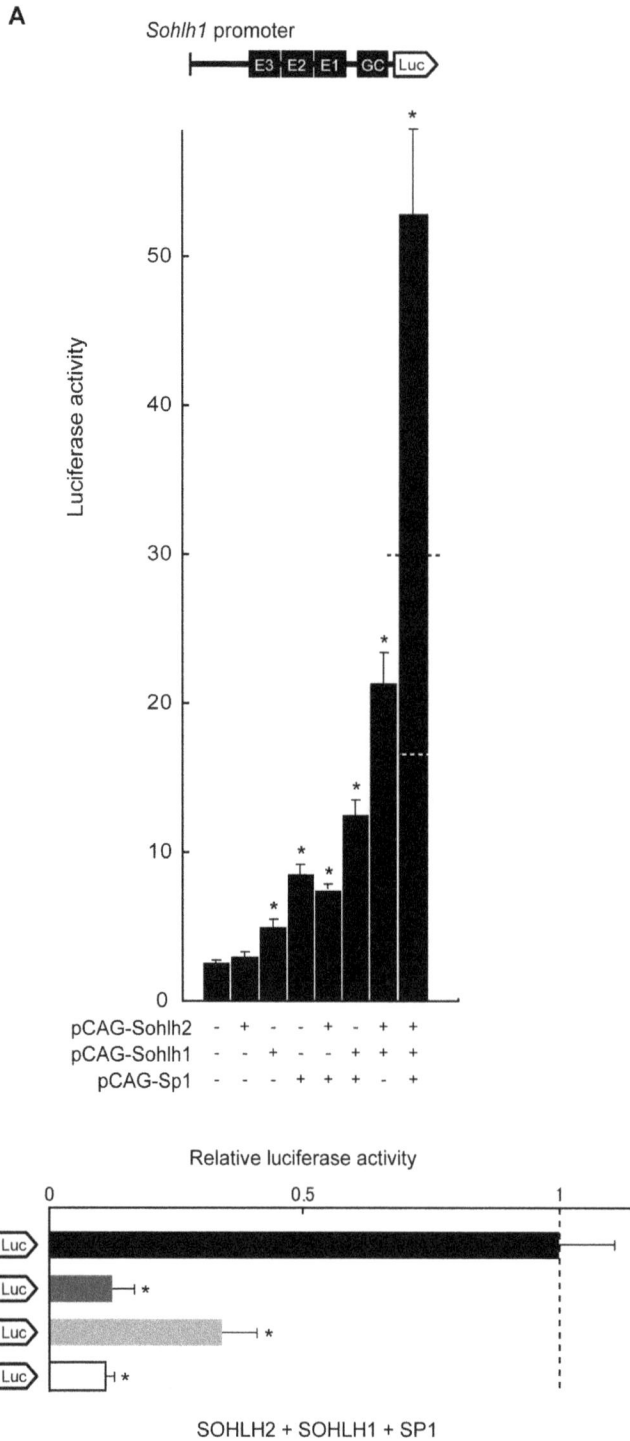

Figure 5. Transcriptional synergy between SP1 and the SOHLH proteins. (A) Reporter assays using the pGL3-Basic vector with the −1036 bp intact *Sohlh1* promoter (200 ng), a pRL-CMV normalization vector (0.1 ng), and expression vectors (pCAG-Sohlh2 (40 ng), pCAG-Sohlh1 (40 ng), and pCAG-Sp1 (40 ng)). Results show the *Sohlh1* promoter-driven firefly luciferase activity relative to that of CMV promoter-driven Renilla luciferase. The white dashed line indicates the sum of the individual transcriptional activities of SOHLH2, SOHLH1, and SP1. The black dashed line indicates the sum of the transcriptional activities of SOHLH2+SOHLH1 and SP1. *P* values were calculated by Mann-Whitney test. *$P<0.05$. (B) Reporter assays using the pGL3-Basic vector containing the −1036 bp *Sohlh1* promoter with various mutations in the E-boxes and/or GC-box (200 ng), a pRL-CMV normalization vector (0.1 ng), and expression vectors (pCAG-Sohlh2 (40 ng), pCAG-Sohlh1 (40 ng), and pCAG-Sp1 (40 ng)). Results show the firefly/Renilla luciferase activity relative to that of the intact −1036 bp *Sohlh1* promoter (black bar), which was arbitrarily set at 1. *P* values were calculated by Welch's *t*-test. *$P<0.05$. Error bars represent the S.E.M. of the means of 3–8 separate experiments done in triplicate.

A

B

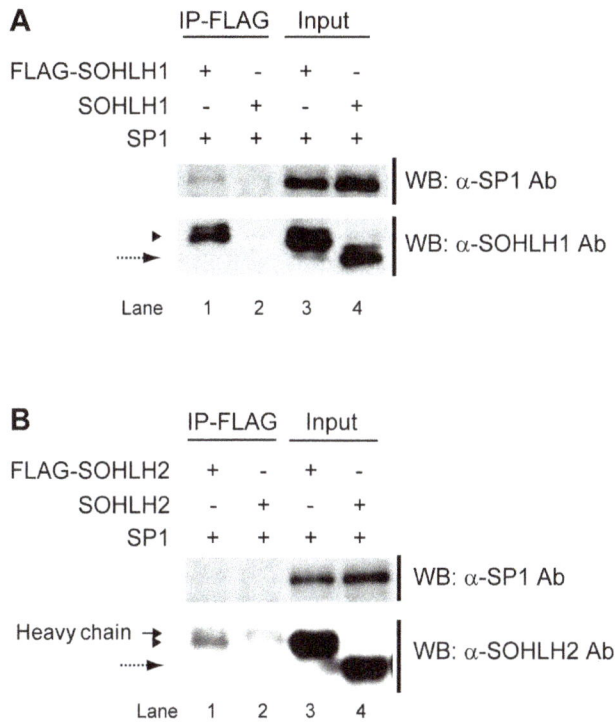

Figure 6. Interaction of SP1 to the SOHLH proteins. (A) Lysates of HEK293 cells overexpressing SP1 and FLAG-SOHLH1 or SOHLH1 were immunoprecipitated with anti-FLAG agarose beads and analyzed by western blotting using anti-SOHLH1 or anti-SP1 antibodies. Arrowhead: FLAG-SOHLH1. Dashed arrow: SOHLH1. (B) Lysates of HEK293 cells overexpressing SP1 and FLAG-SOHLH2 or SOHLH2 were immunoprecipitated with anti-FLAG agarose beads and analyzed by western blotting using anti-SOHLH2 or anti-SP1 antibodies. Arrowhead: FLAG-SOHLH2. Dashed arrow: SOHLH2. Arrow: IgG heavy chain. Pre-immunoprecipitation lysate was used as input.

A

B

Figure 7. Binding of SOHLH1 and SP1 to the *Sohlh1* promoter region *in vivo*. Binding of SOHLH1 and SP1 to the *Sohlh1* promoter region *in vivo* was examined by ChIP assay using P8 testes. The *Sohlh1* promoter region (-371 to -284) and the control region far upstream of the *Sohlh1* promoter (-8946 to -8834) were quantitated by Real-time PCR from chromatin fractions immunoprecipitated with anti-SOHLH1 antibody, anti-SP1 antibody, or control rabbit IgG. Fold enrichment represents the quantity of the region immunoprecipitated with anti-SOHLH1 (A) or anti-SP1 (B) relative to that immunoprecipitated with control rabbit IgG. Values are expressed as means ±S.E.M. of three technical replicates. P values were calculated by Student's t-test. *P< 0.03.

plasmid vectors for SOHLH2, SOHLH1, or SP1, alone or in combination, into HEK293 cells. Expressing SOHLH2, SOHLH1, or SP1 alone with the wild-type *Sohlh1* promoter vector increased the reporter activity by1.1-, 1.8-, or 2.9-fold above the basal level (the level of the reporter vector alone) (**Figure 5A**). Expressing SOHLH2 and SP1 or SOHLH1 and SP1 with the wild-type *Sohlh1* promoter vector enhanced the reporter activity by 2.5-, or 4.0-fold above the basal level, respectively (**Figure 5A**). However, expressing SOHLH2, SOHLH1, and SP1 together increased the reporter activity by approximately 19.2-fold above the basal level, far exceeding the sum of the activation levels obtained with each factor individually (**Figure 5A**). Thus, SOHLH2, SOHLH1, and SP1 transactivated the *Sohlh1* promoter synergistically.

We next examined whether this synergistic transcriptional activation of the *Sohlh1* gene by SOHLH2, SOHLH1, and SP1 required the binding sites we had postulated for these factors. In addition to the promoter with mutations in all three E-boxes, we produced an *Sohlh1* promoter with a mutation in the GC-box (GGGGCGGGGC to GAAGCTTGTC). We introduced these E-box or GC-box mutants, alone or in a combined, double-mutant reporter plasmid, into HEK293 cells along with SOHLH2, SOHLH1, and SP1 expression vectors (**Figure 5B**), and found that reporter activity decreased significantly, to 12.5%, 34.1%, and 10.9% of the activity of the intact promoter, with the E-box, GC-box, and double E- and GC-box mutations, respectively. These

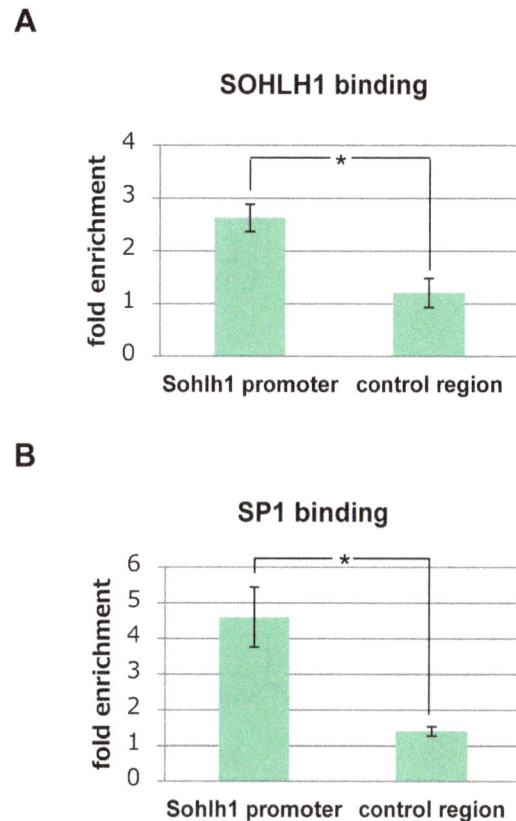

observations suggested that the SOHLH proteins, in cooperation with SP1, transactivate *Sohlh1* through juxtaposed E- and GC-boxes.

SOHLH1 binds to SP1

To determine whether SOHLH proteins interact physically with SP1, we co-expressed SP1 and FLAG-tagged SOHLH1 or SOHLH2 in HEK293 cells. The FLAG-tagged SOHLH proteins immunoprecipitated from cell lysates with an anti-FLAG antibody were then analyzed by western blotting with an anti-SP1 antibody. These experiments showed that FLAG-SOHLH1 could associate with co-expressed SP1 (**Figure 6A**). On the other hand, the co-expression of FLAG-SOHLH2 and SP1 did not reveal any detectable association between SOHLH2 and SP1 (**Figure 6B**). Considering the heterodimerization of SOHLH2 and SOHLH1, SOHLH1 might act as a bridge between SOHLH2 and SP1, resulting in the formation of the SOHLH2/SOHLH1/SP1 complex. As it was reported that the DNA-binding domain of SP1 and the HLH domain of MYOGENIN or NEUROD1 mediate protein-protein interactions [14,15], it is possible that the

A

B

C

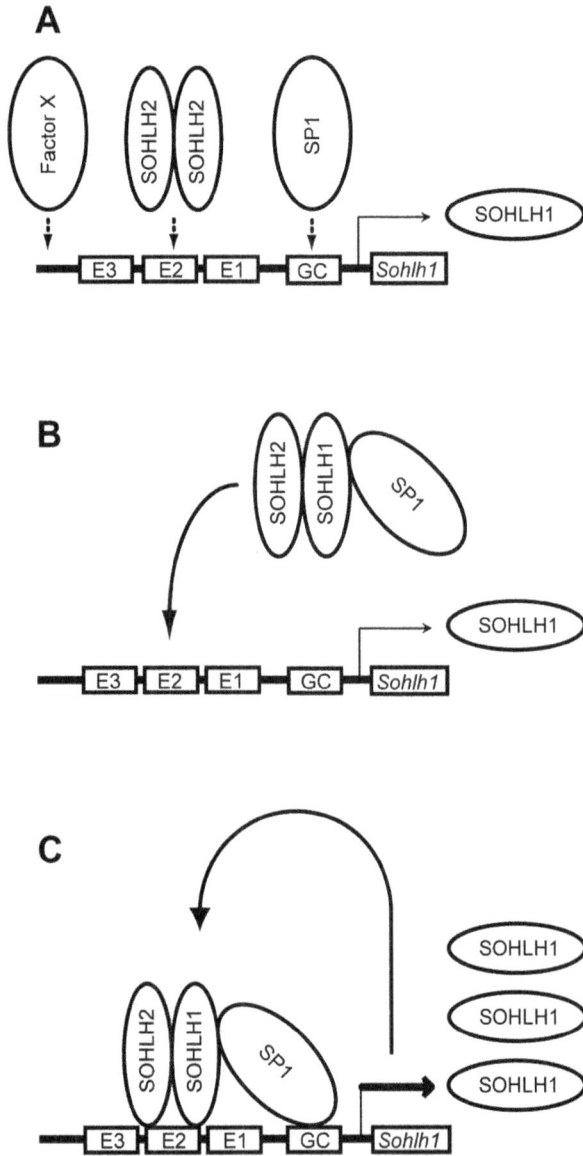

Figure 8. A model of *Sohlh1* gene regulation by the SOHLH2/ SOHLH1/SP1 complex. (A) The SOHLH2 homodimer, SP1, or an unknown Factor X turns on weak *Sohlh1* transcription. (B) The resulting small amounts of SOHLH1 form the SOHLH2/SOHLH1/SP1 complex due to affinity. This complex is recruited to the E- and GC-box regions in the *Sohlh1* promoter. (C) The *Sohlh1* gene is highly upregulated through its auto-regulatory mechanism, which involves the SOHLH2/SOHLH1/SP1 complex.

HLH domain of SOHLH1 also associates with SP1's DNA-binding domain.

SOHLH1 and SP1 are recruited to the *Sohlh1* promoter region *in vivo*

To examine whether SOHLH1 and SP1 are recruited to the *Sohlh1* promoter region *in vivo*, we performed ChIP assay using P8 testes (**Figure 7**). By qPCR, the *Sohlh1* promoter region (-371 to -284) was shown to be significantly enriched in SOHLH1- and SP1-immunoprecipitated chromatin fractions, while a control region far upstream of the *Sohlh1* promoter (-8946 to -8834) was not. These results indicated that the sequences located immedi-

ately upstream of the *Sohlh1* transcription start site are bound by SOHLH1 and SP-1 *in vivo*.

Discussion

In our previous study, we reported that abnormalities in the testes and ovaries of *Sohlh2*-null mice are similar to those seen in *Sohlh1*-null mice, and that *Sohlh1* transcription is downregulated in the gonads of *Sohlh2*-null mice [6]. We also demonstrated that SOHLH2 can form a heterodimer with SOHLH1 [6]. In the current study, we showed that SOHLH2 and SOHLH1 could also form homodimers (**Figure 1C, D**). We further demonstrated that SOHLH2 and SOHLH1, expressed together, upregulated the *Sohlh1* promoter through its E-boxes, while SOHLH2 or SOHLH1 alone upregulated this promoter activity only weakly (**Figure 2A**).

Similar observations have been made for MAX and MYC. MAX, a MYC-family bHLH protein, not only forms homodimers, but also forms heterodimers, preferentially with MYC [21,22]. The MAX/MYC heterodimer binds to the E-box sequence CACGTG with higher affinity than does MAX or MYC alone [21]. MAX homodimers have no transcriptional activity, but the MAX/MYC heterodimer is a principal transcriptional activator [10,11]. In this respect, SOHLH2 and SOHLH1 appear to function similarly to MAX and MYC.

We next demonstrated that the SOHLH proteins and SP1, which are all present in germ cells, are functionally linked. A species-conserved GC-box, which contains the SP1 consensus binding sequence, is adjacent to the *Sohlh1* promoter's E-boxes (**Figure 3**). Some bHLH proteins, including NEUROD1, interact with SP1 to synergistically activate target genes containing juxtaposed E- and GC-boxes [14,15,23]. It has been reported that the synergy between E12/NEUROD1 and SP1 occurs when the bHLH proteins recruit SP1 and stabilize its DNA-binding [15]. The association we observed between SOHLH1 and SP1 suggests that the SOHLH2/SOHLH1 heterodimer might also recruit SP1 and stabilize SP1's binding to the GC-box, thereby synergistically activating the *Sohlh1* promoter activity to its greatest extent (**Figure 8**). Consistent with this notion, ChIP assay using P8 testes showed that SOHLH1 and SP1 bind to the *Sohlh1* promoter region *in vivo* (**Figure 7**).

Of the three E-boxes (E1, E2, and E3), the reporter activity was reduced most greatly by a mutation in E3 (**Figure 4**). The ternary complex of NEUROD1, E12, and SP1 requires proper spacing between the E- and GC-boxes for the strongest promoter activation [15]. The SOHLH2/SOHLH1/SP1 complex is also likely to require a particular spacing between the E- and GC-boxes for maximum promoter activation, with the E3 E-box being at the most appropriate distance from the GC-box for promoter activation by the SOHLH2/SOHLH1/SP1 ternary complex.

In early spermatogenesis and oogenesis, the *Sohlh2* gene is upregulated prior to the *Sohlh1* gene [6]. However, SOHLH2 was barely able to transactivate the *Sohlh1* promoter in the absence of SOHLH1 (**Figure 2A**), so *Sohlh2* upregulation alone may not lead to *Sohlh1*'s transcription *in vivo*. The mechanism that initially activates *Sohlh1*'s transcription remains unknown. It is possible that SOHLH2 homodimers have weak transcriptional activity. Similarly, although MAX homodimers are generally thought to repress transcription, they have been found to activate transcription at low levels in a yeast system [24]. Another possibility is that other factors are responsible for turning on *Sohlh1*'s transcription (**Figure 8A**). Since *Sohlh1* transcription is detected in the gonads of *Sohlh2*-null mice [6], the *Sp1* transcript was consistently observed throughout spermatogenesis and oogenesis (data not shown), and

SP1 alone was able to activate weak but detectable *Sohlh1* transcription (**Figure 5A**), SP1 might be an initiation factor for *Sohlh1* transcription (**Figure 8A**).

In *Sohlh2*-null mice, *Kit* expression is downregulated in both the testis and ovary, and KIT-positive germ-cell differentiation is disturbed [6]. In *Sohlh1*-null mice as well, KIT-positive germ-cell differentiation appears to be disturbed in the testis and ovary [5,8]. KIT is expressed in A_{diff} spermatogonia and in primordial-to-growing oocytes, corresponding to the SOHLH2 and SOHLH1 expression. SP1 is also expressed in spermatogonia [16] and oocytes [17]. Recently, Barrios et al. [25] reported that SOHLH2 and SOHLH1 control the *Kit* expression during postnatal male germ-cell development. The *Kit* proximal promoter is reported to contain E- and GC-boxes [26,27]. These reports suggest that the SOHLH2/SOHLH1/SP1 complex might directly regulate expression of the *Kit* gene *in vivo* through its E- and GC-boxes.

Bioinformatics analyses have predicted that a number of spermatogonia-related genes contain E- and/or GC-boxes in their 5′-cis regulatory elements [28]. The SOHLH2/SOHLH1/SP1 ternary complex might be a key factor in this transcriptional cascade. The self-regulated activity of the *Sohlh1* promoter and the synergistic action of SOHLH2, SOHLH1, and SP1 could promote rapid germ-cell differentiation (**Figure 8**). It was recently shown that retinoic acid and BMP4 are commonly involved in both early spermatogenesis and oogenesis [29–32], and that retinoic acid signaling directly cooperates with SP1 [33]. Further investigation of the interactions among these transcription factors and their signaling pathways should deepen our understanding of the common differentiation mechanisms of early spermatogenesis and oogenesis.

Acknowledgments

The authors are grateful to Mr. Masafumi Ashida for excellent technical assistance. We acknowledge the editorial assistance of Drs. Leslie A. Miglietta and Grace E. Gray.

Author Contributions

Conceived and designed the experiments: ST JiM. Performed the experiments: ST TY JM. Analyzed the data: ST TY JiM. Wrote the paper: ST TY JiM.

References

1. Mithraprabhu S, Loveland KL (2009) Control of KIT signalling in male germ cells: what can we learn from other systems? Reproduction 138: 743–757.
2. Edson MA, Nagaraja AK, Matzuk MM (2009) The mammalian ovary from genesis to revelation. Endocr Rev 30: 624–712.
3. Ballow DJ, Xin Y, Choi Y, Pangas SA, Rajkovic A (2006) *Sohlh2* is a germ cell-specific bHLH transcription factor. Gene Expr Patterns 6: 1014–1018.
4. Pangas SA, Choi Y, Ballow DJ, Zhao Y, Westphal H, et al. (2006) Oogenesis requires germ cell-specific transcriptional regulators Sohlh1 and Lhx8. Proc Natl Acad Sci USA 103: 8090–8095.
5. Ballow D, Meistrich ML, Matzuk M, Rajkovic A (2006) *Sohlh1* is essential for spermatogonial differentiation. Dev Biol 294: 161–167.
6. Toyoda S, Miyazaki T, Miyazaki S, Yoshimura T, Yamamoto M, et al. (2009) Germ cell specific bHLH gene, *Sohlh2*, is required for differentiation of Kit positive spermatogonia and oocytes. Dev Biol 325: 238–248.
7. Hao J, Yamamoto M, Richardson TE, Chapman KM, Denard BS, et al. (2008) Sohlh2 knockout mice are male-sterile because of degeneration of differentiating type A spermatogonia. Stem Cells 26: 1587–1597.
8. Choi Y, Yuan D, Rajkovic A (2008) Germ cell-specific transcriptional regulator sohlh2 is essential for early mouse folliculogenesis and oocyte-specific gene expression. Biol Reprod 79: 1176–1182.
9. Swanson HI, Yang JH (1999) Specificity of DNA binding of the c-Myc/Max and ARNT/ARNT dimers at the CACGTG recognition site. Nucleic Acids Res 27: 3205–3212.
10. Amati B, Dalton S, Brooks MW, Littlewood TD, Evan GI, et al. (1992) Transcriptional activation by the human c-Myc oncoprotein in yeast requires interaction with Max. Nature 359: 423–426.
11. Kretzner L, Blackwood EM, Eisenman RN (1992) Myc and Max proteins possess distinct transcriptional activities. Nature 359: 426–429.
12. Saffer JD, Jackson SP, Annarella MB (1991) Developmental expression of Sp1 in the mouse. Mol Cell Biol 11: 2189–2199.
13. Briggs MR, Kadonaga JT, Bell SP, Tjian R (1986) Purification and biochemical characterization of the promoter-specific transcription factor, Sp1. Science 234: 47–52.
14. Biesiada E, Hamamori Y, Kedes L, Sartorelli V (1999) Myogenic basic helix-loop-helix proteins and Sp1 interact as components of a multiprotein transcriptional complex required for activity of the human cardiac α-actin promoter. Mol Cell Biol 19: 2577–2584.
15. Ray SK, Leiter AB (2007) The basic helix-loop-helix transcription factor NeuroD1 facilitates interaction of Sp1 with the secretin gene enhancer. Mol Cell Biol 27: 7839–7847.
16. Ma W, Horvath GC, Kistler MK, Kistler WS (2008) Expression patterns of SP1 and SP3 during mouse spermatogenesis: SP1 down-regulation correlates with two successive promoter changes and translationally compromised transcripts. Biol Reprod 79: 289–300.
17. Worrad DM, Schultz RM (1997) Regulation of gene expression in the preimplantation mouse embryo: temporal and spatial patterns of expression of the transcription factor Sp1. Mol Reprod Dev 46: 268–277.
18. Yoshimura T, Miyazaki T, Toyoda S, Miyazaki S, Tashiro F, et al. (2007) Gene expression pattern of *Cue110*: a member of the uncharacterized UPF0224 gene family preferentially expressed in germ cells. Gene Expr Patterns 8: 27–35.
19. Niwa H, Masui S, Chambers I, Smith AG, Miyazaki J (2002) Phenotypic complementation establishes requirements for specific POU domain and generic transactivation function of Oct-3/4 in embryonic stem cells. Mol Cell Biol 22: 1526–1536.
20. Suzuki H, Ahn HW, Chu T, Bowden W, Gassei K, et al. (2012) SOHLH1 and SOHLH2 coordinate spermatogonial differentiation. Dev Biol 361: 301–312.
21. Blackwood EM, Eisenman RN (1991) Max: a helix-loop-helix zipper protein that forms a sequence-specific DNA-binding complex with Myc. Science 251: 1211–1217.
22. Prendergast GC, Lawe D, Ziff EB (1991) Association of Myn, the murine homolog of max, with c-Myc stimulates methylation-sensitive DNA binding and ras cotransformation. Cell 65: 395–407.
23. Kyo S, Takakura M, Taira T, Kanaya T, Itoh H, et al. (2000) Sp1 cooperates with c-Myc to activate transcription of the human telomerase reverse transcriptase gene (*hTERT*). Nucleic Acids Res 28: 669–677.
24. Fisher F, Crouch DH, Jayaraman PS, Clark W, Gillespie DA, et al. (1993) Transcription activation by Myc and Max: flanking sequences target activation to a subset of CACGTG motifs *in vivo*. EMBO J 12: 5075–5082.
25. Barrios F, Filipponi D, Campolo F, Gori M, Bramucci F, et al. (2012) SOHLH1 and SOHLH2 control Kit expression during postnatal male germ cell development. J Cell Sci 125: 1455–64.
26. Park GH, Plummer HK 3rd, Krystal GW (1998) Selective Sp1 binding is critical for maximal activity of the human c-*kit* promoter. Blood 92: 4138–4149.
27. Lécuyer E, Herblot S, Saint-Denis M, Martin R, Begley CG, et al. (2002) The SCL complex regulates c-*kit* expression in hematopoietic cells through functional interaction with Sp1. Blood 100: 2430–2440.
28. Lee TL, Alba D, Baxendale V, Rennert OM, Chan WY (2006) Application of transcriptional and biological network analyses in mouse germ-cell transcriptomes. Genomics 88: 18–33.
29. van Pelt AM, van Dissel-Emiliani FM, Gaemers IC, van der Burg MJ, Tanke HJ, et al. (1995) Characteristics of A spermatogonia and preleptotene spermatocytes in the vitamin A-deficient rat testis. Biol Reprod 53: 570–578.
30. Li H, Clagett-Dame M (2009) Vitamin A deficiency blocks the initiation of meiosis of germ cells in the developing rat ovary *in vivo*. Biol Reprod 81: 996–1001.
31. Ding X, Zhang X, Mu Y, Li Y, Hao J (2013) Effects of BMP4/SMAD signaling pathway on mouse primordial follicle growth and survival via up-regulation of *Sohlh2* and c-*kit*. Mol Reprod Dev 80: 70–78.
32. Pellegrini M, Grimaldi P, Rossi P, Geremia R, Dolci S (2003) Developmental expression of BMP4/ALK3/SMAD5 signaling pathway in the mouse testis: a potential role of BMP4 in spermatogonia differentiation. J Cell Sci 116: 3363–3372.
33. Kumar P, Garg R, Bolden G, Pandey KN (2010) Interactive roles of Ets-1, Sp1, and acetylated histones in the retinoic acid-dependent activation of guanylyl cyclase/atrial natriuretic peptide receptor-A gene transcription. J Biol Chem 285: 37521–37530.

11

Reversible Suppression of Cyclooxygenase 2 (COX-2) Expression *In Vivo* by Inducible RNA Interference

Anne K. Zaiss[1,2], Johannes Zuber[3¤a], Chun Chu[1,2], Hidevaldo B. Machado[1,2], Jing Jiao[1,2],
Arthur B. Catapang[1,2], Tomo-o Ishikawa[1,2], Jose S. Gil[1,2], Scott W. Lowe[3], Harvey R. Herschman[1,2*¤b]

1 Department of Medical and Molecular Pharmacology, David Geffen School of Medicine, University of California Los Angeles, Los Angeles, California, United States of America, **2** Department of Biological Chemistry, David Geffen School of Medicine, University of California Los Angeles, Los Angeles, California, United States of America, **3** Cold Spring Harbor Laboratory and Howard Hughes Medical Institute, New York, New York, United States of America

Abstract

Prostaglandin-endoperoxide synthase 2 (PTGS2), also known as cyclooxygenase 2 (COX-2), plays a critical role in many normal physiological functions and modulates a variety of pathological conditions. The ability to turn endogenous COX-2 on and off in a reversible fashion, at specific times and in specific cell types, would be a powerful tool in determining its role in many contexts. To achieve this goal, we took advantage of a recently developed RNA interference system in mice. An shRNA targeting the *Cox2* mRNA 3′untranslated region was inserted into a microRNA expression cassette, under the control of a tetracycline response element (TRE) promoter. Transgenic mice containing the COX-2-shRNA were crossed with mice encoding a CAG promoter-driven reverse tetracycline transactivator, which activates the TRE promoter in the presence of tetracycline/doxycycline. To facilitate testing the system, we generated a knockin reporter mouse in which the firefly luciferase gene replaces the *Cox2* coding region. *Cox2* promoter activation in cultured cells from triple transgenic mice containing the luciferase allele, the shRNA and the transactivator transgene resulted in robust luciferase and COX-2 expression that was reversibly down-regulated by doxycycline administration. *In vivo*, using a skin inflammation-model, both luciferase and COX-2 expression were inhibited over 80% in mice that received doxycycline in their diet, leading to a significant reduction of infiltrating leukocytes. In summary, using inducible RNA interference to target COX-2 expression, we demonstrate potent, reversible *Cox2* gene silencing *in vivo*. This system should provide a valuable tool to analyze cell type-specific roles for COX-2.

Editor: Junming Yue, The University of Tennessee Health Science Center, United States of America

Funding: This work was supported by National Institutes of Health grant P50 CA086306-08 (HRH). AKZ is the recipient of an American Society of Hematology (ASH) Scholar Award. SWL is an investigator of the Howard Hughes Medical Institute. The funders had no role in study design, data collection and analysis, decision to publish, or preparation of the manuscript.

Competing Interests: The authors have read the journal's policy and have the following competing interests: SWL and JZ are members of the Scientific Advisory Board and hold equity in Mirimus Inc., a company that has licensed some of the technology reported in this manuscript.

* Email: hherschman@mednet.ucla.edu

¤a Current address: Research Institute of Molecular Pathology (IMP), Vienna, Austria
¤b Current address: Memorial Sloan-Kettering Cancer Center, New York, New York, New York, United States of America

Introduction

Prostanoids modulate a number of complex biological processes, including inflammation and immunity, pregnancy and parturition, cardiovascular function, temperature regulation, neurodegeneration and tumor progression [1]. Prostanoid production is initiated by the cyclooxygenases (COX); enzymes that convert arachidonic acid to the unstable intermediate prostaglandin H2 (PGH_2). PGH_2 is then further converted, in cell-specific pathways, to a variety of products that include a number of prostaglandins (PGE_2, PGF_{2a}, PGD_2, etc.), prostacyclin (PGI_2) and thromboxane (TXA_2).

Vertebrate species have two *Cox* genes, *Cox1* and *Cox2*. The COX-1 and COX-2 enzymes are isoforms, similar in structure, that catalyze identical reactions. However, they differ in the regulation of their expression and their biological roles. Most cells constitutively express the COX-1 isoform. Consequently, COX-1 is thought to play a role in homeostatic functions. In contrast, COX-2 is not constitutively expressed in most tissues, but is rapidly induced in many cell types in response to a wide range of mitogens, cytokines and other stimuli, and is often present in elevated levels at sites of inflammation and epithelial cancers [2].

The cyclooxygenases are the targets of nearly all commonly used non-steroidal anti-inflammatory drugs (NSAIDs; e.g., ibuprofen and aspirin). These drugs inhibit COX-1 and COX-2 enzymatic activity, thereby preventing prostanoid production. Prostaglandins are recognized as important modulators of the immune system; COX-2 has been associated with an orchestrating role in both the induction and resolution of acute inflammation [3]. COX-2 is also expressed in sterile inflammation, and is induced by pathogens, but its role in anti-pathogen immune responses is not well understood [4,5].

The therapeutic effects of NSAIDs and their side effects, as well as studies with COX-2 selective inhibitors (e.g., celecoxib) and with conventional *Cox2* knockout mice suggest roles for COX-2 in a variety of diseases beyond the symptomatic control of pain and fever. Examples of suggested roles for COX-2 aberrant expression include problems in female fertility and parturition [6] and in neurodegenerative diseases such as Alzheimer's disease and Parkinson's disease [7]. Considerable research has also been devoted to understanding the role of COX-2 in cancer. COX-2 dependent prostanoids contribute to cell proliferation, and increased COX-2 expression occurs in a wide variety of epithelial cancers [8]. In particular, the beneficial effects of NSAIDs have suggested a major role for COX-2 in colon cancer, where this enzyme is suggested to play modulatory and even causal roles [9,10].

At what times, and in which subsets of cells, COX-2 expression is required to either promote or suppress tumorigenesis, infection and other biological processes is still unclear. A model system in which spatial, temporal and reversible loss of COX-2 function could be achieved would, therefore, be of great value in understanding the roles COX-2 plays in normal and pathophysiological conditions.

One way cells achieve targeted modulation of gene expression is through RNA interference (RNAi). RNAi is an endogenous mechanism in which short hairpin RNA (shRNA) species modulate targeted transcripts. In mammals, naturally occurring shRNAs are usually embedded within a microRNA (miRNA) backbone [11]. The miRNA is processed to a short hairpin RNA consisting of a sequence of 21–29 nucleotides, a short loop region and the reverse complement of the 21–29-nucleotide region. This short RNA molecule folds back on itself to form a hairpin structure, which is cleaved into double-stranded RNAs by 'Dicer', an endogenous nuclease. After nuclease digestion the RNA silencing complex recognizes the shRNA and guides it to complementary mRNA(s), thereby targeting them for destruction [12,13]. In this manner, RNAi regulates sequence-specific mRNA destruction.

Investigator-initiated experimental RNAi gene silencing has advantages over traditional gene targeting methods, since RNAi agents act without modifying the target gene. Cullen et al. [14] demonstrated they could use a scaffold to embed and subsequently express a synthetic shRNA sequence targeting a gene of interest. Using the microRNA30 (miR30) precursor RNA as a template, they substituted miR30 stem sequences with designed shRNAs, and showed effective target gene inhibition [15]. When shRNAs are expressed from a tetracycline response element (TRE) promoter in cells, along with a tetracycline transactivator (tTA or rtTA) protein, the shRNA can be reversibly expressed in response to doxycycline (DOX) presence or absence [16,17]. Both a 'tet-off' system, in which the tet-transactivator (tTA) protein activates the TRE promoter but is inhibited by the administration of DOX and a 'tet-on' system, in which the reverse tet-transactivator (rtTA) is latent until activated by DOX, are in common usage for reversible regulation of gene expression [18]. Many mouse lines have been developed that express the tTA (tet off) or rtTA (tet-on) transactivators from tissue specific promoters, allowing temporal and reversible control of gene expression in specific cell types [19].

The ability to reversibly control the location, timing and levels of COX-2 expression would facilitate the elucidation of the many and varied COX-2 roles in normal physiology and disease. In this study we describe the development of a mouse model in which DOX-regulated expression of a COX-2 targeted shRNA makes it possible to inactivate endogenous *Cox2* gene expression, providing cell-specific, reversible elimination of COX-2 expression and function.

Materials and Methods

COX-2 shRNA design and cloning

Potential gene silencing short hairpin RNAs (shRNAs) targeting COX-2 were designed as described [20]. Predicted shRNAs were screened against a series of sensor exclusion criteria and cross-checked against a transcript database to exclude sequences with similarity to 'off-target' genes [20]. Four 22-mer predicted shRNAs; Cox2.284, Cox2.1082, Cox2.2058 and Cox2.3711 were identified; numerical designations reflect the first nucleotide position of the targeted mouse *Cox2* mRNA sequence (Fig. 1A). These shRNA sequences, and their corresponding sense strand predictions, were synthesized as 97 mers and cloned into the miR30 shRNA backbone as described previously [21]. These sequences comprise the common and gene-specific stem and 19 bp loop of the miR30-context to create miR30-adapted shRNAs specific for *Cox2*.

Retrovirus vector preparation and shRNA testing

The LMP retroviral vector, a murine stem cell virus (MSCV)-based vector contains unique *Xho*I and *Eco*RI sites within a miR30-shRNA expression cassette, driven by the viral 5'LTR promoter ([17,20] and Fig. 1A). The vector also encodes PGK promoter-driven puromycin resistance and green fluorescent protein sequences, separated by an internal ribosome entry site (IRES). The four *Xho*I/*Eco*RI COX-2-shRNAs were cloned into the LMP vector, and retroviral stocks for the four COX-2-shRNAs and a retrovirus containing a control shRNA targeting the firefly luciferase coding sequence were generated by three plasmid co-transfection using a VSV-G envelope plasmid in HEK293 cells (ATCC).

To test the ability of the four COX-2 shRNAs to block COX-2 induction at single copy integrated levels, NIH3T3 mouse embryonic fibroblast cells and RAW264.7 (ATCC) mouse monocytic leukemia cells were transduced with a viral multiplicity of infection (MOI) resulting in <1% GFP-positive cells. Cells selected with puromycin (2.5 µg/ml) were used to test COX-2 induction 3–5 days later. NIH3T3 cells were cultivated in DMEM-supplemented with 10% fetal bovine serum (FBS) and 1% penicillin-streptomycin at 37°C in 5% CO_2. RAW264.7 cells were maintained in RPMI 1640 medium with 10% FBS and 1% penicillin-streptomycin. COX-2 was induced in RAW264.7 cells by adding bacterial lipopolysaccharide (LPS, Sigma) for four hours. NIH3T3 cells were starved with 1% FBS over night before shifting to medium with 20% FBS for six hours to induce COX-2 expression. Cells were washed with PBS and lysed directly in the culture dish with Passive Lysis Buffer (Promega). For prostaglandin E2 (PGE_2) analyses culture supernatants were harvested, snap frozen in liquid nitrogen, and assayed collectively by ELISA (Cayman Chemical).

Generation of the TRE-shRNA transgenic mouse

Animal experiments were conducted according to guidelines of the UCLA Animal Care Committee and the Cold Spring Harbor Laboratory Animal Facility. Protocols were subject to ethical review and approval by the UCLA Animal Research Committee (Protocol No. 2010-074-01). Mice were generated using the Flp/FRT recombinase-mediated cassette exchange (RMCE) strategy, using 'KH2' C57BL/6; SJL ES cells that contain an frt-hygro-pA 'homing' cassette downstream of the *ColA1* gene on mouse chromosome 11 [22–24]. Site-specific integration into the homing

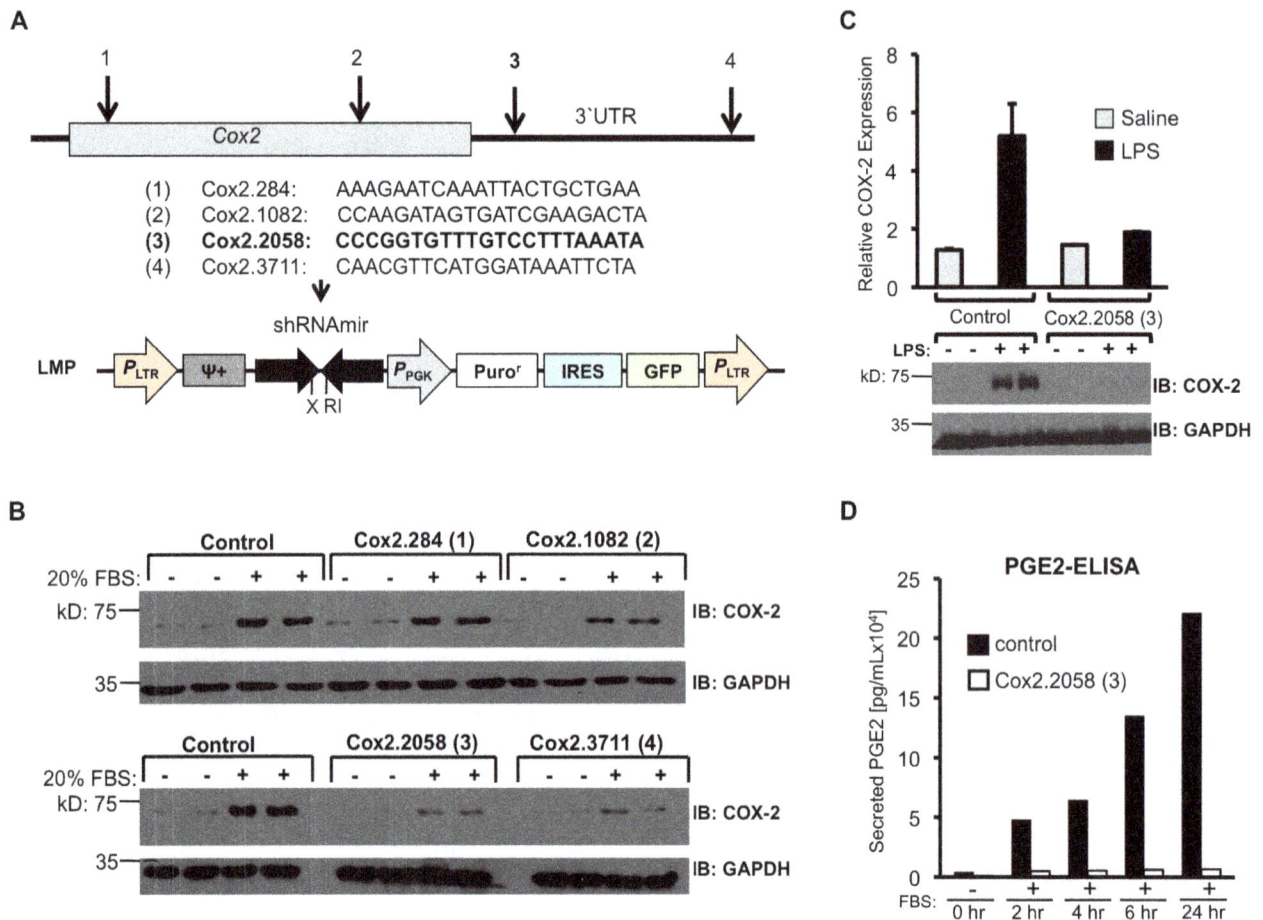

Figure 1. Identifying small inhibitory RNAs for COX-2. (**A**) Schematic representation of the *Cox2* transcript, indicating target areas of four designed shRNAs. The COX-2 protein-coding region is indicated as a filled box, 5′ and 3′ untranslated mRNA as solid lines. Four 22-nucleotide shRNAs for *Cox2* were designed; (1) Cox2.284, (2) Cox2.1082, (3) Cox2.2058 and (4) Cox2.3711, the number reflecting the first nucleotide position of the target sequence in the mRNA transcript. The shRNA sequences were then converted into cloning templates and ligated into the LMP retrovirus vector. This vector contains an *XhoI/Eco*R1 cloning site for shRNAs within a miR30 backbone (shRNAmir). The LMP construct is shown as it appears after integration; shRNAs are constitutively expressed from the 5′LTR promoter. The LMP retrovirus also encodes a puromycin-resistance gene for selection and GFP as a fluorescent marker. (**B**) NIH3T3 cells were transduced at a low MOI with each of the four LMP vectors containing miR30-based shRNAs that target the *Cox2* transcript, or with a control luciferase-targeted shRNA. Retrovirus-transduced cells were selected for integrated provirus by culturing in puromycin. Puromycin-selected cell populations were shifted overnight to 1% serum, then treated with 20% serum for 6 hours. Cell extracts were immunoblotted for COX-2 and GAPDH. (**C**) RAW264.7 cells stably transduced either with the luciferase shRNA LMP vector or with the LMP vector encoding *Cox2*-targeted Cox2.2058 shRNA were stimulated with LPS (50 ng/mL) or with saline for four hours. Cell extracts were prepared and analyzed for COX-2 protein and GAPDH. The quantified COX-2 signal was normalized against GAPDH. (**D**) PGE$_2$ accumulation in the media of serum-stimulated NIH3T3 cells expressing shRNA Cox2.2058 or the control luciferase shRNA. NIH3T3 cells expressing the two shRNAs were shifted from media containing 1% FBS to 20% FBS and, at times shown, media samples were assayed for PGE$_2$ levels.

cassette is achieved by FLPe recombinase-mediated recombination between the FRT-sites in the *ColA1* locus on chromosome 11 and in the targeting vector. The targeting vector, pCol-TGM, contains a GFP open reading frame immediately downstream of the TRE promoter, followed by the miR30-based shRNA expression cassette. The Cox2.2058 shRNA in the LMP shRNA expression cassette was cloned into the miR30 backbone of this targeting vector at the single *Xho*I/*Eco*RI site. The pCol-TGM targeting vector containing the Cox2.2058 shRNA and a plasmid expressing Flpe-recombinase (pCAGs-Flpe) were then co-electroporated into KH2 ES cells. Flpe-mediated recombination confers hygromycin resistance. Sequencing confirmed Cox2.2058 shRNA in Hygromycin-resistant ESC clones. Transgenic mice were then generated using tetraploid embryo complementation, to give rise to mice that are derived directly from the targeted ES cells [25–27].

To elicit doxycycline (DOX) inducible expression from the TRE promoter we used the CAGGs-rtTA3 (4288) transactivator mouse line [24], which expresses rtTA3 from the chicken beta-actin/CMV (CAGGs) promoter. CAGGs-rtTA3 (termed "CAG-rtTA3" or "C3") and TG-Cox2.2058 (termed "shCox-2") mice were crossed to produce double transgenic TG-Cox2.2058/CAG-rtTA3 (shCox2/C3) mice. Offspring were genotyped by genomic DNA PCR for the shCox2 [ColArev45 primer [20] and shCox2 forward primer: 5′-AAA GGA CAA ACA CCG GAT GC-3′] and C3 alleles [24]. To guarantee that all shCox2/C3 mice carry only a single copy of the shCox2 or C3 transgenes, mouse strains were maintained separately and only crossed when generating double transgenic mice.

Generation of a targeted luciferase knockin mouse with the *Cox2* 3'untranslated region (3'UTR)

We followed a procedure similar to that used in the construction of the *Cox2*^*fluc* knockin allele, in which the firefly luciferase (ffLuc) coding region and an SV40 3'UTR are substituted for the COX-2 coding region and 3'UTR [28]. The targeting vector contains a 6.7 kb genomic region upstream of the *Cox2* ATG initiation codon, the firefly luciferase coding region placed at the *Cox2* gene initiating ATG codon, a floxed neo cassette for positive selection, a 1 kb genomic region downstream of the *Cox2* termination codon, and the diphtheria toxin cassette for negative selection (Fig. 2A). This plasmid was generated by recombineering, using pCOXLuc-DT created previously [28], which contains the 5' *Cox2* genomic region and the ffLuc gene, and pCOXlucNeo^floxed COX_3UTR DT which contains the remaining elements described above. The final targeting vector was confirmed by DNA sequencing. To generate *Cox2* knockin mice, the targeting vector was linearized by *Not*I digestion and electroporated into LW1 embryonic stem (ES) cells. ESC clones containing the *Cox2*^*luc-floxed-neo* allele were screened by PCR using the primer pair PGKRa (5'-CTAAAGCG-CATGCTCCAGACT-3' targeting the PGK promoter) and NeoPCR-REV (5'-TAAAGTGACCACGAGAAACGGA-3' targeting the *neoB* gene). Homologous recombination was verified by Southern blotting using a 175 bp SacI/ApaI fragment within the Cox2 5'-UTR to probe the *Pvu*II/*Nhe*I-digested mouse genomic DNA. ESC clones were injected into C57BL/6 blastocysts to generate chimeric mice. One germline transmission founder was obtained. This mouse was crossed with a Cre transgenic mouse to generate *Cox2* heterozygous mice. Complete excision of the floxed neo cassette was confirmed by PCR (Luc-F4 5'-GCT GGG CGT TAA TCA GAG AG-3', Lox-R2 5'-CCA AGC TAT CGA ATT CCT GC-3') and Southern blot. The targeted *Cox2* allele is referred to as *Cox2*^*tm2Luc* ('*Cox2* targeted mutation 2 Luciferase') or 'Luc'.

The *Cox2*^*tm2Luc* allele was detected in cells and in mice by PCR: Luc1 primer, 5'-CCA GGG ATT TCA GTC GAT GT-3' and Luc2 primer; 5'-CGC AGT ATC CGG AAT GAT TT-3'). *Cox2* wild type primers: CoxF1, 5'-AAT TAC TGC TGA AGC CCA CC-3'; E4R2: 5'-AGA AGG CTT CCC AGC TTT TGT AAC C-3'.

Isolation and culture of mouse primary bone marrow macrophages and fibroblasts

Macrophages. Bone marrow macrophages were isolated from the tibia, femur and pelvis, using standard procedures. Cells were washed, depleted of red blood cells, plated into RPMI 1640 medium (Gibco/Invitrogen) supplemented with 10% FBS, 1% Penicillin/Streptomycin, 10 mM Hepes and 0.01 μg/mL mouse macrophage colony stimulating factor (M-CSF, Gibco/Invitrogen), and cultured for 5 days at 37°C in 10% CO_2. All experiments were performed in the following three days. For *Cox2* gene activation, cells were treated with LPS (50 ng/mL) for 4 hours in the presence or absence of DOX (Clontech), as described in Results and Figure Legends.

Fibroblasts. Primary lung and skin fibroblasts were isolated as previously described [29]. Tissues were collected, cut into small pieces, treated with liberase (Roche) for 60 minutes, and then cultured at 37°C in 5% CO_2 for 5 days in DMEM/F12 medium (Gibco/Invitrogen) supplemented with 15% FBS and 1% Penicillin/Streptomycin, to allow fibroblasts to migrate from the fragments. Remaining tissue pieces were removed and the medium was changed to EMEM (Gibco/Invitrogen) supplemented with 15% FBS, 1% Penicillin/Streptomycin, non-essential

amino acids and sodium pyruvate. COX-2 and luciferase expression was induced and analyzed as described above for retrovirus transduced NIH3T3 cells. DOX was added to the media at times, concentrations and durations indicated in Results and Figure Legends. DOX was changed daily.

Cox2 gene activation *in vivo*

Cox2 gene activation by systemic interferon gamma and endotoxin. Systemic *Cox2* activation by IFN-γ and LPS for triple transgenic shCox2/C3/Luc+ mice was performed as described previously [28]. Control mice were injected with saline. Six hours after LPS injection the mice were euthanized. Tissues were rapidly removed, placed in culture dishes and used for luciferase-dependent bioluminescence imaging. The tissues were then snap-frozen, and used subsequently for *in vitro* luciferase enzyme activity quantification and for COX-2 Western blotting.

Cox2 gene activation in the mouse paw by zymosan. Triple transgenic shCox2/C3/Luc+ mice were fed a DOX-free diet or a DOX (625 mg/kg)-containing diet (Harlan Tek) for 12 days. Inflammation was induced by sub-plantar Zymosan injection into the left hind paw, as described previously [28]. Saline injection into the right hind paw was used as a control. At the indicated time points mice were anesthetized with isoflurane and imaged non-invasively for GFP-dependent fluorescence as surrogate for shRNA expression and for luciferase-dependent bioluminescence.

Cox2 gene activation in skin by 12-O-tetradecanoylphorbol-13-acetate (TPA). Mice were anesthetized briefly by isoflurane inhalation. A small area of the dorsal skin was exposed by shaving and TPA (LC Laboratories, 5 μg in acetone) was applied. Twenty-four hours later mice were imaged non-invasively for GFP-dependent fluorescence to analyze shCox2 expression and for luciferase-dependent bioluminescence. Alternatively, mice were euthanized and skin was collected and processed for Western blotting.

Bioluminescence and fluorescence imaging

For bioluminescence *in vivo* imaging, mice were anesthetized with isoflurane, shaved if applicable, injected i.p. with D-luciferin (125 mg/kg) and placed into the imaging chamber of an IVIS imaging system (Xenogen). For *ex vivo* tissue imaging, mice were euthanized and tissues were rapidly excised, placed on culture dishes, covered with D-luciferin (15 mg/mL) solution and imaged. Bioluminescence emissions were collected over 3 minute periods at 5-minute intervals. Whole body and *ex vivo* tissue images were acquired repeatedly until the maximum peak of photon number was confirmed during the 3 minute scans. Data at the 3-minute time point that gave the highest photon number were used for further analysis, using Living Image software (Perkin Elmer). For bioluminescence quantification, a region of interest (ROI) was drawn manually and bioluminescence was recorded as radiance (peak photon/sec/cm^2/sr).

In vivo fluorescent images were obtained at 465 nm excitation and 520 nm emission wavelengths. Fluorescence was quantified by drawing ROIs and displayed as mean fluorescent intensity.

Luciferase activity assays

Cultured cells and tissues were washed with PBS and homogenized in Passive Lysis Buffer (Promega). Debris was removed by centrifugation and luciferase activity was determined in extracts, using the Luciferase Assay System (Promega) and Luminat LB9501 instrumentation. Luciferase activity in cell/tissue extracts was normalized to protein content determined by

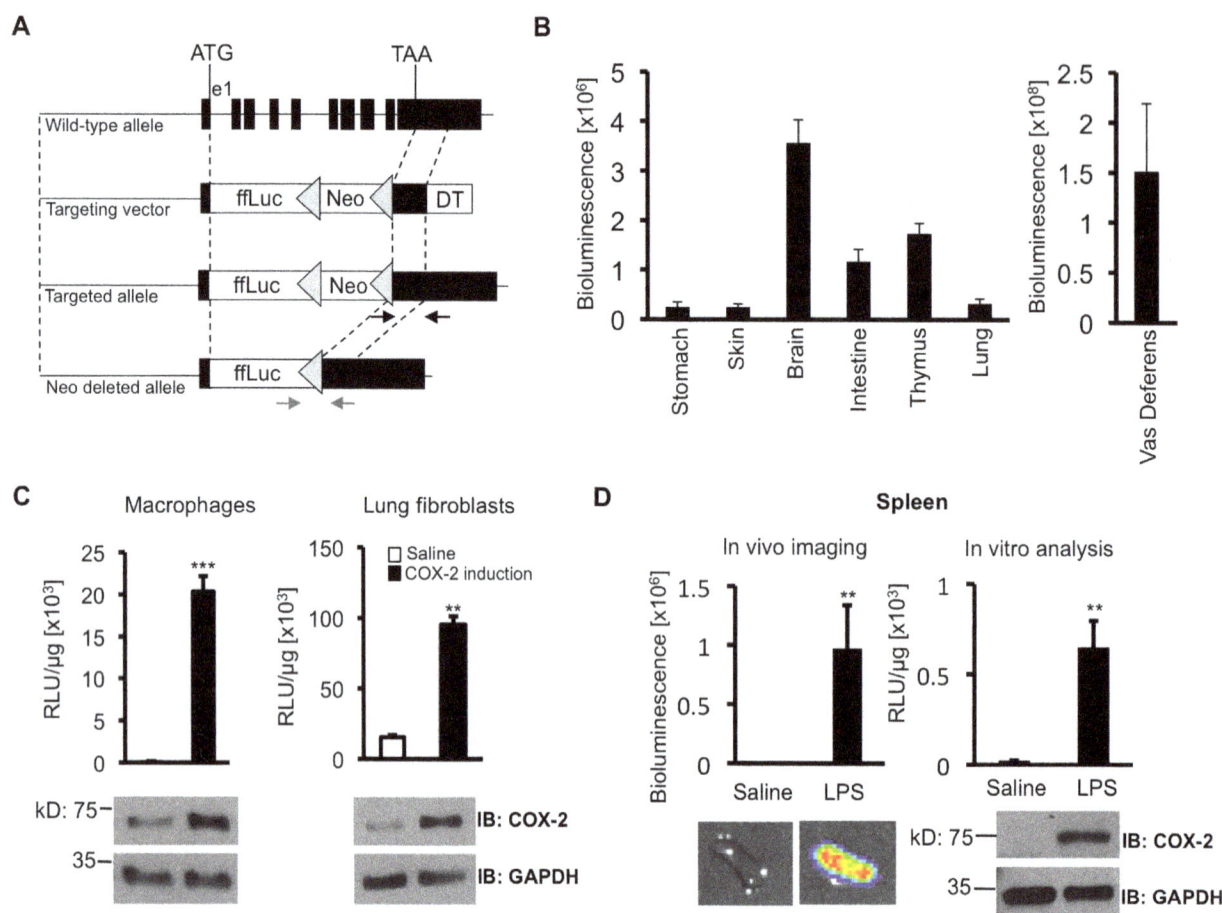

Figure 2. Construction of *Cox2^tm2Luc/+*, a mouse strain in which firefly luciferase replaces the *Cox2* coding region. (**A**) Schematic representation of the wild-type *Cox2* allele and the targeting strategy to create the *Cox2^tm2Luc* knockin allele. The firefly luciferase coding region (ffLuc), PGK-neo (neo) selection cassette, and PGK-DT (DT) selection cassette in the targeting vector are shown as open boxes. Grey triangles depict loxP sites. Homologous recombination was confirmed by PCR (black arrows) and Southern blot analysis in ES cells. The neomycin-resistance cassette was deleted by Cre recombinase expression, resulting in the 'neo-deleted' allele. Deletion was confirmed by PCR (grey arrows). *Cox2* gene sequences are replaced by the firefly luciferase coding region between the ATG translational start site located at the end of exon 1 (e1) and the TAA *Cox2* stop codon located on exon 10. There are no modifications of the untranslated 5′UTR and 3′UTR either upstream of the ATG or downstream from the TAA. (**B**) Unstimulated luciferase activity in isolated *Cox2^tm2Luc/+* tissues. Luciferase activity was quantified by *ex vivo* bioluminescent imaging. Data are means +/− SD (n = 4). (**C**) COX-2 and luciferase induction in primary cells isolated from *Cox2^tm2Luc/+* mice. Bone marrow macrophage cultures were stimulated with LPS (50 ng/mL) for four hours. Lung fibroblast cultures were stimulated with 20% serum for six hours. Cell extracts were analyzed for luciferase enzymatic activity and COX-2 protein. Luciferase activity is displayed as relative light units (RLU) per microgram protein. Data are means +/− SD (**, $p < 0.01$, ***, $p < 0.001$, n = 3). (**D**) Interferon gamma and endotoxin (IFNγ/LPS) COX-2 and luciferase induction in the spleens of heterozygous *Cox2^tm2Luc/+* mice. Four mice were injected i.p. with IFNγ, (1 μg/mouse) and two hours later with LPS (3 mg/kg) or saline. After 6 hours, mice were euthanized, spleens were excised and luciferase bioluminescence was quantified by bioluminescent imaging (left panel). Luciferase enzymatic activity and COX-2 protein levels were measured in extracts (right panel). Data are means +/− SD (**, $p < 0.01$).

Bradford assay, and displayed as relative light units per μg protein (RLU/μg).

Western blotting

The following antibodies were used for Western blotting: GFP, rabbit polyclonal IgG, Abcam, ab290, 1:2000; glyceraldehyde 3-phosphate dehydrogenase (GAPDH), rabbit polyclonal IgG, Santa Cruz, sc-25778, 1:1000; COX-2 (in skin), rabbit polyclonal IgG, Cayman, 160106, 1:100; COX-2 (all other tissues and cells), rabbit polyclonal IgG, Santa Cruz, sc-1747-R, 1:1000. Extracts of cells or tissues that were previously lysed in Passive Lysis Buffer were mixed with SDS sample buffer and boiled prior to separation on SDS-PAGE gels. For Western blot analysis of skin, tissue was homogenized in RIPA buffer containing a proteinase inhibitor cocktail (Santa Cruz, sc-24948), sonicated and incubated at 4°C

for 20 minutes on a rocking platform. Cell debris was removed by centrifugation and protein content was determined by Bradford assay.

Proteins (40–80 μg) were separated on 10% SDS-PAGE gels and transferred onto nitrocellulose membranes. The membranes were blocked with 4% milk protein in PBS/0.1% Tween-20, probed with primary antibodies in the same buffer over night at 4°C, then incubated with anti-rabbit HRP-conjugated secondary antibody (Santa Cruz, sc-2004, 1:5000) for one hour at room temperature. Proteins were visualized on autoradiographic film using ECL reagent (Pierce). For quantification, membranes were scanned with Typhoon instrumentation. Results were graphed as volume densities relative to the GAPDH signal.

Histology and immunohistochemical staining

For immunohistochemistry of COX-2 and GFP, skin samples were fixed overnight in 10% buffered formalin, then embedded in paraffin. Skin sections (5 μm) were de-paraffinated and treated for antigen retrieval with the citrate buffer heat method. Endogenous peroxidases were quenched with 3% hydrogen peroxide and sections were blocked in 5% normal donkey serum/PBS for one hour before overnight incubation with rabbit anti-COX-2 antibody (1:25; Cayman, 160106) or rabbit anti-GFP antibody (1:5000; Abcam, ab290) at 4°C. Slides were washed and incubated with biotin-labeled goat anti-rabbit secondary antibody (1:250; Jackson Laboratories). Signal was developed using the ABC and DAB vector kits (Vector Laboratories). Sections were counter-stained with hematoxylin.

To measure leukocyte infiltration, sections were stained with hematoxylin and eosin and the numbers of infiltrating leukocytes were counted in six fields per section in a blinded fashion. Images were obtained with an inverted microscope (Nikon, Eclipse 2000 TE) using NIS Elements imaging software.

Statistics

Data are expressed as means $+/-$ S.D. (standard deviation). Data were compared between groups with the unpaired Student's t test or one-way analysis of variance as appropriate. p values less than 0.05 were considered significant. All experiments were performed a minimum of three times with a minimum of triplicate samples.

Results

Identifying small hairpin RNAs that can block COX-2 expression

Using improved prediction methods for the design of miR30-based shRNAs [20], we identified four 22-mer guide strand sequences; Cox2.284 (1), Cox2.1082 (2), Cox2.2058 (3) and Cox2.3711 (4) (Fig. 1A), complementary to the *Cox2* coding region (1 and 2) or the *Cox2* 3′-UTR sequence (3 and 4). Guide and complementary sense strand sequences, were embedded into cloning templates as described [20]. Appropriate products carrying the *Xho*I/*Eco*RI restriction sites at their ends and comprising the common and *Cox2*-specific stem sequences and the 19 bp loop were used to create miR30-adapted shRNAs.

Since shRNA transgenic mice will harbor only one copy of the shRNA expression cassette, it is necessary to identify shRNAs that efficiently silence COX-2 expression when expressed from a single genomic locus. To identify appropriate shRNAs, each cloning template containing a COX-2 shRNA sequence was ligated into LMP, a retroviral miR30-shRNA expression vector in which miRNA-based shRNA (shRNAmir) expression is driven from the viral 5′LTR promoter (Fig. 1A). The vector also contains a PGK promoter-driven puromycin resistance gene and a green fluorescence protein (GFP) gene to select and identify transduced cells (Fig. 1A, lower illustration).

To test the candidate shRNAs for suppression of COX-2 expression when expressed from a single gene copy, NIH3T3 cells were transduced with the four LMP vectors at a very low multiplicity of infection (MOI). Flow cytometry of the transduced cell populations demonstrated that only ~1% of the cells expressed GFP (Supplementary Fig. 1A), suggesting transduced cells contained only a single viral genome. Subsequent puromycin selection ensured expansion only of transduced cells, resulting in a mixed cell population of single integrants. Southern blot analysis (Fig. S1B–D) confirmed multiple integration sites in retrovirus transduced, antibiotic-selected cell populations.

The *Cox2* gene can be induced in NIH3T3 cells by increasing the serum concentration in the medium. The abilities of the four *Cox2* specific shRNAs to block COX-2 protein expression were examined by stimulating the four retrovirus-transduced NIH3T3 cell populations with medium containing 20% serum for six hours, then analyzing cell extracts for COX-2 protein content (Fig. 1B). As expected, cells transduced with a vector encoding a control shRNA against luciferase demonstrated substantial COX-2 protein induction. Cells transduced with COX-2 shRNAs 1 and 2 also expressed substantial COX-2 protein in response to 20% serum; these shRNAs did not have a significant effect on COX-2 expression at single-copy conditions. In contrast, cells transduced with COX-2 shRNAs 3 and 4 expressed only small amounts of COX-2 protein in response to serum stimulation. We chose COX-2 shRNA 3 (Cox2.2058) for further studies, as its target location in the mRNA should make it effective in silencing COX-2 expression from both the short and long *Cox2* gene transcripts [30].

To confirm COX-2 knockdown by Cox2.2058 in a different cell type, LMP vector transduction and puromycin selection were repeated in RAW264.7 cells, a murine monocyte cell line. RAW264.7 cells transduced with the control luciferase shRNA vector express substantial COX-2 protein in response to bacterial lipopolysaccharide (LPS, 50 ng/mL), compared to saline-treated cells (Fig. 1C). In contrast, RAW264.7 cells transduced with Cox2.2058 did not express COX-2 protein above baseline values when treated with LPS. Western blot data quantification indicated over 90% reduction of LPS-induced COX-2 expression (Fig. 1C).

To demonstrate functional inactivation of the *Cox2* transcript by Cox2.2058, prostaglandin E2 (PGE$_2$) production was examined in serum-stimulated NIH3T3 cells expressing either the control luciferase shRNA or Cox2.2058. Serum-stimulated cells expressing the control shRNA secrete continuously increasing amounts of PGE$_2$ into the culture medium (Fig. 1D). In contrast, *Cox2*-targeted Cox2.2058 expression completely eliminated serum-induced PGE$_2$ accumulation. In summary, shRNA 3 (Cox2.2058) was highly effective in suppressing COX-2 expression and activity when expressed at a single copy per cell.

A reporter mouse in which the endogenous *Cox2* coding region is replaced with the firefly luciferase coding region. To test the effects of the COX-2 shRNA Cox2.2058 on COX-2 expression it would be useful to have a reporter system that could be non-invasively and repeatedly monitored in individual mice. We previously constructed a knock-in mouse, $Cox2^{fLuc/+}$, in which firefly luciferase is expressed from the endogenous *Cox2* gene [28], and have used this mouse to monitor expression from the *Cox2* gene in a variety of contexts [28,31–33]. However, the $Cox2^{fLuc}$ allele contains an SV40 3′UTR substituted for the endogenous *Cox2* 3′UTR. Because Cox2.2058 targets the 3′UTR of the *Cox2* mRNA, we constructed a new knock-in mouse in which the luciferase coding region is also driven by the endogenous *Cox2* promoter, but in which the luciferase coding sequence is followed by the endogenous *Cox2* 3′UTR. Using this mouse, the efficacy of shRNA-mediated *Cox2* knockdown, and its reversibility can be evaluated by noninvasive imaging of *Cox2* promoter-driven luciferase activity.

We employed a knockin strategy whereby the firefly luciferase (ffluc) coding region and a PGK-neomycin selection cassette in the targeting vector were inserted into the *Cox2* locus by homologous recombination in ES cells (Fig. 2A). Immediately downstream from the neo selection cassette for selection of ES cells containing the targeted allele, the targeting vector contains the *Cox2* translational stop codon (TAA), followed by the *Cox2* gene 3′UTR of exon ten and a diphtheria toxin cassette. After selection of ES cells containing the targeted allele, the neo selection cassette

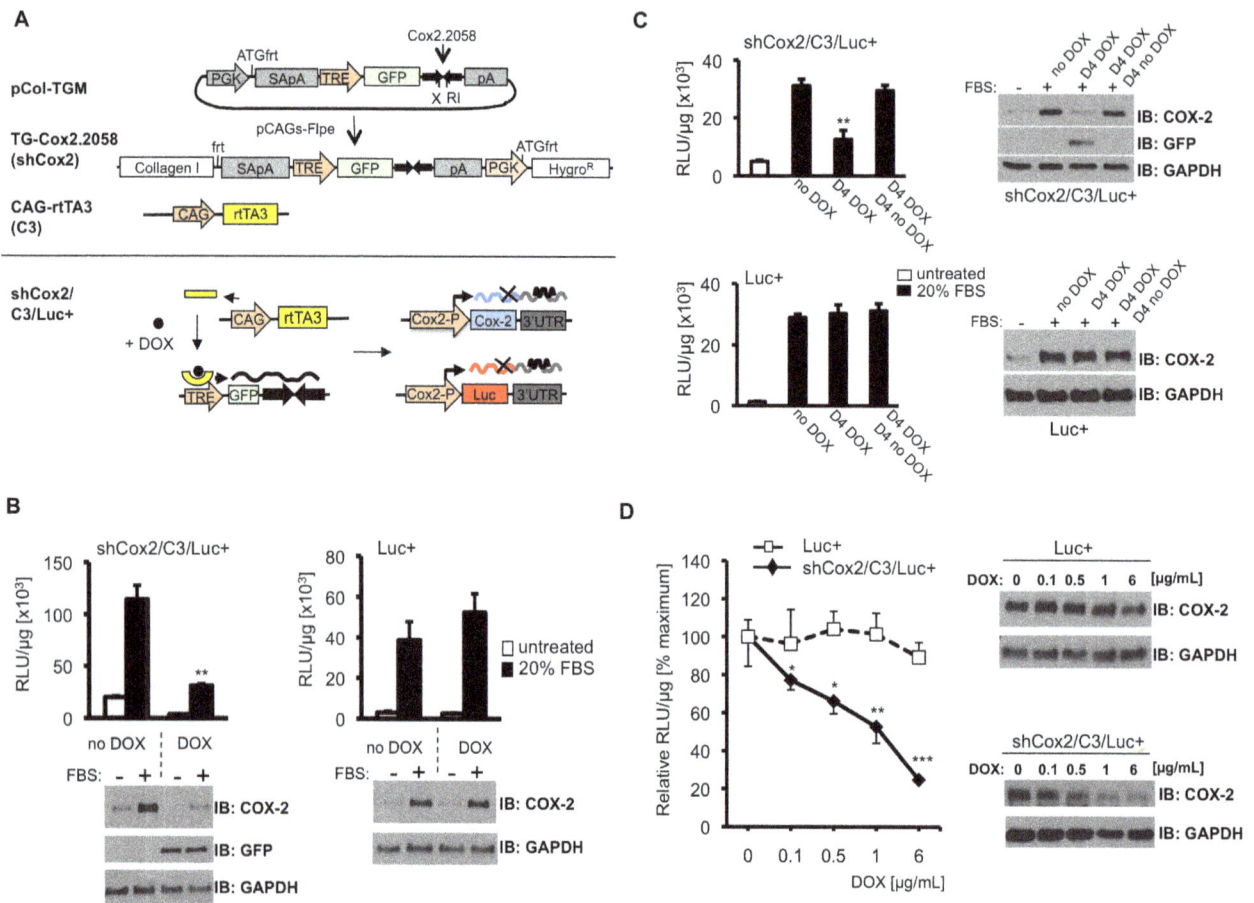

Figure 3. Inducible COX-2 shRNA expression suppresses *Cox2*-driven gene expression in cells cultured from triple transgenic mice. (**A**) Upper panel: The diagram shows the pCol-TGM targeting construct encoding GFP and COX-2 shRNA, expressed from a TRE promoter. Co-electroporation, with pCAGS-Flpe recombinase, into KH2 ES cells results in integration of the construct into the *ColA1* locus. ShCox2 mice are crossed with tet-transactivator mice (CAG-rtTA3/C3) to create double trangenics. Lower panel; in triple transgenic shCox2/C3/Luc+ mice, DOX-rtTA3 activates the TRE promoter, driving GFP and shCox2 expression; shCox2 blocks COX-2 and luciferase expression. Without DOX the TRE promoter is quiescent; COX-2 and luciferase are expressed normally. (**B**) Inducible suppression of *Cox2* gene expression in fibroblasts. Skin fibroblasts from triple transgenic shCox2/C3/Luc+ and Luc+ mice were cultured for four days in the absence or presence of DOX (1 μg/mL), shifted to 1% serum overnight, then stimulated with 20% serum for 6 hours. Luciferase activity was measured in extracts, COX-2 and GFP protein were analyzed by Western blot. Luciferase activity was normalized to protein content. (**C**) Reversible suppression of *Cox2* gene expression in skin fibroblasts. Cells were untreated (no DOX), treated for 4 days with DOX (D4 DOX), or treated for 4 days with DOX followed by 4 days without DOX (D4 DOX, D4 no DOX). Cells were stimulated with medium containing 20% FBS; luciferase activity, COX-2 expression and GFP expression were analyzed. (**D**) Bone marrow macrophages from shCox2/C3/Luc+ and control Luc+ mice were cultured in the indicated concentrations of DOX overnight, then stimulated for 4 hours with LPS (50 ng/mL) to induce COX-2. Cell extracts were analyzed for luciferase activity and COX-2 protein. Luciferase activities are normalized to LPS stimulation in the absence of DOX. Data are means +/− SD. Statistics compare DOX-treated cultures with cells not receiving DOX (*, $p<0.05$; **, $p<0.01$; ***, $p<0.001$).

was excised by transient Cre expression, resulting in the final neo-deleted, ffLuc-expressing allele. This second *Cox2*-targeted firefly luciferase knock-in allele was termed 'targeted mutation 2 luciferase' (tm2Luc).

To characterize $Cox2^{tm2Luc/+}$ mice we first determined which organs demonstrate detectable luciferase gene expression in untreated mice. Luciferase expression was examined by biolumi-nescent imaging of tissues dissected from $Cox2^{tm2Luc/+}$ mice, and quantified from the optical imaging data. The highest constitutive luciferase expression from the *Cox2* gene occurred in the vas deferens (Fig. 2B, right panel), consistent with what has been reported previously [34]. Substantial luciferase expression was also observed in the brain, throughout the gastrointestinal tract, in the thymus, and to lesser extent in stomach, skin and lungs (Fig. 2B, left panel). Luciferase activity was low or undetectable in all other

organs examined, including pancreas, kidney, liver, spleen, and heart. The luciferase expression pattern confirms published results for COX-2 expression obtained by immunohistochemistry and Western blotting, and by our previous results with the $Cox2^{fLuc/+}$ mouse [28,31,34,35].

To further characterize luciferase and COX-2 expression in tissues from heterozygous $Cox2^{tm2Luc/+}$ mice, bone marrow macrophages and lung fibroblasts from mice were harvested, cultured and appropriately stimulated to induce *Cox2* expression. Cell extracts were assayed both for luciferase enzymatic activity and for COX-2 protein (Fig. 2C). LPS (50 ng/mL)-treated macrophages showed parallel increases in luciferase activity and COX-2 protein expression. Serum-stimulated lung fibroblasts also showed similar concomitant luciferase enzymatic activity and COX-2 protein accumulation.

The final set of experiments validating the $Cox2^{tm2Luc/+}$ mouse evaluated coordinated luciferase and COX-2 expression *in vivo*. Systemic interferon gamma (IFN-γ) and endotoxin (LPS) injection induces robust COX-2 induction in spleen, lung, heart and liver [28]. $Cox2^{tm2Luc/+}$ mice were injected first with IFNγ 1 µg/mouse), and two hours later with LPS (3 mg/kg). Six hours after LPS injection organs were removed, exposed to luciferin substrate, and imaged to measure bioluminescence. Substantially enhanced bioluminescence was observed in spleen, liver, heart and lung tissues of [IFN-γ + LPS] injected $Cox2^{tm2Luc/+}$ mice, compared to saline injected control mice (Fig. 2D, left panel and Fig. S2). After imaging, tissues were homogenized and extracts were analyzed both for luciferase enzymatic activity and for COX-2 expression (Fig. 2D, right panel and Fig. S2). IFN-γ+LPS induced both luciferase and COX-2 protein in the tissues examined; demonstrating luciferase induction from the $Cox2^{tm2Luc}$ allele can be used to determine COX-2 expression in heterozygous $Cox2^{tm2Luc/+}$ mice.

Generation of a tetracycline-inducible short hairpin RNA for COX-2 inhibition in mice. To develop a single-copy transgenic mouse in which Cox2.2058 can be expressed in an inducible and reversible manner, we subcloned the Cox2.2058 sequence into the targeting vector (pCol-TGM, [24]) (Fig. 3A, upper panel). pCol-TGM contains a miR30-based expression cassette regulated by an inducible tetracycline response element (TRE) promoter. The pCol-TGM vector also contains, between the TRE promoter and the shRNA cloning site, a GFP coding region, which both enhances knockdown of target genes and enables easy tracking and isolation of shRNA-expressing cells [24]. Transgenic mice were generated using the Flp/FRT recombinase mediated cassette exchange (RMCE) strategy [22] to target the Cox2.2058 containing pCol-TGM sequence into the frt-hygro-pA 'homing' cassette of KH2 ES cells [20,23,24]. Fig. 3A (upper panel) shows the targeting vector (pCol-TGM) and the integrated construct after homologous recombination to create the TG-Cox2.2058 transgene allele. After selection and confirmation of ES cell clones with correct shRNA sequence, transgenic mice were generated using tetraploid embryo complementation [25–27].

Triple transgenic mice containing the CAG-rtTA3 (C3) reverse transactivator, the TG-Cox2. 2058 shRNA (shCox2) and the $Cox2^{tm2Luc}$ allele (Luc+) knockin. The TRE promoter in TG-Cox2.2058 (shCox2) transgenic mice is activated only in the presence of a tetracycline-regulatable transactivator. We crossed the shCox2 mice to the CAG-rtTA3 (C3) reverse transactivator mouse line. The CAG promoter is particularly well expressed in skin epidermis [24]. Moreover, quantification of luciferase expression *in vivo* by non-invasive bioluminescent imaging is most sensitive and efficient in skin [36]. Consequently, to analyze the ability of shCox2 to reversibly suppress COX-2 expression we crossed $Cox2^{tm2Luc/+}$ (Luc+) reporter mice to the shCox2/C3 double transgenic mouse line. Triple transgenic mice carrying the CAG-rtTA3 transactivator (C3), the Cox2.2058 shRNA transgene (shCox2) and the $Cox2^{tm2Luc}$ allele (Luc+) develop normally. In subsequent experiments their luciferase expression is compared with single transgene heterozygous $Cox2^{tm2Luc/+}$ (Luc+) littermate controls.

Tet-regulated COX-2-shRNA reversibly suppresses expression from the *Cox2* gene in cultured cells. CAG-rtTA3 is a 'tet-on' transactivator, inactive unless a tetracycline analog such as doxycycline (DOX) is present. In the presence of DOX, rtTA3 should activate the TRE promoter, leading to shCox2 expression and subsequent down-regulation of *Cox2* transcripts that contain the targeted *Cox2* gene 3'UTR sequence (Fig. 3A lower panel). DOX withdrawal should inactivate the

rtTA3 transactivator, and TRE-mediated transcription should cease.

CAG-rtTa3 driven gene expression is high in mouse skin [24]. We established primary skin fibroblast cultures both from triple transgenic shCox2/C3/Luc+ mice and from Luc+ control mice, to test DOX-regulated *Cox2* gene silencing. To examine whether DOX-induced shCox2 expression results in reduced expression from the *Cox2* gene locus, primary shCox2/C3/Luc+ skin fibroblasts and fibroblasts from Luc+ mice, cultured either in the presence or absence of DOX, were stimulated with 20% FBS, and assayed for luciferase activity. Skin fibroblasts from both mouse strains cultured in the absence of DOX demonstrate a substantial increase in FBS-stimulated luciferase activity in cell extracts over unstimulated cells (Fig. 3B). DOX reduced luciferase production in FBS-treated triple transgenic fibroblasts by ~75% (Fig. 3B, left panel). COX-2 induction in FBS-stimulated triple transgenic fibroblasts was also substantially reduced in the presence of DOX (Fig. 3B, left panel). GFP expression in triple transgenic fibroblasts was observed only in DOX-treated cells, indicating that shCox2 production was DOX-dependent (Fig. 3B, Western blot). In contrast to results for fibroblasts from the triple transgenic cells, both luciferase activity and COX-2 protein expression in Luc+ cells were unaffected by DOX, ruling out non-specific side effects by DOX treatment (Fig. 3B, right panels).

To test the reversibility of DOX-dependent, shCox2-regulated COX-2 knockdown, fibroblasts from triple transgenic shCox2/ C3/Luc+ and Luc+ mice were divided into three treatment groups. Cells were either treated with DOX for four days (D4 DOX), treated with DOX for 4 days followed by 4 days without DOX (D4 DOX, D4 no DOX) or never treated with DOX (no DOX). Cells were then treated with 20% FBS. DOX treatment (D4 DOX) resulted in significant reduction of luciferase and COX-2 expression in response to FBS stimulation in cells from triple transgenic (shCox2/C3/Luc+) mice, but not in FBS-stimulated cells from Luc+ mice (Fig. 3C). Moreover, FBS-stimulated luciferase activity and COX-2 protein expression following DOX withdrawal (D4 DOX, D4 no DOX), were identical to that observed in cells that were never exposed to DOX (no DOX), demonstrating that the DOX-regulated, shCox2-dependent inhibition of COX-2 expression and luciferase activity were fully reversible. GFP expression correlated strongly, in reciprocal fashion, with *Cox2* gene knockdown (Fig. 3C, upper right panel), confirming its role as a reliable biomarker of shCox2 expression.

Luciferase and COX-2 were strongly induced by LPS in primary bone marrow macrophages from both triple transgenic and Luc+ mice (Fig. 3D). Macrophages from both mouse strains were cultured overnight with increasing DOX concentrations, then stimulated with LPS. We observed dose-dependent reductions in luciferase activity and COX-2 expression in shCox2/C3/Luc+ macrophages, but not in Luc+ macrophages. These experiments demonstrate that co-expression of CAG-rtTA3 (C3) and TG-Cox2.2058 (shCox2) results in DOX-dependent, reversible inhibition of *Cox2* gene expression.

Doxycycline-regulated shCox2 expression suppresses *Cox2* gene expression in zymosan-induced inflammation in the mouse paw. Zymosan-induced paw inflammation in rodents is a classic localized inflammation model [37,38]. We previously demonstrated that zymosan inflammation is accompanied by a robust, self-resolving COX-2 induction [28] that can be measured by repeated non-invasive optical imaging in the $Cox2^{fLuc/+}$ mouse [28]. Consequently, the paw inflammation model should also be very useful for initial characterization of the triple transgenic shCox2/C3/Luc+ mouse.

To determine the dynamics of DOX-induced, shCox2-mediated, longitudinal silencing of the *Cox2* gene, three triple transgenic shCox2/C3/Luc+ mice received a DOX-containing diet for 12 days, while three additional triple transgenic mice were fed a control DOX-free diet. To analyze whether the shCox2 shRNA was expressed, and to obtain luciferase activity levels at baseline, the mice were imaged non-invasively for GFP fluorescence and for luciferase-dependent bioluminescent activity prior to zymosan administration (Fig. 4A and 4B, 0 time point). GFP fluorescence from the paws of mice that received the DOX-containing diet (+ DOX) was greater than 10-fold increased when compared to paws of mice that did not receive DOX, indicating that shCox2 expression was strongly induced by DOX (Fig. 4A).

A 2% zymosan suspension in saline (30 μL) was then injected sub-plantar into the left rear paws of all six shCox2/C3/Luc+ mice, and saline was injected into the right rear (contralateral) paws. Luciferase expression from the $Cox2^{tm2Luc}$ allele was measured by bioluminescence imaging at 6, 12, and 24 hours post zymosan injection (Fig. 4B). Zymosan injection resulted in rapid luciferase accumulation compared to luciferase activity at baseline, prior to zymosan injections; zymosan-induced luciferase activity peaked between 6 and 12 hours, then decreased to near-baseline values within 24 hours (Fig. 4B). *Cox2* gene-directed luciferase activity was significantly higher throughout the induction period in mice that received a DOX-free diet (no DOX) compared with mice fed a DOX-containing diet. Thus DOX pretreatment resulted in significant inhibition of *Cox2* gene-driven luciferase expression during zymosan-induced paw inflammation.

Zymosan-induced luciferase activity in Luc+ control mice fed a DOX-containing diet was not significantly different from luciferase activity observed in shCox2/C3/Luc+ mice on a DOX-free diet, confirming that the reduction in luciferase activity observed in shCox2/C3/Luc+ mice on DOX was not simply due to the presence of DOX (Fig. S3A).

Doxycycline-regulated shCox2 reversibly suppresses *Cox2* gene expression in TPA-treated mouse skin. Topical TPA administration induces an inflammatory reaction in mouse skin [39–42]. We used repeated TPA skin treatment of shCox2/C3/Luc+ triple transgenic mice, in combination with repeated noninvasive imaging, to evaluate reversible inactivation of *Cox2*-driven luciferase expression by DOX-inducible shCox2 expression in individual mice.

Three triple transgenic shCox2/C3/Luc+ mice and two control Luc+ littermate control mice received a DOX-containing diet for 12 days. A small area on the back of the mice was shaved and expression from the *Cox2* gene was induced by applying TPA. Twenty-four hours later GFP, as a measure of shCox2 expression, and luciferase activity were measured by respective non-invasive optical fluorescent and bioluminescent imaging (Figs. 5A and B, top rows). After imaging, the DOX diet was replaced with a DOX-free diet. Twelve days later mice were again subjected to TPA treatment on their back and again imaged 24 hours later for GFP and luciferase expression (Figs. 5A and B, middle rows). After imaging, the mice were switched back to a DOX-containing diet for 12 days, treated again with TPA, and imaged a third time for DOX-mediated GFP and *Cox2* gene-mediated luciferase expression (Fig. 5A and B, third rows) to determine whether *Cox2* shRNA-mediated knockdown of *Cox2* gene-driven transcripts is reversible.

DOX induced a ~10-fold GFP fluorescence induction in skin of triple transgenic shCox2/C3/Luc+ mice when compared to fluorescence from Luc+ control mice, which do not contain the GFP-linked shRNA transgene (Fig. 5A, +DOX, top row). Twelve days after the DOX diet was replaced with a DOX-free diet, skin

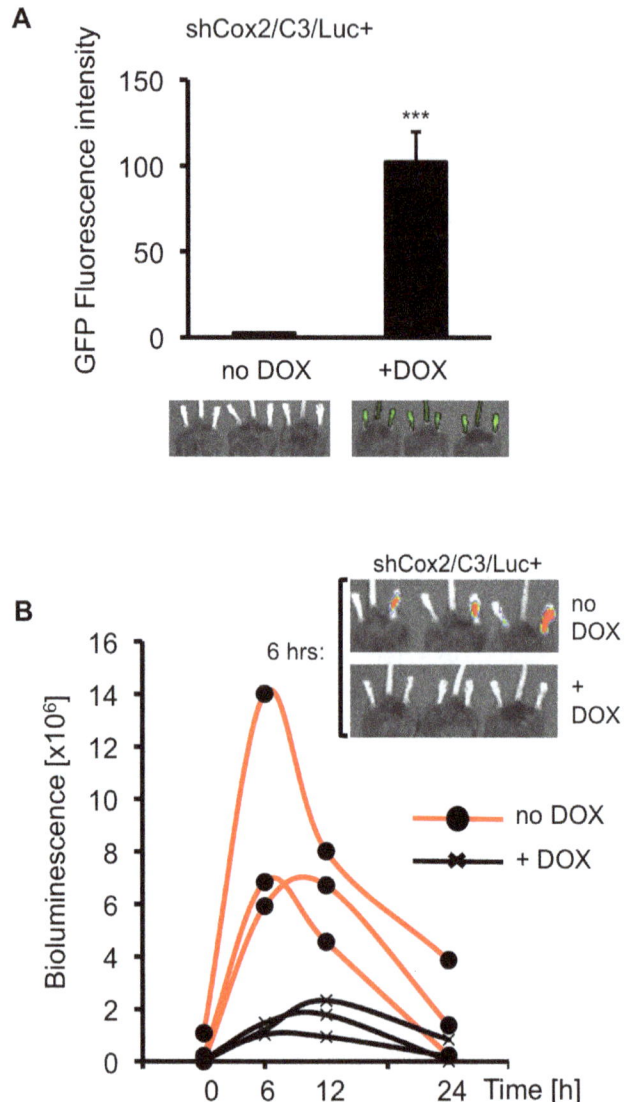

Figure 4. DOX-dependent suppression of *Cox2* driven luciferase expression in triple transgenic mice during zymosan-induced paw inflammation. (**A**) Fluorescent *in vivo* GFP imaging of shCox2/C3/Luc+ triple transgenic mice fed either a control diet (no DOX) or a DOX-containing diet (+DOX) for 12 days. GFP fluorescence was quantified with Living Image software. Data are means +/− SD, (n =3; ***, p<0.001). (**B**) ShCox2/C3/Luc+ mice were injected intraplantarly in the left hind paw with zymosan (30 μL, 2% w/v) and with saline in the contralateral hind paw. *In vivo* luciferase bioluminescence was non-invasively and repeatedly measured and quantified at the indicated time points. Each curve represents data from an individual animal. To control for individual animal variability the bioluminescence in the contralateral control saline-infected paw was subtracted from the bioluminescence value of the zymosan-injected paw for each observation.

GFP expression returned to baseline (Fig. 5A, middle row), suggesting that the COX-2-shRNA is no longer expressed. When shCox2/C3/Luc+ mice were returned to a DOX-containing diet for 12 days, they again showed a ~10-fold induction of GFP expression (Fig. 5A, bottom row). These data demonstrate that shCox2 expression can be reversibly up-and-down-regulated within the same mouse, in response to the presence or absence of DOX in the diet.

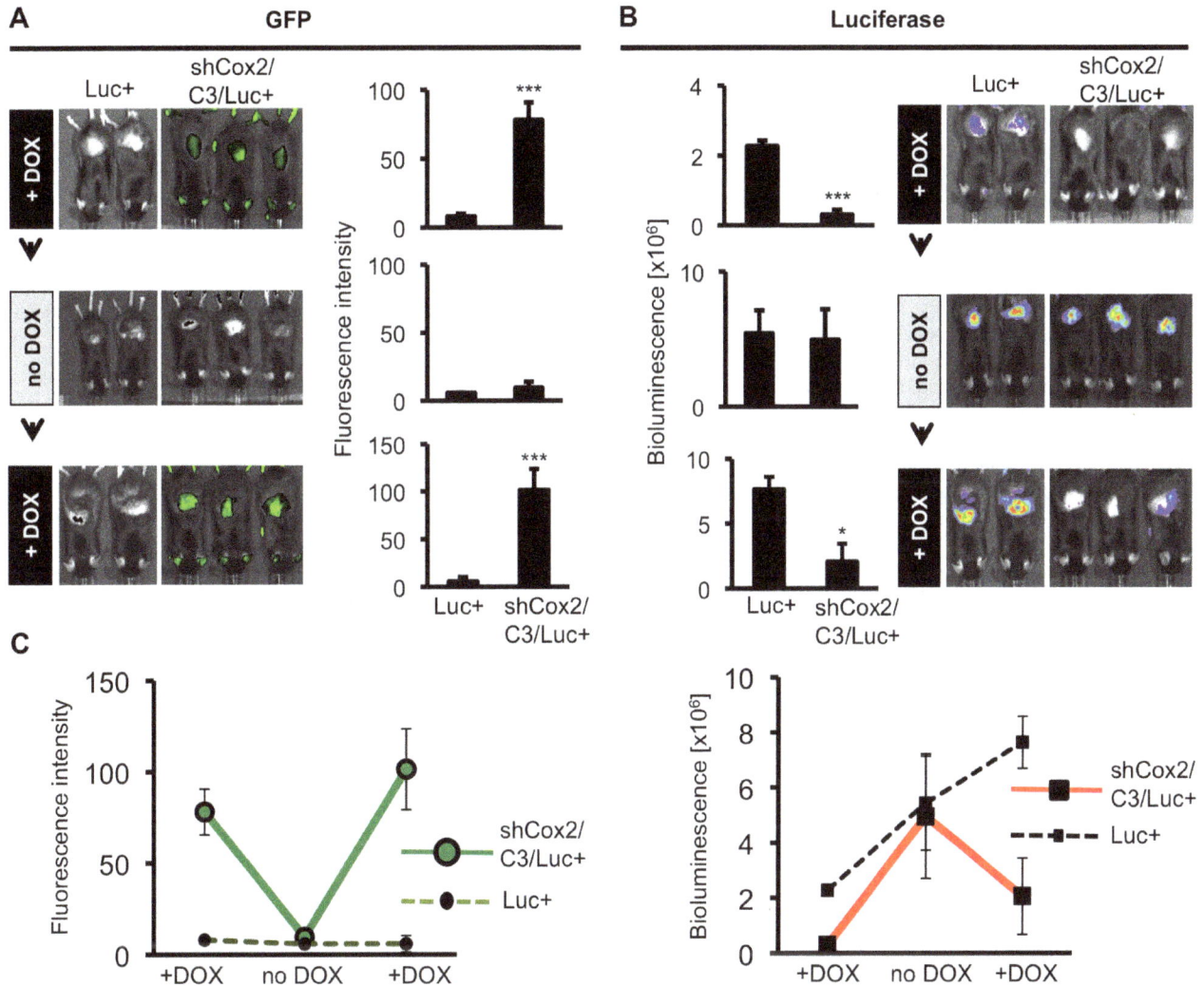

Figure 5. Reversible DOX-dependent suppression of *Cox2*-driven luciferase expression in skin of TPA-treated triple transgenic mice.
Luc+ mice and triple transgenic mice were subjected to the following diet, TPA skin application and imaging schedule: Mice were placed on a DOX-supplemented diet (+DOX) for 12 days followed by skin TPA application. Mice were imaged 24 hours later for GFP fluorescence and luciferase bioluminescence. The mice were then shifted to a DOX-free diet (no DOX) for 12 days and the skin TPA application, GFP fluorescence and luciferase bioluminescence analyses were repeated a second time. The mice were then shifted back to a DOX-supplemented diet (+DOX) for 12 days and the skin TPA application, GFP fluorescence and luciferase bioluminescence analyses were repeated a third time. (**A**) GFP fluorescence, indicating shCox2 expression. Data are means +/− S.D. (***p<0.001). After the DOX-diet was removed (middle panel, no DOX), GFP fluorescence returned to baseline (p>0.05, ns). (**B**) Luciferase bioluminescence, indicating *Cox2* gene-driven luciferase expression. TPA-induced luciferase expression is reversibly reduced, in the presence of DOX, in triple transgenic mice. Data are means +/− S.D (*p<0.05, ***p<0.001). (**C**) Summary of GFP fluorescent intensity values (left panel) and bioluminescence (right panel) in Luc+ mice (dashed lines) and shCox2/C3/Luc+ mice (solid lines) for the three successive non-invasive imaging analyses. Error bars show S.D.

In the presence of DOX in the diet, when DOX-dependent TG-Cox22058 (shCox2) is strongly expressed (Fig. 5A, top row), TPA elicited little or no *Cox2* gene-driven luciferase expression in triple transgenic shCox2/C3/Luc+ mice (Fig. 5B, top row). In contrast, TPA painting of Luc+ mice, which do not contain the shCox2 transgene, elicits robust luciferase expression. Luciferase bioluminescence in shCox2/C3/Luc+ mice was reduced by ~80% relative to luciferase expression in Luc+ mice when the mice are on a DOX-containing diet (Fig. 5B, top row). However, when DOX is removed from the diet of shCox2/C3/Luc+ mice, a condition under which there is no expression of DOX-dependent shCox2, indicated by a lack of GFP expression (Fig. 5A, middle row), TPA induced equivalent levels of *Cox2* gene-dependent luciferase expression in shCox2/C3/Luc+ mice and the control

Luc+ mice (Fig. 5B, middle row). When the shCox2/C3/Luc+ mice are returned to a DOX-containing diet (when DOX-dependent shCox2 is once again expressed; Fig. 5A, bottom row), TPA-induced luciferase expression is once again reduced in shCox2/C3/Luc+ mice (by 75%) relative to luciferase expression in Luc+ mice (Fig. 5B, bottom row). Luc+ mice on DOX-containing and DOX-free diets expressed identical luciferase activities in response to TPA treatment; DOX treatment alone did not influence luciferase induction by TPA in skin (Fig. S3B).

In Fig. 5C we plot the quantification of the successive GFP/shCox2 expression (left panel) and successive *Cox2* gene-driven luciferase expression (right panel) in shCox2/C3/Luc+ mice and control Luc+ mice. In the presence of DOX, GFP/shCox2 is strongly induced, both initially and again following a period of

DOX withdrawal; however, DOX withdrawal reduces shCox2 expression in the triple transgenic mice to the same background level observed in Luc+ mice (Fig. 5C, left panel). The inverse is true for *Cox2* gene-driven luciferase in the triple transgenic shCox2/C3/Luc+ mice; in the presence of DOX, when shCox2 is expressed, luciferase expression is strongly inhibited; in the absence of shCox2 expression (in the absence of DOX) there is no inhibition of *Cox2* gene-driven luciferase expression relative to Luc+ mice. Repeated DOX administration once again suppresses TPA-induced luciferase induction in the shCox2/C3/Luc+ mice. The increasing luciferase activity observed in Luc+ mice with successive TPA treatments (Fig. 5C, right panel, dotted line) reflects the hyperalgesic response observed for COX-2 expression with repeated inflammatory insult [43–46]. In summary, these results show that shCox2 expression can be reversibly regulated *in vivo* by DOX and that reversible regulation of shCox2 expression results in reciprocal downregulation of *Cox2* transcripts.

Doxycycline-regulated shCox2 expression can suppress induction of COX-2 protein and leukocyte infiltration in mice homozygous for the wild-type *Cox2* gene. Heterozygous $Cox2^{tm2Luc/+}$ mice, expressing luciferase from one allele and COX-2 protein from the other allele, are very useful in assessing reversible and/or prolonged *Cox2* gene silencing by repeated non-invasive imaging of the same animal. However, to employ this system to study the reversible, cell-specific role of COX-2 in biological contexts, it is essential to demonstrate that the single copy of the COX-2 shRNA present in double transgenic TG-Cox2.2058/CAG-rtTA3 (shCox2/C3) mice can, when responding to DOX, suppress COX-2 expression in mice with two functional *Cox2* alleles.

TPA treatment induced COX-2 expression in the skin of double transgenic shCox2/C3 mice on a DOX-free diet (Fig. 6A, left panel). In contrast, COX-2 expression was not induced above baseline in TPA-treated double transgenic mice on the DOX-supplemented diet, in which shCox2 was expressed (Fig. 6A, left panel). Quantification of Western blot signals from three independent experiments showed that the average knockdown of COX-2 protein accumulation in the skin of DOX-treated mice following TPA treatment was ~80–90%, corresponding to near baseline expression (Fig. 6A, right panel).

To analyze where in the skin COX-2 expression occurs in response to TPA treatment and to visualize the consequences of DOX-induced shCox2 expression, COX-2 and GFP expression in TPA-treated shCox2/C3 double transgenic mice were analyzed by immunohistochemistry (Fig. 6B). TPA-induced COX-2 expression was visible predominantly in the basal layer of the skin epithelium in mice on a DOX-free diet (Fig. 6B, white arrow). In contrast, no COX-2 staining could be observed in the skin of TPA-treated shCox2/C3 mice either on a DOX-supplemented diet or in untreated control mice (Fig. 6B). GFP staining reflecting shCox2 expression was visible throughout the epithelial layer in mice that received a DOX-containing diet, but was not present in mice on a DOX-free diet (Fig. 6B, lower panel). These observations confirm the Western blot data, demonstrating that single copy shCox2 expression from the DOX-activated TRE locus is sufficient to suppress TPA-induced COX-2 expression in skin of mice homozygous for the wild-type *Cox2* gene.

Our final experiment was to demonstrate inhibition of a COX-2 dependent phenotypic function as a consequence of RNA interference. TPA-treatment of mouse skin results in a COX-2 dependent increase in leukocyte infiltration [39]. To determine whether shCox2-mediated suppression of TPA-induced COX-2 expression in skin results in functional COX-2 attenuation, we analyzed the consequences of DOX treatment on leukocyte infiltration following TPA stimulation in skin of double transgenic shCox2/C3 mice. On a DOX-free diet, we observed an increased number of leukocytes in the dermis of mice treated with TPA compared to untreated control mice (36.3+/−5.1 vs. 84.3+/−6.9 cells per field) (Fig. 6C). In contrast, on a DOX-supplemented diet, TPA-induced leukocyte infiltration was significantly reduced in the dermis of TPA-treated shCox2/C3 mice compared to TPA-treated mice on a control diet (58.9+/−10.9 vs. 84.3+/−6.9) (Fig. 6C).

In summary, we have established a mouse strain that expresses a tet-regulatable, miR30-based shRNA targeting the *Cox2* transcript, and have demonstrated reversible and functional DOX-mediated suppression of *Cox2* gene expression.

Discussion

In this report we describe the creation of a mouse model in which COX-2 expression can be cell-specifically and reversibly regulated by a doxycycline-inducible shRNA that targets the *Cox2* transcript. In addition, we created a luciferase reporter knockin mouse ($Cox2^{tm2Luc}$), in which luciferase replaces the coding region of the endogenous *Cox2* gene, while retaining all *Cox2* regulatory regions (promoter, 5′UTR, 3′UTR). Because shRNA Cox2.2058 targets a sequence in the *Cox2* gene 3′UTR, *Cox2* gene-driven luciferase activity can be used to monitor noninvasively Cox2.2058 shRNA-mediated COX-2 knockdown. Using the $Cox2^{tm2Luc/+}$ heterozygous reporter mouse, investigators can quickly evaluate the feasibility of the system for tissue or cell-specific applications.

Although the heterozygous $Cox2^{tm2Luc/+}$ mouse provides a model in which to quickly monitor *Cox2* gene expression, many aspects of COX-2 biology are gene-dosage dependent [28,47,48]. Since the TG-Cox2.2058 transgene is present in only a single copy, it is necessary to demonstrate substantial DOX-dependent COX-2 suppression in mice homozygous for the *Cox2* wild type allele and, more importantly, to confirm that COX-2 knockdown can result in functional suppression. Both criteria are met for TPA-induced skin inflammation (Fig. 6), a classic model of COX-2 induction [49]. ShCox2 silencing of COX-2 expression resulted in a significant reduction of leukocytes present in the dermis following TPA treatment (Fig. 6C). This finding is consistent with previous studies that show that COX-2 plays a role in leukocyte chemotaxis [39,50]; Nakamura et al [39] found a ~40% reduction in leukocyte infiltration following two TPA applications in the presence of the COX-2 inhibitor nimesulide [39].

In our *in vivo* experiments (Figs. 4–6) we supply and withdraw DOX in 12-day intervals. The literature suggests that DOX induction and de-induction occurs more rapidly; e.g. 2–4 days [24]. However, when monitoring GFP as a surrogate indicator for shCox2 expression, we observed residual fluorescence for 10–12 days after DOX withdrawal. Additional experiments will be needed to determine the relationship between the stability of the GFP reporter and the stability of Cox2.2058 following DOX removal, to optimize minimal times required for reversible, Cox2.2058-mediated COX-2 expression knockdown and recovery.

Measuring *Cox2*-driven luciferase activity confirmed that DOX-regulated shCox2-dependent COX-2 silencing exhibited full reversibility in the skin inflammation system (Fig. 5). The ability to reversibly and repeatedly knock down *Cox2* gene expression is one of the greatest assets of the shCox2 gene silencing protocol; the model has the enormous advantage of not permanently disrupting the *Cox2* gene. This property allows reversible manipulation of COX-2 expression at any time after disease onset, permitting – for example – investigation of the complex roles of COX-2 expression

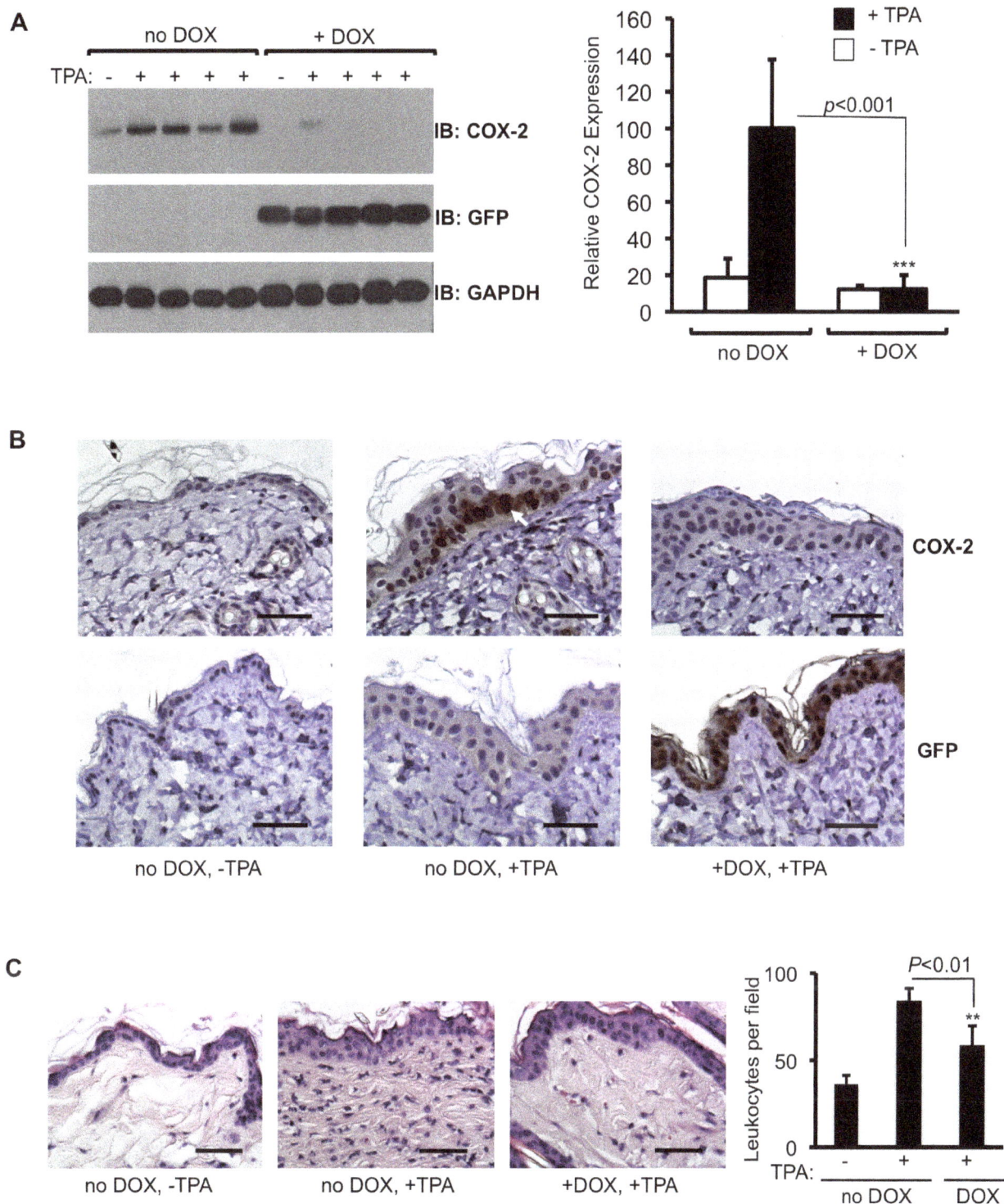

Figure 6. DOX-dependent suppression of TPA-induced COX-2 expression by shCox2 in mice homozygous for the *Cox2* gene. (A) Double transgenic shCox2/C3 mice, which have two wild type *Cox2* alleles, were maintained on either a control (no DOX) or a DOX-supplemented diet (+DOX) for 12 days, then painted with TPA on the back. 24 hours after TPA administration the mice were euthanized and skin extracts were assayed for GFP and COX-2 protein by Western blotting. Quantification is from three independent experiments. The COX-2 signal was normalized to the GAPDH loading control. Data are means +/− S.D. (***p<0.001). **(B)** COX-2 (upper panels) and GFP (lower panels) immunohistochemistry in skin of untreated shCox2/C3 mice (left panels), 24 hours after TPA administration to shCox2/C3 mice maintained on a DOX-free diet (center panels) and 24 hours after TPA treatment of shCox2/C3 mice maintained for 12 days on a DOX-supplemented diet (right panels). COX-2 staining is visible as brown patches (white arrow) in the basal epithelium. **(C)** H&E stain to visualize leukocytes in skin sections from double transgenic shCox2/C3 mice treated with TPA and/or DOX as in B. The graph depicts quantification of leukocytes in the dermis. Quantification is from two independent experiments. Data are means +/− S.D. (**p<0.01). The scale bar indicates 50 μm.

in alternative cells in inflammatory responses to pathogens, by initiating knockdown in specific cell types at different time points after infection. In contrast, conventional *Cox2* gene disruption permanently eliminates global COX-2 expression at all times.

Although conditional Cre recombinase expression can also be induced at distinct time points by using the inducible Cre-ERT/ tamoxifen system, recombination irreversibly disrupts the target locus. Moreover, Cre expression can, in some instances, show significant toxicity [51]. In addition, shCox2 gene silencing requires only two genes; a single TG-Cox2.2058 transgene and the appropriate rtTA or tTA transactivator transgene; Cre-based conditional deletion requires three alleles (two *Cox2^flox* alleles and the targeted Cre recombinase); consequently, shCox2 targeting can be more easily and rapidly combined with existing mouse disease models. The combination of easier genetic modeling and repeated, reversible gene silencing should permit insights that go beyond what can be obtained through conditional knockout approaches. Another advantage of inducible shRNA gene silencing which distinguishes it from the binary Cre-loxP system is the possibility of graded inhibition of gene expression by varying DOX dosage.

Suppression by shCox2 in distinct cell types requires appropriate cell/tissue-specific rtTA/tTA transgenic mice. The list of such mice is already extensive (ww.tetsystems.com), and growing. In addition, a CAG promoter-driven lox-stop-lox conditional rtTA transactivator that enables use of a Cre driver to convert transgenic mice to a cell-specific rtTA setting is currently under development (S. Lowe, submitted for publication).

Despite their great utility, RNAi-based gene regulation systems have their limitations. Tetracycline analogues can have side effects [52]. We included Luc+ control animals, to verify that COX-2 inhibition was due to shRNA expression, and not to DOX side effects (Fig. S3). It is also important to ensure that DOX efficiently reaches the target tissue of interest; DOX in the diet is more effective than DOX supplied in the drinking water [53].

Adequate rtTA expression in the target tissue/cell type is a key, rate-limiting step for DOX-regulated gene silencing. We chose the CAG-rtTA3 transactivator, which shows particularly strong expression in skin [24], to facilitate initial testing of the Cox2.2058 shRNA. The CAG promoter is considered by many to be ubiquitous. Because of our interest in inflammatory roles of COX-2, we also analyzed luciferase activity in spleens and lungs of IFN-γ + LPS-treated triple transgenic mice. Luciferase expression in the spleen or lung was not reduced significantly in IFN-γ + LPS treated mice that received a DOX-supplemented diet (Fig. S4). Analysis of DOX-dependent shRNA expression in target cells is facilitated by GFP co-expression. Consistent with previous reports [24], we observed less than two-fold shRNA-coupled GFP induction in spleen and lung by DOX, in contrast to the greater than 10-fold GFP induction in skin. These results suggest that CAG-driven rtTA3 transactivator expression may not be sufficient to block COX-2 expression in all tissues, or that silencing of the TRE-driven transgene may occur in some tissues, emphasizing the importance of systematic validation of the system for each application. Finally, it is important to keep in mind that, although recent advances in design of the microRNA backbone have significantly increased the effectiveness of single copy RNAi molecules [54], tet-regulated RNAi can rarely produce the null phenotype in individual cells achievable by gene deletion strategies.

Inducible *Cox2* gene knockdown using the tet-regulated shCox2 expression system provides a complementary approach to global and conditional *Cox2* gene knockout strategies to study COX-2 roles *in vivo*. Because COX-2 is inducible by a wide range of

stimuli and plays an important role in many disease states, the double transgenic shCox2/C3 mouse and the triple transgenic shCox2/C3/Luc+ mouse should become useful tools for the study of the many aspects of COX-2 biology.

Supporting Information

Figure S1   (A) Percent of GFP expressing NIH 3T3 cells, two days after LMP retrovirus vector transduction at a low multiplicity of infection (MOI). Less than 1% of the cells are GFP positive. The cells were harvested, washed with PBS, fixed with 0.01% paraformaldehyde and analyzed by flow cytometry to determine the percentage of GFP expression. Mock; untransduced cells, indicating background fluorescence. Control; cells transduced with a LMP vector encoding a control shRNA against luciferase. Cox2.284 (1) – Cox2.3711 (4); cells transduced with LMP vectors encoding *Cox2* specific shRNAs 1–4. **(B)** Schematic representation of the LMP retrovirus vector construct, indicating the probe used for Southern blot (red) and restriction enzyme target sites (X, *Xho*I. RI, *Eco*RI. HIII, *Hind*III. NI, *Nco*I). **(C–D)** Southern blot of LMP vector transduced NIH 3T3 cells. Cells were transduced at a low MOI. Two days later the cells were treated with 2.5 μg/mL puromycin to select for LMP transduced cells. Total DNA was harvested using the DNeasy kit (Qiagen). Cell DNA was digested with restriction enzymes overnight, before being subjected to Southern blot analysis. **(C)** Southern blot of *Hind*III digested DNA from mock transduced NIH 3T3 cells (m) or NIH 3T3 cells transduced with the LMP retrovirus vector encoding control (c) or *Cox2* specific shRNAs (1–4). *Hind*III digestion results in a 1100 bp fragment within the vector backbone, confirming the integrated LMP vector in cell genomic DNA. **(D)** Southern blot of *Nco*I digested DNA from mock-transduced (m), LMP control vector-transduced (c) or LMP vector encoding Cox2.5078 shRNA (3)-transduced NIH3T3 cells. *Nco*I has a single target site close to the 3′ end of the LMP vector, resulting in a ~1000+ bp fragment depending on the location of the closest *Nco*I site in the cellular genome. The characteristic smear on the Southern blot confirms multiple-sized DNA fragments, indicating divers integration sites.

Figure S2   COX-2 and luciferase induction by interferon gamma and endotoxin (IFN-γ/LPS) in the hearts, livers and lungs of heterozygous *Cox-2^tm2Luc/+* mice. Four mice were injected i.p. with IFNγ followed two hours later with LPS, or with saline. After 6 hours, mice were euthanized, tissues were removed and luciferase bioluminescence was quantified by bioluminescence imaging (left panels). Tissue extracts were then prepared and luciferase enzymatic activity was measured. COX-2 protein expression in tissue extracts was also analyzed by immunoblotting (right panels). Data are means +/− SD (**, p< 0.01; *, p<0.05).

Figure S3   (A) Luc+ mice that receive a DOX containing diet (indicated in blue) have similar luciferase expression in the zymosan treated paws at 6 hours compared to triple transgenic (shCox2/C3/Luc+) mice that do not receive a DOX containing diet (indicated in red), confirming that DOX treatment alone does not significantly reduce luciferase expression. Data are means +/− SD, *p<0.05. **(B)** DOX in the diet does not affect TPA-induced *Cox2*-driven luciferase expression in skin. Luc+ mice were maintained either on a DOX-free diet or on a DOX-supplemented diet for 12 days prior to TPA administration to the skin. *In vivo* luciferase bioluminescence was measured 24 hours later. Bioluminescence in response to TPA painting is not significantly different in Luc+ control mice on DOX-free and DOX-supplemented diets. Data are individual values and averages from two mice.

Figure S4 **(A) GFP fluorescence imaging of spleen and lung tissue from triple transgenic shCox2/C3/Luc+ mice that received a DOX containing diet (+DOX) or a control diet (no DOX) for 12 days.** GFP fluorescence is increased significantly, but not to a great extent in the spleens. In the lungs, there was no significant difference in GFP fluorescence between shCox2/C3/Luc+ mice on a DOX containing diet and on a DOX free diet. **(B)** Luciferase induction by interferon gamma and endotoxin (IFN-γ/LPS) in the spleens and lungs of triple transgenic shCox2/C3/Luc+ mice that were fed a DOX containing diet (+DOX) to induce shRNA expression or a control diet (no DOX) for 12 days. Mice were injected i.p. with IFNγ

followed two hours later with LPS, or with saline (mock). After 6 hours, mice were euthanized, tissues were removed and luciferase bioluminescence was quantified by bioluminescence imaging. Luciferase expression was not significantly different in mice receiving the DOX diet versus controls. Data are means +/− SD. (*, $p < 0.05$, *ns*, $p > 0.05$, n = 4).

Acknowledgments

We thank Dr. Sotirios Tetradis for helpful comments and advice.

Author Contributions

Conceived and designed the experiments: AKZ JZ TI SWL HRH. Performed the experiments: AKZ CC. Analyzed the data: AKZ JZ SWL HRH. Contributed reagents/materials/analysis tools: JZ SWL CC HBM JJ ABC JSG. Contributed to the writing of the manuscript: AKZ HRH.

References

1. Herschman HR, Talley JJ, DuBois R (2003) Cyclooxygenase 2 (COX-2) as a target for therapy and noninvasive imaging. Mol Imaging Biol 5: 286–303.
2. Vane JR, Bakhle YS, Botting RM (1998) Cyclooxygenases 1 and 2. Annu Rev Pharmacol Toxicol 38: 97–120.
3. Harris SG, Padilla J, Koumas L, Ray D, Phipps RP (2002) Prostaglandins as modulators of immunity. Trends Immunol 23: 144–150.
4. Hirata T, Narumiya S (2012) Prostanoids as regulators of innate and adaptive immunity. Adv Immunol 116: 143–174.
5. Steer SA, Corbett JA (2003) The role and regulation of COX-2 during viral infection. Viral Immunol 16: 447–460.
6. Lim H, Paria BC, Das SK, Dinchuk JE, Langenbach R, et al. (1997) Multiple female reproductive failures in cyclooxygenase 2-deficient mice. Cell 91: 197–208.
7. Minghetti L (2004) Cyclooxygenase-2 (COX-2) in inflammatory and degenerative brain diseases. J Neuropathol Exp Neurol 63: 901–910.
8. Fischer SM (1997) Prostaglandins and cancer. Front Biosci 2: d482–500.
9. Garcia Rodriguez LA, Cea-Soriano L, Tacconelli S, Patrignani P (2013) Coxibs: pharmacology, toxicity and efficacy in cancer clinical trials. Recent Results Cancer Res 191: 67–93.
10. Dixon DA, Blanco FF, Bruno A, Patrignani P (2013) Mechanistic aspects of COX-2 expression in colorectal neoplasia. Recent Results Cancer Res 191: 7–37.
11. Bartel DP (2009) MicroRNAs: target recognition and regulatory functions. Cell 136: 215–233.
12. Hannon GJ (2002) RNA interference. Nature 418: 244–251.
13. Dykxhoorn DM, Novina CD, Sharp PA (2003) Killing the messenger: short RNAs that silence gene expression. Nat Rev Mol Cell Biol 4: 457–467.
14. Zeng Y, Wagner EJ, Cullen BR (2002) Both natural and designed micro RNAs can inhibit the expression of cognate mRNAs when expressed in human cells. Mol Cell 9: 1327–1333.
15. Zeng Y, Cai X, Cullen BR (2005) Use of RNA polymerase II to transcribe artificial microRNAs. Methods Enzymol 392: 371–380.
16. Stegmeier F, Hu G, Rickles RJ, Hannon GJ, Elledge SJ (2005) A lentiviral microRNA-based system for single-copy polymerase II-regulated RNA interference in mammalian cells. Proc Natl Acad Sci U S A 102: 13212–13217.
17. Dickins RA, Hemann MT, Zilfou JT, Simpson DR, Ibarra I, et al. (2005) Probing tumor phenotypes using stable and regulated synthetic microRNA precursors. Nat Genet 37: 1289–1295.
18. Dickins RA, McJunkin K, Hernando E, Premsrirut PK, Krizhanovsky V, et al. (2007) Tissue-specific and reversible RNA interference in transgenic mice. Nat Genet 39: 914–921.
19. Kistner A, Gossen M, Zimmermann F, Jerecic J, Ullmer C, et al. (1996) Doxycycline-mediated quantitative and tissue-specific control of gene expression in transgenic mice. Proc Natl Acad Sci U S A 93: 10933–10938.
20. Dow LE, Premsrirut PK, Zuber J, Fellmann C, McJunkin K, et al. (2012) A pipeline for the generation of shRNA transgenic mice. Nat Protoc 7: 374–393.
21. Fellmann C, Zuber J, McJunkin K, Chang K, Malone CD, et al. (2011) Functional identification of optimized RNAi triggers using a massively parallel sensor assay. Mol Cell 41: 733–746.
22. Seibler J, Schubeler D, Fiering S, Groudine M, Bode J (1998) DNA cassette exchange in ES cells mediated by Flp recombinase: an efficient strategy for repeated modification of tagged loci by marker-free constructs. Biochemistry 37: 6229–6234.
23. Beard C, Hochedlinger K, Plath K, Wutz A, Jaenisch R (2006) Efficient method to generate single-copy transgenic mice by site-specific integration in embryonic stem cells. Genesis 44: 23–28.

24. Premsrirut PK, Dow LE, Kim SY, Camiolo M, Malone CD, et al. (2011) A rapid and scalable system for studying gene function in mice using conditional RNA interference. Cell 145: 145–158.
25. Nagy A, Rossant J, Nagy R, Abramow-Newerly W, Roder JC (1993) Derivation of completely cell culture-derived mice from early-passage embryonic stem cells. Proc Natl Acad Sci U S A 90: 8424–8428.
26. Eakin GS, Hadjantonakis AK (2006) Production of chimeras by aggregation of embryonic stem cells with diploid or tetraploid mouse embryos. Nat Protoc 1: 1145–1153.
27. Zhao XY, Lv Z, Li W, Zeng F, Zhou Q (2010) Production of mice using iPS cells and tetraploid complementation. Nat Protoc 5: 963–971.
28. Ishikawa TO, Jain NK, Taketo MM, Herschman HR (2006) Imaging cyclooxygenase-2 (Cox-2) gene expression in living animals with a luciferase knock-in reporter gene. Mol Imaging Biol 8: 171–187.
29. Seluanov A, Vaidya A, Gorbunova V (2010) Establishing primary adult fibroblast cultures from rodents. J Vis Exp.
30. Ristimaki A, Narko K, Hla T (1996) Down-regulation of cytokine-induced cyclooxygenase-2 transcript isoforms by dexamethasone: evidence for post-transcriptional regulation. Biochem J 318 (Pt 1): 325–331.
31. Kirkby NS, Zaiss AK, Urquhart P, Jiao J, Austin PJ, et al. (2013) LC-MS/MS confirms that COX-1 drives vascular prostacyclin whilst gene expression pattern reveals non-vascular sites of COX-2 expression. PLoS One 8: e69524.
32. Kirkby NS, Zaiss AK, Wright WR, Jiao J, Chan MV, et al. (2013) Differential COX-2 induction by viral and bacterial PAMPs: Consequences for cytokine and interferon responses and implications for anti-viral COX-2 directed therapies. Biochem Biophys Res Commun 438: 249–256.
33. Ishikawa TO, Jain NK, Herschman HR (2010) Cox-2 gene expression in chemically induced skin papillomas cannot predict subsequent tumor fate. Mol Oncol 4: 347–356.
34. Lazarus M, Munday CJ, Eguchi N, Matsumoto S, Killian GJ, et al. (2002) Immunohistochemical localization of microsomal PGE synthase-1 and cyclooxygenases in male mouse reproductive organs. Endocrinology 143: 2410–2419.
35. Oshima M, Oshima H, Taketo MM (2005) Hypergravity induces expression of cyclooxygenase-2 in the heart vessels. Biochem Biophys Res Commun 330: 928–933.
36. Contag PR, Olomu IN, Stevenson DK, Contag CH (1998) Bioluminescent indicators in living mammals. Nat Med 4: 245–247.
37. Calhoun W, Chang J, Carlson RP (1987) Effect of selected antiinflammatory agents and other drugs on zymosan, arachidonic acid, PAF and carrageenan induced paw edema in the mouse. Agents Actions 21: 306–309.
38. Tarayre JP, Delhon A, Aliaga M, Barbara M, Bruniquel F, et al. (1989) Pharmacological studies on zymosan inflammation in rats and mice. 2: Zymosan-induced pleurisy in rats. Pharmacol Res 21: 385–395.
39. Nakamura Y, Kozuka M, Naniwa K, Takabayashi S, Torikai K, et al. (2003) Arachidonic acid cascade inhibitors modulate phorbol ester-induced oxidative stress in female ICR mouse skin: differential roles of 5-lipoxygenase and cyclooxygenase-2 in leukocyte infiltration and activation. Free Radic Biol Med 35: 997–1007.
40. Kadoshima-Yamaoka K, Goto M, Murakawa M, Yoshioka R, Tanaka Y, et al. (2009) ASB16165, a phosphodiesterase 7A inhibitor, reduces cutaneous TNF-alpha level and ameliorates skin edema in phorbol ester 12-O-tetradecanoyl-phorbol-13-acetate-induced skin inflammation model in mice. Eur J Pharmacol 613: 163–166.
41. Puignero V, Queralt J (1997) Effect of topically applied cyclooxygenase-2-selective inhibitors on arachidonic acid- and tetradecanoylphorbol acetate-induced dermal inflammation in the mouse. Inflammation 21: 431–442.

42. Gabor M (2003) Models of acute inflammation in the ear. Methods Mol Biol 225: 129–137.

43. Araldi D, Ferrari LF, Lotufo CM, Vieira AS, Athie MC, et al. (2013) Peripheral inflammatory hyperalgesia depends on the COX increase in the dorsal root ganglion. Proc Natl Acad Sci U S A 110: 3603–3608.

44. Doyle T, Chen Z, Muscoli C, Obeid LM, Salvemini D (2011) Intraplantar-injected ceramide in rats induces hyperalgesia through an NF-kappaB- and p38 kinase-dependent cyclooxygenase 2/prostaglandin E2 pathway. FASEB J 25: 2782–2791.

45. Jain NK, Ishikawa TO, Spigelman I, Herschman HR (2008) COX-2 expression and function in the hyperalgesic response to paw inflammation in mice. Prostaglandins Leukot Essent Fatty Acids 79: 183–190.

46. Hay C, de Belleroche J (1997) Carrageenan-induced hyperalgesia is associated with increased cyclo-oxygenase-2 expression in spinal cord. Neuroreport 8: 1249–1251.

47. Tiano HF, Loftin CD, Akunda J, Lee CA, Spalding J, et al. (2002) Deficiency of either cyclooxygenase (COX)-1 or COX-2 alters epidermal differentiation and reduces mouse skin tumorigenesis. Cancer Res 62: 3395–3401.

48. Dinchuk JE, Car BD, Focht RJ, Johnston JJ, Jaffee BD, et al. (1995) Renal abnormalities and an altered inflammatory response in mice lacking cyclooxygenase II. Nature 378: 406–409.

49. Kujubu DA, Fletcher BS, Varnum BC, Lim RW, Herschman HR (1991) TIS10, a phorbol ester tumor promoter-inducible mRNA from Swiss 3T3 cells, encodes a novel prostaglandin synthase/cyclooxygenase homologue. J Biol Chem 266: 12866–12872.

50. Menezes GB, Rezende RM, Pereira-Silva PE, Klein A, Cara DC, et al. (2008) Differential involvement of cyclooxygenase isoforms in neutrophil migration in vivo and in vitro. Eur J Pharmacol 598: 118–122.

51. Loonstra A, Vooijs M, Beverloo HB, Allak BA, van Drunen E, et al. (2001) Growth inhibition and DNA damage induced by Cre recombinase in mammalian cells. Proc Natl Acad Sci U S A 98: 9209–9214.

52. Ahler E, Sullivan WJ, Cass A, Braas D, York AG, et al. (2013) Doxycycline alters metabolism and proliferation of human cell lines. PLoS One 8: e64561.

53. Cawthorne C, Swindell R, Stratford IJ, Dive C, Welman A (2007) Comparison of doxycycline delivery methods for Tet-inducible gene expression in a subcutaneous xenograft model. J Biomol Tech 18: 120–123.

54. Fellmann C, Hoffmann T, Sridhar V, Hopfgartner B, Muhar M, et al. (2013) An optimized microRNA backbone for effective single-copy RNAi. Cell Rep 5: 1704–1713.

12

Consumption of *Bt* Rice Pollen Containing Cry1C or Cry2A Protein Poses a Low to Negligible Risk to the Silkworm *Bombyx mori* (Lepidoptera: Bombyxidae)

Yan Yang[1,2,3], **Yue Liu**[1], **Fengqin Cao**[1], **Xiuping Chen**[2], **Lisheng Cheng**[3], **Jörg Romeis**[2,4], **Yunhe Li**[2]*****, **Yufa Peng**[2]

1 College of Environment and Plant Protection, Hainan University, Haikou, China, **2** State Key Laboratory for Biology of Plant Diseases and Insect Pests, Institute of Plant Protection, Chinese Academy of Agricultural Sciences, Beijing, China, **3** Qiongtai Teachers College, Haikou, China, **4** Agroscope, Institute for Sustainability Science ISS, Zurich, Switzerland

Abstract

By consuming mulberry leaves covered with pollen from nearby genetically engineered, insect-resistant rice lines producing Cry proteins derived from *Bacillus thuringiensis* (*Bt*), larvae of the domestic silkworm, *Bombyx mori* (Linnaeus) (Lepidoptera: Bombyxidae), could be exposed to insecticidal proteins. Laboratory experiments were conducted to assess the potential effects of Cry1C- or Cry2A-producing transgenic rice (T1C-19, T2A-1) pollen on *B. mori* fitness. In a short-term assay, *B. mori* larvae were fed mulberry leaves covered with different densities of pollen from *Bt* rice lines or their corresponding near isoline (control) for the first 3 d and then were fed mulberry leaves without pollen. No effect was detected on any life table parameter, even at 1800 pollen grains/cm^2 leaf, which is much higher than the mean natural density of rice pollen on leaves of mulberry trees near paddy fields. In a long-term assay, the larvae were fed *Bt* and control pollen in the same way but for their entire larval stage (approximately 27 d). *Bt* pollen densities ≥150 grains/cm^2 leaf reduced 14-d larval weight, increased larval development time, and reduced adult eclosion rate. ELISA analyses showed that 72.6% of the Cry protein was still detected in the pollen grains excreted with the feces. The low exposure of silkworm larvae to Cry proteins when feeding *Bt* rice pollen may be the explanation for the relatively low toxicity detected in the current study. Although the results demonstrate that *B. mori* larvae are sensitive to Cry1C and Cry2A proteins, the exposure levels that harmed the larvae in the current study are far greater than natural exposure levels. We therefore conclude that consumption of *Bt* rice pollen will pose a low to negligible risk to *B. mori*.

Editor: Juan Luis Jurat-Fuentes, University of Tennessee, United States of America

Funding: The study was supported by the National GMO New Variety Breeding Program of PRC (2014ZX08011-02B, 2014ZX08011-001). The funders had no role in study design, data collection and analysis, decision to publish, or preparation of the manuscript.

Competing Interests: The authors have declared that no competing interests exist.

* Email: yunhe.li@hotmail.com

Introduction

Rice, *Oryza sativa* L., is a staple food for more than half of the world's population and for over 65% of the Chinese people [1,2]. To feed a growing population worldwide, rice production will have to increase by more than 40% by the year 2030 [3]. Similarly, China will need to increase its rice production by at least 20% by 2030 in order to meet its domestic needs [4]. Rice production, however, is constrained by many factors, and insect pests are among the most important [5].

Insect pests that can substantially reduce rice production in China include the following lepidopteran species. Recent research has confirmed that genetic engineering of rice is an efficient strategy for insect pest control. Multiple insect-resistant genetically engineered (IRGE) rice lines have been developed that produce Cry toxins derived from the bacterium *Bacillus thuringiensis* (*Bt*), and these IRGE rice lines are very effective against these lepidopteran pests [5–7].

Before a novel GE variety is commercialized, its potential risks to the environment and animal and human health must be extensively evaluated, and an important component of the risk assessment concerns the potential effects of IRGE crops on non-target organisms [8,9]. Many laboratory and field studies have demonstrated that *Bt* rice represents a negligible threat to non-target arthropods belonging to orders that differ from that of the target pests, i.e., *Bt* proteins produced in current IRGE rice lines only affect lepidopterans [5,9–14]. On the other hand, non-target lepidopterans could be affected by *Bt* rice and therefore warrant special attention in the risk assessment of IRGE crops [15]. A non-target lepidopteran of particular concern in China is the silkworm *Bombyx mori* Linnaeus (Lepidoptera: Bombycidae).

B. mori is an economically and culturally important insect in China, which is a world center of silk production [16]. *B. mori* larvae feed exclusively on mulberry (*Morus atropurpurea* Roxb.) leaves. In southeast China, mulberry trees are typically planted near or around rice fields in a planting system that is referred to as mulberry-mixed cropping [17]. Thus, once *Bt* rice is commercially grown in China, mulberry leaves may be covered with *Bt* rice pollen. It follows that *B. mori* larvae could be exposed to Cry proteins if they consume mulberry leaves covered with *Bt* rice pollen and if Cry proteins are produced in the pollen [18–22]. Because *B. mori* belongs to the same order as the target pests, i.e., the Lepidoptera, it may be sensitive to lepidopteran-active Cry proteins produced by the current *Bt* rice lines. Thus, before *Bt* rice lines are approved for commercial use, their potential effects on *B. mori* should be assessed [5].

In the current study, we developed and used a rice pollen-feeding assay to assess the potential effects of *Bt* rice pollen containing Cry2A or Cry1C protein on *B. mori* larvae.

Results

Bt protein contents in rice pollen

No *Bt* protein was detected in pollen from the control rice (Minghui 63). The mean ($\pm$SE) content of Cry2A was 28.15 ± 1.19 µg/g dry weight (DW) in T2A-1 pollen, which was more than 11-fold higher than the content of Cry1C in T1C-19 pollen (2.40 ± 0.08 µg/g DW).

Pollen consumption by *B. mori*

Pollen consumption per larva increased as the density of pollen on the mulberry leaf squares increased (one-way ANOVA; $P < 0.01$ for each type of rice pollen). Pair-wise comparisons by Tukey HSD tests showed significant differences between any two pollen doses in both bioassays except for the two lowest doses in the short-term bioassay (all $P < 0.001$) (Fig. 1A&B).

In the short-term bioassay, the number of pollen grains consumed was independent of pollen type (Cry1C, Cry2A, and control rice pollen) (Dunnett test; all $P > 0.1$ at any pollen density) (Fig. 1A). In the long-term assay, however, consumption was lower with Cry1C pollen than with control pollen ($P < 0.05$ for any pollen density) but did not differ ($P > 0.05$ for any pollen density) between Cry2A pollen and control pollen ($P > 0.05$) (Fig. 1B).

Effects of pollen consumption on larval weight

In both bioassays, 14-d larval weight was not affected by increased pollen consumption regardless of the source of the pollen (one-way ANOVA for the short-term assay; control: $F = 0.36$, df $= 17$, $P = 0.832$; Cry1C: $F = 0.16$, df $= 17$, $P = 0.96$; Cry2A: $F = 0.33$, df $= 17$, $P = 0.86$; one-way ANOVA for the long-term assay; control: $F = 0.20$, df $= 17$, $P = 0.93$; Cry1C: $F = 2.79$, df $= 17$, $P = 0.07$; Cry2A: $F = 1.96$, df $= 17$, $P = 0.16$). In the short-term assay at any pollen density, 14-day larval weight did not differ significantly between either *Bt* treatment and the control (Dunnett's test; $P > 0.1$). In the long-term assay, however, larval weight was significantly lower with *Bt* rice pollen than with control pollen at the pollen dose of 150 (Cry1C: $P = 0.015$; Cry2A: $P = 0.018$), 450 (Cry1C: $P = 0.015$; Cry2A: $P = 0.012$), and 1800 grains/cm^2 leaf (Cry1C: $P = 0.033$; Cry2A: $P = 0.033$) (Table 1).

Effects of pollen consumption on larval development

Larval developmental time was not significantly affected by the increased pollen consumption of any rice pollen in the short-term assay (Kruskal-Wallis test; control: $\chi^2 = 4.77$, $P = 0.31$; Cry1C: $\chi^2 = 7.88$, $P = 0.10$; Cry2A: $\chi^2 = 2.94$, $P = 0.57$) (Table 1). In the long-term assay, developmental time was not affected by an increase in the density of control pollen ($\chi^2 = 3.63$, $P = 0.46$) but was significantly increased by an increase in the density of both kinds of *Bt* pollen ($\chi^2 = 14.378$, $P = 0.006$ for Cry1C, and $\chi^2 = 13.769$, $P = 0.008$ for Cry2A) (Table 1). In the short-term

assay, pollen type did not significantly affect larval developmental time (Dunnett's test, all $P > 0.10$). In the long-term assay, however, larval developmental time was significantly longer with Cry1C pollen than with control pollen at doses of 450 and 1800 grains/cm^2 (both $P < 0.05$). Larval developmental time did not differ between Cry2A and control pollen at any pollen density (all $P > 0.05$) (Table 1).

Effects of pollen consumption on pupation and eclosion rate

In the short-term assay, the pupation rate was not significantly affected by pollen density on the mulberry leaves (one-way ANOVA; control: $F = 2.06$, df $= 17$, $P = 0.15$; Cry1C: $F = 0.47$, df $= 17$, $P = 0.76$; Cry2A: $F = 2.70$, df $= 17$, $P = 0.08$) (Table 1), although the pupation rate tended to drop with the highest density of Cry2A pollen. In the long-term assay, the pupation rate gradually decreased with the increase of pollen density; the decrease was marginally significant with control pollen ($F = 3.12$, df $= 17$, $P = 0.05$), but was significant with both Cry1C pollen ($F = 6.58$, df $= 17$, $P = 0.004$) and Cry2A pollen ($F = 6.77$, df $= 17$, $P = 0.004$).

In both bioassays, eclosion rate of the silkworm significantly decreased as pollen density increased (one-way ANOVA; $P < 0.001$ for any pollen type). In the short-term assay, the eclosion rate was similar with *Bt* pollen and control pollen (Dunnett's test; $P > 0.2$ for any pollen density). In the long-term assay, however, the eclosion rate was significantly lower with Cry1C pollen than with control pollen at 450 and 1800 grains/cm^2 leaf ($P = 0.001$ and 0.011, respectively) or with Cry2A pollen than with control pollen at 1800 grains/cm^2 leaf ($P = 0.016$) (Table 1).

Fate of Cry1C contained in *Bt* rice pollen after silkworm gut passage

Pollen grains from *Bt* rice T1C-19 contained 27.4% less Cry1C protein after passage through the digestive system of *B. mori* larvae. This difference, however, was not statistically significant (Student's-t test, $t = 1.96$, df $= 4$, $P = 0.12$) (Table 2). Observations under the microscope revealed that the majority of the pollen grains were only partly digested.

Discussion

The risk represented by a *Bt* crop for a non-target organism depends on the organism's sensitivity to the *Bt* protein and on the probability that it is exposed to harmful concentrations of that protein in the field [8,23]. Therefore, when dietary assays are used to assess the effects of *Bt* pollen on a non-target species, selection of appropriate pollen doses is important and should be based on the pollen densities that the species may encounter in the field [24]. Fan et al. [17] reported that the density of rice pollen deposited on the leaves of mulberry trees near rice fields ranged from 13–199 grains/cm^2 with an average density of 93 grains/cm^2. Yao et al. [20] reported that the maximum density of rice pollen on mulberry leaves was 1636 grains/cm^2, although the probability of that density occurring in the field was only 0.2%. In the current study, the pollen doses tested ranged from 0 to 1800 grains/cm^2 of mulberry leaf, which covered the potential pollen densities to which silkworms may be exposed in the field. In addition, both a short-term feeding assay (3 d) and a long-term feeding assay (27 d, covering the entire larval stage) were conducted. In the field, rice anthesis usually lasts 10 to 15 d [20,21] and pollen is typically shed at a high rate for less than 1 week (Yunhe Li et al., unpublished data). In addition, the silkworm larval period may not totally overlap with rice anthesis, and environmental factors such as rain

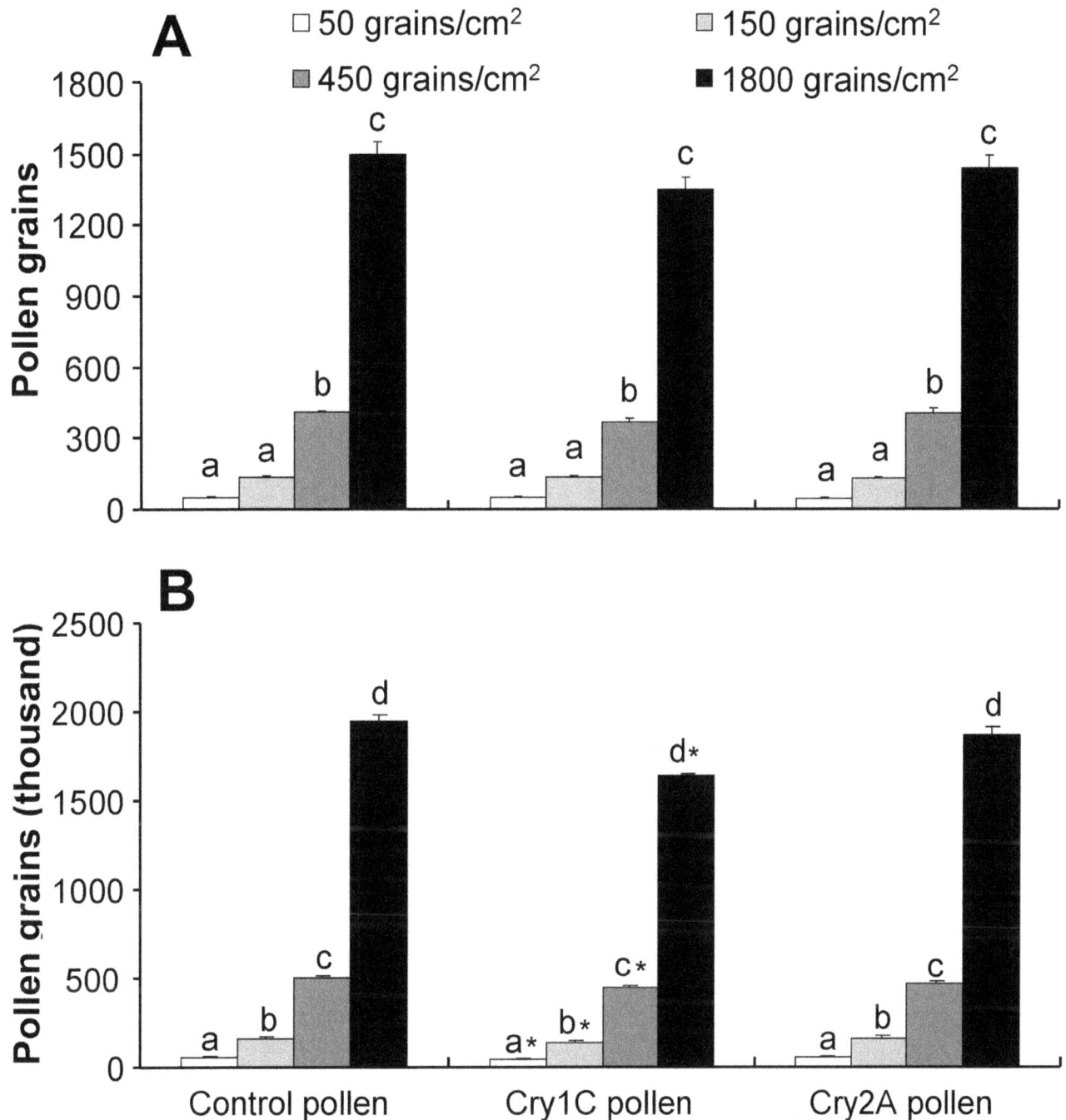

Figure 1. Rice pollen consumption per *Bombyx mori* larva. Lavae were fed mulberry leaves covered with different doses of pollen from Cry1C- or Cry2A-expressing *Bt* rice or pollen from the corresponding non-transformed varieties for (A) 3 d (short-term assay) and for (B) the entire larval stage (long-term assay). For each pollen type within each assay, bars with different letters are significantly different (one-way ANOVA with Tukey test), while asterisks indicate significant differences between the *Bt* pollen treatment and the corresponding non-*Bt* pollen treatment at the same pollen dose (Dunnett test). Values are means + SE, n = 3.

and wind will also affect rice pollen deposition on mulberry leaves. Therefore the short-term assay may represent an realistic exposure scenario and the long-term assay represents a worst-case scenario. Thus, the assays used in this study are useful for assessing the risk of that *Bt* rice pollen represents to the silkworm, *B. mori*.

As expected, the number of pollen grains consumed by *B. mori* larvae increased as the density of rice pollen on mulberry leaves increased. Interestingly, the larvae in the long-term assay

consumed significantly less Cry1C pollen than control pollen even at the low dose of 50 grains/cm^2 and even though this density of Cry1C pollen did not affect survival or development. This suggests that the reduced consumption of Cry1C pollen was not caused by harm to the larvae and that Cry1C protein may have antifeedant activity towards *B. mori* larvae. This was not the case with Cry2A pollen in the long-term assay in that the larvae consumed similar quantities of Cry2A and control pollen.

Table 1. Life table parameters of *Bombyx mori* larvae when fed pollen from Cry1C- or Cry2A-expressing *Bt* rice or pollen from the corresponding non-transformed varieties.

Parameter	No. of pollen grains/cm² leaf	Short-term assay[1]			Long-term assay[2]		
		Control pollen	Cry1C pollen	Cry2A pollen	Control pollen	Cry1C pollen	Cry2A pollen
Larval weight (mg)	0	400.91±14.70 a	400.91±14.70 a	400.91±14.70 a	400.91±14.70 a	400.91±14.70 a	400.91±14.70 a
	50	394.98±13.42 a	399.64±17.07 a	389.49±22.92 a	404.18±12.18 a	380.03±5.31 a	384.13±22.84 a
	150	410.48±10.80 a	398.88±4.62 a	383.40±10.56 a	415.31±9.41 a	366.86±8.41 a*	371.34±6.98 a*
	450	396.30±12.12 a	388.54±23.64 a	375.96±25.52 a	410.00 ±11.49 a	356.02±2.64 a*	359.17±5.18 a*
	1800	385.45±4.09 a	386.24±18.18 a	388.89±4.21 a	412.40±8.82 a	347.25±17.51 a*	354.32±4.69 a*
Larval developmental time (d)	0	26.99±0.37 a	26.99±0.37 a	26.99±0.37 a	26.99±0.37 a	26.99±0.37 a	26.99±0.37 a
	50	26.67±0.49 a	27.00±0.21 a	27.41±0.16 a	27.33±0.24 a	27.70±0.12 a	27.27±0.12 a
	150	26.97±0.28 a	27.57±0.20 a	27.67±0.15 a	27.47±0.19 a	27.83±0.09 a	28.07±0.30 b
	450	27.60±0.12 a	27.83±0.13 a	27.27±0.19 a	27.57±0.24 a	28.90±0.15 b*	28.37±0.15 b
	1800	27.30±0.15 a	27.53±0.07 a	27.45±0.23 a	27.90±0.40 a	29.63±0.15 b*	29.17±0.35 b
Pupation rate (%)	0	86.67±3.57 a	86.67±3.57 a	86.67±3.57 a	86.67±3.57 a	86.67±3.57 a	86.67±3.57 a
	50	90.00±0.00 a	85.00±2.89 a	81.67±1.67 a	80.00±2.89 a	80.00±2.89 ab	85.00±2.89 a
	150	83.33±1.67 a	83.33±4.41 a	83.33±1.67 a	81.67±1.67 a	73.33±1.67 ab	71.67±3.33 ab
	450	78.33±3.33 a	80.00±5.00 a	81.67±6.01 a	76.67±1.67 a	73.33±1.67 ab	75.00±2.89 ab
	1800	80.00±0.00 a	81.67±3.33 a	70.00±2.89 a	71.67±3.33 a	66.67±1.67 b	65.00±2.89 b
Eclosion rate (%)	0	73.33±2.47 a	73.33±2.47 a	73.33±2.47 a	73.33±2.47 a	73.33±2.47 a	73.33±2.47 a
	50	48.33±1.67 b	53.33±1.67 b	48.33±4.41 b	45.00±2.89 b	38.33±6.01 b	41.67±4.41 b
	150	41.67±4.41 b	35.00±2.89 c	28.33±3.33 c	31.67±4.41 bc	18.33±6.01 c	28.33±4.41 bc
	450	41.67±6.01 b	30.00±5.00 c	31.67±4.41 c	30.00±2.89 bc	6.67±1.67 c*	25.00±2.89 c
	1800	33.33±1.67 b	30.00±2.89 c	-	23.33±6.01 c	1.67±1.67 c*	3.33±1.67 d*

Values are means ± SE, n = 3.

[1]Neonates of *B. mori* were fed mulberry leaves covered with rice pollen for the first 3 d and were then fed mulberry leaves without rice pollen until pupation.

[2]Neonates of *B. mori* were fed mulberry leaves covered with rice pollen for their entire larval stage.

"-" denotes lost data.

For each parameter, means in a column followed by different letters are significantly different (one-way ANOVAs with Tukey tests for larval weight, pupation and eclosion rate; Kruskal-Wallis Tests followed by Mann-Whitney U-Tests for larval development time).

An asterisk denotes a significant difference between the *Bt* pollen treatment and the corresponding non-*Bt* pollen treatment at the same pollen dose (Dunnett test).

Table 2. Cry1C protein content of rice pollen grains before and after passage through the digestive system of *Bombyx mori* larvae.

Sample	Cry1C concentration (µg/g dry weight)	No. of pollen grains per mg	Cry1C content per grain (pg) ($c_x = a_x/b_x \times 10^3$)	Lost rate of Cry1C protein (%) ($R_{Cry} = (c_1-c_2)/c_1 \times 100$)
Fresh pollen	3.62 ± 0.10 (a_1)	43106.62 ± 1140.13 (b_1)	0.084 ± 0.002 (c_1)	27.40
Feces	0.08 ± 0.03 (a_2)	1026.01 ± 68.66 (b_2)	0.061 ± 0.011 (c_2)	

Insects were fed with mulberry leaves covered with T1C-19 rice pollen, and the feces were collected; n = 3.
Values are means ± SE in columns 2 and 4.

In both bioassays, the larval weight, development time, and pupation rate were not negatively affected by consumption of control rice pollen even at the highest pollen dose. It seems that consumption of control rice pollen does not affect the normal development of B. mori larvae. Effects seen in the Bt pollen treatments can thus be attributed to the Bt Cry toxins. Surprisingly, however, the eclosion rate of larvae was significantly decreased by consumption of control or Bt pollen even at the lowest dose of 50 grains/cm² leaf in both bioassays. The biological mechanism underlying this effect is unclear. These results suggest that dietary effects can be specific to certain life table parameters. It follows that as many parameters as possible should be observed in such dietary bioassays.

No adverse effect was detected for Bt pollen in the short-term assay, even at the highest pollen density of 1800 grains/cm² leaf. In the long-term assay, however, Cry1C or Cry2A pollen negatively affected all of the tested B. mori life table parameters with an exception of pupation rate. Even at a dose of only 150 grains/cm² leaf, Bt pollen significantly reduced larval weight. This is consistent with Wang et al. [18], who reported that B. mori larval weight but not survival was reduced when the larvae were fed Bt rice pollen containing Cry1Ab toxin. Although we did not statistically compare Cry1C pollen and Cry2A pollen treatments, the data indicate that consumption of Cry1C pollen was more harmful than consumption of Cry2A pollen. For example, B. mori larval developmental time was significantly increased by feeding on Cry1C pollen at 450 and 1800 grains/cm² leaf but was not increased by feeding on Cry2A pollen even at 1800 grains/cm² leaf. Our ELISA determination indicated that the Cry2A content in T2A-1 rice pollen was 11-times higher than the Cry1C content in T1C-19 rice pollen. This shows that Cry1C is much more toxic than Cry2A to B. mori larvae, which is also the case for other lepidopterans including the stem borer C. suppressalis, a target pest of the Bt rice lines. In sensitive-insect bioassays, neonate larvae of C. suppressalis were fed for 7 d with artificial diet containing a range of Cry protein concentrations. The EC_{50} (toxin concentration resulting in 50% weight reduction compared to the control) was 18 ng/mL diet for Cry1C [25] and 1310 ng/mL diet for Cry2A [12]. Previous research has demonstrated that different Cry proteins can have significantly different insecticidal spectra even if they have high homology. For example, Cry1Aa, Cry1Ab, and Cry1Ac have similar structures and belong to the same taxonomic class [26]. In assays with B. mori, however, Cry1Aa was 17-times more toxic than Cry1Ab [27] and 400-times more toxic than Cry1Ac [28]. Based on the EC_{50} values reported for Cry1C and Cry2A, we would have expected a higher mortality in the silkworm larvae fed with high doses of Bt rice pollen. A likely explanation for the relatively low toxicity is that the rice pollen grains were only partly digested in the larval gut and thus the larvae were only exposed to a fraction of the Cry protein. This was confirmed by the fact that 72.6% of the Cry protein was detected in the pollen grains excreted with the feces.

That cry1C- and cry2A-expressing Bt rice pollen are toxic to the lepidopteran B. mori is not surprising because Cry1 and Cry2 proteins are specifically toxic to lepidopterans. Our results do not indicate, however, that the growing of these Bt rice lines will pose a significant risk to the silk industry for the following reasons. First and as discussed earlier, our long-term pollen exposure assay represents a worst-case scenario, and B. mori larvae are unlikely to be exposed to rice pollen for such a long period [20,21,29]. Second, the negative effects were detected only at the relatively high pollen densities of ≥150 grains/cm² mulberry leaf, which rarely occur in the field. Yao et al. [20] reported a mean of 62.3 grains/cm² of mulberry leaf at a distance of 0 m from the paddy field edge, and the density steeply declined to 4.0 grains/cm² of leaf at a distance of 10 m. Third, the mulberry leaves that are fed to B. mori larvae are picked from trees and transferred to a building where the larvae are fed. Thus, the quantity of rice pollen deposited on the leaves would likely be reduced during transport and handling [20]. Fourth, the Bt rice pollen used in our assays was fresh and protected from sunlight and other environmental factors, which would not be the case for Bt rice pollen in the field; in the field, the activity of the Bt proteins in rice pollen may be partially reduced by exposure to rainfall and sunlight before the mulberry leaves are picked and fed to the larvae [20,30]. Considering all four reasons and the results of our short-term bioassay, we conclude that the impact of T1C-19 and T2A-1 rice pollen on the silkworm, B. mori, is probably minimal.

Although exposure of B. mori larvae to Bt rice pollen is likely to be limited under natural conditions, B. mori has received substantial attention in the risk assessment of Bt rice because of its economic importance [18–22,31]. Wang et al. [18] found that consumption of Bt rice pollen containing Cry1Ab protein at a density of 110 grains/cm² mulberry leaves during the whole larval development did not affect the survival of the silkworms, but significantly reduced larval weight. Yao et al. [20] reported that Bt rice pollen containing a fusion Cry1Ab/Ac protein had no negative effect on B. mori larvae, when their neonates were exposed to Bt pollen at the density of 3395 grains/cm² mulberry leaves for 48 h. Likewise, two subsequent studies did not find detrimental effects of Cry1Ab-containing rice pollen from Bt rice lines KMD1 and B1 on B. mori larvae [19,21]. That different studies have reported differences in the toxicity of Bt pollen containing Lepidoptera-active Cry proteins to B. mori larvae can be explained as follows: i) the studies used rice pollen that contained different types of Bt proteins; ii) the concentrations of Bt proteins in the rice pollen differed; and especially iii) the Bt protein exposure differed among studies [21]. Although Bt rice pollen was found to be toxic to B. mori larvae in some studies, previous researchers have generally concluded that Bt rice pollen poses a negligible risk to this domesticated lepidopteran [5,21].

Ours is the first study to assess the effects of Cry1C- and Cry2A-containing rice pollen on B. mori larvae. Although the results indicate that the larvae are sensitive to Cry1C and Cry2A proteins contained in T1C-19 and T2A-1 rice pollen, the results also indicate that the *Bt* rice lines probably represent a low to negligible risk to *B. mori* larvae because of the limited exposure of the larvae to *Bt* rice pollen under natural conditions. To guarantee the safety of the silk industry, however, we recommend that mulberry leaves on trees that are near paddy fields planted with *Bt* rice lines should not be used to feed *B. mori* larvae or should be washed before they are fed to *B. mori* larvae.

Materials and Methods

Ethics statement

No specific permits were required for the described field studies. The rice fields from which rice pollen were collected were owned by the author's institute (Institute of Plant Protection, Chinese Academy of Agricultural Sciences, CAAS). These field studies did not involve endangered or protected species.

Insects

A hybrid of *B. mori*, Liangguang 1, was used in this study. Eggs of *B. mori* were purchased from the Hainan Silk Development Co., Ltd. (Qiongzhong County, Hainan, China) and were kept in a climatic chamber at $27 \pm 0.5°C$, $75 \pm 5\%$ RH, and 12:12 h L:D photoperiod. Newly hatched larvae (<12 h after hatching) were used for all experiments.

Plant materials

Two transgenic rice varieties, T1C-19 and T2A-1, and their corresponding non-transformed near isoline Minghui 63 were used for the experiments. T2A-1 plants express a synthesized modified *cry2A* gene and T1C-19 plants express a modified *cry1C* gene targeting lepidopteran rice pests. Minghui 63 is an elite indica restorer line for cytoplasmic male sterility in China. All rice seeds were kindly provided by Prof. Yongjun Lin (Huazhong Agricultural University, Wuhan).

The rice lines were simultaneously planted in three adjacent plots at the experimental field station of the Institute of Plant Protection, CAAS, near Langfang city, Hebei Province, China (39.5°N, 116.4°E). Each plot was approximately 0.1 hectare, and plots were separated by a 1-m ridge. The rice seeds were sown in a seeding bed on 6 May 2012, and the seedlings were transplanted to the experimental plots on 14 June 2012 when the seedlings were at the four-leaf stage. The plants were cultivated according to the common local agricultural practices but without pesticide sprays.

During rice anthesis from 3 to 13 September 2012, rice pollen was collected daily by shaking the rice tassels in a plastic bag. The collected pollen was air dried at room temperature for 48 h and subsequently passed through a screen with 0.125-mm openings to remove anthers and contaminants. Pollen collected from each rice line was pooled and stored at $-80°C$ until used.

The leaves of mulberry were collected from a mulberry garden at Qiongzhong County, Hainan Province, China. The freshly collected mulberry leaves were washed in water, air dried at room temperature, and stored at 4°C. The leaves were used within 4 days of collection.

Bt protein content in rice pollen

The concentrations of Cry1C and Cry2A proteins in pollen were measured with double-antibody sandwich enzyme-linked immunosorbent assay (DAS-ELISA) kits from EnviroLogix Inc. (Portland, ME, USA; Catalog No. AP 007and kit lot 140433N for Cry1C, and AP 005 and 202482 for Cry2A). Five samples (5–10 mg) of *Bt* or control rice pollen were lyophilized and then homogenized in 1 ml of PBST with a micro-mortar and pestle on ice. After centrifugation and appropriate dilution of the supernatants, ELISA was performed according to the manufacturer's instructions. The optical density (OD) values were read with a microplate spectrophotometer (PowerWave XS2, BioTek, USA). The concentrations of Cry1C and Cry2A were calculated by calibrating the OD values to a range of concentrations of standard Cry1C and Cry2Aa samples provided with the kit.

Bioassays

Two bioassays were conducted in a climate chamber at $27 \pm 0.5°C$, $75 \pm 5\%$ RH, and a 12:12 h L:D photoperiod. In a short-term assay, neonates of *B. mori* were fed cut mulberry leaves (described later in this paragraph) that were covered with rice pollen for the first 3 d and then were fed with cut mulberry leaves without rice pollen for the remainder of the larval period. In a long-term assay, *B. mori* were fed cut mulberry leaves with rice pollen for their entire larval period. For both assays, a scissors was used to cut mulberry leaves into squares of different sizes. Squares that were 1, 4, 30, and 50 cm^2 were used for feeding the first, second, third, and fourth and fifth instar larvae, respectively. Each bioassay included three main treatments: mulberry leaf squares with control rice pollen, Cry1C pollen, or Cry2A pollen. Each main treatment included five doses of rice pollen so that each cm^2 of mulberry leaf square contained 0 or about 50, 150, 450, or 1800 pollen grains.

To obtain the different pollen doses on the mulberry leaf squares, the mean weight of a single rice pollen grain was estimated using the method described in Li et al. [32]. Based on the individual weight of a rice pollen grain, the appropriate quantity of rice pollen grains was weighed and placed in a plastic Petri dish. The Petri dish was shaken by hand after a single wet leaf square was placed in the dish. Examination of leaf squares with a stereo-microscope (50×) confirmed that the actual densities of pollen grains that adhered to the leaf surfaces were very similar to the expected doses, and that the pollen grains were relatively evenly distributed.

Plastic boxes ($12 \times 7 \times 6$ cm for 1st to 3rd instar larvae and $35 \times 25 \times 20$ for 4th and 5th instar larvae) with small holes in the lids were used for both feeding assays. A single treated leaf square was placed on a filter paper on the bottom of a box, and 20 randomly selected *B. mori* neonates were placed on the leaf square. Three replicates and a total of 60 insects were tested for each pollen dose. The number of alive larvae was recorded daily. When a leaf square was almost completely consumed, it was replaced with a new pollen-treated or untreated leaf square. When leaf squares were changed, the uneaten leaf area was calculated using the method of Lang and Vojtech [33]; this information was needed to estimate the mean amount of pollen grains consumed by each larva. When larvae developed into 5th instars and stopped eating, a net was introduced into the plastic dish for larval cocooning. The assays were terminated when all of the insects had developed into adults or died. The following variables were determined: pupation rate, eclosion rate, larval development time, and 14-day larval weight.

Fate of Cry1C contained in *Bt* rice pollen after silkworm gut passage

To estimate the degree at which silkworm larvae are exposed to Cry protein when *Bt* rice pollen grains pass through their gut, the mean Cry1C content in pollen grains before ingestion or from pollen grains in the feces of the silkworm larvae was compared. Neonates of *B. mori* were fed mulberry leaves until the third instar and then starved for 24 h to empty their gut. Subsequently, the larvae were placed in 3 plastic boxes (10 insects per box) and provided with mulberry leaves covered with *Bt* rice pollen of T1C-19 at a density of >1800 grains/cm². After 12 h, the silkworm larvae were transferred to new boxes and received the same food. Subsequently, fresh fecal pellets were collected three times at a 2-h interval. All feces collected from each box were pooled as one sample, resulting a total of 3 samples. Meanwhile 3 samples of fresh T1C-19 rice pollen were also obtained. All samples were stored at −80°C. After lyophilization, the concentrations of Cry1C were measured using ELISA as described above.

The total digestion rate of Cry1C in pollen cannot be determined from *Bt* protein concentrations (Cry protein per dry weight) because the digestion process reduces both the amount of Cry protein and the weight of the pollen grains, and the feces also contained mulberry leaf residues. Therefore, the mean Cry protein content of a single pollen grain before ingestion or present in the feces was assessed. The number of pollen grains in 1.0 mg (dw) rice pollen or feces was estimated. Lyophilized fresh pollen or feces (1.0 mg) were mixed with 300 μl fuchsin acid solution. The pollen grains were counted in each of three 5 μl aliquots of the suspension with a microscope at 50× magnification. The mean number of pollen grains in the aliquots was multiplied by 60 to obtain the number in the whole sample. This procedure was repeated seven to 10 times. Based on the mean number of pollen grains in 1.0 mg per fresh rice pollen and feces and the Cry1C protein concentrations in fresh rice pollen and feces, the Cry1C content in single pollen grains was calculated [13].

Statistical analysis

Student's *t*-tests were conducted to compare Cry1C and Cry2A contents in rice pollen. For the pollen feeding bioassays, one-way ANOVAs followed by Tukey HSD tests were used to determine how the nature of rice pollen (from non-*Bt* rice or from *Bt* rice producing Cry1C and Cry2A) and pollen dose affected pollen consumption, pupation and eclosion rate, and 14-day weight. Because the assumptions for parametric analyses were not met for larval development time (days to pupae), the data were analyzed by Kruskal-Wallis tests, and pair-wise comparisons were further conducted using Mann-Whitney U-tests if significant differences were detected. The Bonferroni correction was applied to correct for 10 pair-wise comparisons leading to an adjusted $\alpha = 0.005$. At each pollen dose, comparisons were made between each *Bt* pollen treatment (Cry1C and Cry2A) and the control (non-*Bt* pollen treatment) using Dunnett tests. The mean Cry1C concentrations in rice pollen grains before and after gut passage were compared using Student's *t* test.

SPSS 13.0 for Windows was used for all statistical analyses.

Acknowledgments

We thank Yanan Wang for technical assistance and Prof. Yongjun Lin (Huazhong Agricultural University) for kindly providing transgenic rice seeds.

Author Contributions

Conceived and designed the experiments: Y. Li YY. Performed the experiments: YY Y. Liu. Analyzed the data: Y. Li YY. Contributed reagents/materials/analysis tools: XC FC LC YP. Contributed to the writing of the manuscript: YY Y. Li JR YP.

References

1. Food and Agriculture Organization of the United Nations (FAO) (2001) Human energy requirements, Report of a Joint FAO/WHO/UNU Expert Consultation, FAO Food and Nutrition Technical Report Series No 1, Rome. http://www.fao.org/docrep/007/y5686e/y5686e00.htm. Accessed 30 Mar 2014.
2. Zhang XF, Wang DY, Fang FP, Zhen YK, Liao XY (2005) Food Safety and Rice Production in China. Reserch of Agricultural Modernization 2: 85–88.
3. Gurdev SK (2005) What it will take to feed 5.0 bill ion rice consumers in 2030. Plant Molecular Biology 59: 1–6.
4. Peng S, Tang Q, Zou Y (2009) Current status and challenges of rice production in China. Plant Production Science 12: 3–8.
5. Chen M, Shelton A, Ye GY (2011) Insect-resistant genetically modified rice in China: From research to commercialization. Annual Review of Entomology 56: 81–101.
6. Zhang YJ, Li YH, Zhang Y, Chen Y, Wu KM, et al. (2011) Seasonal expression of Cry1Ab and Cry1Ac proteins in transgenic rice lines and their resistance against striped rice borer *Chilo suppressalis* (Walker). Environmental Entomology 40: 1323–1330.
7. Wang YN, Zhang L, Li YH, Liu YM, Han LZ, et al. (2014) Expression of Cry1Ab protein in a marker-free transgenic Bt rice line and its efficacy in controlling a target pest, *Chilo suppressalis* (Lepidoptera: Crambidae). Environmental Entomology 43: 528–536.
8. Romeis J, Bartsch D, Bigler F, Candolfi MP, Gielkens MMC, et al. (2008) Assessment of risk of insect-resistant transgenic crops to nontarget arthropods. Nature Biotechnology 26: 203–208.
9. Li YH, Peng YF, Hallerman EM, Wu KM (2014) Safety management and commercial use of genetically modified crops in China. Plant Cell Reports 33: 565–573.
10. Wang YY, Li YH, Romeis J, Chen XP, Zhang J, et al. (2012) Consumption of *Bt* rice pollen expressing Cry2Aa does not cause adverse effects on adult *Chrysoperla sinica* Tjeder (Neuroptera: Chrysopidae). Biological Control 61: 245–251.
11. Li YH, Wang YY, Romeis J, Liu QS, Lin KJ, et al. (2013) *Bt* rice expressing Cry2Aa does not cause direct detrimental effects on larvae of *Chrysoperla sinica*. Ecotoxiology 22: 1413–1421.
12. Li YH, Romeis J, Wu KM, Peng YF (2014) Tier-1 assays for assessing the toxicity of insecticidal proteins produced by genetically engineered plants to non-target arthropods. Insect Science 21: 125–134.
13. Zhang XJ, Li YH, Romeis J, Yin XM, Wu KM, et al. (2014) Use of a pollen-based diet to expose the ladybird beetle *Propylea japonica* to insecticidal proteins. PLoS ONE 9(1): e85395.

14. Li YH, Hu L, Romeis J, Wang YN, Han LZ, et al. (2014) Use of an artificial diet system to study the toxicity of gut-active insecticidal compounds on larvae of the green lacewing *Chrysoperla sinica*. Biological Control 69: 45–51.
15. Romeis J, Raybould A, Bigler F, Candolfi MP, Hellmich RL, et al. (2013) Deriving criteria to select arthropod species for laboratory tests to assess the ecological risks from cultivating arthropod-resistant transgenic crops. Chemosphere 90: 901–909.
16. Liu Y, Li Y, Li X, Qin L (2010) The origin and dispersal of the domesticated Chinese oak silkworm, *Antheraea pernyi*, in China: A reconstruction based on ancient texts. Journal of Insect Science 180: 1–10.
17. Fan L, Wu Y, Pang H, Wu J, Shu Q, et al. (2003) *Bt* rice pollen distribution on mulberry leaves near rice fields. Acta Ecologica Sinica 23: 826–833.
18. Wang Z, Ni X, Xu M, Shu Q, Xia Y (2001) The effect on the development of silkworm larvae of transgenic rice pollen with a synthetic *cry1Ab* gene from *Bacillus thuringiensis*. Hereditas (Beijing) 23: 463–466.
19. Wang Z, Shu Q, Cui H, Xu M, Xie X, et al. (2002) The effect of *Bt* transgenic rice flour on the development of silkworm larvae and the sub-micro-structure of its midgut. Scientia Agricultura Sinica 35: 714–718.
20. Yao H, Ye G, Jiang C, Fan L, Datta K, et al. (2006) Effect of the pollen of transgenic rice line, TT9-3 with a fused *cry1Ab/cry1Ac* gene from *Bacillus thuringiensis* Berliner on non-target domestic silkworm, *Bombyx mori* Linnaeus (Lepidoptera: Bombyxidae). Applied Entomology and Zoology 41: 339–348.
21. Yao H, Jiang C, Ye G, Hu C, Peng Y (2008) Toxicological assessment of pollen from different *Bt* rice lines on *Bombyx mori* (Lepidoptera: Bombyxidae). Environmental Entomology 37: 825–837.
22. Yuan Z, Yao H, Ye G, Hu C (2006) Survival analysis of the larvae from different hybrids of silkworm, *Bombyx mori* to *Bt* rice pollen. Bulletin of Sericulture 3: 23–27.
23. Raybould A, Caron-Lormier G, Bohan DA (2011) Derivation and interpretation of hazard quotients to assess ecological risks from the cultivation of insect-resistant transgenic crops. Journal of Agricultural and Food Chemistry 59: 5877–5885.
24. Sears MK, Hellmich RL, Stanley-Horn DE, Oberhauser KS, Pleasants JM, et al. (2001) Impact of *Bt* corn pollen on monarch butterfly populations: A risk assessment. Proceedings of the National Academy of Sciences, USA 98: 11937–11942.
25. Li YH, Chen XP, Hu L, Romeis J, Peng YF (2014) Bt rice producing Cry1C protein does not have direct detrimental effects on the green lacewing *Chrysoperla sinica* (Tjeder). Environmental Toxicology and Chemistry 33: 1391–1397.

26. Crickmore N, Zeigler DR, Feitelson J, Schnepf E, Van Rie J, et al. (1998) Revision of the nomenclature for the *Bacillus thuringiensis* pesticidal crystal proteins. Microbiology and Molecular Biology Reviews 62: 807–813.

27. Ihara H, Kuroda E, Wadano A, Himeno M (1993) Specific toxicity of δ-endotoxins from *Bacillus thuringiensis* to *Bombyx mori*. Bioscience Biotechnology and Biochemistry 57: 200–204.

28. Ge AZ, Shivarova NI, Dean DH (1989) Location of the *Bombyx mori* specificity domain on a *Bacillus thuringiensis* delta-endotoxin protein. Proceedings of the National Academy of Sciences, USA 86: 4037–4041.

29. Julia C, Dingkuhn M (2012) Variation in time of day of anthesis in rice in different climatic environments. European Journal of Agronomy 43: 166–174.

30. Pusztai M, Fast P, Gringorten L, Kaplan H, Lessard T, et al. (1991) The mechanism of sunlight-mediated inactivation of *Bacillus thuringiensis* crystals. Journal of Biochemistry 273(Pt 1): 43–47.

31. Niu L, Ma Y, Mannakkara A, Zhao Y, Ma W, et al. (2013) Impact of single and stacked insect-resistant *Bt*-cotton on the honey bee and silkworm. PLoS ONE 8(9): e72988.

32. Li YH, Meissle M, Romeis J (2010) Use of maize pollen by adult *Chrysoperla carnea* (Neuroptera: Chrysopidae) and fate of Cry proteins in *Bt*-transgenic varieties. Journal of Insect Physiology 56: 157–164.

33. Lang A, Vojtech E (2006) The effects of pollen consumption of transgenic *Bt* maize on the common swallowtail, *Papilio machaon* L. (Lepidoptera, Papilionidae). Basic and Applied Ecology 7: 296–306.

13

Microarray Expression Analysis of the Main Inflorescence in *Brassica napus*

Yi Huang[1], Jiaqin Shi[1], Zhangsheng Tao[1], Lida Zhang[2], Qiong Liu[1,3], Xinfa Wang[1], Qing Yang[1], Guihua Liu[1], Hanzhong Wang[1]*

1 Oil Crops Research Institute, Chinese Academy of Agricultural Sciences, Wuhan, Hubei, P. R. China, **2** Plant Biotechnology Research Center, School of Agriculture and Biology, Shanghai Jiao Tong University, Shanghai, P. R. China, **3** School of Life Science and Technology, Hubei University, Wuhan, P. R. China

Abstract

The effect of the number of pods on the main inflorescence (NPMI) on seed yield in *Brassica napus* plants grown at high density is a topic of great economic and scientific interest. Here, we sought to identify patterns of gene expression that determine the NPMI during inflorescence differentiation. We monitored gene expression profiles in the main inflorescence of two *B. napus* F_6 RIL pools, each composed of nine lines with a low or high NPMI, and their parental lines, Zhongshuang 11 (ZS11) and 73290, using a *Brassica* 90K elements oligonucleotide array. We identified 4,805 genes that were differentially expressed ($\geq$1.5 fold-change) between the low- and high-NPMI samples. Of these, 82.8% had been annotated and 17.2% shared no significant homology with any known genes. About 31 enriched GO clusters were identified amongst the differentially expressed genes (DEGs), including those involved in hormone responses, development regulation, carbohydrate metabolism, signal transduction, and transcription regulation. Furthermore, 92.8% of the DEGs mapped to chromosomes that originated from *B. rapa* and *B. oleracea*, and 1.6% of the DEGs co-localized with two QTL intervals (*PMI10* and *PMI11*) known to be associated with the NPMI. Overexpression of *BnTPI*, which co-localized with *PMI10*, in *Arabidopsis* suggested that this gene increases the NPMI. This study provides insight into the molecular factors underlying inflorescence architecture, NPMI determination and, consequently, seed yield in *B. napus*.

Editor: Jinfa Zhang, New Mexico State University, United States of America

Funding: This work was financially supported by grants from the Hubei Provincial Key Natural Science Foundation (No.: 2011CDA073), National High-tech R&D Program (No.: 2012AA101107), The National System of Modern Technology for Rapeseed Industry in China (No.: CARS-13), and Hubei Agricultural Science and Technology Innovation Center. The funders had no role in study design, data collection and analysis, decision to publish, or preparation of the manuscript.

Competing Interests: The authors have declared that no competing interests exist.

* Email: wanghz@oilcrops.cn

Introduction

The inflorescence architecture of plants is a key agronomic factor determining seed yield, and is thus a major target of crop domestication and improvement [1,2]. Variations in agricultural productivity and reproductive success are largely determined by differences in shoot architecture, and reproductive shoots known as inflorescences exhibit tremendous diversity across flowering plants in both branch and flower number [3–5]. The inflorescence architecture depends on developmental decisions at the inflorescence meristem [6,7].

Brassica napus is one of the most important cash crops in the *Brassicaceae* family. The transition from vegetative to reproductive growth is particularly important for successful seed production. After the floral transition, the main inflorescence (MI) meristem either acquires a floral meristem identity or produces lateral meristems, which further iterate the MI meristem pattern (Benlloch et al., 2007; Bradley et al., 1997; Prusinkiewicz et al., 2007; Thompson and Hake, 2009; Wang and Li, 2008). Since the number of pods on the main inflorescence (NPMI) influences the productivity of *B. napus*, especially under high-density growth conditions, plant breeders throughout the world are interested in identifying the factors that determine the initiation and formation of pods on the MI. Therefore, understanding the genetic basis of inflorescence architecture will not only elucidate this intriguing evolutionary mechanism, but may also be used to improve crop grain yield [1,2,8]. Although the link between the NPMI and yield is well established, surprisingly little research has examined the molecular basis of the differentiation and development of pods on the MI in *B. napus*. Such information is essential for improving molecular breeding techniques and developing genetic modification strategies.

Several lines of genetic evidence suggest that inflorescence architecture is controlled by multiple genes and their interactions [9–12]. Transcriptome analysis provides a valuable tool for identifying the network of genes underlying MI development. Microarray analysis and transcriptomic sequencing have been used to identify genes involved in numerous biological processes in several species [13–19]. Transcriptomic sequencing is widely used for genome-wide expression pattern analysis in species that have been fully sequenced, such as *Arabidopsis* and rice. Conversely, microarray analysis is suitable for identifying genes in species such as *B. napus*, for which little reference genome information is available. Recently, an oligonucleotide array of *Brassica* harboring

90K genes was developed in CombiMatrix and used to identify genes in *B. napus* siliques subjected to drought stress [20] and to decipher the biology of seed coats in canola [21].

To gain insight into the mechanism underlying inflorescence shaping, the differentiation and development of pods on the MI, and high yield formation in *B. napus*, we adopted a microarray-based approach to study large-scale gene expression changes in the inflorescence. Specifically, we investigated the transcriptomic variation at the pistil-stamen primordial differentiation stage of two *B. napus* F_6 RIL pools, each composed of nine lines with an extremely low or high NPMI, and their elite parental lines, Zhongshuang 11 (ZS11) and 73290, which exhibit marked differences in the NPMI, using the 90K oligonucleotide array. We sought to identify patterns of gene expression in the MI during pod differentiation, determine biological processes associated with pod differentiation and determination, and identify potential candidate genes that are involved in inflorescence development, and especially in pod determinacy, by identifying genes that are both differentially expressed in the MI and co-localized with QTL intervals associated with the NPMI. Furthermore, we planned to demonstrate the functional roles of some representative potential DEGs in increasing the NPMI by overexpressing these DEGs in *Arabidopsis* wild type plants.

Materials and Methods

Plant growth and sample collection

Two elite *Brassica napus* L. lines, cv. Zhongshuang 11 (ZS11), which was used for genome sequencing, and 73290, which was used for re-sequencing, developed by the Rapeseed Biotechnology Breeding Unit, Oil Crops Research Institute of the Chinese Academy of Agriculture Sciences (Wuhan, China), were used to generate an $F_{2:3}$ population harboring 183 offspring for QTL mapping of yield traits, as previously described [22]. Subsequently, an F_6 inbred line (RIL) population was obtained from the $F_{2:3}$ population by single seed descent. Twelve lines with an extremely high number of pods on the main inflorescence (NPMI) and 11 with an extremely low NPMI were selected from the above F_6 RIL population. Four of these lines flowered too late to harvest and one was partially sterile, leaving nine high- and nine low-NPMI lines. The ZS11, 73290, and 18 RIL lines were used for gene expression profiling analysis in the main inflorescence (MI).

Seeds were sown on May 17, 2011 in the experimental station of Qinghai University (Xining, Qinghai Province, China), which is managed by local people. The owner provided full access permission to the experimental site. The field trails did not range over any protected or endangered species, and no vertebrates were involved in this study. Individual plants in each line that did not exhibit any obvious phenotypic differences were selected at several developmental stages, to avoid the erroneous inclusion of plants not belonging to the line. The development of the MI was observed under a microscope (Olympus, Japan). The inflorescence primordia of three to five plants for each line were examined every three days under an anatomical microscope, beginning at the eighth leaf stage. The MI primordia of five plants per line in which the first flower on the MI was in the pistil-stamen primordia initiation stage were harvested in the morning, immediately frozen in nitrogen, and then kept at $-80°C$ for total RNA isolation. Ten mature plants of each line were harvested and the inflorescence traits were recorded.

Seeds of *Arabidopsis thaliana* ecotype Col-0, transgenic *D35S::BnTPI* and *pBnTPI::GUS*, and empty vector (EV) lines were germinated on MS (Murashige & Skoog) medium in Petri dishes for 4 days at 4°C in the dark, and then transferred to a growth room at 21°C under a regime of 16-h light/8-h dark at a light intensity of $\sim 150~\mu E\,m^{-2}\,s^{-1}$. After 7 to 10 days, the seedlings were transplanted to pots of soil (Peilei Co., Jiangsu, China) and grown under the same conditions as described above. Transgenic and control lines were grown under identical conditions in different pots to avoid cross-pollination.

RNA extraction

Using an RNeasy Mini Kit (Cat. 74124, Qiagen, Mississauga, ON) according to the manufacturer's recommendations, the total RNA of inflorescence primordia was extracted in biological triplicate, each consisting of three main inflorescence sections taken from three independent plants. Trace amounts of DNA were removed using DNase I (Cat. 18068-015, Invitrogen, USA). The RNA yield and purity were determined spectrophotometrically with a NanoDrop 1000 spectrophotometer (Thermo Fisher Scientific, USA), and the intactness of RNA was verified by electrophoresis on a 1% agarose gel with $1 \times TBE$ running buffer at 70 V for 40 min. Purified total RNA was precipitated and re-suspended in DEPC-treated water to a final concentration of about 500 ng/μL. For the high- and low-NPMI lines, the total RNA of inflorescence primordia was extracted individually and pooled at equal concentrations within each extreme-NPMI group. Nine lines were used for each biological replicate in the reverse transcription analysis.

Total RNA of *Arabidopsis* transgenic and control seedlings was extracted using an RNAprep Pure Plant Kit (TIANGEN, China) under the manufacturer's recommendations.

Microarray analysis

1) ***Brassica* 90K array.** A *Brassica* genomics resource was developed at the Plant Biotechnology Institute, National Research Council Canada (PBI-NRC) and Agriculture and Agri-Food Canada (AAFC) in collaboration with other institutes. About 95,418 unique sequences were assembled using 781,826 EST sequences mainly from three species: *B. napus*, *B. rapa*, and *B. oleracea*. These unique sequences were submitted to CombiMatrix for the development of a 35–40mer oligonucleotide microarray. Probes were synthesized *in situ* on electrodes of a microchip. The customized CombiMatrix *Brassica* array comprised 90,500 probes and was named the Brassica 90K array.

2) **aRNA in vitro synthesis, dye coupling, and hybridization to the *Brassica* 90K array.** cDNAs were synthesized from total RNA extracted from the main inflorescence of ZS11 and 73290, and from the high- and low-NPMI pools. About 3.0 μg of total RNA for each sample was reverse transcribed into amino allyl-modified aRNA using a Message II aRNA Amplification Kit (Cat: 1753, Ambion, Austin, TX, USA), according to the manufacturer's protocol. The amplification products were analyzed by agarose gel electrophoresis and ethidium bromide staining, and the aRNA yield was quantified using a NanoDrop 1000 spectrophotometer. The aRNA was selected as template for fluorescent target preparation for microarray experiments.

About 5.0 μg of aRNA was labeled with mono-reactive NHS esters of Cy5 in the dark for 30 min at room temperature, and the samples were purified and used as targets for hybridization. The pre-hybridization and hybridization procedures were performed precisely according to the protocol of CombiMatrix (www. combimatrix.com). Cy5 dye-labeled RNA populations from individual tissue samples were hybridized to the 90K array and three biological replicates were performed.

3) **Data collection and analysis.** Hybridized arrays were scanned using a LuxScan 10K Scanner (CapitalBio Corporation,

Beijing, China) at 5 μm resolution, 100% laser power, and different PMT values to obtain a similar overall intensity between slides. The scans were saved as TIF files. Raw spot fluorescence intensities were collected using LuxScan version 2.0 (CapitalBio Corporation, Beijing, China) and saved as LSR files. The microarray dataset generated in this study was deposited in NCBI's Gene Expression Omnibus and is accessible through GEO series accession number GSE57886 (http://www.ncbi.nlm.nih. gov/geo/query/acc.cgi?acc = GSE57886).

Before normalization, basic pre-processing was performed. Samples that were originally negative after background subtraction and those with an overall signal intensity of ≤100 were filtered out. Array features annotated as "Empty", "Blank" and those internal controls were flagged and excluded from the analysis. Twelve LSR files containing the raw probe intensity values were imported into R and the signal intensities were subjected to quantile normalization to normalize between arrays using the R-package LIMMA [23] with a 5% false discovery rate (FDR≤0.05). Probes with at least one missing value within triplicates were removed from subsequent analysis. The background-corrected signal intensity of each probe on the array was combined by averaging three biological replicates. A gene was considered to be differentially expressed when its fold-change in expression between two samples was ≥1.5 or ≤0.67.

Probe annotation analysis was performed using BLASTx at an E-value of $\leq 10^{-5}$ or a BLASTn score of ≥100 against the TAIR10 database (http://www.arabidopsis.org). The functional categories were calculated using bootstrap analysis with 100 replicates of the TAIR10 Arabidopsis genome annotation, at a 95% confidence level ($p<0.05$), and MapMan was used as the classification source [24].

Isolation and cloning of BnTPI

1) Construct preparation and transformation. The entire open reading frame (ORF) of *BnTPI* was amplified by RT-PCR using a pair of gtpi primers from the cDNA of ZS11, resulting in the *pD1301S::BnTPI* construct with a 765-bp insert. The *pBnTPI::GUS* construct harbored a 1,706-bp fragment of the *BnTPI* ORF that had been amplified with ptpi primers from the cDNA of ZS11. PCR was carried out with a C1000 PCR system (BIO-RAD) using an *Easy Taq* DNA Polymerase Kit (Cat: E51117, TransGen, China) and the following amplification cycle: 30 cycles of 94°C for 30 s, 58°C for 30 s, and 72°C for 40 s. The amplicons were collected and cloned using TA-overhangs into the pMD18-T vector (Takara). The integrity of cloned inserts was confirmed by sequencing with a pair of M13 primers. These coding regions of *BnTPI* in pMD18-T were respectively transferred to generate expression plasmids *pD1301S::BnTPI* and *pBnTPI::GUS*, which were confirmed by PCR using a pair of primers based on the cauliflower mosaic virus 35S constitutive promoter (35s-f) and gtpi-r, and the pC1301GT (cgt-r) and ptpi-f primers, respectively. The sequences of all the PCR primers are listed in Table S3.

Agrobacterium tumefaciens strain GV3101 harboring *pD1301S::BnTPI* or *pBnTPI::GUS* was transformed into *Arabidopsis thaliana* wild-type Col-0 plants by floral dip as previously described [25]. Transformants were germinated and screened based on hygromycin B resistance on Murashige and Skoog medium.

2) Real-time PCR. Total RNA of transgenic and control seedlings was extracted using the RNAprep Pure Plant Kit (TIANGEN, China) under the manufacturer's recommendations. First-strand cDNA was synthesized from 10.0 μL of total RNA using Oligo (dT) and 200 units of M-MLV reverse transcriptase (Promega) and incubated at 70°C for 5 min, in an ice-bath for 5 min, and at 42°C for 60 min. For comparative PCR, different cDNAs were normalized using *ACTIN2* (AT3G18780)-specific primers (Table S3). Reactions were performed in a final volume of 20.0 μL containing 10.0 μL of 2×SYBR Green Master Mix (TOYOBO), 0.5 mM of each primer, and 2.0 μL of cDNA, using a real-time PCR system (CFX96, Bio-RAD). PCR conditions were as follows: 95°C for 2 min, followed by 40 cycles of 95°C for 15 s, 58°C for 15 s, 72°C for 25 s, and 25°C for 30 s. *ACTIN2* expression was used to normalize the transcript level in each sample. Data were analyzed using sequence detector software (Bio-RAD, Version 2.1).

3) GUS Staining. Nine-day-old, light-grown seedlings expressing the *pBnTPI::GUS* fusion were vacuum infiltrated for 5 min and then incubated overnight at 37°C in reaction buffer containing 50.0 mM sodium phosphate (pH 7.0), 0.5 mM ferricyanide, 0.5 mM ferrocyanide, 0.05% Triton X-100, and 1.0 mM X-Gluc [26]. Plantlets were depigmented with 70% ethanol and GUS staining patterns were documented using a digital camera [27].

Results

Morphological differences in the inflorescences of B. napus lines

The MIs of plants of line 73290 were much larger than those of ZS11 (Figures 1A and 1B). Furthermore, line 73290 had more differentiated flower buds and a higher NPMI than ZS11 and a longer MI. The NPMI is determined by the duration and rate of differentiation. We found that line 73290 had a longer period of differentiation than did ZS11. A Student's t-test ($p<0.05$) showed that the difference in the NPMI between ZS11 (80~90) and 73290 (140~150) was significant. Consequently, we constructed genetic linkage maps using SSR markers to detect QTLs associated with multiple yield-related traits from 183 lines within the $F_{2:3}$ population derived from the cross between ZS11 and 73290 (unpublished).

Based on the variation in the NPMI in lines within the F_6 RIL populations, we selected a subset of 23 F_6 RIL lines that had an extremely low or high NPMI. We counted the NPMI in the ZS11, 73290, and F_6 lines using 10 mature plants per accession. Differences between ZS11 and all accessions except line OG1805-4 were significant (Student's t-test; $p<0.05$) (Figure 1C), as were those between 73290 and all other accessions examined.

We selected 9 RILs with a high NPMI and 9 with a low NPMI to construct high- and low-NPMI RNA pools for gene expression profiling of the MI using the Brassica 90K oligonucleotide array. Hereafter, we refer to ZS11 and the low-NPMI RNA pool as the low-NPMI sample, and 73290 and the high-PMI RNA pool as the high-NPMI sample.

Gene expression profiles of the main inflorescence

As previously described [21], each oligonucleotide probe on the array was designed based on non-overlapping *Brassica* EST contigs and singletons that ideally represent unique genes. For ease of description, we will refer to each of the 90K probes on the array as a gene, and provide AGI locus numbers of the *Arabidopsis* sequences with the highest level of identity for each cDNA or contig that was used as a source for each array probe. A total of 84970 genes produced hybridization signals in all four RNA samples within the biological triplicates and these signals were subjected to data processing. We sought to identify differences in gene expression that led to differences in the MI and NPMI between the low-NPMI and high-NPMI samples, by comparing the expression of genes in the ZS11 vs. 73290 or high-NPMI RNA pool and the low-NPMI RNA pool vs. 73290 or the high-NPMI

Figure 1. Morphological features of the main inflorescence (MI). (A) Top view of young floral buds on the MI of ZS11 and 73290 at four development stages, namely sepal initiation (I), pistil-stamen initiation (II), petal initiation (III), and pistil-stamen elongation (IV). (B) The intact MI of ZS11 and 73290 before harvesting. (C) Histogram indicating differences in the NPMI between ZS11, 73290, and the F_6 RIL lines. Significance was determined using Student's t-test (n = 30). Asterisks above the columns indicate significant differences compared to ZS11. *, $p < 0.05$; **, $p < 0.01$.

RNA pool. We identified 1051 and 2136 genes with 207 overlaps that were differentially expressed between ZS11 and 73290, as well as between ZS11 and the high-PMI RNA pool. Similarly, 907 and 1269 genes, including 97 common genes, showed expression differences between the low-NPMI and high-NPMI RNA pool or between the low-NPMI RNA pool and 73290, respectively. Furthermore, 1051 and 1269 genes, with 92 overlaps, displayed

expression differences between ZS11 vs. 73290 and between the low-NPMI RNA pool vs. 73290, respectively, and 2136 and 907 genes, with 114 overlaps, exhibited expression difference between ZS11 vs. the high-NPMI RNA pool and between the low-NPMI RNA pool vs. the high-NPMI RNA pool, respectively (Figure 2A). Ten genes were differentially expressed both in ZS11 vs. 73290 and in the low-NPMI vs. the high-NPMI RNA pool. In sum, we

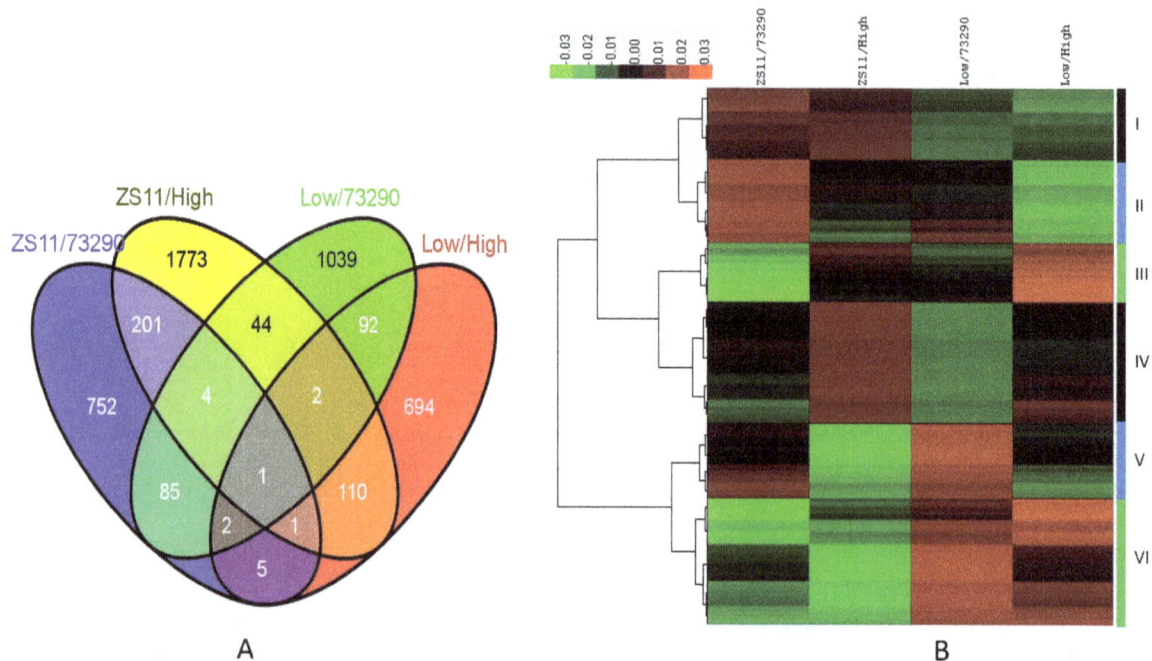

Figure 2. Transcriptomic analysis of the MI of ZS11, 73290, and the two F_6 pools. (A) Venn diagram of differentially expressed genes (DEGs) between the low-NPMI and high-NPMI samples ($p < 0.05$, absolute fold change of ≥ 1.5). (B) Hierarchical cluster analysis of genes that were detected as being differentially expressed in at least one of the low-NPMI versus high-NPMI pairs. Red indicates up-regulation and green denotes down-regulation. I–VI indicate the six clusters, and the color bars on the right denote the range of each cluster. Low and High indicate the Low-NPMI and High-NPMI RNA samples, respectively.

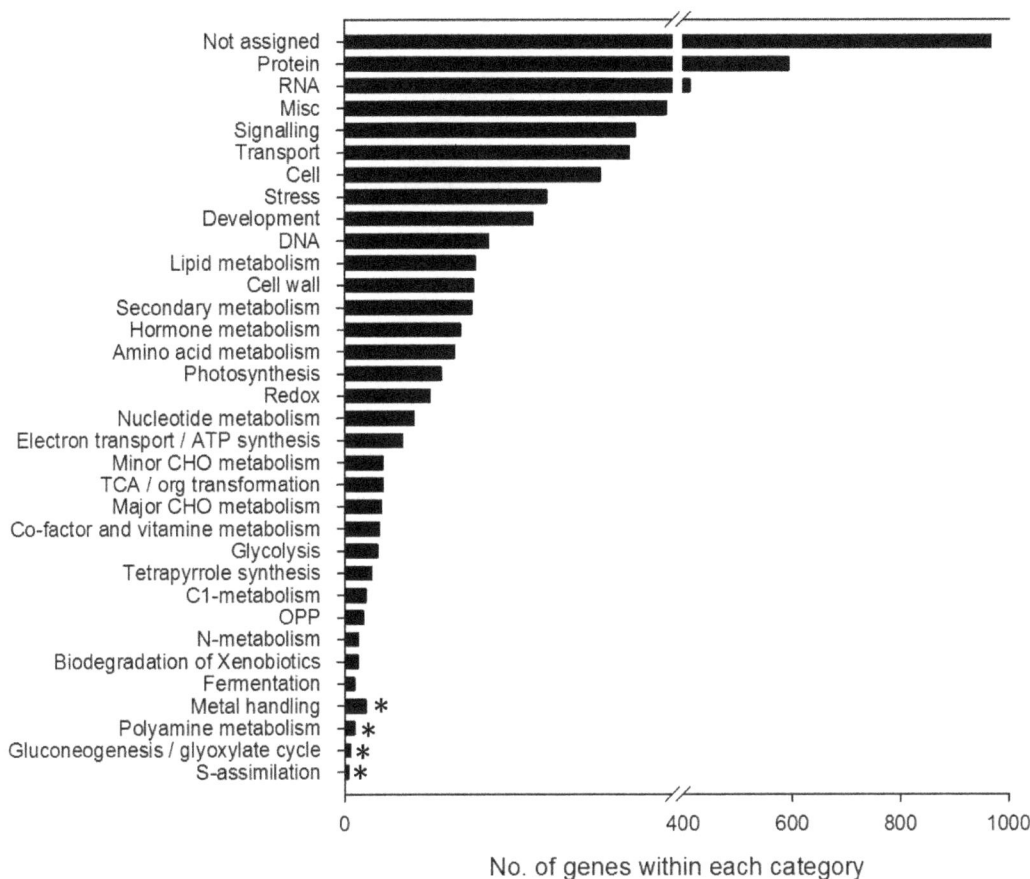

Figure 3. Functional categories of differentially expressed genes are significantly enriched using MapMan as a classification source. GO terms are sorted based on P value (p<0.05). * indicates the unenriched GO terms.

identified 4805 genes that were significantly differentially expressed among the four samples (Table S1).

Furthermore, probe annotation analysis against TAIR10 indicated that 3981 of the 4805 DEGs matched at least one gene annotated in TAIR10, and 824 shared no significant homology with any Arabidopsis accession. The functional categories were determined with 100 bootstrap replicates against TAIR10 using MapMan as the classification source and produced a total of 31 enriched clusters (Figure 3). Among the enriched categories, we detected highly enriched clusters with strong confidence levels (p< 0.05), such as carbohydrate synthesis and metabolism, development regulation, hormone responses, signal transduction, stress and redox, and transcription regulation (Table S1).

Expression patterns of DEGs in the main inflorescence of B. napus

Next, we monitored the expression patterns of these 4805 DEGs in the MI using hierarchical cluster analysis. This revealed six major clusters (Figure 2B). Genes were subjected to functional category enrichment analysis in each cluster compared with their distribution among all genes in *Arabidopsis*. About 17.94% (862) of the DEGs occurred in cluster I and were up-regulated in the MI of ZS11 but down-regulated in that of the low-NPMI RNA pool, 73290, and high-NPMI RNA pool, while the mRNA abundance of these genes was similar in the MIs of 73290 and the high-NPMI RNA pool. Amongst these DEGs, several functional categories, such as protein and amino acid metabolism, nucleotide metabo-

lism, cell organization, photosynthesis and mitochondrial electron transport, DNA duplication and repair, and transport, were significantly enriched.

In cluster II, which accounted for 22.33% (1073) of the DEGs, genes were upregulated in the MIs of ZS11and partially in the low-NPMI RNA pool relative to 73290 and the high-NPMI RNA pool, with the expression in the MI of the high-NPMI RNA pool being greater than that in 73290. These DEGs were mainly involved in protein and amino acid metabolism, transcription regulation, photosynthesis and mitochondria electron transport, carbohydrate metabolism, stress response, and transport, or had not been assigned to a functional group.

About 10.92% (525) of genes fell in cluster III and displayed significantly higher expression in line 73290 than in the high-NPMI RNA pool. In contrast, in cluster V, which accounts for 15.98% (768) of the DEGs, genes were expressed at higher levels in the high-NPMI RNA pool and lower levels in line 73290. Furthermore, the expression level of genes in clusters III and V was similar in the MI of ZS11 and the low-NPMI RNA pool. These genes were enriched in categories such as protein metabolism, transcription regulation, signaling, and transport.

Similarly, 449 (9.34%) of the DEGs were classified into cluster IV and these genes had lower expression in the high-NPMI RNA pool and line 73290 than in line ZS11 and the low-NPMI RNA pool, with expression being lowest in the high-NPMI pool. The genes in this category were mainly related to protein metabolism, transcription regulation, signal transduction, and lipid metabolism.

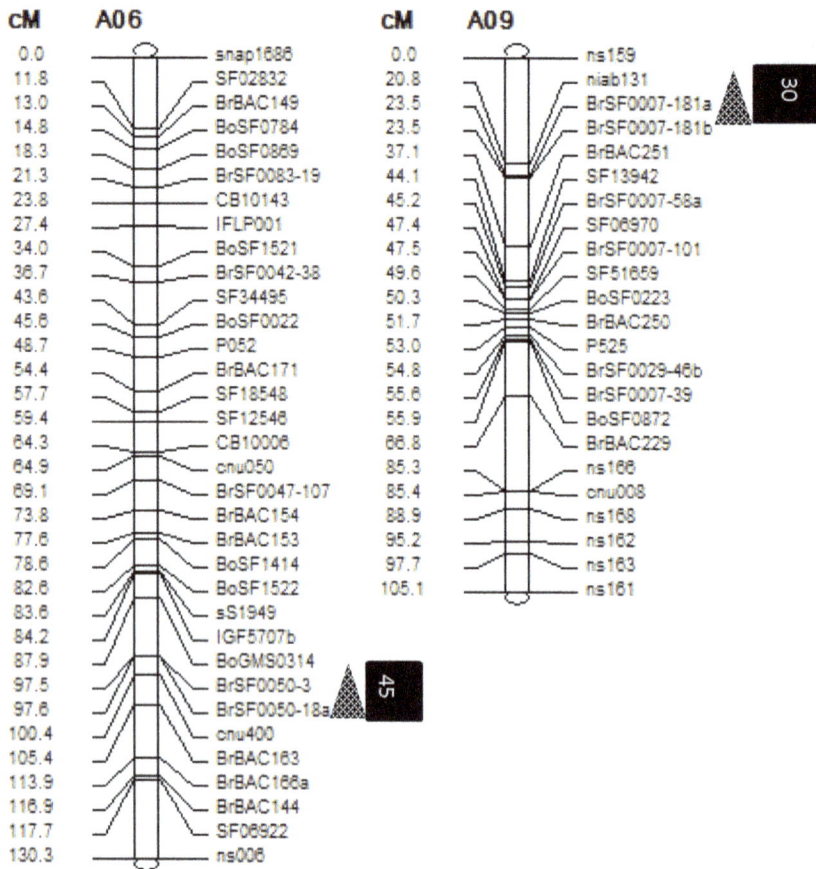

Figure 4. Mapped QTLs associated with the NPMI and DEGs that localized within the QTL intervals. Triangles indicate the mapped QTLs for NPMI and black squares with embedded digits denote the position and number of co-localized DEGs. A06 and A09 denote chromosomes A06 and A09 of *B. napus*, respectively.

Cluster VI was the largest, consisting of 1128 genes and accounting for 23.48% of the DEGs. The expression of genes in this cluster was greatest in the low-NPMI RNA pool and weakest in ZS11, and was similar in 73290 and the high-NPMI lines. We found that 41.8% of the genes in this cluster had not been assigned functional roles or had no hits against TAIR10, and the remaining genes were involved in photosynthesis, cell wall formation, lipid metabolism, secondary metabolism, the stress response and other categories.

Taken together, these results suggest that most of the DEGs identified in the experiment are likely to be involved in MI and floral organ differentiation and development, and that most of the observed expression changes are a consequence of allele-specific gene expression among the two parents and their F_6 RILs.

Localization of DEGs in QTL-containing intervals

Although the genomic sequence of *B. napus* is currently incomplete, the genome of *B. rapa* has been sequenced [28] and that of *B. oleracea* is almost complete in OCRI-CAAS (Shengyi Liu et al., 2014). Since *B. rapa* (AA, $2n = 20$) and *B. oleracea* (CC, $2n = 18$) are the diploid ancestors of amphidiploid *B. napus* (AACC, $2n = 38$), the 4805 DEGs identified in this study can be localized on chromosomes of *B. napus* by referencing the genome information of *B. rapa* and *B. oleracea* using BLASTn analysis at an E-value of $<10^{-5}$. Consequently, we established that 4363 of the 4805 DEGs are specifically co-localized on *B. rapa* chromosomes and 4300 are on *B. oleracea* chromosomes. Furthermore,

4460 genes are distributed over all 19 linkage groups of *B. napus*, and the remaining 345 are currently homeless, and do not localize to any chromosomes of *B. napus*.

A genetic map (whole length 1869.7 cM) containing 519 markers (average distance = 3.6 cM) was constructed using the $F_{2:3}$ population derived from the cross ZS11*73290. We detected two QTLs associated with the NPMI on chromosomes A06 in 2010 (PMI10) and A09 in 2011 (PMI11), with a 17.6% and 13.9% contribution rate, respectively (Table 1 and Figure 4). We found that 45 and 30 DEGs identified in the current study were co-localized to the QTL intervals *PMI10* and *PMI11* (Table S2), suggesting that these genes regulate pod differentiation and development on the MI in *B. napus*. Ten of these 75 mapped DEGs are novel, with no homology to sequences in *Arabidopsis*, and the remaining 65 have orthologs in *Arabidopsis* that are involved in several biological processes, including transcription regulation, cell division and organization, and carbohydrate metabolism (Table S2). Several DEGs that co-localized to QTL intervals and were up- or down-regulated in both ZS11 and the low-NPMI pool were selected for quick functional analysis by overexpression in *Arabidopsis*.

Functional analysis of *BnTPI* in *Arabidopsis thaliana*

Some of the DEGs transformed into *Arabidopsis* caused morphological variations in inflorescence traits. For instance, MI52894, a DEG that co-localized to a QTL interval on chromosome A6, was strongly up-regulated in the MI of both

Table 1. Identified QTLs for pods on the main inflorescence and co-localized DEGs.

Date	QTLs	Chrom.	Interval	QTL position (cM)	LOD	Additive effect	Dominant effect	VE (%)	No. of co-localized genes
2010	PMI10	A06	97.6–105.4	99.7	9.8	−5.31	2.21	17.6	45
2011	PMI11	A09	20.8–23.5	23.5	4.6	−12.56	−9.03	13.9	30

the high-NPMI pool and 73290. This gene had 96% amino acid identity with *Arabidopsis AtTPI* (AT3G55440), which encodes triosephosphate isomerase (TPI). We thus named this gene *BnTPI* (Figure 5A). The full-length ORF (765 bp) of *BnTPI* was isolated and sequenced in both ZS11 and 73290. *BnTPI* was composed of nine exons, and its CDSs from ZS11 and 73290 were identical (Figure 5B). The expression level of *BnTPI* in the MI was determined by RT-PCR (Figure 5C), and found to be up-regulated both in the high-NPMI pool and 73290 compared to the low-NPMI pool and ZS11.

Subsequently, we generated the *D35S::BnTPI* construct and transformed this construct and its corresponding empty vector (EV) into wild type *Arabidopsis* Col-0. Nineteen positive homozygous transgenic lines (T_4) and their EV and wild type Col-0 controls were planted under identical conditions in a growth room and harvested for phenotype investigation. We found that 14 of the 19 *BnTPI* overexpression lines (T_4) had 4 to 18 more pods on the MI than did the EV and Col-0 controls, and 10 transgenic lines produced MIs that were 0.5 to 9.3 cm longer than those on the EV control (Figures 6A, 6D, and 6E). Simultaneously, we examined the expression level of *BnTPI* in 9-day-old seedlings of the positive transgenic, EV, and Col-0 lines using real-time PCR with a pair of gtpi primers (Table S3). We found that the transcript abundance of *BnTPI* (relative to the expression level of *ACTIN2* (AT3G18780)) was significantly higher in the positive transgenic line than in the Col-0 and EV plants (Figure 6B). Moreover, the expression level of *BnTPI* was positively correlated with the length of the MI and the NPMI (Figures 6C, 6D and 6E), while no obvious correlation occurred between *BnTPI* expression and pod length on the MI (Figures 6C and 6F).

Furthermore, we cloned and sequenced the *BnTPI* promoter from ZS11 and 73290, but did not detect any differences between the two. We monitored the tissue-specific activity of the *BnTPI* promoter by generating a translational fusion construct *pBnTPI::-GUS* and transforming it into *Arabidopsis* Col-0 plants. Twelve independent *pBnTPI::GUS*-positive transgenic lines were obtained, and the GUS activity was found to vary throughout development in a tissue-specific manner. In nine-day-old transgenic seedlings, GUS staining was clearly visible in the shoot apex meristem (SAM), node-joint, and roots (Figure 7A), whereas in 24-day-old transgenic plants, it was mainly confined to the axil and the SAM (Figures 7B and 7C), and at flowering, weak GUS staining was detectable in the SAM and floral buds (Figures 7D and 7E).

Discussion

The *B. napus* inflorescence is a raceme in which the apical meristem is able to grow indefinitely, generating a continuous main axis that laterally produces floral meristems. The MI plays a determinant role in the reproductive success of *B. napus* [29]. In the current study, in an attempt to identify the genes underlying pod differentiation and development on the MI in *B. napus*, we analyzed the gene expression profiles in the MIs of two F_6 RIL populations with great differences in pod number and their corresponding elite parental cultivars using an oligonucleotide array. The analysis was performed when the first flower on the MI was in the pistil-stamen primordial differentiation state, as the NPMI is directly correlated with the number of differentiated floral primordia, and it is easy to synchronize the development of different plants. The expression of about 5% of transcripts differed significantly between the low-NPMI and high-NPMI samples, and these transcripts were involved in multiple pathways. This analysis provided a considerable amount of information that can be used to assess patterns of gene expression that may be relevant for

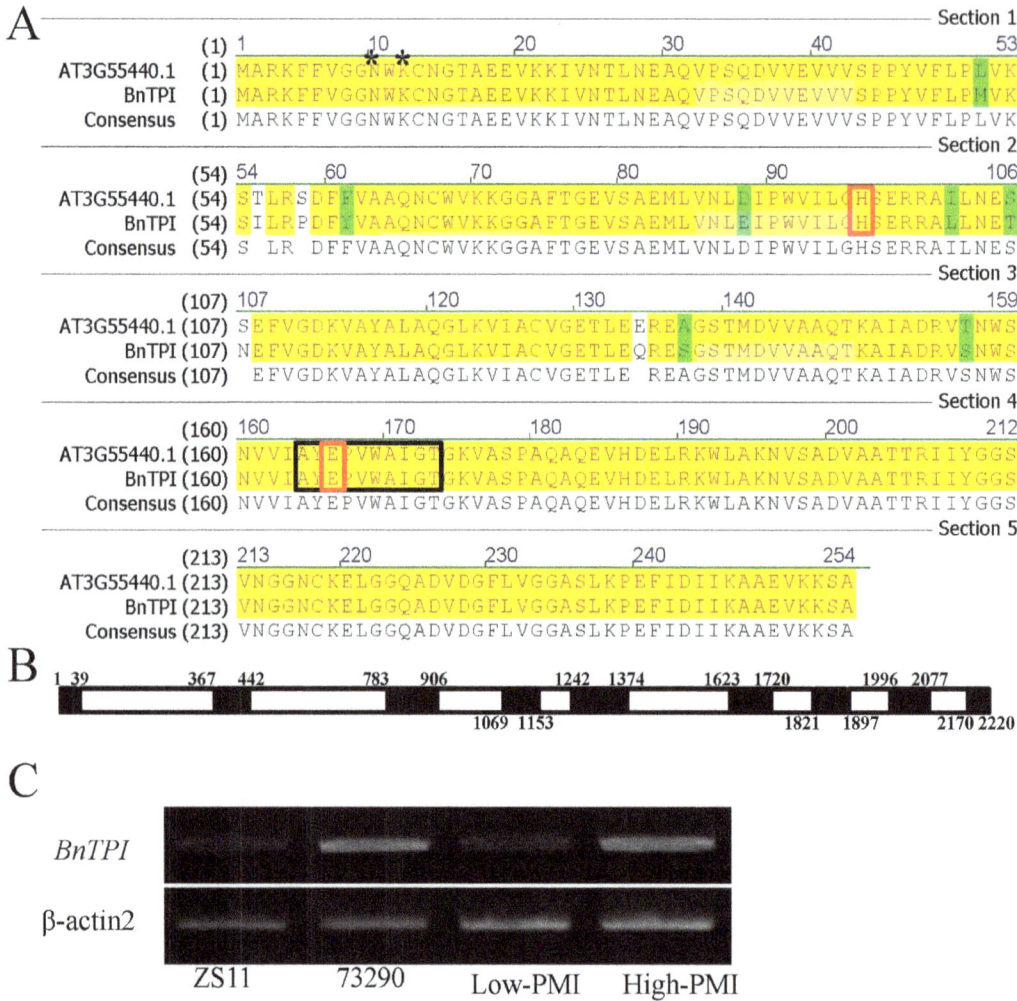

Figure 5. Alignment of the BnTPI and AtTPI amino acid sequences, gene structure prediction, and expression level validation using RT-PCR. (A) BnTPI displays a highly conserved domain (164–174 AA, black rectangular box), two binding sites (N10 and K12, black star), two active sites (H96 and E166, red box), an electrophilic site, and a proton acceptor point. Yellow and green shading indicate conserved and mutant amino acids between BnTPI and the *Arabidopsis* ortholog AT3G55440.1, respectively. (B) Schematic representation of *BnTPI*. Exons are shown as black boxes and introns as white boxes. Numbers above the boxes indicate the nucleobases (bp). (C) The expression level of *BnTPI* in the MI of the ZS11, 73290, and F_6 RIL lines was determined by RT-PCR. *β-actin2* expression was used to normalize the transcript level in each sample.

determining pod differentiation and development on the MI of *B. napus*.

A remarkable outcome of this study was the finding that a considerable proportion of the genes in the four RNA samples exhibited significant differences in expression level. These differences were displayed in several forms: the genes may be highly expressed in one of the parents, but expressed at low levels in the other parent and F_6 lines, or highly expressed in both low-NPMI samples, but expressed at low levels in the high-NPMI samples and vice versa, or highly expressed in both extreme F_6 line pools, but expressed at low levels in either or both parents. The expression level of some genes was altered among the four RNA samples, but not at a significant level (i.e., <1.5-fold), and we would like to examine the potential involvement of these genes in pod differentiation and development as well. The growth and development of organisms is widely believed to be spatially and temporally regulated at the transcriptional level. This study shows that a gene may be strongly expressed in the MI of one genotype, but very weakly expressed in another, resulting in large differences

in the level of gene expression among the genotypes. These differences in gene expression provide insight into the mechanisms that regulate pod differentiation and development.

Recent studies showed that quantitative variations in gene expression levels were controlled genetically and were regulated by both *cis*- and *trans*-acting loci, with *trans*-acting elements playing the dominant roles [30,31]. We thus selected DEGs as candidate regulators of NPMI and rapidly validated the functions of ten gene through overexpression analysis in *Arabidopsis*. Our analysis of DEGs between the two parental lines revealed that sequence differences in the CDS, promoter, or both regions are common, and that the sequence differences of several DEGs result in obvious phenotypic variation in the NPMI and other floral organs in *Arabidopsis* (data no shown), suggesting that these phenotypic variations are attributable to *cis*-acting regulation or to variations in the CDSs of genes between the two genotypes. These findings are consistent with previous studies in other organisms [32–35]. In addition, we identified some DEGs, such as *BnTPI*, which exhibited no sequence differences in either the CDS or promoter

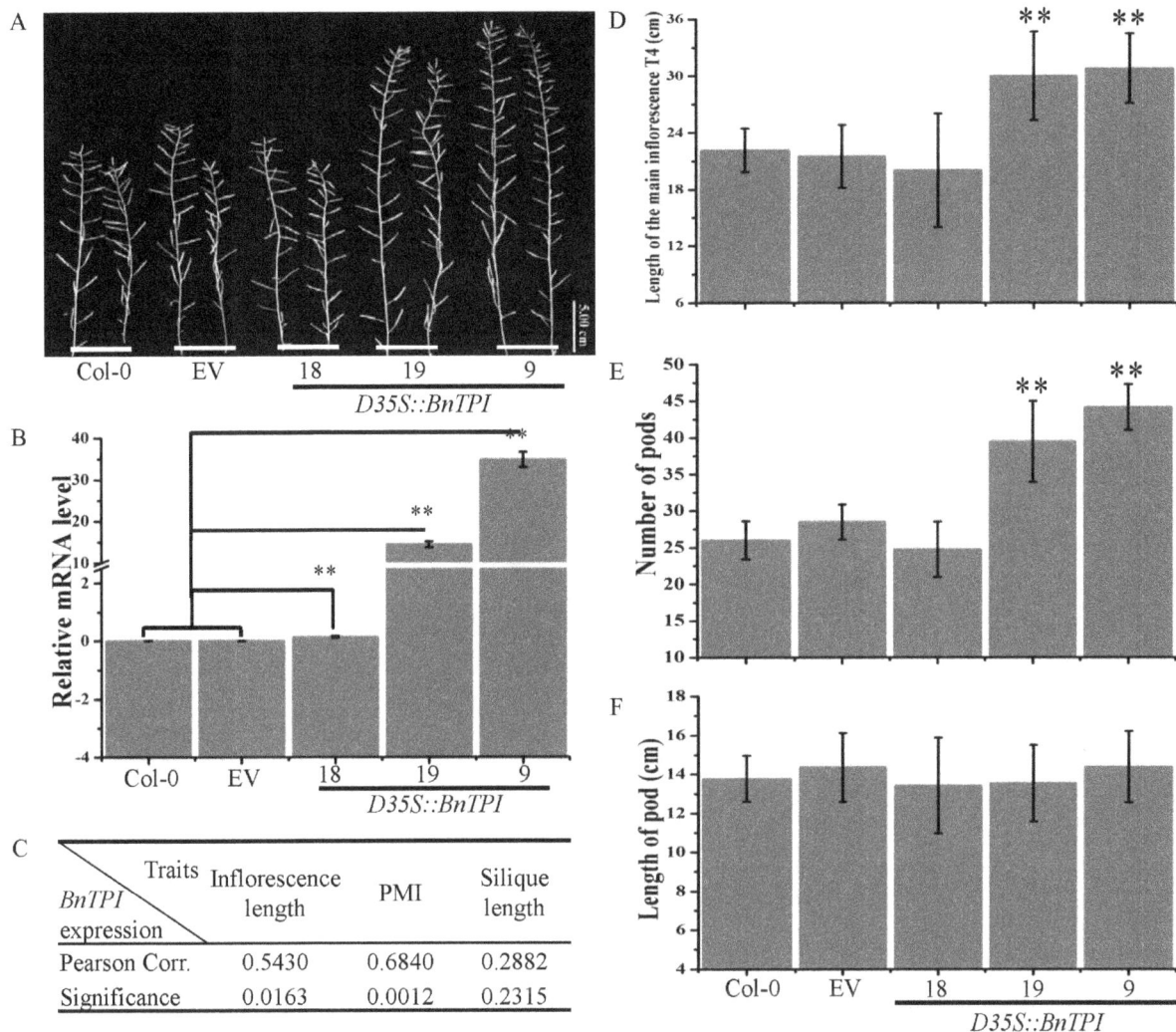

Figure 6. Features of the MI of wild type Col-0, EV, and *D35S::BnTPI* transgenic lines. (A) Morphological variations in the MIs and pods of the wild type Col-0, EV, and *D35S::BnTPI* transgenic lines. (B) Relative RNA abundance of *BnTPI* in the MI of wild type Col-0, EV, and *D35S::BnTPI* transgenic lines. The error bar is based on biological triplicates. (C) Pearson's correlation coefficient between *BnTPI* expression and inflorescence traits, including MI length, NPMI, and silique length; (D), (E), and (F) transgenic *D35S::BnTPI* and Col-0 and EV lines display significant differences in MI length, NPMI, and pod length, as determined by Student's *t*-test at the $p<0.05$ level. Phenotype variation of twenty plants per line was investigated.

regions, but displayed obvious phenotypic variation in the NPMI and other floral organs, indicating that *trans*-acting elements play important roles during pod differentiation in *B. napus*. We established that 5% of *B. napus* genes were differentially expressed among the four RNA samples. We plan to determine the cause of the observed expression differences by characterizing more of the DEGs, as this would enhance our fundamental understanding of mechanisms underlying differences in the NPMI in *B. napus*.

It was suggested that gene expression data together with QTL analysis may provide an avenue for identifying candidate genes for traits of interest [36]. In this analysis, we therefore also attempted to establish a link between the DEGs and potential QTLs related to the NPMI. We identified two QTLs associated with the NPMI using an $F_{2:3}$ population derived from a cross between the parents used in this study, and 4460 of the 4805 DEGs were mapped to 19 chromosomes of *B. napus* referencing genome sequences of both *B. rapa* and *B. oleracea*. Among the mapped DEGs, 45 and 30 were localized to the two QTLs for NPMI spanning 7.8 cM and 2.7 cM, respectively (Table 1). As the ORF regions of *Arabidopsis*

thaliana and *B. napus* exhibit an estimated 85% sequence identity [37], it would be helpful to identify the homologs of *Arabidopsis* genes known to control floral and inflorescence differentiation in *B. napus*. The functional roles of a representative DEG were further investigated, with overexpression in *Arabidopsis* resulting in an increased NPMI and longer inflorescences. We identified *BnTPI* as a factor in pod determination on the MI and, based on the function of the *Arabidopsis* homolog, suggest that energy harvesting, reservation, and conversion contribute to inflorescence differentiation, elongation, and floral bud procreation during flowering in *B. napus*. Thus, this study identified several likely candidate genes that affect the NPMI, and these genes warrant further studies.

It should be pointed out that the current study only established the landscape of gene expression in the *B. napus* MI using an oligonucleotide array that was developed based on non-overlapping *Brassica* EST contigs and singletons that ideally represent unique gene models. Along with the full-scale genome sequencing of *B. napus* and *B. oleracea* and advances in micro-technologies, it

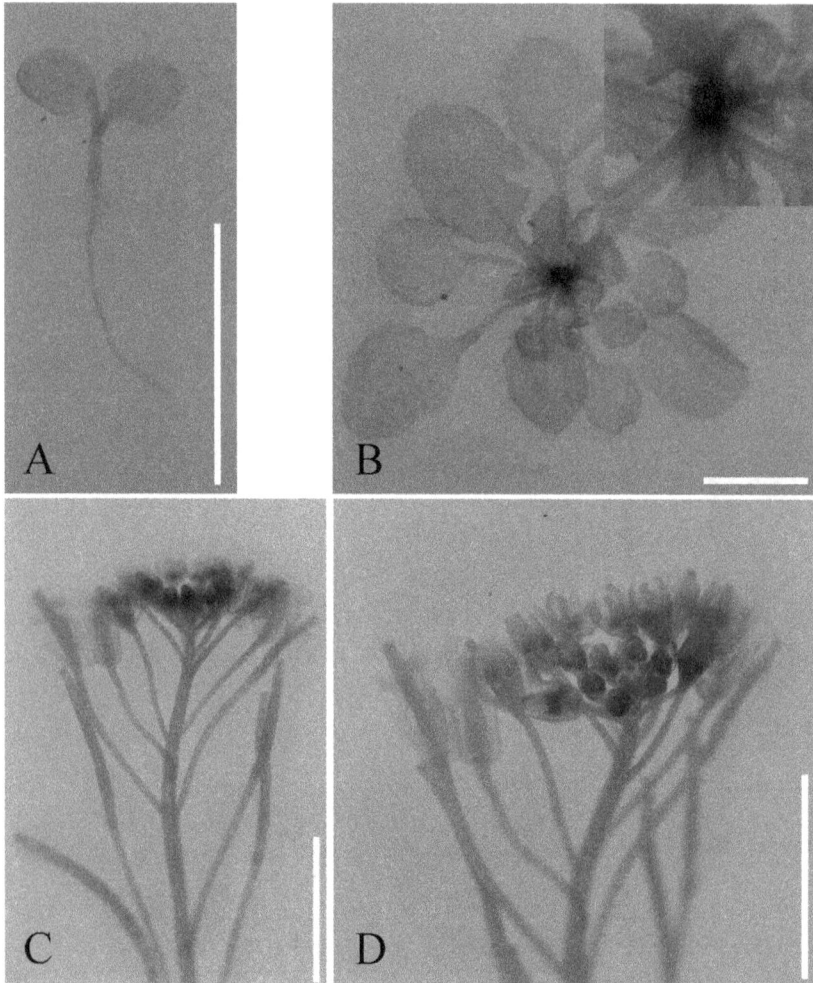

Figure 7. Histochemical staining of GUS activity in *pBnTPI::GUS* transgenic *Arabidopsis* lines. (A) In nine-day-old seedlings, GUS activity was detectable at the shoot apex meristem (SAM), node-joint, and roots. (B) GUS activity in the rosette axils of 24-day-old plants. The insets on the top right corner indicate the megascopic rosette. (D) Weak GUS activity in the SAM and floral buds at flowering, and (E) a zoomed-in view of the stained floral buds.

would be helpful to decipher all potential genes that affect the NPMI in the genomes of *B. rapa*, *B. oleracea*, and *B. napus* and to place the entire set of *Brassicaceae* genes on a single chip. This comprehensive chip would provide an effective approach for establishing extensive expression profiles throughout the life cycle of *B. napus*. All genes that affect the NPMI and other inflorescence traits would certainly be identified using such a chip, and the roles of these genes could be gradually determined using functional assays.

Conclusions

We report here the first comprehensive survey of gene expression profiles of the *B. napus* MI. The datasets created in this study exhibited variation at the gene expression level that is intricate, but coordinated. Several expression patterns and prominent processes and pathways in the MI were identified as increasing pod quantity and the duration of inflorescence differentiation in *B. napus*. The comparison of DEGs with NPMI QTLs suggested a number of likely candidates that warrant further studies. Our initial analysis of the functions of *BnTPI* and other interesting genes in *Arabidopsis* provides a starting point for further

investigations into the regulation of pod primordia initiation, maintenance of SAM duration, and even control of seed yield in *B. napus*. These findings contribute to our understanding of the mechanisms underlying pod and inflorescence differentiation and development, and pave the way for developing molecular breeding strategies for improved yield *B. napus* and other crops.

Acknowledgments

We thank Jingyin Yu (Oil Crops Research Institute of Chinese Academy of Agricultural Sciences, Wuhan, China) for partial data processing. We are grateful to Dr. Kathleen L. Farquharson from The Plant Cell for suggestions and language editing.

Author Contributions

Conceived and designed the experiments: YH HZW. Performed the experiments: YH ZST QL QY XFW. Analyzed the data: YH LDZ. Contributed reagents/materials/analysis tools: JQS GHL. Contributed to the writing of the manuscript: YH ZST HZW.

References

1. Zhu Z, Tan L, Fu Y, Liu F, Cai H, et al. (2013) Genetic control of inflorescence architecture during rice domestication. Nat Commun 4: 2200.
2. Eveland AL, Goldshmidt A, Pautler M, Morohashi K, Liseron-Monfils C, et al. (2014) Regulatory modules controlling maize inflorescence architecture. Genome Res 24: 431–443.
3. MacAlister CA, Park SJ, Jiang K, Marcel F, Bendahmane A, et al. (2012) Synchronization of the flowering transition by the tomato TERMINATING FLOWER gene. Nat Genet 44: 1393–1398.
4. Kawamura K, Hibrand-Saint Oyant L, Crespel L, Thouroude T, Lalanne D, et al. (2011) Quantitative trait loci for flowering time and inflorescence architecture in rose. Theor Appl Genet 122: 661–675.
5. Park SJ, Eshed Y, Lippman ZB (2014) Meristem maturation and inflorescence architecture–lessons from the Solanaceae. Curr Opin Plant Biol 17: 70–77.
6. Liu C, Teo ZW, Bi Y, Song S, Xi W, et al. (2013) A conserved genetic pathway determines inflorescence architecture in Arabidopsis and rice. Dev Cell 24: 612–622.
7. Iwata T, Nagasaki O, Ishii HS, Ushimaru A (2012) Inflorescence architecture affects pollinator behaviour and mating success in Spiranthes sinensis (Orchidaceae). New Phytol 193: 196–203.
8. Upadyayula N, da Silva HS, Bohn MO, Rocheford TR (2006) Genetic and QTL analysis of maize tassel and ear inflorescence architecture. Theor Appl Genet 112: 592–606.
9. Larson SR, Kellogg EA, Jensen KB (2013) Genes and QTLs controlling inflorescence and stem branch architecture in Leymus (Poaceae: Triticeae) Wildrye. J Hered 104: 678–691.
10. Harder LD, Prusinkiewicz P (2013) The interplay between inflorescence development and function as the crucible of architectural diversity. Ann Bot 112: 1477–1493.
11. Koes R (2008) Evolution and development of virtual inflorescences. Trends Plant Sci 13: 1–3.
12. Benlloch R, Berbel A, Serrano-Mislata A, Madueno F (2007) Floral initiation and inflorescence architecture: a comparative view. Ann Bot 100: 659–676.
13. Koia J, Moyle R, Botella J (2012) Microarray analysis of gene expression profiles in ripening pineapple fruits. BMC Plant Biology 12: 240.
14. Abu-Abied M, Szwerdszarf D, Mordehaev I, Levy A, Stelmakh OR, et al. (2012) Microarray analysis revealed upregulation of nitrate reductase in juvenile cuttings of Eucalyptus grandis, which correlated with increased nitric oxide production and adventitious root formation. Plant J 71: 787–799.
15. Carlsson J, Lagercrantz U, Sundstrom J, Teixeira R, Wellmer F, et al. (2007) Microarray analysis reveals altered expression of a large number of nuclear genes in developing cytoplasmic male sterile Brassica napus flowers. Plant J 49: 452–462.
16. Huang Y, Chen L, Wang L, Vijayan K, Phan S, et al. (2009) Probing the endosperm gene expression landscape in Brassica napus. BMC genomics 10: 256.
17. Jansen L, Hollunder J, Roberts I, Forestan C, Fonteyne P, et al. (2013) Comparative transcriptomics as a tool for the identification of root branching genes in maize. Plant Biotechnol J 11: 1092–1102.
18. Oono Y, Kawahara Y, Yazawa T, Kanamori H, Kuramata M, et al. (2013) Diversity in the complexity of phosphate starvation transcriptomes among rice cultivars based on RNA-Seq profiles. Plant Mol Biol 83: 523–537.
19. Hart T, Komori HK, LaMere S, Podshivalova K, Salomon DR (2013) Finding the active genes in deep RNA-seq gene expression studies. BMC Genomics 14: 778.
20. Zhu Y, Cao Z, Xu F, Huang Y, Chen M, et al. (2012) Analysis of gene expression profiles of two near-isogenic lines differing at a QTL region affecting oil content at high temperatures during seed maturation in oilseed rape (Brassica napus L.). Theor Appl Genet 124: 515–531.
21. Jiang Y, Deyholos MK (2010) Transcriptome analysis of secondary-wall-enriched seed coat tissues of canola (Brassica napus L.). Plant cell reports 29: 327–342.
22. Sun M, Hua W, Liu J, Huang S, Wang X, et al. (2012) Design of new genome- and gene-sourced primers and identification of QTL for seed oil content in a specially high-oil Brassica napus cultivar. PLoS One 7: e47037.
23. Smyth GK, Michaud J, Scott HS (2005) Use of within-array replicate spots for assessing differential expression in microarray experiments. Bioinformatics 21: 2067–2075.
24. Provart NJ, Zhu T (2003) A browser-based functional classification superviewer for arabidopsis genomics. Currents Comput Mol Biol 271–272.
25. Clough SJ, Bent AF (1998) Floral dip: a simplified method for Agrobacterium-mediated transformation of Arabidopsis thaliana. Plant J 16: 735–743.
26. Jefferson RA, Kavanagh TA, Bevan MW (1987) GUS fusions: β-Glucuronidase as a Sensitive and Versatile Gene Fusion Marker in Higher Plants. EMBO Journal 6: 39001–33907.
27. Krouk G, Lacombe B, Bielach A, Perrine-Walker F, Malinska K, et al. (2010) Nitrate-regulated auxin transport by NRT1.1 defines a mechanism for nutrient sensing in plants. Dev Cell 18: 927–937.
28. Wang X, Wang H, Wang J, Sun R, Wu J, et al. (2011) The genome of the mesopolyploid crop species Brassica rapa. Nat Genet 43: 1035–1039.
29. Teo ZW, Song S, Wang YQ, Liu J, Yu H (2014) New insights into the regulation of inflorescence architecture. Trends Plant Sci 19: 158–165.
30. Schadt EE, Monks SA, Drake TA, Lusis AJ, Che N, et al. (2003) Genetics of gene expression surveyed in maize, mouse and man. Nature 422: 297–302.
31. Emilsson V, Thorleifsson G, Zhang B, Leonardson AS, Zink F, et al. (2008) Genetics of gene expression and its effect on disease. Nature 452: 423–428.
32. Rodriguez-Suarez C, Atienza SG, Piston F (2011) Allelic variation, alternative splicing and expression analysis of Psy1 gene in Hordeum chilense Roem. et Schult. PLoS One 6: e19885.
33. Shao G, Wei X, Chen M, Tang S, Luo J, et al. (2012) Allelic variation for a candidate gene for GS7, responsible for grain shape in rice. Theor Appl Genet 125: 1303–1312.
34. Yan F, Di S, Rojas Rodas F, Rodriguez Torrico T, Murai Y, et al. (2014) Allelic variation of soybean flower color gene W4 encoding dihydroflavonol 4-reductase 2. BMC Plant Biol 14: 58.
35. Lo HS, Wang Z, Hu Y, Yang HH, Gere S, et al. (2003) Allelic variation in gene expression is common in the human genome. Genome Res 13: 1855–1862.
36. Hitzemann R, Malmanger B, Reed C, Lawler M, Hitzemann B, et al. (2003) A strategy for the integration of QTL, gene expression, and sequence analyses. Mamm Genome 14: 733–747.
37. Cavell AC, Lydiate DJ, Parkin IA, Dean C, Trick M (1998) Collinearity between a 30-centimorgan segment of Arabidopsis thaliana chromosome 4 and duplicated regions within the Brassica napus genome. Genome 41: 62–69.

14

Mutational Analysis of the Ve1 Immune Receptor That Mediates *Verticillium* Resistance in Tomato

Zhao Zhang[1,2]**, Yin Song**[1]**, Chun-Ming Liu**[2]**, Bart P. H. J. Thomma**[1]*

1 Laboratory of Phytopathology, Wageningen University, Wageningen, The Netherlands, **2** Key Laboratory of Plant Molecular Physiology, the Chinese Academy of Sciences, Beijing, China

Abstract

Pathogenic *Verticillium* species are economically important plant pathogens that cause vascular wilt diseases in hundreds of plant species. The *Ve1* gene of tomato confers resistance against race 1 strains of *Verticillium dahliae* and *V. albo-atrum*. *Ve1* encodes an extracellular leucine-rich repeat (eLRR) receptor-like protein (RLP) that serves as a cell surface receptor for recognition of the recently identified secreted *Verticillium* effector Ave1. To investigate recognition of Ave1 by Ve1, alanine scanning was performed on the solvent exposed β-strand/β-turn residues across the eLRR domain of Ve1. In addition, alanine scanning was also employed to functionally characterize motifs that putatively mediate protein-protein interactions and endocytosis in the transmembrane domain and the cytoplasmic tail of the Ve1 protein. Functionality of the mutant proteins was assessed by screening for the occurrence of a hypersensitive response upon co-expression with Ave1 upon *Agrobacterium tumefaciens*-mediated transient expression (agroinfiltration). In order to confirm the agroinfiltration results, constructs encoding Ve1 mutants were transformed into Arabidopsis and the transgenes were challenged with race 1 *Verticillium*. Our analyses identified several regions of the Ve1 protein that are required for functionality.

Editor: Boris Alexander Vinatzer, Virginia Tech, United States of America

Funding: These authors have no support or funding to report.

Competing Interests: The authors have declared that no competing interests exist.

* E-mail: bart.thomma@wur.nl

Introduction

In order to activate immune responses that ward off invading microorganisms, plants utilize various types of receptors that recognize pathogen(-induced) ligands of various nature [1,2]. Appropriate recognition of these ligands by the immune receptors is crucial for the activation of immune responses. These immune receptors are either extracellular cell surface receptors that detect (conserved) pathogen-associated molecular patterns (PAMPs) or damage-associated modified self-patterns, or cytoplasmic receptors that recognize highly specific pathogen effectors either directly, or indirectly through recognition of their activities [3,4]. Both types of receptors may activate an hypersensitive response (HR), which is a rapid cell death surrounding the infection site that is thought to prevent further pathogen invasion [5].

The *Verticillium* genus comprises vascular pathogens that cause Verticillium wilt diseases in over 200 plant species worldwide [6,7]. In tomato, immunity against *Verticillium* wilt is governed by the immune receptor Ve1 that recognizes the secreted *Verticillium* effector Ave1 [8,9]. *Ve1* encodes a putative plasma membrane-localized extracellular leucine-rich repeat (eLRR)-containing cell surface receptor of the receptor-like protein (RLP) class [10]. Typically, the amino acid sequence of RLPs is composed of a signal peptide (SP), an eLRR domain that is shielded by N-terminal and C-terminal eLRR-caps, a single-pass transmembrane (TM) domain, and a short cytoplasmic tail that lacks obvious motifs for intracellular signaling. In some cases, an acidic domain is present between the eLRR domain and the TM domain.

Furthermore, the eLRR domain can be subdivided into three domains in which a non-eLRR island or C2 domain interrupts the C1 and C3 eLRR regions [11,12]. As RLPs lack an obvious domain for intracellular signaling, they presumably form a complex with other proteins, such as receptor-like kinases, to respond to ligand binding and initiate an immune response [11]. Indeed, it was recently demonstrated that interaction of Ve1 with the SUPPRESSOR OF BIR1 (SOBIR1) receptor-like kinase is required for Ve1-mediated immunity [13,14].

It is conceivable that the eLRR domain of cell surface receptors acts as ligand sensor [15]. This similarly holds true for the eLRRs of Toll-like receptors (TLRs) that act in animal innate immunity [16]. The typical plant eLRR consensus motif comprises 24 amino acids, xxLxxLxxLxxLxLxxNxLt/sGxIP, where (x) represents any amino acid and (L) is sometimes substituted by other hydrophobic residues. For plants, the first eLRR protein crystal structures were resolved for a polygalacturonase-inhibiting protein (PGIP) [17], the brassinosteroid receptor brassinosteroid-insensitive 1 (BRI1) [18–20] and the flagellin receptor flagellin-sensitive 2 (FLS2) [21]. These studies revealed that successive eLRRs align in parallel to form a curved, slightly twisted "horseshoe-like" structure, in which parallel core β-strands (xxLxLxx) form the concave (inner) side of the protein and various helices, short β-strands and additional connecting residues form the convex (outer) side [15]. The concave side of the eLRR is thought to serve for ligand binding, where the hydrophobic (L) residues in the β-sheet (xxLxLxx) are involved in the framework that determines the overall shape of the protein, and the five variable, solvent exposed residues (x) of the β-strands determine ligand binding specificity [15]. Crystallographic analysis of PGIP demonstrated that the solvent exposed residues on the concave β-sheet surface determine the interaction with polygalacturonases [17]. Furthermore, the recently released crystal structure of BRI1 showed that the brassinosteroid hormone binds to a groove in between the concave β-sheet surface and the island domain [18–20]. Similarly, the conserved N-terminal epitope of bacterial flagellin (flg22) binds to the inner concave surface of the FLS2 LRR solenoid [21].

In the majority of studied eLRR receptors, ligand specificity is determined by the C1 domain [11]. We recently carried out domain swaps between Ve1 and its non-functional homolog Ve2, and demonstrated that the chimeras in which the first thirty eLRRs of Ve1 were replaced with those of Ve2 remained able to activate *Verticillium* resistance [13]. However, the C3 domain and C-terminus of Ve2 appeared not to be functional [13]. Potentially, the non-functional Ve2 receptor still interacts with the Ave1 elicitor in the C1 domain, but fails to activate immune signaling due to a non-functional C3 domain and C-terminus. Nevertheless, similar to Ve1, Ve2 still interacts with the receptor-like kinase SOBIR1 [13]. To further determine the role of eLRRs of Ve1 in ligand specificity and signal transduction, we employed a high-throughput alanine scanning mutagenesis strategy to mutate solvent exposed residues on the concave surface of each eLRR repeat of Ve1 in this study.

Results

Alanine scanning of the concave side of the Ve1 eLRR domain

Considering the large size of the Ve1 eLRR domain and avoiding the potential inefficiency of random mutagenesis, a site-directed mutagenesis strategy was performed to identify functional regions of the Ve1 eLRR domain which contains 37 imperfect eLRRs. To this end, solvent exposed residues in the β-strand of each eLRR repeat were mutated. In total, 37 mutant *Ve1* alleles were engineered, named M1–M37 respectively, in which two of the five variable solvent exposed residues in the xxLxLxx consensus of a single eLRR were mutated such that they were substituted by alanines (Figure 1). To generate mutant alleles, the *Ve1* coding sequence was cloned into pDONR207 (Invitrogen, Carlsbad, California) through a Gateway BP reaction to generate entry vector pDONR207::Ve1. Using pDONR207::Ve1 as template, and inverse PCR was performed to establish alanine substitutions by changing wild type codons in the primer sequence. The mutated *Ve1* variants were sequenced and subsequently cloned into an expression construct driven by the constitutive CaMV35S promoter.

C1 domain eLRRs 1 to 8 and 20 to 23 are required for Ve1 functionality

We previously suggested that ligand recognition is determined by the Ve1 eLRRs 1 to 30 [13]. To determine which eLRRs of the C1 domain are required for Ve1 functionality in more detail, tobacco leaves were co-infiltrated with 1:1 mixture of *Agrobacterium tumefaciens* cultures carrying *Ave1* and *Ve1* alleles that encode mutants in the C1 domain (M1–M31). Intriguingly, agroinfiltration in at least three independent experiments revealed that expression of mutant alleles M1, M3 to M8, and M20 to M23 together with Ave1 showed significantly compromised HR at five days post infiltration (dpi; Figure 2; Figure 3A). In contrast, co-expression of Ave1 with the mutant alleles M2, M9–M19, and M24–M31 resulted in full HR. To exclude the possibility that co promised HR is the result of the expression of unstable receptor

proteins rather, the Ve1 mutants that failed to induce full HR were C-terminally tagged with a green fluorescent protein (GFP), and protein stability was verified by immunoblotting (Figure S1). Similar to the discrepancies have previously been reported for Ve1, Ve2 and other eLRR proteins, the estimated sizes of the Ve1-GFP proteins exceeded the calculated sizes, likely due to N-glycosylation of the proteins [13,22,23]. Importantly, most of the GFP-tagged Ve1 mutants accumulated to similar levels as GFP-tagged wild type Ve1 protein or GFP-tagged Ve1 mutant M2 that are able to induce full HR. Only mutant M1-GFP could not be detected by western blotting, indicating that this LRR are essential for Ve1 protein stability (Figure S1).

To further assess functionality of the mutant alleles, all mutant constructs were transformed into Arabidopsis [24]. For each mutant, three independent transformants were challenged with race 1 *V. dahliae*. As expected based on the occurrence of HR in tobacco, transgenic plants carrying the non-functional mutant alleles M1, M3–M8 and M20–M23 displayed Verticillium wilt symptoms that were comparable to those on inoculated non-transgenic control plants (Figure 2; Figure S2). In contrast, expression of functional mutant alleles M2, M9–M19 and M24–M31 in Arabidopsis resulted in complete *Verticillium* resistance, as the transgenes showed few to no symptoms upon inoculation when compared to non-transgenic control plants (Figure 2; Figure S2). The differential symptom display correlated with the amount of *Verticillium* biomass, when compared with the *Verticillium* biomass in inoculated wild type plants and Ve1-expressing plants (Figure 2). Collectively, these results show that the LRR region between eLRR1 and eLRR8, as well as between eLRR20 and eLRR23, is required for Ve1-mediated resistance.

The island (C2) domain is required for Ve1 function

To test the contribution of the island domain, the non-LRR region (C2) that separates the two LRR-containing domains (C1 and C3) in the extracellular domain of Ve1, to Ve1 function, two alanine substitutions were introduced into the predicted island domain to engineer mutant allele MIS (Figure 1). Agroinfiltraion revealed that the mutant allele can still activate an HR upon co-expression with Ave1, as the complete infiltrated sectors became fully necrotic (Figure 2; Figure 3A). Similarly, expression of the mutant allele in Arabidopsis resulted in *Verticillium* resistance, as the transgenes showed few to no symptoms of disease and significantly less fungal biomass accumulated upon inoculation with race 1 *V. dahliae* when compared with wild-type plants (Figure 2; Figure S2). Previously, Wang et al. [25] demonstrated that deletion of the island domain from CLV2 does not affect its functionality in plant development. We thus designed the deletion construct Ve1_ΔIS, in which the complete island domain of Ve1 was removed. In contrast to mutant allele MIS, co-expression of the deletion construct with Ave1 did not induce an HR in tobacco (Figure 3B), suggesting that the island domain is required for Ve1 functionality. Importantly, the Ve1_ΔIS-GFP mutant accumulates to detectable levels (Figure S1).

Alanine scanning reveals functionally important solvent-exposed residues in the β-strands of the C3 domain

Based on domain swaps between Ve1 and Ve2, we previously demonstrated that the C3 domain and C-terminus of Ve2 are not able to activate immune signaling [13]. To further determine the role of solvent exposed residues in the β-strands of the C3 domain, tobacco leaves were co-infiltrated with *A. tumefaciens* cultures carrying mutant *Ve1* alleles in the region that encodes the C3 domain (M32-M37) and Ave1. Intriguingly, five of the six Ve1 mutants that were generated in the C3 domain resulted in

```
                    MKMMATLYFLWLLLIPSFQILSGYHIFLV          SP
                              xxLxLxx
   SSQCLDDQKSLLLQFKGSLQYDSTLSKKLAKWND                      1
   MTSECCNWNGVTCNLFGHVIALELDDETISSGIE                      2
           NSSALFSLQYLESLNLADNMFNVGIP                      3
            VGIANLTNLKYLNLSNAGFVGQIP                       4
             ITLSRLTRLVTLDLSTILPFFDQP                      5
   LKLENPNLSHFIENSTELRELYLDGVDLSSQRT                       6
       EWCQSLSLHLPNLTVLSLRDCQISGPLD                        7
          ESLSKLHFLSFVQLDQNNLSSTVP                         8
          EYFANFSNLTTLTLGSCNLQGTFP                         9
          ERIFQVSVLESLDLSINKLLRGSIP                       10
           IFFRNGSLRRISLSYTNFSGSLP                        11
          ESISNHQNLSRLELSNCNFYGSIP                        12
          STMANLRNLGYLDFSFNNFTGSIP                        13
          YFRLSKKLTYLDLSRNGLTGLLS                         14
          RAHFEGLSELVHINLGNNLLSGSLP                       15
          AYIFELPSLQQLFLYRNQFVGQVD                        16
       EFRNASSSPLDTVDLTNNHLNGSIP                          17
          KSMFEIERLKVLSLSSNFFRGTVP                        18
          LDLIGRLSNLSRLELSYNNLTVDAS                       19
   SSNSTSFTFPQLNILKLASCRLQKFPD                            20
          LKNQSWMMHLDLSDNQILGAIP                          21
       NWIWGIGGGGLTHLNLSFNQLEYVEQ                          22
          PYTASSNLVVLDLHSNRLKGDLLIP                       23
          PCTAIYVDYSSNNLNNSIPTDIG                         24
            KSLGFASFFSVANNGITGIIP                         25
          ESICNCSYLQVLDFSNNALSGTIP                        26
          PCLLEYSTKLGVLNLGNNKLNGVIP                       27
          DSFSIGCALQTLDLSANNLQGRLP                        28
          KSIVNCKLLEVLNVGNNRLVDHFP                        29
          CMLRNSNSLRVLVLRSNKFYGNLM                        30
   CDVTRNSWQNLQIIDIASNNFTGVLN                             31
   AEFFSNWRGMMVADDYVETGRNHIQ                              IS
   YEFLQLSKLYYQDTVTLTIKGMELELVKI                          32
          LRVFTSIDFSSNRFQGAIP                             33
          DAIGNLSSLYVLNLSHNALEGPIP                        34
          KSIGKLQMLESLDLSTNHLSGEIP                        35
          SELASLTFLAALNLSFNKLFGKIP                        36
   STNQFQTFSADSFEGNSGLCGLPLNNSCQSNGSA                     37

              SESLPPPTPLPDSDDEWE                          AC

            FIFAAVGYIVGAANTISVVWF                         TM

            YKPVKKWFDKHMEKCLLWFSRK                        CT
```

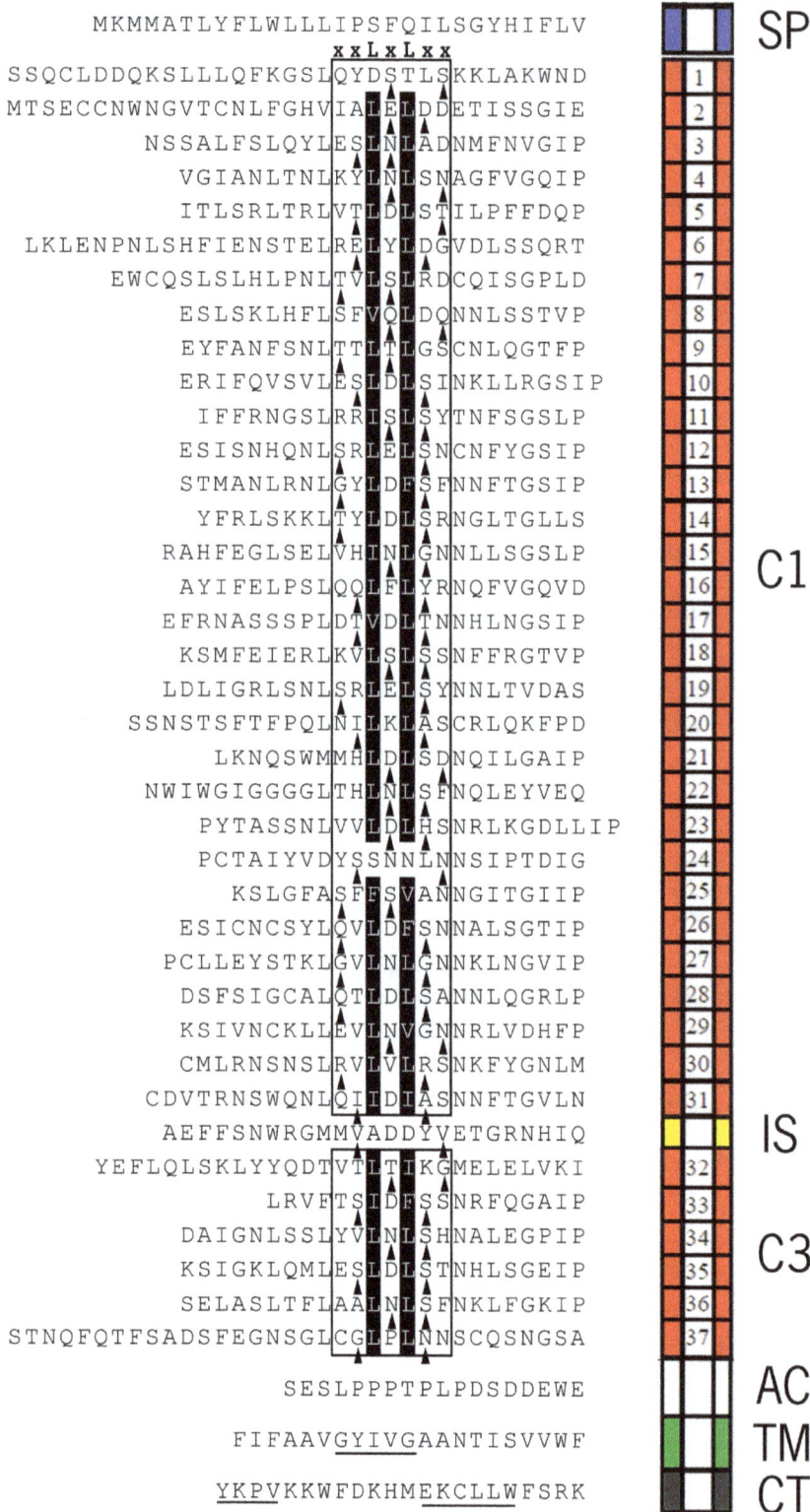

Figure 1. Primary structure of the Ve1 protein. Alignment of the amino acid sequence of Ve1 with a schematic representation of the protein structure. Ve1 is composed of a signal peptide (SP), eLRR region C1 (C1), island domain (IS), eLRR region C3 (C3), acidic domain (AC), transmembrane domain (TM) and cytoplasmic tail (CT). Double alanine scanning was performed on the solvent exposed β-strand residues across the Ve1eLRR domain. The putative parallel β-strands (xxLxLxx) on the concave surface are boxed, and the conserved hydrophobic residues on the concave β-sheet surface are indicated with black shading. Triangles represent solvent-exposed amino acid residues (x) subjected to alanine substitution for each of the repeats. Only one eLRR was mutated per mutant allele. The putative GxxxG motif and endocytosis signals are underlined.

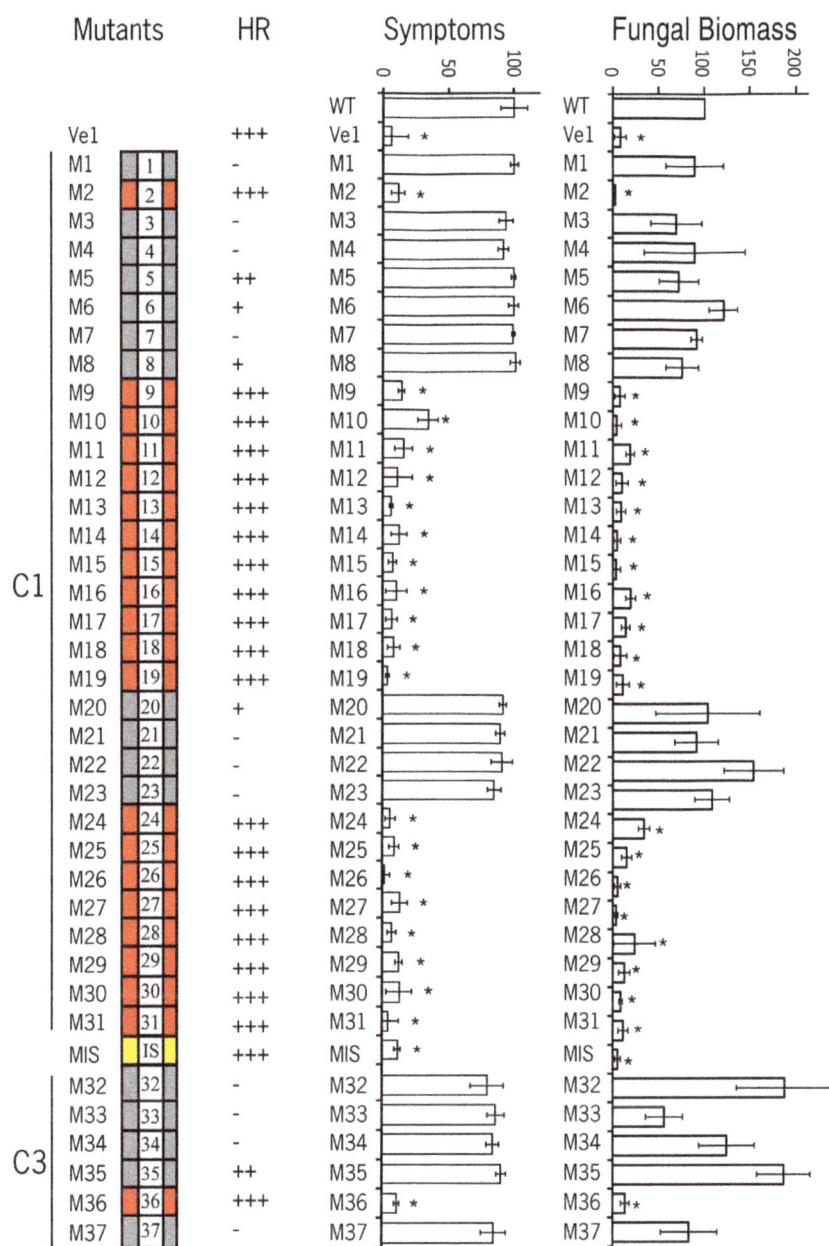

Figure 2. Double alanine scanning reveals eLRRs required for Ve1 functionality. A schematic representation of the Ve1 eLRR domain is shown with a summary of the functionality of the double alanine scanning mutant alleles. Grey boxes indicate mutant alleles that compromise Ve1 functionality while red boxes indicate mutants that remain fully functional. The occurrence of HR upon co-expression of Ve1 mutant alleles with Ave1 is provided, where +++ corresponds to an HR that is similar to the HR induced by wild type Ve1; ++ corresponds to an HR that is reduced when compared with the HR induced by wild-type Ve1; + corresponds to a limited HR; and - corresponds to absence of a detectable HR. Quantification of Verticillium wilt symptoms in wild type (WT) and transgenic lines is indicated. Bars represent quantification of symptoms presented as percentage of diseased rosette leaves with standard deviation with WT set to 100%. Asterisks indicate significant differences when compared with WT (P<0.001). Quantification of Verticillium biomass in Arabidopsis expressing Ve1 mutantconstructs is shown. Fungal biomass is determined by real-time qPCR in wild-type (WT) Arabidopsis and transgenic lines, and the fungal biomass in WT plants is set to 100%. For qPCR, Verticillium internal transcribed spacer (ITS) transcript levels were determined relative to Arabidopsis RuBisCo transcript levels for equilibration. Bars represent an average Verticillium quantification of three independent transgenic lines. Error bars represent standard deviations of qPCR results from three independent transgenic lines. Asterisks indicate significant differences when compared with WT (P<0.05). Data from a representative experiment are shown.

abolished or significantly compromised HR in tobacco leaves at five dpi, as only mutant (M36) still activated full HR (Figure 2; Figure 3A). The nonfunctional mutants were C-terminally tagged with GFP, and protein stability was tested by immunoblotting (Figure S1). GFP-tagged mutant proteins M32-GFP, M35-GFP and M37-GFP were found to accumulate to similar levels as non-mutated

Ve1-GFP protein or the functional mutant protein M36-GFP, whereas the M32-GFP and M34-GFP mutant constructs did not lead to detectable protein levels, suggesting that these LRRs are essential for Ve1 protein stability (Figure S1). As expected based on the agroinfiltration results, expression of M36 resulted in Verticillium resistance in Arabidopsis, while plants expressing the other C3

Figure 3. Typical appearance of tobacco leaves transiently co-expressing Ave1 with Ve1 mutant alleles. (A) Occurrence of HR upon co-expression of Ave1 and Ve1 double alanine scanning mutant alleles. **(B)** Co-expression of the island domain deletion construct Ve1_ΔIS with Ave1. All pictures were taken at 5 days post infiltration and are representative of at least three independent experiments.

domain mutant alleles displayed typical Verticillium wilt symptoms that were comparable to wild type plants (Figure 2; Figure S2). Collectively, as expected based on the domain swaps experiments [13], these alanine scanning assays confirm that the C3 region (eLRR32-eLRR37) is critical for Ve1 functionality.

The C3 domain of Cf-9 is required for functionality

Previous comparison of eLRR-RLP sequences of Arabidopsis, rice and tomato has shown that the C3 domains of these proteins are relatively conserved. Based on this finding it was suggested that the conserved C3 region may be involved in interaction with common factors, such as (a) co-receptor(s) [11–13,26]. To prove

that the C3 domain of Cf-9 is functionally important similar to that of Ve1, we performed site-directed mutagenesis on the C3 domain of Cf-9, which has four eLRRs. The alanine substitutions are made at the same sites of the concave surface that were used for the mutagenesis of Ve1 (Figure 4). Intriguingly, co-expression of Avr9 with Cf-9 mutants M24, M25 and M27 resulted in compromised HR, whereas co-expression with mutant M26 did not show compromised HR. Collectively, these results demonstrate that the C3 region is required for Cf-9 function, as was similarly demonstrated for Ve1.

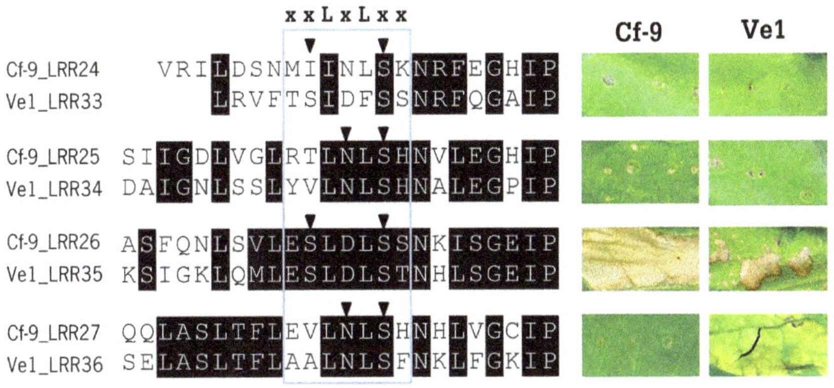

Figure 4. The C3 domain of Cf-9 is required for functionality. A sequence alignment of the C3 domain of Cf-9 and Ve1 is shown, with identical and similar residues indicated with black shading. The putative parallel β-strands (xxLxLxx) on the concave surface are boxed. Triangles represent solvent-exposed amino acid residues subjected to alanine substitution. Functional characterization of the mutants is shown on the right. Photographs illustrate typical appearance of tobacco leaves upon co-expression of Cf-9 mutants with Avr9, or Ve1 mutants with Ave1. Pictures were taken at 5 days post infiltration and are representative of at least three independent experiments.

Alanine scanning of putative functional motifs in the C-terminus of Ve1

In addition to the eLRR domain, the domain swaps between Ve1 and Ve2 also pointed towards a function of the transmembrane region and cytoplasmic tail of Ve1 [13]. A GxxxG motif that has been implicated in protein-protein interactions is found in the transmembrane domain of many membrane proteins [27,28], including Ve1 and other eLRR-containing cell surface receptors such as Cf-2, Cf-4, Cf-9, EFR and HrcVf [11]. Interestingly, a mutation in the second glycin of GxxxG motif abolished the function of Cf-9, which was thought to be due to disruption of the interaction with a co-receptor that associates through the GxxxG motif [29]. Similar mutations in Arabidopsis AtRLP51 and AtRLP55 resulted in constitutively activated defense [30]. Furthermore, endocytosis of membrane proteins is often associated with presence of a Yxxφ or E/DxxxLφ consensus motif in the cytoplasmic domains of such proteins, where φ is a hydrophobic residue and x is any amino acid [31,32]. Both Yxxφ and E/DxxxLφ motifs are present in the cytoplasmic domain of Ve1. To further determine the role of the GxxxG, E/DxxxLφ and Yxxφ motifs in Ve1 function, we employed alanine scanning mutagenesis.

The putative transmembrane GxxxG motif is not required for Ve functionality

All five residues in the Ve1 putative GxxxG domain were selected for mutagenesis and subjected to alanine substitution (G1 to G5; Figure 5A). Co-expression of the mutants with Ave1 in tobacco showed that the mutations did not affect Ve1 functionality, as full HR was still observed (Figure 5A). Next, Arabidopsis plants were transformed with the mutant alleles, and the resulting transgenes were challenged with V. dahliae. As expected, all mutant Ve1 alleles still mediated Verticillium resistance as the transgenic plants showed few to no symptoms upon inoculation and accumulated significantly less fungal biomass when compared with non-transgenic wild type plants (Figure 5C; Figure S3).

Putative C-terminal endocytosis motifs are not required for Ve1 functionality

To investigate whether the putative C-terminal E/DXXXLφ endocytosis motif is involved in Ve1 functionality, we generated six Ve1 mutant alleles, E1 to E6, in which each amino acid of the

E/DXXXLφ motif was replaced by an alanine (Figure 5B). Expression of none of the mutant alleles resulted in reduced HR upon co-expression with Ave1 by agroinfiltration in tobacco (Figure 5B). Also in this case, Arabidopsis transgenes expressing the mutant alleles were resistant against Verticillium (Figure 5; Figure S3). Similarly, we generated alanine substitution construct Y4 in which the conserved Tyr1032 of the putative Yxxφ endocytosis motif was mutated. However, co-infiltration with Ave1 showed that also this mutation does not affect Ve1 functionality (Figure 5B). Collectively, although our data do not show whether or not endocytosis of the Ve1 immune receptor takes place as part of the immune signaling process, we show that the two putative endocytosis motifs in the Ve1 C-terminus are not required for Ve1 functionality.

Discussion

The plant eLRR-containing cell surface receptors encompass many members that were shown to play important roles in either development or pathogen immunity. Since solved structures of receptor-ligand co-crystals often are not readily available, thus far, knowledge about the functioning of plant eLRR receptors is mainly based on domain swaps, domain deletions, gene shuffling analyses and site-directed mutagenesis. We previously swapped domains of Ve1 with homologous domains of its non-functional homolog Ve2, and analysis of the chimeras suggested that Ve2 may still detect the (activity of the) Ave1 effector in the C1 eLRR domain, but that its C3 domain and C-terminus are not able to activate defense signaling. Here, we employed a site-directed mutagenesis strategy to further dissect functional determinants of Ve1.

Previously, site-directed mutagenesis has been employed for functional analysis of eLRR-containing cell surface receptors. For example, van der Hoorn et al [23] analyzed a number of site-directed mutants of Cf-9 and demonstrated that conserved Trp and Cys residues present in the N- and C-terminal eLRR flanking regions are important for Cf-9 activity. Similarly, recently reported site-directed mutations proved that the Cys residues in the N-terminal flanking region of the FLS2 eLRRs are required for protein stability and function [33]. However, as these Trp or Cys residues are conserved in many other plant eLRR proteins as well, they likely contribute to the conformation and stability of the protein rather than to ligand specificity. In addition, another site-directed

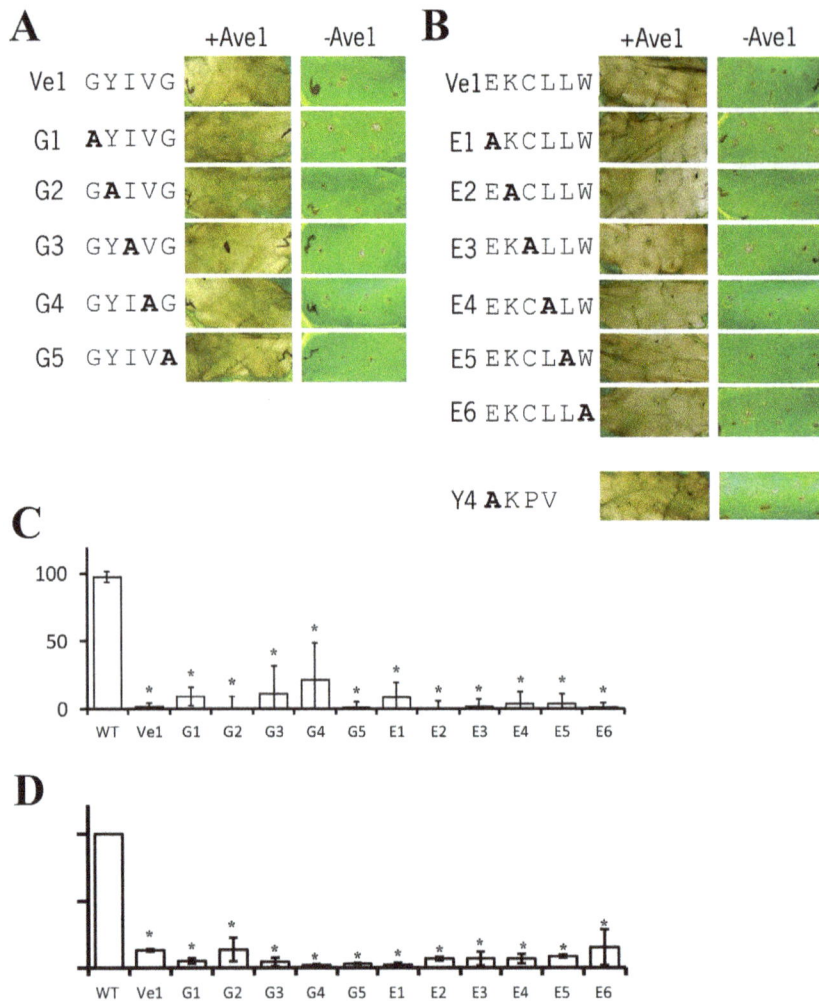

Figure 5. The putative transmembrane GxxxG motif and C-terminal endocytosis motifs are not required for Ve1 functionality. (A) Typical appearance of tobacco leaves transiently expressing wild type Ve1 and Ve1 mutants in presence or absence of Ave1 for the GxxxG motif (**A**) or the C-terminal endocytosis motifs (**B**). Pictures were taken at 5 days post infiltration and are representative of at least three independent experiments. (**C**) Quantification of *Verticillium* wilt symptoms in wild type (WT) and transgenic lines. Bars represent quantification of symptoms presented as percentage of diseased rosette leaves with standard deviation. WT is set to 100%. Asterisks indicate significant differences when compared with WT (P<0.001). (**D**) Quantification of *Verticillium* biomass in Arabidopsis expressing Ve1 mutants in the GxxxG motif and the C-terminal endocytosis motifs. Fungal biomass determined by real-time qPCR in wild-type (WT) Arabidopsis and transgenic lines, and the fungal biomass in WT plants is set to 100%. For qPCR, *Verticillium* internal transcribed spacer (ITS) transcript levels are shown relative to Arabidopsis RuBisCo transcript levels (for equilibration). Bars represent an average *Verticillium* quantification of three independent transgenic lines. Error bars represent standard deviations of qPCR results from three independent transgenic lines. Asterisks indicate significant differences when compared with WT (P<0.05).

mutagenesis strategy focused on putative *N*-linked glycosylation sites, which frequently occur in the eLRR domain of cell surface receptors. Through Asn to Asp substitution, van der Hoorn et al [23] demonstrated that four glycosylation sites contribute to Cf-9 functionality. These four sites are located in putative α-helixes that are exposed at the convex surface of the Cf-9 eLRR domain and are also conserved in many plant eLRR proteins [23]. Glycosylation may contribute to protein conformation, facilitate interactions with the cell wall [34], or protect proteins from degradation [35]. However, it seems unlikely that these putative glycosylation sites contribute to ligand specificity of Cf-9 [23]. Most of the Ve1 glycosylation sites are located at convex face of the eLRR domain (18 of 21 for Cf-9 and 15 of 18 for Ve1), and thus they were not specifically targeted in our study. To the best of our knowledge, no examples of ligand perception at convex side of the eLRR domain have been reported [11]. Moreover, *N*-linked glycosylation was

determined to make only subtle quantitative contributions to FLS2 functionality [33]. In contrast, alanine scanning mutagenesis on the concave β-sheet surface across the Arabidopsis FLS2 eLRR domain identified eLRR9-eLRR15 as contributors to flagellin perception [36]. To identify eLRRs that are required for Ve1 ligand recognition, we focused our attention on the concave β-sheet surface and evaded conserved hydrophobic leucine residues in β-sheets that are likely involved in framework of protein. A double-alanine scanning was performed in which two of the five variable, solvent exposed residues in a single eLRR repeat were mutated. Mutagenesis of two non-adjacent amino acids increases the chance of substituting functionally important residues.

In this study, we showed that mutant alleles that reveal compromised Ve1 function are restricted to three consecutive eLRR regions, eLRR1-eLRR8, eLRR20-eLRR23 and eLRR32-eLRR37. This is consistent with previously studies, in which eLRR function was found to be determined by solvent-exposed residues in clustered LRRs of the concave β-sheet surface. For example, domain swaps of tomato Cfs revealed that eLRR13-eLRR16 of Cf-4 contribute to ligand specificity [37], while ligand specificity of Cf-9 is determined by eLRR10-eLRR16 [38]. In addition, photoaffinity labelling showed that BAM1 directly interacts with the small peptide ligand CLE9 at the eLRR6–eLRR8 region [39]. Finally, the crystal structure of PGIP showed that the concave surface of eLRR4-eLRR8 is involved in polygalacturonase binding [17]. Similarly, crystallographic studies revealed that brassinosteroid binds to a hydrophobic groove of BRI1 in between the island domain and the concave β-sheet surface of eLRR20-eLRR25 [18,19]. Significantly, crystal structure analysis showed that flg22 binds to the concave surface of FLS2 eLRR3 to eLRR16 [21]. This similarly holds true for the eLRR domain of mammalian TLRs, for example, a crystal structure of the TLR4–MD-2–LPS complex demonstrated that the TLR4 interaction with cofactor MD-2 is restricted to the concave β-sheet surface of two eLRR clusters, eLRR2-eLRR5 and eLRR8-eLRR10 [40].

Because ligand specificity is often determined by the C1 domain, we previously suggested that this may similarly be true for Ve1 [13]. Therefore, the two regions eLRR1-eLRR8 and eLRR20-eLRR23 are proposed to contribute to ligand binding. However, most of the mutant alleles in the C3 domain (eLRR32-eLRR37) also abolished Ve1 function. This finding is consistent with previous domain swap experiments between Ve1 and Ve2, which demonstrated that the C3 domain of Ve2 is not able to activate successful immune signaling [13]. Similar to Ve1, alanine scanning of the C3 domain of Cf-9, which is rather conserved when compared with the C3 domain of Ve1, compromised its functionality. This is also consistent with previous mutagenesis studies on Cf-9, where Wulff et al [29] showed that the Ser675Leu mutation in the solvent-exposed resides of the concave side of the Cf-9 eLRR24 in the C3 domain abolished functionality. Similarly, van der Hoorn et al [23] proved that Cf-9 function is compromised upon Asp substitution of Asn697, which is located on the concave side of eLRR25. In addition, a Glu662Val mutation in Cf-4 similarly showed the importance of concave side of the eLRR C3 domain [29]. It has previously been demonstrated that the C3 domains of the Cf-4 and Cf-9 receptors, that perceive sequence-unrelated effector proteins Avr4 and Avr9, respectively, is identical, supporting a role in immune signaling rather than in ligand perception [37].

The eLRR domain has recently been shown to be involved in hetero-dimerization of receptor molecules [41-43]. Possibly, the relatively conserved C3 domain [11,13,26] is involved in the interaction with downstream signaling partners such as (a) common co-receptor(s) [13]. BRASSINOSTEROID INSENSITIVE 1-ASSOCIATED KINASE 1 (BAK1) is such a common co-receptor and forms a heteromerization with FLS2 for activation of plant immunity. Interestingly, although FLS2 do not carry a non-eLRR island domain that interrupts its 28 eLRRs into the C1 and C3 regions, recent crystallographic analysis on FLS2-BAK1-flg22 co-crystals reveals that flg22 ligand binds to the N-terminus of FLS2 (eLRR3-eLRR16), whereas BAK1 binds to concave surface of the C-terminal eLRRs of FLS2 (eLRR18-eLRR25) [21]. Previously, BAK1 was shown to be genetically involved in Ve1-mediated immunity [9,24]. Other common co-receptor candidates for both Ve1 and Cf proteins have recently been identified as SOBIR1 and SOMATIC EMBRYOGENESIS RECEPTOR-LIKE KINASE 1 (SERK1), which both encode an eLRR-RLK with a short eLRR domain [20,43]. It was demonstrated that tomato SOBIR1 physically interacts with various eLRR-RLPs, including Cf-9, Cf-4 and Ve1, irrespective of ligand binding

[13,14], while SERK1 was shown to be genetically required for both Ve1- and Cf-4-mediated immune signaling [9,24]. Although it remains unknown how various eLRR-RLPs interact with SOBIR1 and SERK1, the relatively high conservation of the C3 domain suggests that this region may be involved.

Overall, this study identified exposed concave β-sheet surfaces with a functional role in Ve1-mediated resistance. This extensive analysis of Ve1 provides fuel for our understanding of eLRR protein function and brings novel leads for further research on eLRR protein function in plants.

Materials and Methods

Plant materials

Tobacco (*Nicotiana tabacum* cv. Petite Havana SR1) and Arabidopsis (*Arabidopsis thaliana*) plants were grown in the greenhouse at 21°C/19°C during 16/8 hours day/night periods, respectively, with 70% relative humidity and 100 W•m^{-2} supplemental light when the light intensity dropped below 150 W•m^{-2}. After agroinfiltration, plants were grown in the climate room at 22°C/19°C during 16/8 hours day/night periods, respectively, with 70% relative humidity. Arabidopsis transformations were performed as described [44]. Homozygous single insert transgenic lines were selected by analyzing the segregation of antibiotic resistance.

Generation of constructs for over-expression of Ve1 and Cf-9

The tomato *Ve1* coding sequence was PCR amplified from *pMOG800::Ve1* [9] using primers attB-Ve1-F and attB-Ve1-R containing AttB1 and AttB2 sites for Gateway-compatible cloning. The tomato Cf-9 coding sequence was PCR amplified from *pMOG800::Cf-9* [45] using primers attB-Cf9-F and attB-Cf9-R. The resulting PCR product was cleaned from 1% agarose gel using the QIAquick Gel Extraction Kit (Qiagen, Valencia, California) and transferred into donor vector *pDONR207* using Gateway BP Clonase II enzyme mix (Invitrogen, Carlsbad, California) to generate entry vector *pDONR207::Ve1* and *pDONR207::Cf-9*, respectively. The entry constructs *pDONR207::Ve1* and *pDONR207::Cf-9* were subsequently cloned into Gateway destination vector using Gateway LR Clonase II enzyme mix (Invitrogen, Carlsbad, California) to generate expression constructs driven by the CaMV35S promoter. The expression constructs were transformed into *E. coli* and transformants were checked by colony PCR analysis using primers AttB1F and AttB2R. The expression constructs were subsequently sequenced and transformed into *Agrobacterium tumefaciens* strain GV3101 by electroporation.

Alanine scanning mutagenesis

For the alanine scanning mutagenesis, inverse PCR was performed to introduce alanine substitutions. Primers to introduce mutations (Table S1) were designed according to user manual of GeneTailor site-directed mutagenesis kit (Invitrogen, Carlsbad, California). PCR reactions were performed in a total volume of 30 μL with 23 μL water, 3 μL 10x PCR buffer, 1 μL dNTPs, 1 μL of each primer, 1 μL Pfu DNA polymerase (Promega, Madison, Wisconsin) and 1 μL of *pDONR207::Ve1* or *pDONR207::Cf-9*. The PCR consisted of an initial denaturation step of 5 minutes at 95°C, followed by denaturation for 30 sec at 95°C, annealing for 30 sec at 45°C to 55°C, and extension for 14 min at 72°C for 20 cycles, and then a final extension for 20 min at 72°C. The product was purified by QIAquick PCR Purification Kit (Qiagen, Valencia, California), treated with *Dpn*I endonuclease kinase (New England Biolabs, Ipswich, UK), and transformed into DH5α chemically competent cells. Mutant plasmid DNA was extracted and sequenced to verify the mutations, and recombined with the Gateway-compatible destination vector to generate an expression construct driven by the constitutive CaMV35S promoter.

Agrobacterium tumefaciens-mediated transient expression

A. tumefaciens containing expression constructs were infiltrated into tobacco plants as described previously [45–47]. Briefly, an overnight culture of *A. tumefaciens* cells was harvested at OD_{600} of 0.8 to 1 by centrifugation and resuspended to a final OD of 2. *A. tumefaciens* cultures containing constructs to express *Ave1* and mutated *Ve1* proteins were mixed in a 1:1 ratio and infiltrated into leaves of five- to six-week-old tobacco plants. At five days post infiltration (dpi), leaves were examined for necrosis.

Protein extraction and immunoblotting

For detection of Ve1 mutants that showed compromised function, corresponding mutant constructs were C-terminally tagged with the green fluorescent protein (GFP) as described previously [46]. *A. tumefaciens* containing the relevant expression constructs was infiltrated into tobacco plants as described previously [46]. Tobacco leaves were harvested at two days post infiltration, flash frozen and ground to a fine powder in liquid nitrogen. Total proteins were dissolved in extraction buffer (150 mM Tris-HCL pH 7.5, 150 mM NaCl, 10 mM DTT, 10% glycerol, 10 mM EDTA, 1% IGEPAL CA-630, 0.5% polyvinylpyrrolidon and 1% protease inhibitor cocktail [Roche, Basel, CH]). The immunopurifications and immunoblotting were performed as described previously [48].

Verticillium inoculations

Race 1 *V. dahliae* strain JR2 was grown on potato dextrose agar (PDA) at 22°C. *V. dahliae* conidia were harvested from 7- to 14-day-old fungal plates and washed with tap water. The conidia were suspended to a final concentration of 10^6 conidia per milliliter in potato dextrose broth (PDB). For inoculation, 2- to 3-week-old Arabidopsis plants were uprooted, and subsequently the roots were dipped in the conidial suspension for 3 min. As a control, plants were mock-inoculated in PDB without conidia. After inoculation, plants were immediately transplanted to new pots, and disease development was evaluated at 21 days post inoculation (dpi) as described earlier [24]. Fungal biomass quantification in infected Arabidopsis plants was performed with

real-time quantitative PCR (qPCR) as described previously [49]. Briefly, qPCR was conducted on total DNA isolated from *V. dahliae* infected Arabidopsis with primers amplifying *Verticillium* internal transcribed spacer (ITS; ITS1-F and STVe1-R) and the primers amplifying the Arabidopsis RuBisCo gene as endogenous control (AtRub-F3 and AtRub-R3). The qPCR was conducted using an ABI7300 PCR machine (Applied Biosystems, Foster City, California) in combination with the SensiMix SYBR Hi-ROX Kit (Bioline, London, UK). Real-time PCR conditions were as follows: an initial 95°C hot start activation step for 10 min was followed by denaturation for 15 sec at 95°C, annealing and extension for 60 sec at 60°C for 40 cycles.

Supporting Information

Figure S1 Stability of Ve1 mutants that showed compromised HR-inducing capacity. GFP-tagged Ve1 mutants were detected by immunoblotting using GFP antibody (α-GFP). Coomassie-stained blots (CBS) showing the 50 kDa Rubisco band present in the input samples confirm equal loading.

Figure S2 Typical appearance of non-transgenic Arabidopsis (WT) and transgenic Arabidopsis expressing Ve1 mutants, upon mock-inoculation or inoculation with race 1 *V. dahliae*. Pictures were taken at 21 days post inoculation and are representative of three independent experiments.

Figure S3 Typical appearance of non-transgenic Arabidopsis (WT) and transgenic Arabidopsis producing Ve1 mutants in the putative GxxxG motif and the E/DxxxLφ endocytosis motifs, upon mock-inoculation or inoculation with *V. dahliae* race 1. Pictures were taken at 21 days post infiltration and are representative of three independent experiments.

Author Contributions

Conceived and designed the experiments: ZZ YS CML BPHJT. Performed the experiments: ZZ YS. Analyzed the data: ZZ YS BPHJT. Contributed reagents/materials/analysis tools: YS. Contributed to the writing of the manuscript: ZZ BPHJT.

References

1. Boller T, Felix G (2009) A renaissance of elicitors: perception of microbe-associated molecular patterns and danger signals by pattern-recognition receptors. Annu Rev Plant Biol 60: 379–406.
2. Thomma BP, Nürnberger T, Joosten MH (2011) Of PAMPs and effectors: the blurred PTI-ETI dichotomy. Plant Cell 23: 4–15.
3. Jones JD, Dangl JL (2006) The plant immune system. Nature 444: 323–329.
4. Wu Y, Zhou JM (2013) Receptor-like kinases in plant innate immunity. J Integr Plant Biol 55: 1271–1286.
5. Hammond-Kosack KE, Jones JD (1996) Resistance gene-dependent plant defense responses. Plant Cell 8: 1773–1791.
6. Fradin EF, Thomma BP (2006) Physiology and molecular aspects of *Verticillium* wilt diseases caused by *V. dahliae* and *V. albo-atrum*. Mol Plant Pathol 7: 71–86.
7. Klosterman SJ, Atallah ZK, Vallad GE, Subbarao KV (2009) Diversity, pathogenicity, and management of *Verticillium* species. Annu Rev Phytopathol 47: 39–62.
8. de Jonge R, van Esse P, Maruthachalam K, Bolton MD, Santhanam P, et al. (2012) Tomato immune receptor Ve1 recognizes effector of multiple fungal pathogens uncovered by genome and RNA sequencing. Proc Natl Acad Sci U S A 109: 5110–5115.
9. Fradin EF, Zhang Z, Juarez Ayala JC, Castroverde CD, Nazar RN, et al. (2009) Genetic dissection of *Verticillium* wilt resistance mediated by tomato Ve1. Plant Physiol 150: 320–332.
10. Kawchuk LM, Hachey J, Lynch DR, Kulcsar F, van Rooijen G, et al. (2001) Tomato *Ve* disease resistance genes encode cell surface-like receptors. Proc Natl Acad Sci U S A 98: 6511–6515.
11. Zhang Z, Thomma BP (2013) Structure-function aspects of extracellular leucine-rich repeat-containing cell surface receptors in plants. J Integr Plant Biol 55: 1212–1223.
12. Wang G, Fiers M, Ellendorff U, Wang ZZ, de Wit PJGM, et al. (2010) The diverse roles of extracellular leucine-rich repeat-containing receptor-like proteins in plants. Crit Rev Plant Sci 29: 285–299.
13. Fradin EF, Zhang Z, Rovenich H, Song Y, Liebrand TW, et al. (2014) Functional analysis of the tomato immune receptor Ve1 through domain swaps with its non-functional homolog Ve2. PLoS One 9: e88208.
14. Liebrand TW, van den Berg GC, Zhang Z, Smit P, Cordewener JH, et al. (2013) Receptor-like kinase SOBIR1/EVR interacts with receptor-like proteins in plant immunity against fungal infection. Proc Natl Acad Sci U S A 110: 10010–10015.
15. Kobe B, Kajava AV (2001) The leucine-rich repeat as a protein recognition motif. Curr Opin Struct Biol 11: 725–732.
16. Chang JH, McCluskey PJ, Wakefield D (2012) Recent advances in Toll-like receptors and anterior uveitis. Clin Experiment Ophthalmol 40: 821–828.
17. Di Matteo A, Federici L, Mattei B, Salvi G, Johnson KA, et al. (2003) The crystal structure of polygalacturonase-inhibiting protein (PGIP), a leucine-rich

repeat protein involved in plant defense. Proc Natl Acad Sci U S A 100: 10124–10128.

18. Hothorn M, Belkhadir Y, Dreux M, Dabi T, Noel JP, et al. (2011) Structural basis of steroid hormone perception by the receptor kinase BRI1. Nature 474: 467–471.

19. She J, Han Z, Kim TW, Wang J, Cheng W, et al. (2011) Structural insight into brassinosteroid perception by BRI1. Nature 474: 472–476.

20. Jiang J, Zhang C, Wang X (2013) Ligand perception, activation, and early signaling of plant steroid receptor brassinosteroid insensitive 1. J Integr Plant Biol 55: 1198–1211.

21. Sun Y, Li L, Macho AP, Han Z, Hu Z, et al. (2013) Structural basis for flg22-induced activation of the Arabidopsis FLS2-BAK1 immune complex. Science 342: 624–628.

22. Bleckmann A, Weidtkamp-Peters S, Seidel CA, Simon R (2010) Stem cell signaling in Arabidopsis requires CRN to localize CLV2 to the plasma membrane. Plant Physiol 152: 166–176.

23. van der Hoorn RA, Wulff BB, Rivas S, Durrant MC, van der Ploeg A, et al. (2005) Structure-function analysis of Cf-9, a receptor-like protein with extracytoplasmic leucine-rich repeats. Plant Cell 17: 1000–1015.

24. Fradin EF, Abd-El-Haliem A, Masini L, van den Berg GC, Joosten MH, et al. (2011) Interfamily transfer of tomato Ve1 mediates Verticillium resistance in Arabidopsis. Plant Physiol 156: 2255–2265.

25. Wang G, Long Y, Thomma BP, de Wit PJ, Angenent GC, et al. (2010) Functional analyses of the CLAVATA2-like proteins and their domains that contribute to CLAVATA2 specificity. Plant Physiol 152: 320–331.

26. Fritz-Laylin LK, Krishnamurthy N, Tor M, Sjolander KV, Jones JDG (2005) Phylogenomic analysis of the receptor-like proteins of rice and Arabidopsis. Plant Physiol 138: 611–623.

27. Senes A, Gerstein M, Engelman DM (2000) Statistical analysis of amino acid patterns in transmembrane helices: The GxxxG motif occurs frequently and in association with beta-branched residues at neighboring positions. J Mol Biol 296: 921–936.

28. Curran AR, Engelman DM (2003) Sequence motifs, polar interactions and conformational changes in helical membrane proteins. Curr Opin Struct Biol 13: 412–417.

29. Wulff BB, Thomas CM, Parniske M, Jones JD (2004) Genetic variation at the tomato Cf-4/Cf-9 locus induced by EMS mutagenesis and intralocus recombination. Genetics 167: 459–470.

30. Zhang YX, Yang YA, Fang B, Gannon P, Ding PT, et al. (2010) Arabidopsis snc2-1D activates receptor-like protein-mediated immunity transduced through WRKY70. Plant Cell 22: 3153–3163.

31. Murphy AS, Bandyopadhyay A, Holstein SE, Peer WA (2005) Endocytic cycling of PM proteins. Annu Rev Plant Biol 56: 221–251.

32. Geldner N, Robatzek S (2008) Plant receptors go endosomal: A moving view on signal transduction. Plant Physiol 147: 1565–1574.

33. Sun W, Cao Y, Jansen Labby K, Bittel P, Boller T, et al. (2012) Probing the Arabidopsis flagellin receptor: FLS2-FLS2 association and the contributions of specific domains to signaling function. Plant Cell 24: 1096–1113.

34. Leconte I, Carpentier JL, Clauser E (1994) The functions of the human insulin receptor are affected in different ways by mutation of each of the four N-glycosylation sites in the beta subunit. J Biol Chem 269: 18062–18071.

35. Gahring L, Carlson NG, Meyer EL, Rogers SW (2001) Granzyme B proteolysis of a neuronal glutamate receptor generates an autoantigen and is modulated by glycosylation. J Immunol 166: 1433–1438.

36. Dunning FM, Sun W, Jansen KL, Helft L, Bent AF (2007) Identification and mutational analysis of Arabidopsis FLS2 leucine-rich repeat domain residues that contribute to flagellin perception. Plant Cell 19: 3297–3313.

37. van der Hoorn RA, Roth R, De Wit PJ (2001) Identification of distinct specificity determinants in resistance protein Cf-4 allows construction of a Cf-9 mutant that confers recognition of avirulence protein Avr4. Plant Cell 13: 273–285.

38. Wulff BBH, Heese A, Tomlinson-Buhot L, Jones DA, de la Pena M, et al. (2009) The major specificity-determining amino acids of the tomato Cf-9 disease resistance protein are at hypervariable solvent-exposed positions in the central leucine-rich repeats. Mol Plant Microbe Interact 22: 1203–1213.

39. Shinohara H, Moriyama Y, Ohyama K, Matsubayashi Y (2012) Biochemical mapping of a ligand-binding domain within Arabidopsis BAM1 reveals diversified ligand recognition mechanisms of plant LRR-RKs. Plant J 70: 845–854.

40. Kim HM, Park BS, Kim JI, Kim SE, Lee J, et al. (2007) Crystal structure of the TLR4-MD-2 complex with bound endotoxin antagonist Eritoran. Cell 130: 906–917.

41. Jaillais Y, Belkhadir Y, Balsemao-Pires E, Dangl JL, Chory J (2011) Extracellular leucine-rich repeats as a platform for receptor/coreceptor complex formation. Proc Natl Acad Sci U S A 108: 8503–8507.

42. Li J (2011) Direct involvement of leucine-rich repeats in assembling ligand-triggered receptor-coreceptor complexes. Proc Natl Acad Sci U S A 108: 8073–8074.

43. He K, Xu S, Li J (2013) BAK1 directly regulates brassinosteroid perception and BRI1 activation. J Integr Plant Biol 55: 1264–1270.

44. Clough SJ, Bent AF (1998) Floral dip: a simplified method for Agrobacterium-mediated transformation of Arabidopsis thaliana. Plant J 16: 735–743.

45. van der Hoorn RAL, Laurent F, Roth R, de Wit PJGM (2000) Agroinfiltration is a versatile tool that facilitates comparative analyses of Avr9/Cf-9-induced and Avr4/Cf-4-induced necrosis. Mol Plant Microbe Interact 13: 439–446.

46. Zhang Z, Fradin E, de Jonge R, van Esse HP, Smit P, et al. (2013) Optimized agroinfiltration and virus-induced gene silencing to study Ve1-mediated Verticillium resistance in tobacco. Mol Plant Microbe Interact 26: 182–190.

47. Zhang Z, Thomma B (2014) Virus-Induced gene silencing and Agrobacterium tumefaciens-mediated transient expression in Nicotiana tabacum. Methods Mol Biol 1127: 173–181.

48. Liebrand TW, Smit P, Abd-El-Haliem A, de Jonge R, Cordewener JH, et al. (2012) Endoplasmic reticulum-quality control chaperones facilitate the biogenesis of Cf receptor-like proteins involved in pathogen resistance of tomato. Plant Physiol 159: 1819–1833.

49. Ellendorff U, Fradin EF, de Jonge R, Thomma BP HJ (2009) RNA silencing is required for Arabidopsis defence against Verticillium wilt disease. J Exp Bot. 60: 591–602.

15

Gene-Splitting Technology: A Novel Approach for the Containment of Transgene Flow in *Nicotiana tabacum*

Xu-Jing Wang[1♦]**, Xi Jin**[1♦]**, Bao-Qing Dun**[2♦]**, Ning Kong**[1]**, Shi-Rong Jia**[1]**, Qiao-Ling Tang**[1]**, Zhi-Xing Wang**[1]*

1 Biotechnology Research Institute, Chinese Academy of Agricultural Sciences, Beijing, China, **2** National Key Facility for Crop Gene Resources and Genetic Improvement, Institute of Crop Sciences, Chinese Academy of Agricultural Sciences, Beijing, China

Abstract

The potential impact of transgene escape on the environment and food safety is a major concern to the scientists and public. This work aimed to assess the effect of intein-mediated gene splitting on containment of transgene flow. Two fusion genes, *EPSPSn-In* and *Ic-EPSPSc*, were constructed and integrated into *N. tabacum*, using *Agrobacterium tumefaciens*-mediated transformation. *EPSPSn-In* encodes the first 295 aa of the herbicide resistance gene 5-enolpyruvyl shikimate-3-phosphate synthase (EPSPS) fused with the first 123 aa of the Ssp DnaE intein (In), whereas *Ic-EPSPSc* encodes the 36 C-terminal aa of the Ssp DnaE intein (Ic) fused to the rest of EPSPS C terminus peptide sequences. Both *EPSPSn-In* and *Ic-EPSPSc* constructs were introduced into the same *N. tabacum* genome by genetic crossing. Hybrids displayed resistance to the herbicide N-(phosphonomethyl)-glycine (glyphosate). Western blot analysis of protein extracts from hybrid plants identified full-length EPSPS. Furthermore, all hybrid seeds germinated and grew normally on glyphosate selective medium. The 6-8 leaf hybrid plants showed tolerance of 2000 ppm glyphosate in field spraying. These results indicated that functional EPSPS protein was reassembled *in vivo* by intein-mediated trans-splicing in 100% of plants. In order to evaluate the effect of the gene splitting technique for containment of transgene flow, backcrossing experiments were carried out between hybrids, in which the foreign genes *EPSPSn-In* and *Ic-EPSPSc* were inserted into different chromosomes, and non-transgenic plants NC89. Among the 2812 backcrossing progeny, about 25% (664 plantlets) displayed glyphosate resistance. These data indicated that transgene flow could be reduced by 75%. Overall, our findings provide a new and highly effective approach for biological containment of transgene flow.

Editor: Mario Soberón, Instituto de Biotecnología, Universidad Nacional Autónoma de México, Mexico

Funding: This work was supported by the Major Project of China on New Varieties of GMO Cultivation (2013ZX08010-003) and the National Natural Science Foundation of China (31100408). The funders had no role in study design, data collection and analysis, decision to publish, or preparation of the manuscript.

Competing Interests: The authors have declared that no competing interests exist.

* E-mail: wangcotton@126.com

♦ These authors contributed equally to this work.

Introduction

Along with the rapid development and commercialization of genetic modified crops worldwide [1], the potential impact of transgene flow mediated by pollen dispersal on the environment and food safety has become a major concern to the scientific community as well as the public. To date, spatial or temporal isolation is generally applied to control transgene flow, but these methods generally show limited efficacy. To further minimize and hopefully eliminate transgene flow, a series of biological containment strategies have been developed, including transgene excision, chloroplast transformation, cytoplasmic male sterility and restorer genes, cleistogamy, etc. [2–11]. Although a steady progress has been achieved in recent decades, Hüsken et al. (2010) concluded that no single containment strategy would result in 100% reduction of gene flow, suggesting that combinations of complementary containment systems are required [12].

Generally, transgenic plants express full length genes encoding the active proteins of interest. Alternatively, target traits may be established by engineering functional proteins that result from reassembly of separately expressed inactive precursor peptides. The method, referred to as gene splitting, can be useful in controlling transgene flow.

The discovery of split inteins provides a useful tool for gene splitting studies. Inteins, referred to as "protein introns", are internal protein elements that undergo self-splicing resulting in the ligation of flanking sequences (exteins) through a peptide bond to form a new mature protein [13]. Inteins were first discovered by two research groups in the yeast *Saccharomyces cerevisiae* [14,15]. To date, more than 600 inteins have been described in organisms from all three domains of life [16]. Split inteins are capable of protein *trans*-splicing. An inactive target protein N-fragment (N-extein) fused to the N-terminal intein fragment (N-intein) and another inactive protein encompassing the C-terminal intein fragment (C-intein) fused to the target protein C-fragment (C-extein) could reassemble into a functional mature target protein, through intein mediated *trans*-splicing. *Ssp* DnaE, the first described split intein, was identified in the *Synechocystis* sp. strain PCC6803 [17]. *Ssp* DnaE is able to cyclize and trans-splice proteins in plants, such as tobacco [18], Arabidopsis [19] and wheat [20].

The G_2-*aroA* gene (GenBank accession No.: EF155478) was identified from the G_2 strain of *Pseudomonas fluorescens* isolated from glyphosate polluted area. This gene encodes the 445 aa EPSPS protein, which confers glyphosate resistance. The transgenic

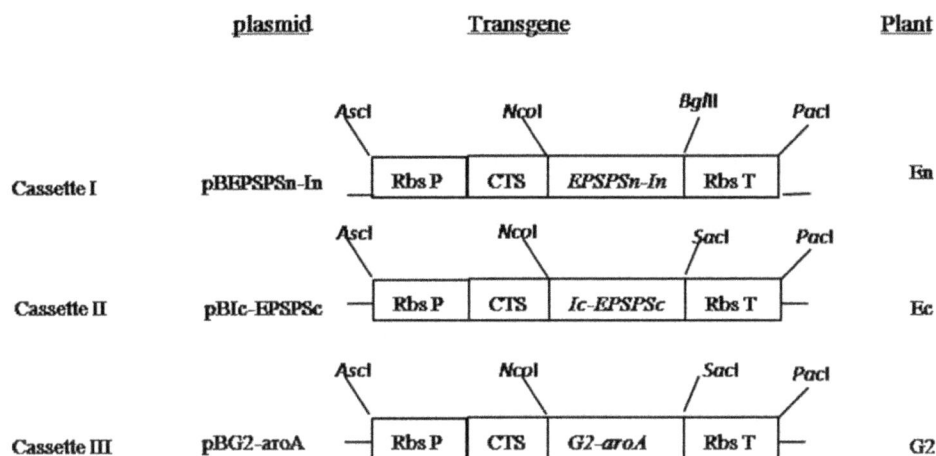

Figure 1. Summary of transgene constructs used for tobacco transformation. Construct names, gene expression cassettes, and names of transgenic plants are presented. Rbs P represents the promoter of *chrysanthemum* Rubisco small unit gene (*RbcS*); CTS is the chloroplast signal peptide containing a natural intron from the *RbcS* gene of *chrysanthemum*; and Rbs T represents the terminator of *chrysanthemum RbcS*.

tobacco containing *G2-aroA* gene can tolerate 1.6% (the working concentration for control annual weed) isopropylamine salt of glyphosate (Roundup, 41.0% (W/V)) [21]. Dun et al. (2007) identified a suitable splitting site in *G2-aroA* named F295/T296. Using *Ssp* DnaE trans-splicing strategy, reassembly of the full-length and functional EPSPS protein in tobacco and *E. coli* was achieved [22]. In this study, we used *G2-aroA* as target gene and tobacco as model plant to assess the effect of gene splitting on transgene flow control. We confirmed the reconstitution of the functional EPSPS protein by *Ssp* DnaE intein mediated trans-splicing when the two gene split fragments were introduced into the same tobacco plant genome by genetic crossing. In addition, we demonstrated that successful reassembly was achieved in 100% cross hybrid plants. Furthermore, we found that gene splitting reduced the transgene flow by more than 75%. This is the first report on reassembly efficiency and effectiveness of transgene flow containment by gene splitting.

Materials and Methods

Genes and Germplasm

The *G₂-aroA* gene encoding the glyphosate-resistant EPSPS protein was obtained from Lin lab of Biotechnology Research Institute of Chinese Academy of Agricultural Science. Pint-n (In) and Pint-c (Ic), the oligonucleotide sequences encoding the N-terminal (123 aa; Genebank accession no. AF545504) and C-terminal (36 aa; Genebank accession no. AF545505) domains of the *Ssp* DnaE intein, respectively, were kindly provided by Professor Thomas C. Evans, Jr. The vectors pImpactvector1.4 and pBinPLUS were purchased from Plant Research International (Netherlands). *E. coli* strains and other reagents were purchased from Takara company. Tobacco (*Nicotiana tabacum* NC89) seeds were conserved by our lab.

Construction of expression vectors

DNAs encoding *EPSPS* segments, *Ic* and *In* were amplified with PCR using the following primers (restriction enzyme sites of *Nco*I (CCATGG), *Bgl*II (AGATCT) and *Sac*I (GAGCTC) are underlined):

EPSPSn   forward: 5′-GGCCATGGATGGCGTGTTTG-CCTGATGA-3′
reverse: 5′-GAAGTCCTGCGCGGCTACGC-3′

In   forward: 5′-GCGCAGGACTTCAAATTTGCT-GAATATTGCCT-3′
reverse: 5′-GGCAGATCTTTATTTAATTGT-CCCAGCGTCAAG-3′
Ic   forward: 5′-GGCCATGGATGGTTAAAGTTAT-CGGTCG-3′
reverse: 5′-GGATATGTTAAAGCAGTTAG-3′
EPSPSc   forward: 5′-TTTAACATATCC-ACCCAGCCC-GACGCCAAGGC-3′
reverse: 5′-CCGGAGCTCTCAGTCGTTTAG-GTGAACGCCCAGG-3′

EPSPSn-In was then amplified with nested PCR using previous PCR products for *EPSPSn* and *In* as substrates, and EPSPSn forward and In reverse primers. The fusion gene *EPSPSn-In* was inserted into pImpactvector1.4 to generate the intermediate vector pIV1.4EnIn. The *EPSPSn-In* expression cassette was subcloned into the plant expression vector pBinPLUS to yield pBEPSPSn-In. Similar techniques were employed to construct the plant expression vectors pBIc-EPSPSc and pBG2-aroA.

Transformation of tobacco

The three final plant expression vectors pBIc-EPSPSc, pBG2-aroA and pBEPSPSn-In were mobilized into *Agrobacterium tumefaciens* strain LBA4404 by the freeze-thaw method. Transformed bacteria were grown on YEB medium containing 100 mg/L kanamycin at 28°C and 150–250 rpm overnight. Cultures were diluted 1:1 with YEB and allowed to grow to absorbance (measured at 550 nm) of ≈0.8. NC89 tobacco leaf discs from approximately 4-week-old shoot cultures were used for transformation with *A. tumefaciens*. After infection, leaf discs were incubated on a co-cultivation medium (1×MS salts, 3% sucrose, 2 mg/L 6-benzylaminopurine and 0.1 mg/L α-naphthalene acetic acid) at 28°C in the dark for 3–4 days and then selected on co-cultivation medium containing 500 mg/L cephalosporin and 100 mg/L kanamycin. The selected transgenic plantlets were then grown on media containing 1×MS salts, 3% sucrose, 100 mg/L kanamycin and 500 mg/L cephalosporin.

Transgene insertion number analysis

T1 seeds of transgenic plants were germinated on a selective medium containing MS salt, 3% sucrose and 100 mg/L kanamycin. Pale and moribund seedlings were defined as kanamycin - susceptible (Kan^S) plants, while healthy and green seedlings were considered kanamycin- resistant (Kan^R). Numbers of Kan^R and Kan^S seedlings for each transformation event were analyzed by the χ-squared test to identify plants with a single copy insertion. Furthermore, real-time quantitative PCR was used to assess copy number of the inserted *nptII* gene in transgenic plants. The primer pairs Ef (CTATCAGGACATAGCGTTGG)/Er (GCTCAGAA-GAACTCGTCAAG) and Rf (GACGAAGCTTACTGAG-GAAC)/Rr (CCAACAATCTATCAGCCACG) were designed according to gene sequences of *nptII* and *mr2*, which encode neomycin phosphotransferase (the most frequently used marker in plant transformation experiments) and ribonucleotide reductase (endogenous reference), respectively. Real-time PCR was carried out individually with genomic DNA from single transgenic plants as templates on an AB 7500 Real Time PCR System (Applied Biosystems, USA) with the following reaction conditions: 30 sec at 95°C, followed by for 45 cycles of 5 sec at 95°C, 34 sec at 52.8°C, and 40 sec at 70°C. The initial *nptII* and *mr2* template copy numbers were derived from CT values, and the inserted gene's copy number was estimated by the ratio of initial template copy number of *nptII* to that of *mr2*.

Selection of homozygous transgenic tobacco

Ten T1 seedlings were grown in soil, and T2 seeds were collected from individual plants and germinated on selective medium. The homozygous transgenic plants were recognized by healthy seedlings after three successive selections.

Genetic crossing of tobacco plants

For cross-fertilization, pollen was collected from fully opened flower of homozygous male parent plants and dusted onto the stigma of homozygous female parent plants prepared from unopened buds.

Western blot analysis

Soluble proteins were extracted from transgenic plant leaves using a Plant Protein Purification kit (Beijing CoWin Biotech Co., Ltd., China) according to manufacturer's instructions. Western blot detection of EPSPS peptides was carried following standard procedure with polyclonal antibodies raised in mice against EPSPSn-In and intact EPSPS (kindly provided by Lin's lab, Biotechnology Research Institute of Chinese Academy of Agricultural Science) at 1000 and 10000 dilutions, respectively.

Figure 2. Growth of homozygous transgenic plant lines homEn-33 (A) and homEc-11 (B). T3 seeds from transformed plants were germinated on selective media containing MS salt, 3% sucrose and 100 mg/L kanamycin. Pale and moribund seedlings represented kanamycin-susceptible (Kan^S) plants, while healthy and green seedlings were kanamycin- resistant (Kan^R). All homEn-33-5 and homEc-11 seedlings grew healthy and displayed kanamycin resistance.

Evaluation of protein splicing reassembly efficiency

Hybrid seeds from genetic crossing plants were sterilized and inoculated onto medium containing MS salt, 3% sucrose and 33.8 mg/L glyphosate for germination. Seedlings with green leaves were considered glyphosate-resistant plant (gly^R) and otherwise defined as glyphosate-susceptible (gly^S). Seedlings were evaluated by leaf color (green or yellow) after 20 days of culture. Resistant and susceptible seedling amounts were analyzed by the χ-squared test to assess the efficiency of protein splicing reassembly. In the leaf spraying experiment, 6 to 8-leaf-stage transgenic plants grown in the greenhouse were sprayed with 41.0% Roundup (isopropylamine salt of glyphosate as active ingredient) at indicated concentrations. The survival of the plants was evaluated after one week.

Analysis of transgene insertion site

Tail-PCR was carried out to analyze the flanking sequences at the insertion sites of transgenic tobacco plants. A Genome Walking Kit (TaKaRa, Japan) was used to amplify the flanking sequences at target gene insertion sites. The specific primers F-1 (5'-GGACAGGTCGGTCTTGACAAAAAGAACCGG-3'), F-2 (5'-GTGCCCAGTCATAGCCGAATAGCCTCTCC-3'), F-3 (5'-CCTGCGTGCAATCCATCTTGTTCAATCATGCG-3'), and F-4 (5'- CGAGATAGGGTTGAGTGTTGTTCCAG -3') were designed and synthesized based on the *nptII* gene sequence. Subsequently, three nested PCRs were carried out using primer pairs containing a specific primer and compound annex primers (AP1, AP2, AP3 and AP4) provided with the kit, with the genomic DNA as template. The obtained sequences were analyzed by comparison with the GenBank and tobacco genome databases.

Evaluation of the effect of gene splitting on transgene flow control

Hybrids harboring both *EPSPSn-In* and *Ic-EPSPSc* were selfed and artificially back-crossed to the non-transgenic tobacco line NC89. The resulting seeds were sterilized and inoculated onto media containing MS salt, 3% sucrose and 33.8 mg/L glyphosate to assess glyphosate resistance of the hybrid progeny. The gly^R/gly^S segregation ratio was calculated, and statistical analyses were carried out using the chi squared test.

Results

Fusion between EPSPS and Ssp DnaE intein

According to the suitable and effective splitting site (F295/T296) reported by Dun et al. [22], divided *G$_2$-aroA* gene segments were fused to *Ssp DnaE* segments. The resulting constructs encoded two fusion proteins, EPSPSn-In and Ic-EPSPSc. EPSPSn-In is an in-frame fusion between the first 295 amino acid residues of the EPSPS protein and the 123 amino acid residues of *Ssp* DnaE intein-N. Likewise, Ic-EPSPSc is an in-frame fusion between the 36 amino acid moieties of Ssp DnaE intein-C and the remaining 150 EPSPS amino acid residues.

The fused genes were cloned into ImpactVector1.4 vector. The vector contains a Rubisco small subunit promoter from which the target genes were transcribed, and a Rubisco small subunit terminator (RbcS1 T) from *Asteraceous chrysanthemum* at the 3' end[23]. The vector also contains a signal peptide of the first 11 amino acids from *Chrysanthemum morifolium* Rubisco small subunit protein fused at the N terminus to deliver target proteins into chloroplast stroma[23]. The expression cassettes were then subcloned into T-regions of plant expression vector pBinPLUS. This resulted in three expression plasmids: pBEPSPSn-In, pBIc-EPSPSc and pBG2-aroA. The T-regions in all plasmids harbored

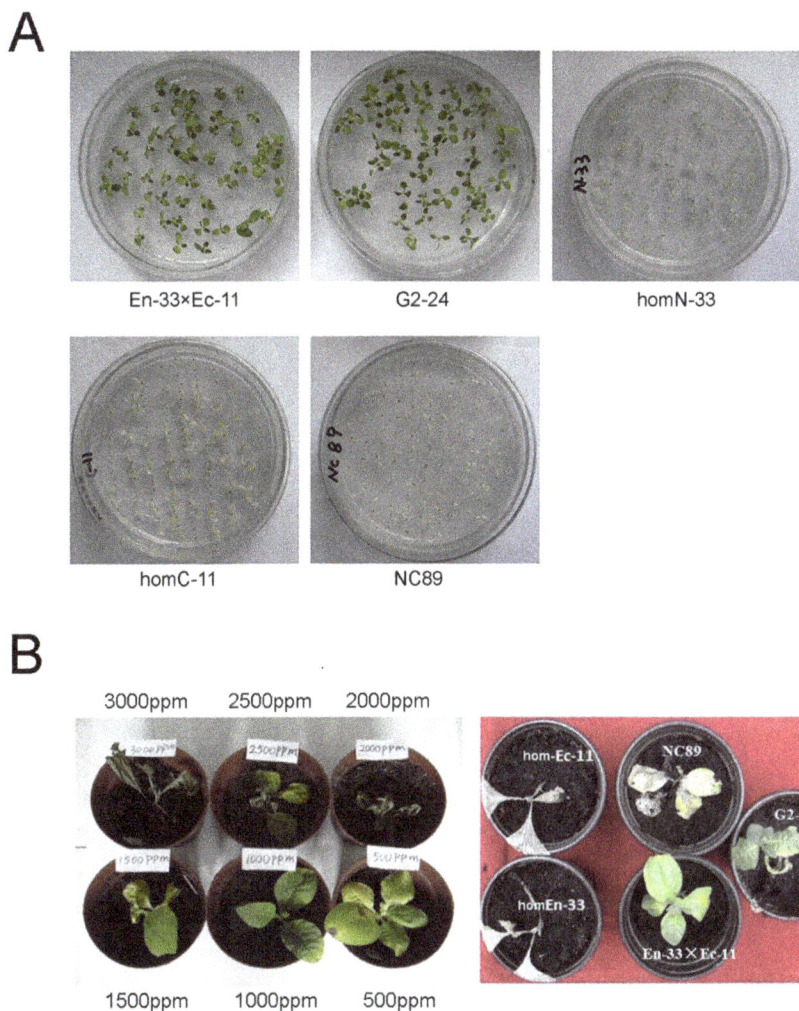

Figure 3. Analysis of glyphosate resistance in the transgenic tobacco plants. A. Growth of different transgenic tobacco lines on glyphosate selective medium. Seeds were sterilized and inoculated onto medium containing MS salt, 3% sucrose and 33.8 mg/L glyphosate for germination. The seedlings with green leaves represented glyphosate-resistant plants and otherwise defined as glyphosate-susceptible. B. Grown plants (6 to 8-leaf stage) were tested for glyphosate resistance. Left: NC89 plants one week after spraying different concentrations of 41% Roundup herbicide. Right: transgenic tobacco plants one week after spraying 2500 ppm Roundup.

nosP::nptII::nosT of kanamycin resistance marker. The structures of the three plasmids are summarized in Fig. 1.

In vivo reassembly efficiency of the split EPSPS through intein mediated trans-splicing

Infection with pBEPSPSn-In, pBIc-EPSPSc or pBG2-aroA plasmids resulted in three sets of transformed plant lines, labeled as En, Ec and G2, respectively. Totally 35 of En, 42 of Ec and 60 of G2 transgenic tobacco lines were obtained by agrobacterium mediated transformation. T1 seeds of transgenic tobacco were germinated on 100 mg/L kanamycin selective medium. Analysis of segregation ratio showed that the seeds from plant lines En-33, Ec-11 and G2-24 exhibited Kan^R: Kan^S seedling at 3:1 ratio, as would be predicted from single copy transgenic insertions.

In order to confirm the insertion number of transformed genes, real-time PCR was used to detect the copy numbers of the exogenous *nptII* gene in En-33, Ec-11 and G2-24. The standard curve of the reference gene *mr2* was CT = 33.953–3.069log(cn), with r^2 and amplification efficiency of 0.998 and 111.752%, respectively, where CT is the cycle threshold and cn is the copy number. The standard curve of *nptII* was CT = 35.236–3.313log(cn), with r^2 and amplification efficiency of 0.999 and 100.382%, respectively. The *mr2* and *nptII* copy number (cn) values were derived from the standard curves and CT values obtained in real-time PCR. The cn ratios of *mr2* to *nptII* for En-33, Ec-11 and G2-24 were 1.06, 0.95, and 1.34, respectively, indicating a single copy insertion of *nptII* in the transgenic tobacco lines En-33, Ec-11 and G2-24.

T0 plantlets of En-33 and Ec-11 were planted in a greenhouse. The seeds were collected at harvest and germinated on MS0 medium containing 100 mg/L kanamycin. The kanamycin resistant aseptic seedlings were then again transferred to greenhouse culture. After three rounds of selfing, homozygous lines of En-33 and Ec-11 were obtained and named homEN-33 and homEC-11, respectively. All seeds of homEn-33 and homEc-11 seedlings grew healthy on kanamycin selective medium (Fig. 2). For genetic crossing, homEn-33 was used as pollen donor, whereas homEc-11 was pollen recipient. The resulting hybrids were designed En-33×Ec-11.

A

66KD →
45KD →

Antiserum 1:2000

B

66KD →
45KD →
35KD →

Antiserum, 1:2000
Sample amount of 10μg in No. 2
2μg in others

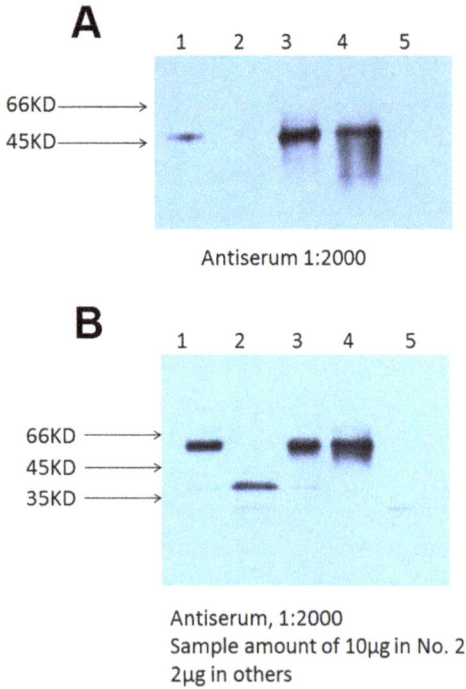

Figure 4. Western blot analysis of EPSPS peptides. Proteins from different plants were blotted with antibodies against EPSPSn-In (A) or full length EPSPS (B).1: En-33; 2: Ec-11; 3: G2-24; 4: En-33×Ec-11; 5: NC89.

To assess the efficiency of protein reassembly through intein mediated trans-splicing, glyphosate resistance of hybrid En-33×Ec-11 was analyzed by seed germination on selective medium containing 33.8 mg/L glyphosate. As controls, seeds of hybridization parent homEn-33, homEc-11 as well as G2-24 and non-transgenic NC89 were germinated on selective and nonselective

media. All En-33×Ec-11 hybrid seedlings grew normally on glyphosate selective medium and displayed similar glyphosate resistance phenotype compared to G2-24, which contained the full length of the glyphosate resistance gene (Fig. 3A). In addition, glyphosate resistance was tested by leaf spraying experiment in 6 to 8-leaf-stage transgenic plants. Plants grown in the greenhouse were sprayed with the 41.0% Roundup (isopropylamine salt of glyphosate as active ingredient) at doses of 2500 ppm. Within one week, the homEn-33, homEc-11 and NC89 plants wilted and turned yellow, gradually dying, whereas the hybrid En-33×Ec-11 and G2-24 transgenic tobacco plants grew normally (Fig. 3B). These results indicated the successful reassembly of a functional EPSPS protein from EPSPSn-In and Ic-EPSPSc by intein mediated protein trans-splicing. These findings were further confirmed by Western blot assays. As shown in Fig. 4, in the hybrid En-33×Ec-11 plants, accumulation of the reassembled full length EPSPS protein with slight larger size than EPSPSn-In was observed. Meanwhile no smaller Ic-EPSPSc was detected in hybrid plants, indicating a highly efficient trans-splicing induced reassembly.

Gene splitting and control of transgene flow

The analysis of En-33 and Ec-11 flanking sequences by thermal asymmetric interlaced–PCR showed that the two transgenic inserts were located in different chromosomes. According to Mendel's laws of inheritance, without possibility of chromosome crossing-over, 25% pollen possess A and B genes at the same time if A and B were located at different chromosomes in a given plant (Fig. 5A). Upon the backcrossing between En-33×Ec-11 hybrid (as pollen donor) and non-transgenic tobacco NC89 (as pollen recipient), 664 of 2812 back-crossed progeny plantlets displayed glyphosate-resistance, a gly^R ratio of 23.61%. While self-crossed, 2328 of 4158 progeny seeds of En-33×Ec-11 germinated on glyphosate selective medium were glyphosate-resistant, a ratio of 55.99%. As estimated by the χ-squared test, both data fit the hypothesis that *EPSPSn-In* and *Ic-EPSPSc* were inserted into different chromosomes according to Mendel's laws of inheritance.

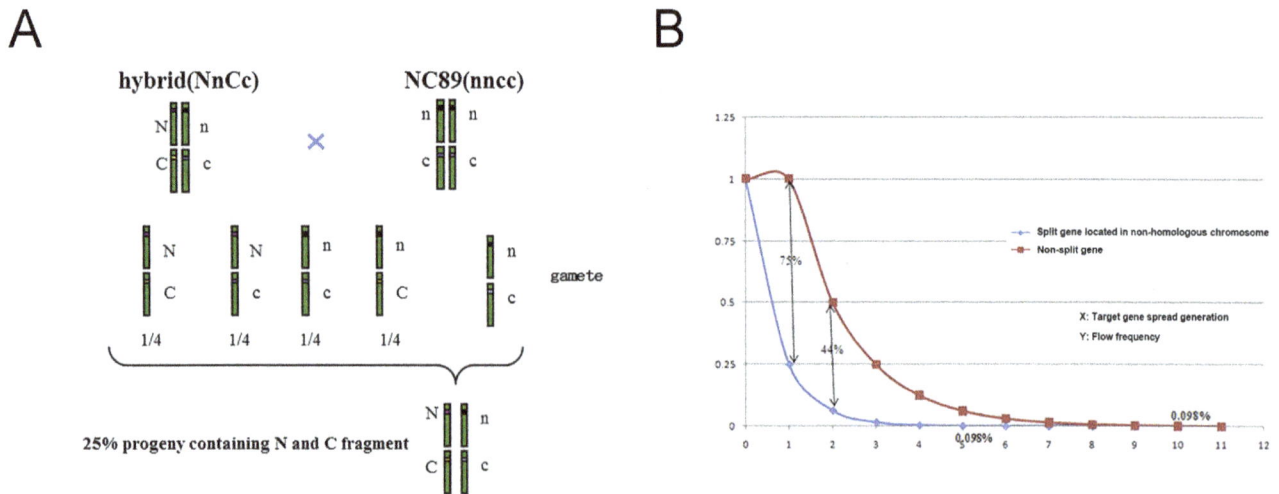

A

hybrid(NnCc)   NC89(nncc)

×

gamete

1/4   1/4   1/4   1/4

25% progeny containing N and C fragment

B

X: Target gene spread generation
Y: Flow frequency

Figure 5. Gene flow frequency with gene splitting strategy. A. Prediction of glyphosate resistant plants percentage in the En-33×Ec-11 backcrossing progeny. N represents the EPSPSn-In fragment; C stands for the Ic-EPSPSc fragment. According to the Mendel's laws of inheritance, if the two genes were inserted into different chromosomes, 25% backcrossing progeny would contain both genes and display glyphosate resistance. B. Percentages of progeny plants resulting from backcrossing with gene splitting or full length transgenic strategy were compared. At F1 generation, 25% of hybrid backcrossing progeny would display the target character, 75% lower than backcrossing progeny plants with full length transgenic and wild type plants. After 5 generations, the ratio of backcrossing transgenic plants using gene splitting strategy would display the target character in less than 0.1% population.

After five generations, lower than 0.1% progeny plants resulting from backcrossing between hybrid and wild type would be expected to contain both splitted gene fragments for the reassembly of full length functional protein (Fig. 5B). The same threshold decrease would be expected after 10 generations if the transgene was carried with full length target gene. Therefore gene splitting technique would significantly reduce transgene flow. At F1 generation, the reduction would be expected to be at least 75%.

Discussion

Transgene flow continues to pose a threat on environment and food safety, and has therefore become a major concern with increasing production of genetically modified organisms [27-29]. Intein mediated protein trans-splicing may limit the environmental impact of a foreign gene by keeping different parts in different chromosomes while assembling gene products in one cell through crossing to achieve the desired function [18,24,25].

Ssp DnaE intein has been used by others to cyclize and trans-splice proteins in various plants, such as tobacco, Arabidopsis and wheat [18–20]. It was recently proposed that gene flow from such transgenic plants to wild or weedy relatives would transmit only a portion of the full-length gene, which should imply reduced environmental impact [18,24,25]. However, to optimally apply the gene splitting technique for containment of transgene flow, it is critical to ensure reassembly efficiency. Few reports on detection and analysis of reassembly efficiency are available. Iwai et al. (2006) estimated the efficiency of ligation by trans-splicing using band intensities after sodium dodecyl sulfate–polyacrylamide gel electrophoresis [26]: the reassembly efficiency was determined by comparing the molar amount of the ligated product and either of the residual N- or C-terminal precursor fragments, a useful method for prokaryotic expression systems. However, this approach is not suitable for plant expression systems due to the laborious purification of target proteins. Therefore, it is critical to design new methods to estimate reassembly efficiency after gene

splitting. In this study, based on the phenotype of glyphosate-resistant hybrids, the reassembly efficiency of the target protein was estimated at the plant level. Our data indicated that 100% En-23×Ec-11 hybrid plants tolerate glyphosate treatment, suggesting perfect functional reassembly efficiency after gene splitting. This is the first report on reassembly efficiency after gene splitting in a plant expression system.

Transgenic plants containing divided target gene incorporated into different chromosomes are with less risk to pass the transgenic products into environment. Our results showed that less than 25% of progeny plants still expressed reassembled functional EPSPS proteins when the hybrid transgenic plants were backcrossed to wild type species. Indeed, the gene splitting technique allows the two gene segments to be located on different chromosomes instead of expressing the full length gene on a single chromosome; according to Mendel's laws of inheritance, the latter situation would result in 100% inheritance in the first hybrid generation, while only 25% should be expected with gene splitting. Importantly, the percentage kept decreasing after passing to more generations.

Overall, our results demonstrate that the gene splitting technique can effectively reduce transgene flow, providing a new biological containment strategy in the biosafety field. It is worth mentioning that a series of biological strategies for transgene flow containment have been devised, each with unique characteristics and suitable application scale. It is difficult to control transgene flow completely using only one strategy. Therefore, future studies should focus on controlling transgene flow by combining two or more strategies.

Author Contributions

Conceived and designed the experiments: XJW ZXW XJ. Performed the experiments: XJW XJ BQD NK. Analyzed the data: XJW ZXW SRJ QLT. Wrote the paper: XJW ZXW SRJ.

References

1. James C (2013) Global status of commercialized biotech/GM crops: 2012. China Biotech 33: 1–8.
2. Ruf S, Karcher D, Bock R (2007) Determining the transgene containment level provided by chloroplast transformation. Proc Natl Acad Sci U S A 104: 6998–7002.
3. Kempken F (2010) Engineered Male Sterility. Biotechnology in Agriculture and Forestry 64: 253–265.
4. Gidoni D, Srivastava V, Carmi N (2008) Site-specific excisional recombination strategies for elimination of undesirable transgenes from crop plants. In Vitro Cellular & Developmental Biology-Plant 44: 457–467.
5. Benitez ER, Khan NA, Matsumura H, Abe J, Takahashi R (2010) Varietal differences and morphology of cleistogamy in soybean. Crop Science 50: 185–190.
6. Munsch M, Camp KH, Stamp P, Weider C (2008) Modern maize hybrids can improve grain yield as plus-hybrids by the combined effects of cytoplasmic male sterility and allo-pollination. Maydica 53: 262–268.
7. Svab Z, Maliga P (2007) Exceptional transmission of plastids and mitochondria from the transplastomic pollen parent and its impact on transgene containment. Proc Natl Acad Sci U S A 104: 7003–7008.
8. Yoshida H, Itoh J, Ohmori S, Miyoshi K, Horigome A, et al. (2007) superwoman1-cleistogamy, a hopeful allele for gene containment in GM rice. Plant Biotechnol J 5: 835–846.
9. Leflon M, Hüsken A, Njontie C, Kightley S, Pendergrast D, et al. (2009) Stability of the cleistogamous trait during the flowering period of oilseed rape. Plant breeding 129: 13–18.
10. Al-Ahmad H, Galili S, Gressel J (2005) Poor competitive fitness of transgenically mitigated tobacco in competition with the wild type in a replacement series. Planta 222: 372–385.
11. Moon HS, Li Y, Stewart CN Jr (2010) Keeping the genie in the bottle: transgene biocontainment by excision in pollen. Trends Biotechnol 28: 3–8.
12. Husken A, Prescher S, Schiemann J (2010) Evaluating biological containment strategies for pollen-mediated gene flow. Environ Biosafety Res 9: 67–73.
13. Elleuche S, Poggeler S (2010) Inteins, valuable genetic elements in molecular biology and biotechnology. Appl Microbiol Biotechnol 87: 479–489.

14. Hirata R, Ohsumk Y, Nakano A, Kawasaki H, Suzuki K, et al. (1990) Molecular structure of a gene, VMA1, encoding the catalytic subunit of H(+)-translocating adenosine triphosphatase from vacuolar membranes of Saccharomyces cerevisiae. J Biol Chem 265: 6726–6733.
15. Kane PM, Yamashiro CT, Wolczyk DF, Neff N, Goebl M, et al. (1990) Protein splicing converts the yeast TFP1 gene product to the 69-kD subunit of the vacuolar H(+)-adenosine triphosphatase. Science 250: 651–657.
16. Perler FB (2002) InBase: the Intein Database. Nucleic Acids Res 30: 383–384.
17. Wu H, Hu Z, Liu XQ (1998) Protein trans-splicing by a split intein encoded in a split DnaE gene of Synechocystis sp. PCC6803. Proc Natl Acad Sci U S A 95: 9226–9231.
18. Chin HG, Kim GD, Marin I, Mersha F, Evans TC Jr, et al. (2003) Protein trans-splicing in transgenic plant chloroplast: reconstruction of herbicide resistance from split genes. Proc Natl Acad Sci U S A 100: 4510–4515.
19. Yang J, Fox GC Jr, Henry-Smith TV (2003) Intein-mediated assembly of a functional beta-glucuronidase in transgenic plants. Proc Natl Acad Sci U S A 100: 3513–3518.
20. Gils M, Rubtsova M, Kempe K (2012) Split-transgene expression in wheat. Methods Mol Biol 847: 123–135.
21. Zhu Y, Yu ZL, Lin M (2003) Bioresistance and biodegradation of glyphosate and construction of transgenic plants. Molecular Plant Breeding 1: 435–441.
22. Dun BQ, Wang XJ, Lu W, Zhao ZL, Hou SN, et al. (2007) Reconstitution of glyphosate resistance from a split 5-enolpyruvyl shikimate-3-phosphate synthase gene in Escherichia coli and transgenic tobacco. Appl Environ Microbiol 73: 7997–8000.
23. Dafny-Yelin M, Chung SM, Frankman EL, Tzfira T (2007) pSAT RNA interference vectors: a modular series for multiple gene down-regulation in plants. Plant Physiol 145: 1272–1281.
24. Khan MS, Khalid AM, Malik KA (2005) Intein-mediated protein trans-splicing and transgene containment in plastids. Trends Biotechnol 23: 217–220.
25. Bock R, Khan MS (2004) Taming plastids for a green future. Trends Biotechnol 22: 311–318.

26. Iwai H, Zuger S, Jin J, Tam PH (2006) Highly efficient protein trans-splicing by a naturally split DnaE intein from Nostoc punctiforme. FEBS Lett 580: 1853–1858.

27. Rieben S, Kalinina O, Schmid B, Zeller SL (2011) Gene flow in genetically modified wheat. PLoS One 6: e29730.

28. Heuberger S, Ellers-Kirk C, Tabashnik BE, Carriere Y (2010) Pollen- and seed-mediated transgene flow in commercial cotton seed production fields. PLoS One 5: e14128.

29. Dyer GA, Serratos-Hernandez JA, Perales HR, Gepts P, Pineyro-Nelson A, et al. (2009) Dispersal of transgenes through maize seed systems in Mexico. PLoS One 4: e5734.

16

Transgenic Tobacco Plants Overexpressing a Grass *PpEXP1* Gene Exhibit Enhanced Tolerance to Heat Stress

Qian Xu[1], Xiao Xu[1], Yang Shi[1], Jichen Xu[1]*, Bingru Huang[2]*

1 National Engineering Laboratory for Tree Breeding, Beijing Forestry University, Beijing, China, **2** Dep. of Plant Biology and Pathology, Rutgers, the State Univ. of New Jersey, New Brunswick, New Jersey, United States of America

Abstract

Heat stress is a detrimental abiotic stress limiting the growth of many plant species and is associated with various cellular and physiological damages. Expansins are a family of proteins which are known to play roles in regulating cell wall elongation and expansion, as well as other growth and developmental processes. The *in vitro* roles of expansins regulating plant heat tolerance are not well understood. The objectives of this study were to isolate and clone an expansin gene in a perennial grass species (*Poa pratensis*) and to determine whether over-expression of expansin may improve plant heat tolerance. Tobacco (*Nicotiana tabacum*) was used as the model plant for gene transformation and an expansin gene *PpEXP1* from *Poa pratensis* was cloned. Sequence analysis showed *PpEXP1* belonged to α-expansins and was closely related to two expansin genes in other perennial grass species (*Festuca pratensis* and *Agrostis stolonifera*) as well as *Triticum aestivum*, *Oryza sativa*, and *Brachypodium distachyon*. Transgenic tobacco plants over-expressing *PpEXP1* were generated through *Agrobacterium*-mediated transformation. Under heat stress (42°C) in growth chambers, transgenic tobacco plants over-expressing the *PpEXP1* gene exhibited a less structural damage to cells, lower electrolyte leakage, lower levels of membrane lipid peroxidation, and lower content of hydrogen peroxide, as well as higher chlorophyll content, net photosynthetic rate, relative water content, activity of antioxidant enzyme, and seed germination rates, compared to the wild-type plants. These results demonstrated the positive roles of *PpEXP1* in enhancing plant tolerance to heat stress and the possibility of using expansins for genetic modification of cool-season perennial grasses in the development of heat-tolerant germplasm and cultivars.

Editor: M. Lucrecia Alvarez, TGen, United States of America

Funding: National Natural Science Foundation of China (#31128016), Beijing Natural Science Foundation (#5112015), and Rutgers Center for Turfgrass Science for funding support. The funders had no role in study design, data collection and analysis, decision to publish, or preparation of the manuscript.

Competing Interests: The authors have declared that no competing interests exist.

* Email: jcxu282@sina.com (JX); huang@aesop.rutgers.edu (BH)

Introduction

Plant cell walls provide structural support, control the shape and size of the cell, and protect cells from external biotic and abiotic stresses, and expansin proteins have been shown to control cell wall extensibility and growth [1]. Expansins are small proteins 25–27 kDa in size which bind to glucan-coated cellulose in the cell wall causing a reversible disruption of hydrogen bonding between cellulose microfibrils and the glucan matrix loosening the cell wall [2,3]. Cho and Kende [4] reported that cell wall extensibility of rice (*Oryza sativa*) coleoptiles was positively correlated with the expression level of an expansin gene, *OsEXP4*. Cho et al. [5] confirmed the role of *OsEXP4* in coleoptile cell wall loosening through sense and antisense expression of *OsEXP4* in rice. Expansins also affect shoot and root elongation [6–10] and leaf morphogenesis [11–13]. Over-expressing an expansin gene, *AtEXP10*, resulted in increased petiole growth in transgenic *Arabidopsis thaliana* whereas antisense suppressed petiole growth [11].

Expansins may play a role in regulating plant tolerance to abiotic stresses [5,14–17]. At low water potential, several expansin genes were up-regulated in the root-elongation zone of maize (*Zea mays*), which was correlated to root cell elongation and turgor pressure [6,16,18]. In rice coleoptiles, *OsEXP2* and *OsEXP4*

mRNA induction increased due to flooding stress and northern hybridization results showed that these two expansin genes were correlated with coleoptile elongation in response to oxygen concentration [15]. Transgenic experiments further proved that *OsEXP4* was involved in seedling growth by mediating cell wall loosening [5]. Li et al. [17,19] identified a drought-related expansin gene *TaEXPB23* and confirmed the functions through over-expression in tobacco (*Nicotiana tabacum*) plants using constitutive and inducible promoters. Limited information is available on expansin regulation of plant tolerance to heat stress. Several studies reported that plant responses to heat stress may involve changes in the expression level of expansin genes. Yang et al. [20] identified an expansin gene homologous to *AtEXP13* which was induced in Chinese cabbage (*Brassica rapa*) seedlings exposed to 37°C for 3 h. Xu et al. [14] identified an expansin gene (*AsEXP1*) from a grass species, which was up-regulated during heat stress in heat-tolerant thermal *Agrostis scabra* but little or no expression in heat-sensitive *A. stolonifera*. Further molecular and physiological analyses revealed that the expression level of *AsEXP1* was positively correlated to whole-plant heat tolerance evaluated as leaf photochemical efficiency and cell-membrane stability in different genotypes of *A. scabra* and *A. stolonifera*. Overexpressing *TaEXPB23* from wheat in tobacco did not result in improvement in heat tolerance [21]. As stated above, expansins are a family of

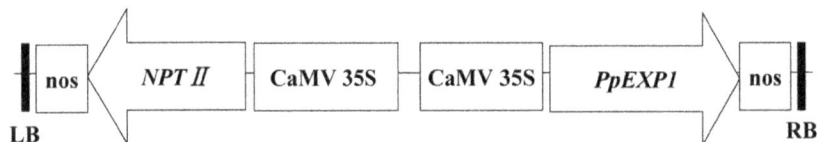

Figure 1. Schematic diagram of the *PpEXP1* over-expression chimeric gene construct, p35S-*NPTII*/35S-*PpEXP1*. *PpEXP1* is under the control of the CaMV 35S promoter and the kanamycin resistance gene *NPTII* is the selection marker.

proteins, which are encoded by multiple genes. The differential effects on heat tolerance of different expansins may be due to differences in specific gene functions. The functions of specific expansin genes in conferring plant tolerance to heat stress require further investigation.

One method to determine the role of candidate genes is to manipulate gene expression through transgenic transformation. Transgenic approaches have been successfully used for determining roles of other genes previously identified to be associated with improved heat tolerance in various plant species, such as heat shock proteins, heat shock transcription factors, and antioxidant enzymes [22–27]. To our knowledge, the *in vivo* roles of expansins regulating plant tolerance to heat stress have not been well documented. Therefore, the objectives of this study were to isolate and clone an expansin gene in a perennial grass species, Kentucky bluegrass (*Poa pratensis*) and to determine whether over-expression of expansin may improve plant tolerance to heat stress. Tobacco (*Nicotiana tabacum*) was chosen for transformation since previous research has shown it is a simple yet robust method and has been successfully utilized in confirming gene function in various studies. Plant tolerance to heat stress was evaluated through plant growth vigor, as well as cellular and physiological responses of tobacco during heat stress.

Materials and Methods

Cloning of a *PpEXP1* gene from *Poa pratensis*

Seeds of Kentucky bluegrass (*Poa pratensis* cv. 'Diva') were sown in plastic pots (25 cm deep and 15 cm in diameter) filled with sandy loamy soil. Seedlings were established for 35 d at 25°C and then exposed to heat stress at 40°C in a growth chamber (HP1500 GS-B; Wuhan Ruihua Instrument & Equipment, Wuhan, China). The other growth chamber conditions were 70% relative humidity, 16 h/8 h light/dark cycle, and photosynthetically active radiation (PAR) of 510 mmol m^{-2} s^{-1} during seedling establishment and subsequent heat stress. After 24 h heat treatment, leaves were collected for RNA extraction. Reverse transcription was performed using an oligo (dT)$_{15}$ primer and Moloney murine leukemia virus reverse transcriptase (Promega) at 42°C for 1 h. A primer set was designed for the upstream and downstream sites of the *AsEXP1* gene in *A. stolonifera* (PL7: 5′-CACATTGCT-TCTCCCGCTTTTGT-3′, UTR4: 5′-TCGGAGTAGCCGAA-GGCCTC-3′), and was used for the homologous gene cloning (*PpEXP1*). The PCR conditions were 94°C for 3 min followed by 32 cycles of 94°C for 1 min, 55°C for 1 min, and 72°C for 1 min. The amplified fragment was cloned into vector pEASY–Blunt (TransGen Biotech). After sequencing, the putative amino acid sequence of the gene *PpEXP1* was deduced using the DNAMAN software (Lynnon Corp. Quebec, Canada). The sequence alignment and phylogenetic tree construction among *PpEXP1* and its homologous genes were analyzed using MEGA 5 [28].

Construction of the gene over-expression vectors

The full coding sequence (including stop codon) was amplified from the recombinant plasmids with the specific primers for the over-expression vector (PpEXP1: 5′- CCCAAGCTTCAATGG-CCTCCTCCAATGC-3′, with added Hind restriction site underlined; PpEXP2: 5′-CCGGAATTCCTAGAACTGGCC-TCCTTCG-3′, with added EcoRI restriction site underlined). The amplified fragments and a binary vector pEZR-(K)-LC [29] were both digested with Hind III and EcoRI enzymes and then ligated. The positive recombinant constructs p35S-nptII/35S-PpEXP1 (Fig. 1) were verified using PCR with primers PpEXP1 and PpEXP2 and restriction enzyme digestion with HindIII and EcoRI and constructs were introduced into *Agrobacterium tumefaciens* strain LBA4404 by electroporation transformation.

Transformation of *PpEXP1* into tobacco plants

Recombinant plasmids were transformed into tobacco plants by *Agrobacterium*-mediated leaf disc infiltration [30]. *Agrobacterium* (LBA4404) carrying the recombinant constructs was cultured in LB medium plus 50 mg L^{-1} kanamycin plates. The plates were incubated for 2–3 days at 28°C. The *Agrobacterium* was then cultured in 30 mL YEP media with appropriate antibiotics at 28°C overnight. The *Agrobacterium* was separated from the YEP medium by centrifuging and the pellets were re-suspended in MS media (OD600 = 0.6~1.0). Leaf disks (0.5 cm in diameter) of 4–5 week old tobacco plants were incubated in the *Agrobacterium* suspension solution for 10 min. Leaf disks were then blotted dry with sterile paper towels, and cultured in MS solid media in the dark at 28°C to generate transgenic plants.

Shoot regeneration was induced on Tobacco Selection Medium (30 g L^{-1} sugar, 4 g L^{-1} phytagel, 3 mg L^{-1} 6-BA, 0.2 mg L^{-1} NAA, 100 mg L^{-1} kanamycin, 250 mg L^{-1} sodium cefotaxime in Murashige and Skoog medium) [31]. After five weeks, the regenerated shoots were transferred to a root-inducing medium (Murashige and Skoog medium with 30 g L^{-1} sugar, 4 g L^{-1} phytagel, 100 mg L^{-1} kanamycin, 250 mg L^{-1} sodium cefotaxime). The positive transformants were verified using PCR with the primer pair PpEXP1/PpEXP2.

Germination test of the transgenic and wild type tobacco seeds with the heat treatment

Seeds of the T0 transgenic line and wild type (WT) plants were wrapped in wet paper towels in petri dishes and incubated at 30, 40, or 50°C for 2 h in an incubator. Seeds were then placed on MS medium and kept at 28°C for germination. The percentage of germinated seeds from the total number of seeds for the transgenic line or the wild type was determined.

Physiological analysis of transgenic and the wild type tobacco plants in heat stress

Positive transgenic tobacco seedlings were planted in pots filled with vermiculite. Plants were irrigated daily and fertilized weekly with half-strength Hoagland's nutrient solution [32]. Two-week

```
1    ATGGCCTCCTCTAATGCTCTGCTCCTGCTCTTCTCGGCCCTCTGCTTCCTTGCCCGCCGG
1     M  A  S  S  N  A  L  L  L  L  L  F  S  A  L  C  F  L  A  R  R
61   GCCGCCGGCGACTACGGCTCGTGGCAGAGCGCCCACGCCACCTTCTATGGCGGCGGCGAT
21    A  A  G  D  Y  G  S  W  Q  S  A  H  A  T  F  Y  G  G  G  D
121  GCGTCCGGCACAATGGGCGGCGCGTGCGGCTACGGGAACCTGTACAGCACGGGCTACGGC
41    A  S  G  T  M  G  G  A  C  G  Y  G  N  L  Y  S  T  G  Y  G
181  ACCAACACGGCAGCGCTGAGCACGGCGTTGTTCAACGAGGGCGCGGCGTGCGGGTCCTGC
61    T  N  T  A  A  L  S  T  A  L  F  N  E  G  A  A  C  G  S  C
241  TACGAGCTCAAGTGCGAGGGGGAGCTACTGCGTGCCGGGCAGCATCATCATCACCGCCACC
81    Y  E  L  K  C  E  G  S  Y  C  V  P  G  S  I  I  I  T  A  T
301  AACCTCTGCCCGCCCAACTACGCGGCTGCCCAACGACGACGGCGGCTGGTGCAACCCGCCG
101   N  L  C  P  P  N  Y  A  L  P  N  D  D  G  G  W  C  N  P  P
361  CGCGCCCACTTCGACATGGCCGAGCCGGCCTACCTCCAGATCGGCGTCTACCGCGCCGGC
121   R  A  H  F  D  M  A  E  P  A  Y  L  Q  I  G  V  Y  R  A  G
421  ATCGTGCCGTCAACTACAGGAGGGTACCCTGCGTGAAGAAGGGCGGCATCAGGTTCACC
141   I  V  P  V  N  Y  R  R  V  P  C  V  K  K  G  G  I  R  F  T
481  ATCAACGGCCACTCCTACTTCAACCTGGTGCTGGTCACCAACGTCGCCGGCGCCGGGGAC
161   I  N  G  H  S  Y  F  N  L  V  L  V  T  N  V  A  G  A  G  D
541  GTGCAGTCCGTCTCGATCAAGGGCTCTAGCACCGGCTGGCAGGCCATGTCACGCAACTGG
181   V  Q  S  V  S  I  K  G  S  S  T  G  W  Q  A  M  S  R  N  W
601  GGCCAGAACTGGCAGAGCAACTCCGACCTCGACGGCCAGAGCCTCTCCTTCAAGGTCACC
201   G  Q  N  W  Q  S  N  S  D  L  D  G  Q  S  L  S  F  K  V  T
661  CTCAGCGACGGCCGAACCATCGTCAGCAACAACGCCGCCCGGCAGGCTGGACGTTCGGC
221   L  S  D  G  R  T  I  V  S  N  N  A  A  P  A  G  W  T  F  G
721  CAGACCTTCGAAGGAGGCCAGTTCTAG
240   Q  T  F  E  G  G  Q  F  *
```

Figure 2. The full sequence of *PpEXP1* gene from *Poa pratensis* with the deduced amino acids.

old seedlings of the transgenic and wild-type (WT) plants were exposed to heat stress (42°C) or the optimal growth temperature (25°C) for 6 d with other environmental conditions the same as previously described. Leaves were destructively sampled for the physiological tests following 6 d of treatment. Several physiological parameters were measured to evaluate heat tolerance in the wild type and transgenic lines, including leaf cell membrane stability, relative water content, chlorophyll content, net photosynthetic rate, membrane lipid peroxidation, content of hydrogen peroxide, and antioxidant activity of superoxide dismutase.

Cellular membrane stability was evaluated by measuring relative electrolyte leakage (EL) of leaves using the methods of Blum and Ebercon [33]. For EL measurement, leaf disks (1.2 cm in diameter) of 0.2 g fresh weight were taken from three leaves per plant and incubated in 40 mL deionized water for 12 h on a conical shaker. The conductance of the solution was measured as the initial level of conductance (C_i) using a conductance meter (Yellow Springs Instrument Co., Ohio, USA). Leaf tissues were

then killed in an autoclave at 121°C for 15 min. After 12 h incubation on a conical shaker, the conductance of the sample was measured again as C_{max}. Leaf EL was calculated as: EL (%) = C_i/C_{max} × 100.

Leaf relative water content was calculated using fresh weight (FW), turgid weight (TW), and dry weight (DW) of leaves according to the equation: (FW-DW)/(TW-DW) × 100 [34]. Fresh weight was measured immediately after leaves were cut off the plant. After FW determination, leaves were soaked in distilled water for 12 h at 4°C until leaves became fully turgid, and then blotted dry to determine TW. Leaf DW was measured after leaves were dried in an oven at 87°C for 72 h.

For leaf chlorophyll content, chlorophyll was extracted from leaf tissue of 0.1 g fresh weight in dimethyl sulfoxide and kept in the dark for 2 d. The absorbance of the solution was determined at 645 and 663 nm using a spectrophotometer. Leaf chlorophyll content was calculated using the equation described in [35].

Leaf net photosynthetic rate was measured with an infrared gas analyzer (LI-6400, LICOR Inc., Lincoln, NB). A single leaf was enclosed in the leaf chamber (2 × 3 cm) and exposed to light (PAR of 650 μmol photon m^{-2} s^{-1}) with a built-in red and blue light source of the LI-6400 and carbon dioxide concentration controlled at 380 μmol m^{-1}). Leaf net photosynthetic rate was measured at 10:00 AM.

Membrane lipid peroxidation was determined by measuring malondialdehyde (MDA) content following the procedure described in [36]. Briefly, fresh leaves (0.2 g) were ground in 10% trichloroacetic acid containing 0.5% thiobarbituric acid. The mixture was heated in a water bath at 95°C for 30 min, quickly cooled on ice, and then centrifuged at 14,000 g for 20 min. The absorbance of the supernatant was measured at 532 and 600 nm using a spectrophotometer. The concentration of MDA was calculated using an extinction coefficient of 155 mM^{-1} cm^{-1} [37].

The content of hydrogen peroxide was determined following the method described in [38] with modification. Leaf samples were extracted in acetone. The supernatants (1 mL) were added in a reaction solution containing 0.1 mL 20% titanium reagent (20% w/v TiCl$_4$ in 12.1 mL HCl) and 0.2 mL 17 M ammonia. The solution with the extractants was centrifuged at 3000 g at 4°C for 10 min and the pallets were dissolved in 3 mL 1 M sulfuric acid. The absorbance of the solutes was measured at 410 nm with a spectrophotometer.

The activity of an antioxidant enzyme, superoxide dismutase (SOD), was measured using the method previously described in [39] with modification. Briefly, 0.2 g of fresh leaf tissue was ground to a fine powder using a mortar and pestle and extracted in 4 mL of extraction buffer at pH 7.8 containing 50 mM potassium

```
AsEXP1 : MASYNALLLLVSAECFLARRAAGDYGSWQSAHATFYGEADASGTMGGACGYGNLYSTGYGTNTAALSTALFNEGAACGSCYELKCLGA
PpEXP1 : MASSNALLLLFSALCFLARRAAGDYGSWQSAHATFYGGGDASGTMGGACGYGNLYSTGYGTNTAALSTALFNEGAACGSCYELKCEC-

AsEXP1 : AGSSCRAGSITITATNLCPPNYALPNDDGGWCNPPRAHFDMAEPAYLQIGIYRAGIVPVNYRRVPCVKKGGIRFTINGHSYFNLVLVT
PpEXP1 : --SYCVPGSIIITATNLCPPNYALPNDDGGWCNPPRAHFDMAEPAYLQIGVYRAGIVPVNYRRVPCVKKGGIRFTINGHSYFNLVLVT

AsEXP1 : NVAGAGDVQAVSIKGSSTGWQAMSRNWGQNWQSNADLDGQALSFKVTISDGRTIISNNAAPAGWQFGQTFEGGQF
PpEXP1 : NVAGAGDVQSVSIKGSSTGWQAMSRNWGQNWQSNSDLDGQSLSFKVTLSDGRTIVSNNAAPAGWTFGQTFEGGQF
```

Figure 3. The protein sequence alignment of *PpEXP1* from *Poa pratensis* and *AsEXP1* from *Agrostis stolonifera*.

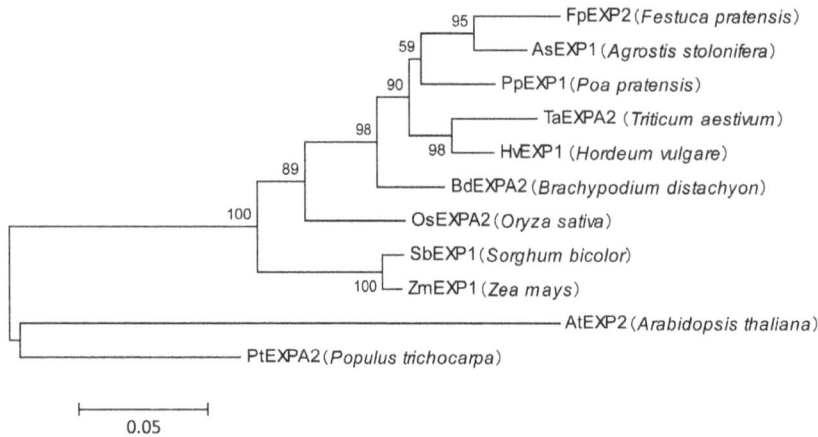

Figure 4. A phylogenetic tree constructed with MEGA5 based on the amino acids of the *PpEXP1* homologous genes from different plant species. Accession numbers of these proteins listed below: FpEXP2 (CAC06433), TaEXPA2 (AAT94292), HvEXP1 (BAK05504), BdEXPA2 (XP_003564501), OsEXPA2 (NP_001044656), SbEXP1 (XP_002456550), ZmEXP1 (NP_001105040), AtEXP2 (XP_002871148), PtEXPA2 (XP_002319409). Sequence of AsEXP1 was from Zhou et al. [47].

phosphate, 1 mM ethylenediaminetetraacetic acid, 1% polyvinyl-pyrrolidone, 1 mM dithiothreitol, and 1 mM phenylmethylsulfo-nyl. The extractants were centrifuged at 15,000 g for 30 min at 4°C, and supernatant was collected for enzyme assay. The SOD (EC 1.15.1.1) activity was measured by recording changes in the rate of nitro blue tetrazolium chloride reduction in absorbance at 560 nm.

Cellular structural changes of transgenic and wide-type tobacco plants under heat stress

The 5 mm leaf segments were fixed by immersion in newly-prepared neutral sodium phosphate (0.1 M) with 5% paraformal-dehyde at 4°C overnight. After washing in phosphate buffer, leaf samples were dehydrated in a series of ethanol solutions at concentration of 30%, 50%, 70%, 85%, 95%, and 100%. The dehydrated tissues were embedded and polymerized in Epon812 resin at 70°C. Leaf samples were then cut horizontally into 2 μm-thin sections and stained by toluidine blue. The images of leaf cross sections were recorded by a camera fitted to a microscope (Nikon, Tokyo, Japan).

Statistical analysis

Data for effects of temperature and transformation on physiology and germination rate were determined using the analysis of variance (ANOVA) with a general linear procedure. Significance between treatments was determined using the least significance test (LSD) at $p = 0.05$ with a statistical program (SAS Version 9.0, Cary, NC).

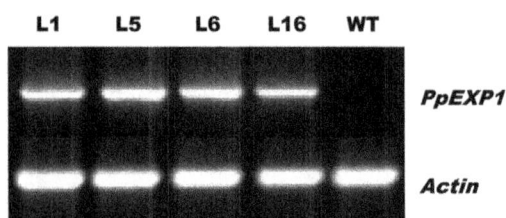

Figure 5. *PpEXP1* gene expression in transgenic (L1, L5, L6, L16) and wild type (WT) tobacco by RT-PCR.

Results

Sequence characteristics and expression of *PpEXP1*

Using the primers designed based on the sequence of *AsEXP1*, the homologous gene fragment was amplified and cloned from the transcript of *Poa pratensis* plants exposed to heat stress and named as *PpEXP1*. Sequencing results showed the gene's open reading frame contained 744 nucleotides (bp) encoding 247 amino acids (Fig. 2). The amino acid alignment shows an identity of up to 92% between *PpEXP1* and *AsEXP1* from *A. stolonifera* (Fig. 3) with a 3 amino acid deduction in *PpEXP1*. Searches in Genebank for homologous genes revealed several matches including 8 monocots and 2 eudicots. A phylogenetic tree was constructed based on their amino acid sequences and shown in Fig. 4. The monocot and eudicot plants were basically distributed into two branches. The C_3 plants (*Triticum aestivum*, *Oryza sativa*, *Festuca pratensis*, *Poa pratensis*, *Agrostis stolonifera*, and *Brachypodium distachyon*) and C_4 plants (*Zea mays* and *Sorghum bicolor*) of the 9 monocots species were classified into two sub-clades. The *PpEXP1* gene was more closely grouped with two other perennial grass species (*F. pratensis* and *A. stolonifera*), *Triticum aestivum*, *Hordeum vulgare*, and *Brachypodium distachyon* based on the gene sequences. The topology structure fits well to the phylogenetic tree constructed by whole chlorophyll genome sequence and molecular markers [40,41].

Transgenic lines of tobacco obtained using agrobacterium-mediated transformation were tested at the RNA level using the specific primers of PpEXP1/PpEXP2. *PpEXP1* expression was confirmed in all transgenic lines using RT-PCR. The expression of *PpEXP1* in four transgenic tobacco lines (L1, L5, L6, and L16) is shown in Fig. 5.

Seed germination rate of transgenic and WT plants as affected by heat stress

The T1 seeds for the WT and four transgenic lines (L1, L5, L6, and L16) were harvested and used for the germination test. Seed germination rate decreased with increasing temperature from 30 to 40 and 50°C in the WT (Table 1). Seed germination rate did not change in any of the four transgenic lines as temperature increased from 30 to 40°C, while the decline was observed at 50°C in L1 and L5. Increasing temperatures did not have significant effects on the seed germination rates of L6 and L16. Seed

Table 1. Germination rate of seeds from transgenic lines and wide-type tobacco plants treated at different temperatures.

Treatment	Germination rate (%)				
	L1	L5	L6	L16	WT
30°C	92.3±1	94.2±1	93.6±0.6	92.3±1	91.7±0.6
40°C	91.0±0.6	92.3±1	92.9±0.6	91.0±0.6	76.3*±0.6
50°C	87.2±0.6	89.1±0.6	90.4±1	88.5±1	60.9*±0.6

All data are mean ± SD of each treatment. Three replicates were performed for each treatment.
"*" indicates significant difference of the WT data from all transgenic lines at 40 or 50°C based on least significance difference test at p = 0.05. Data at 30°C without "*" indicate no significant differences among the WT and transgenic lines.

germination rates of all four transgenic lines were significantly higher than that of the WT at 40 and 50°C.

Plant vigor and cellular structural changes of transgenic and WT plants in response to heat stress

Plants of four transgenic lines (L1, L5, L6, and L16) and the WT were exposed to heat stress to evaluate growth responses. Leaf wilting was observed in the WT plants at 3 d of heat stress, but was not seen in any of the transgenic lines. After 6 d of heat stress, the WT plants exhibited severe leaf wilting, chlorosis of mature leaves, and stunted growth, but the transgenic plants did not exhibit signs of heat damages as seen in the wild type (Fig. 6). Plant vigor was rated based on the level of leaf wilting, leaf color, and canopy height on the scale of 1–9, with 9 being undamaged and 1 being a dead plant. The rating was 4±0.5 (n = 10) for the WT plants and averaged at 8±0.5 (n = 10) for the four transgenic lines.

Cellular damages in transverse sections of leaf tissues from the WT and transgenic plants exposed to the optimal growth temperature and heat stress were examined by microscopy (Fig. 7). Under the optimal growth temperature (25°C), mesophyll cells were intact while palisade and spongy mesophyll cells were arranged orderly with a clearly-defined cellular structure in leaves of WT (Fig. 7A) and four transgnic lines (Fig. 7B–E). After heat treatment at 42°C for 6 d, the WT plants displayed some cellular disruption and disorganization of the spongy mesophyll cells with large intercellular spaces, which could be the result of cell death (Fig. 7F). Heat-stressed transgenic plants still maintained well-defined mesophyll cells and transversal structures of the leaf were similar between leaves exposed to the optimal growth temperature and heat stress (Fig. 7G–J). Plastids were present as granules along the mesophyll cell walls in both the WT and transgenic plants

grown at the optimal growth temperature (Fig. 7A and 7B–E). Under heat stress, majority of plastids disappeared in the WT leaves (Fig. 7F) but remained intact in high density in the transgenic leaves (Fig. 7G–J).

Physiological changes of the transgenic and WT plants in response to heat stress

Under the optimal temperature conditions, no significant differences in any of the physiological parameters (Fig. 8, 9, 10) were detected between the four transgenic lines and the WT plants. At 6 d of heat stress, relative water content (Fig. 8A), chlorophyll content (Fig. 9A), and net photosynthetic rate (Fig. 9B) declined in all plants, to a lesser extent for all three parameters in the four transgenic lines than the WT plants. Leaf EL (Fig. 8B), H_2O_2 content (Fig. 10A), and MDA content (Fig. 10B), as well as SOD activity (Fig. 10 C) increased significantly at 6 d of heat stress in all plants, and the magnitude of increase for EL, H_2O_2, and MDA was less pronounced in the four transgenic lines than in the WT plants. Four transgenic lines maintained significantly higher relative water content, chlorophyll content, and net photosynthetic rate, and SOD activity, but had lower EL, H_2O_2, and MDA content compared to the WT.

Discussion

Expansins include two major types, α –expansins and β-expansins, which are encoded by a large gene family [42]. As shown in Fig. 4, the *PpEXP1* gene was grouped closely with two other perennial grass species (*F. pratensis* and *A. stolonifera*), as well as other monocots including *Triticum aestivum*, *Oryza sativa*, and *Brachypodium distachyon*, but was less related to *Arabidopsis thaliana* based on the gene sequence of the current study. The functional

Figure 6. The transgenic (right) and wild-type (left) tobacco plants were under heat treatment of 42°C for 6 d, which the leaf wilting was rating as 8 and 4, respectively.

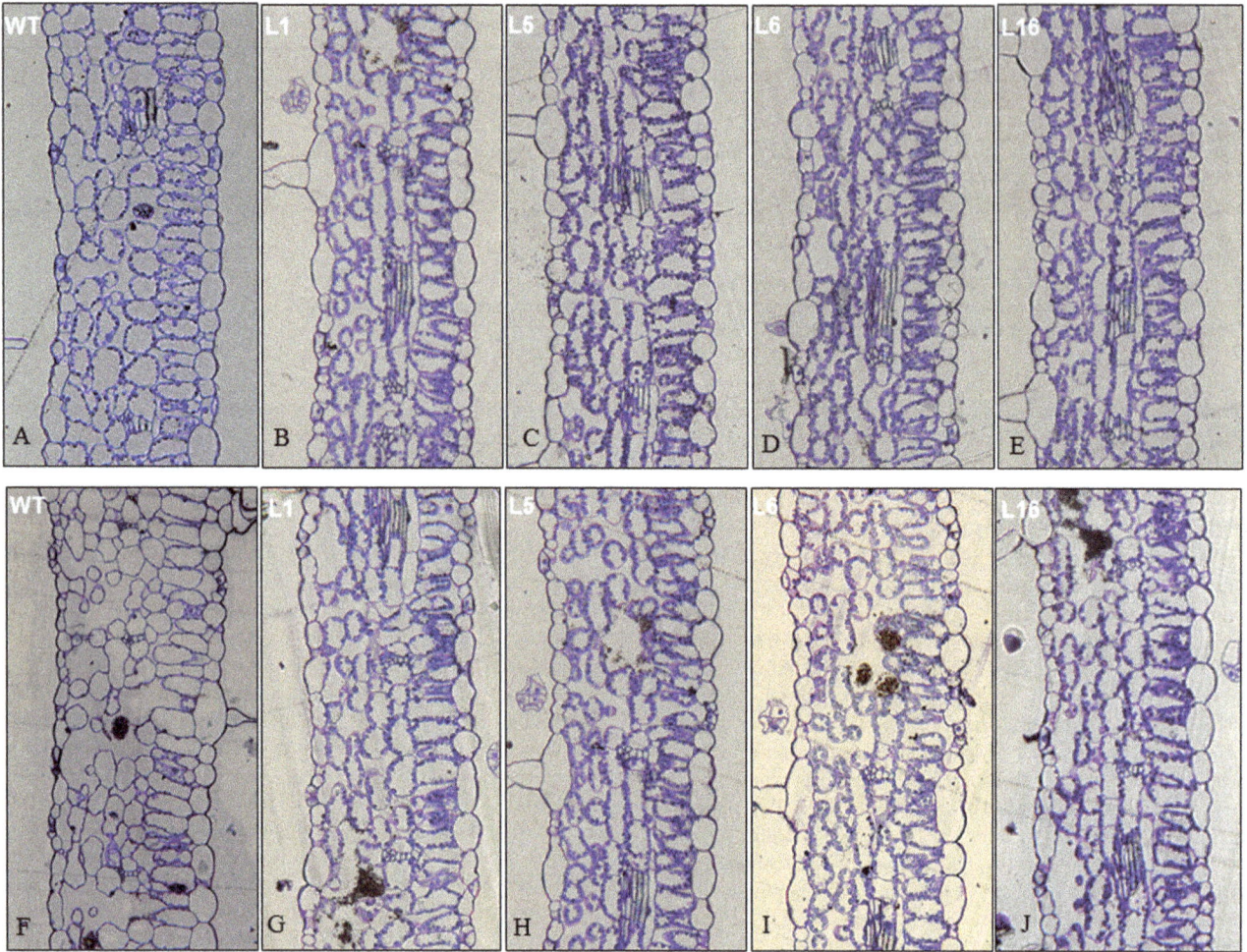

Figure 7. Transverse sections of leaf tissues from WT and transgenic tobacco plants and imaged captured on a camera fitted to a microscope at 20 X. A, WT 25°C; B, 35S::*PpEXP1* 25°C; C, WT 42°C; D, 35S::*PpEXP1* 42°C.

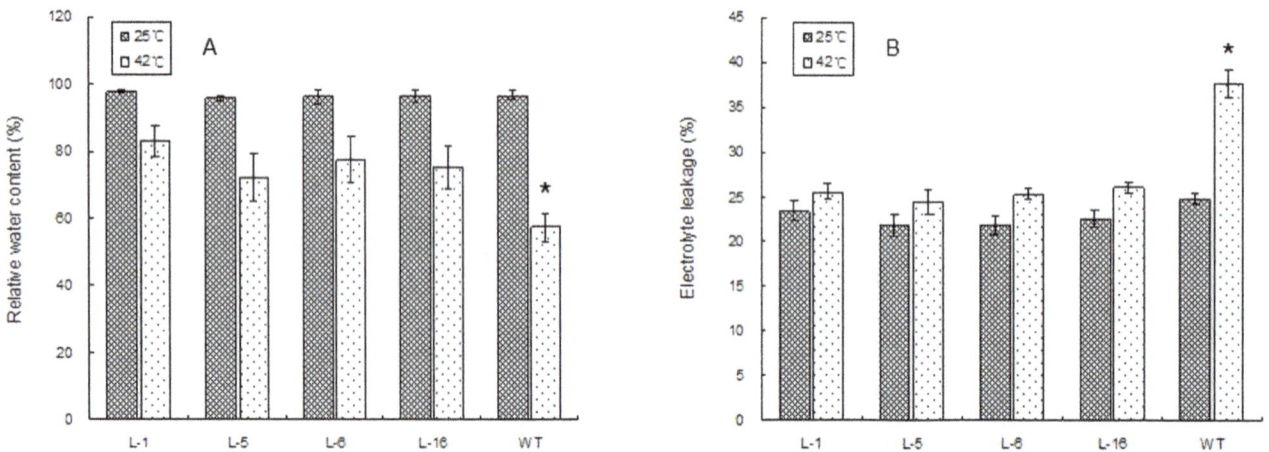

Figure 8. Leaf relative water content (A) and electrolyte leakage (B) in WT and four lines of transgenic tobacco under normal temperature and at 6 d of heat treatment. WT represents the wild type tobacco. L1, L5, L6, and L16 are different transgenic tobacco lines. The vertical bar over each column represents standard error of the mean for three replicates in each treatment. The "*" over the column of the WT indicates significant difference of the WT from all transgenic lines under heat stress based on least significance difference test at p = 0.05. Columns without "*" indicate no differences among the WT and transgenic lines.

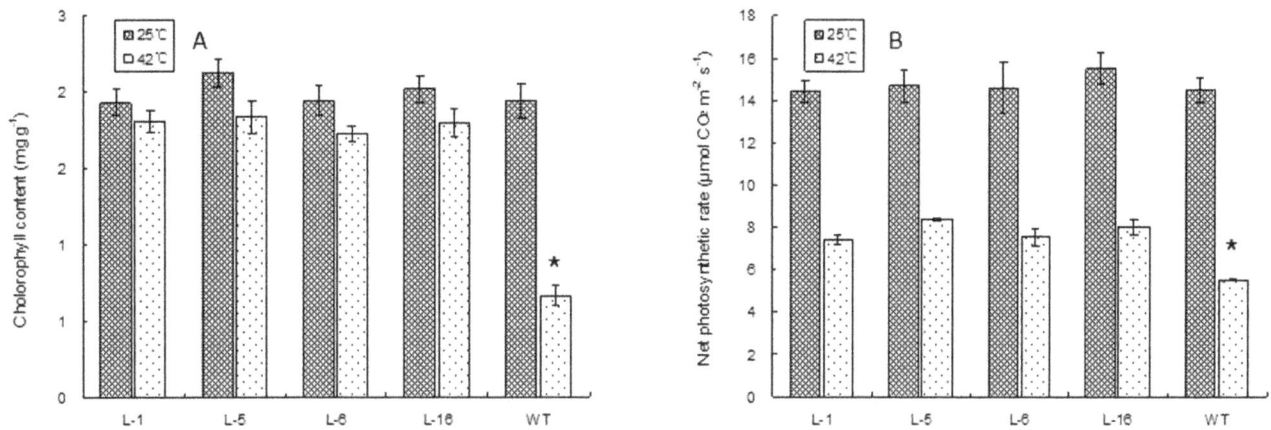

Figure 9. Leaf chlorophyll content (A) and net photosynthetic rate (B) in WT and four lines of transgenic tobacco under normal temperature and at 6 d of heat treatment. WT represents the wild type tobacco. L1, L5, L6, and L16 are different transgenic tobacco lines. The vertical bar over each column represents standard error of the mean for three replicates in each treatment. The "*" over the column of the WT indicates significant difference of the WT from all transgenic lines under heat stress based on least significance difference test at p = 0.05. Columns without "*" indicate no differences among the WT and transgenic lines.

conserved amino acids and motifs specific to α-expansin were found in *PpEXP1* proteins, such as cysteines that form the disulfide bridges, HFD motif in the GH45-like domain and the aromatic residues in the binding-domain, suggesting that *PpEXP1* is belong to α-expansin family. The α-expansins are predominantly found in dicots, and also found in monocots [42]. In addition, the amino acid sequence alignment showed a 92% similarity to the reported *AsEXP1*sequences in *A. stolonifera* [14]. These data suggest that *PpEXP1* is a conserved α-expansin gene in grass species.

Over-expression of *PpEXP1* in tobacco enhanced seed germination during heat stress and suppressed heat-induced leaf damages by maintaining higher leaf chlorophyll content, relative water content, net photosynthetic rate, antioxidant enzyme activity, and lower EL, H_2O_2 and MDA content (Fig. 8, 9, 10). These results suggested that *PpEXP1* could be involved in heat adaptation by promoting leaf photosynthesis and hydration status, and strengthening antioxidant activities to suppress the accumulation of reactive oxygen species and membrane damages. Less cellular disruption during heat stress was seen during microscopic observations of cells in the transgenic plants (Fig. 7). Cell death associated with cell structural damages, such as rupture of cell membranes and cell walls are typical heat stress symptoms in plant cells [43]. Loss of mesophyll cellular structure and presence of large intercellular spaces were shown in the WT plants under heat stress. However, the mesophyll cells in leaves of *PpEXP1* transgenic plants were maintained with well-defined cellular structure and a high density of plastids. These results indicated that over-expression of *PpEXP1* could facilitate cellular survival during heat stress and possibly be linked to the effects on maintaining cell wall extensibility. Previous research has shown that heat treatments inactivate expansins and increase cell wall rigidity [44]. The cellular observations were in line with the physiological data, supporting the positive effects of *PpEXP1* over-expression on heat tolerance. Leaf chlorophyll content, net photosynthetic rate, and water status have been shown to be positively correlated to plant heat tolerance while EL and oxidative damages due to increased H_2O_2 and MDA production are negatively correlated to plant heat tolerance [45,46]. The expression level of *AsEXP1* was previously found to be positively correlated to heat tolerance in two grass species contrasting in heat tolerance. Higher expression

of expansin in the heat-tolerant thermal *A. scabra* relative to heat-sensitive *A. stolonifera* was associated with significantly higher leaf chlorophyll content and lower electrolyte leakage under heat stress in the former compared to the latter species [47]. To our knowledge, this is the first study that demonstrated the positive effects of over-expressing α-expansin, as *PpEXP1*, on improving heat tolerance. Our results suggest that expansins play important roles in protecting membranes and reducing cellular and physiological damages from heat stress, thereby improving plant tolerance to heat stress.

The underlying mechanisms of expansin genes regulating plant tolerance to heat stress are not well understood. Cell wall extensibility regulated by expansin can be affected by external stresses, including temperature [44]. Despite the knowledge of the positive association of expansin expression and plant heat tolerance, the specific effects of high temperatures on cell-wall properties as affected through the regulation of expansins is not clear. Heat treatment inactivates expansins in the cell wall, leading to the reduction in cell wall extensibility and the cell wall extensibility can be recovered by increasing expansin proteins in the heat-treated cell walls [44]. The correlation of expansin and cell wall extensibility has been confirmed in various plant species, including tomato leaves (*Solanum lycopersicum*) [48], oat (*Avena sativa*) coleoptiles [49], rice (*Oryza sativa*) internodes [50], and soybean (*Glycine max*) roots [7]. The abnormal cell growth of *Arabidopsis* plants exposed to high temperature has been associated with the loss of cellulose crystalline and cell wall extensibility [51]. Based on the functions of expansins in the cell wall it could be assumed that *PpEXP1* may be involved in protecting cellular damages from heat stress by loosening cell walls, and maintaining cellular turgor and cell integrity under heat stress. This notion deserves further investigation using the transgenic plants over-expressing *PpEXP1*.

In summary, *PpEXP1* cloned from Kentucky bluegrass was classified as an α-expansin and closely related to expansin genes in other grass species. Over-expressing *PpEXP1* in tobacco mitigated cellular damages and growth inhibition due to heat stress and improved heat tolerance of tobacco as manifested by greater seed germination rate, chlorophyll content, and cell membrane stability, as well as lower level of membrane lipid peroxidation. However the cellular and molecular mechanisms of PpEXP1

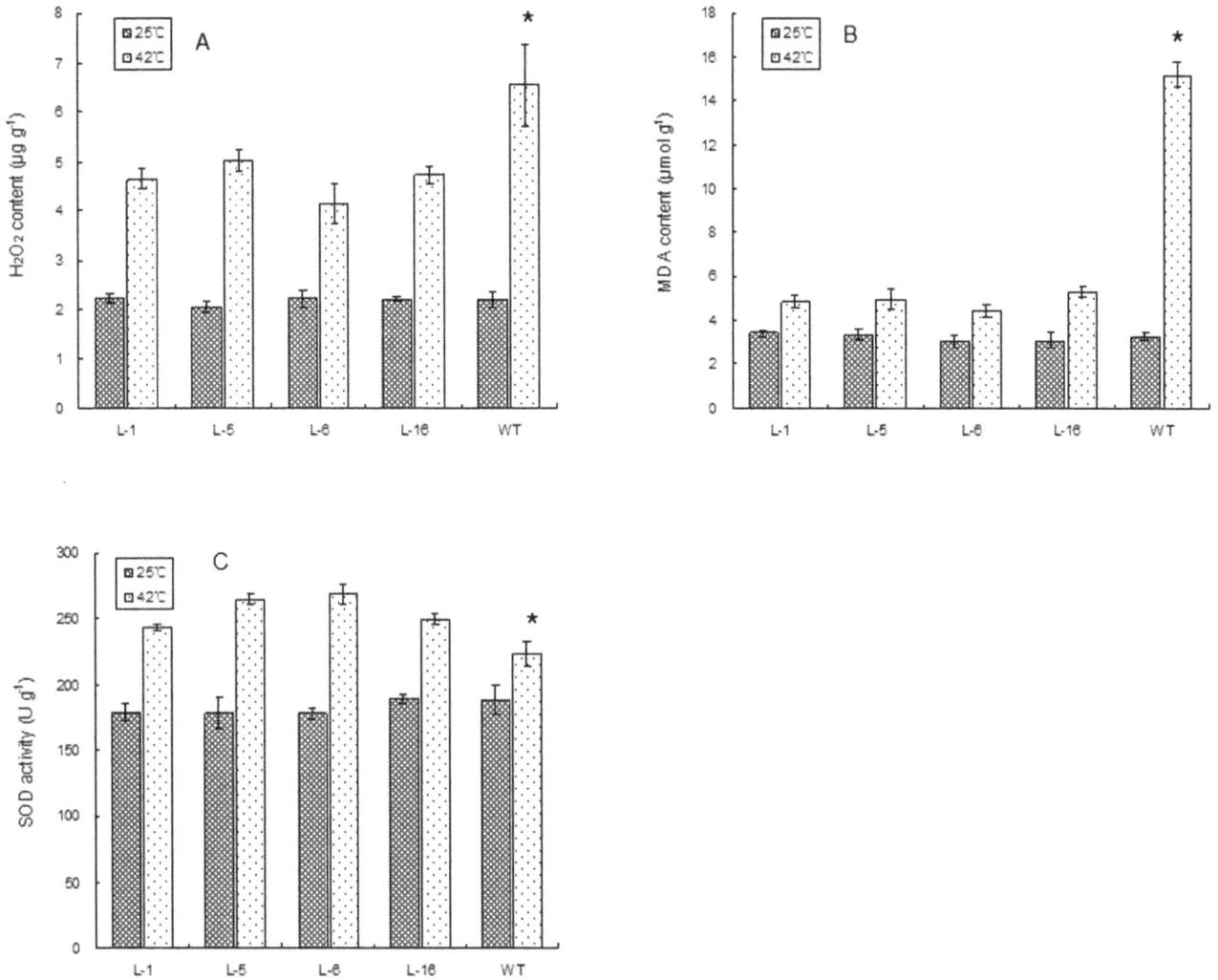

Figure 10. Hydrogen peroxide (H₂O₂) content (A), and malondialdehyde (MDA) content (B), and activity of superoxide dismutase (SOD) (C) in WT and four lines of transgenic tobacco under normal temperature and at 6 d of heat treatment. WT represents the wild type tobacco. L1, L5, L6, and L16 are different transgenic tobacco lines. The vertical bar over each column represents standard error of the mean for three replicates in each treatment. The "*" over the column of the WT indicates significant difference of the WT from all transgenic lines under heat stress based on least significance difference test at $p = 0.05$. Columns without "*" indicate no differences among the WT and transgenic lines.

regulating plant heat tolerance requires further investigation. The effectiveness of *PpEXP1* in grass heat tolerance will be investigated in future research through over-expressing the gene in cool-season perennial grasses that are sensitive to heat stress. *PpEXP1* is a potential candidate gene that could be used to generate heat-tolerant cool-season perennial grass germplasm and cultivars, such as used in forage and turf grasses.

Acknowledgments

Thanks to Patrick Burgess, David Jespersen, and Jillian Keough at Rutgers University for editing and critical review of the manuscript.

Author Contributions

Conceived and designed the experiments: QX JX BH. Performed the experiments: QX XX YS. Analyzed the data: QX BH JX. Contributed reagents/materials/analysis tools: JX BH. Wrote the paper: QX BH JX.

References

1. Cosgrove DJ (2000) Expansive growth of plant cell walls. Plant Physiol Biochem 38: 109–124.
2. McQueen-Mason SJ, Cosgrove DJ (1995) Expansin mode of action on cell walls (analysis of wall hydrolysis, stress relaxation, and binding). Plant Physiol 107: 87–100.
3. Cosgrove DJ (2000) Loosening of plant cell walls by expansins. Nature 407: 321–326.
4. Cho HT, Kende H (1997) Expression of expansin genes is correlated with growth in deepwater rice. Plant Cell 9: 1661–1671.
5. Choi DS, Lee Y, Cho HT, Kende H (2003) Regulation of expansin gene expression affects growth and development in transgenic rice plants. Plant Cell 15: 1386–1398.
6. Lee DK, Ahn JH, Song SK, Choi YD, Lee JS (2003) Expression of an expansin gene is correlated with root elongation in soybean. Plant Physiol 131: 985–997.
7. Kam MJ, Yun HS, Kaufman PB, Chang SC, Kim SK (2005) Two expansins, *EXP1* and *EXPB2*, are correlated with the growth and development of maize roots. Journal of Plant Biol 48: 304–310.
8. Lin C, Choi HS, Cho HT (2011) Root hair-specific *EXPANSIN A7* is required for root hair elongation in Arabidopsis. Mol Cells 31: 393–397.

9. Ookawara R, Satoh S, Yoshioka T, Ishizawa K (2005) Expression of alpha-expansin and xyloglucan endotransglucosylase/hydrolase genes associated with shoot elongation enhanced by anoxia, ethylene and carbon dioxide in arrowhead (*Sagittaria pygmaea* Miq.) tubers. Ann Bot 96: 693–702.

10. Yu Z, Kang B, He X, Lu S, Bai Y, et al. (2011) Root hair-specific expansins modulate root hair elongation in rice. Plant J 66: 725–734.

11. Cho HT, Cosgrove DJ (2000) Altered expression of expansin modulates leaf growth and pedicel abscission in *Arabidopsis thaliana*. Proc Natl Acad Sci U S A 97: 9783–9788.

12. Goh HH, Sloan J, Dorca-Fornell C, Fleming A (2012) Inducible repression of multiple expansin genes leads to growth suppression during leaf development. Plant Physiol 159: 1759–1770.

13. Hu LW, Cui DY, Neill S, Cai WM (2007) *OsEXPA4* and *OsRWC3* are involved in asymmetric growth during gravitropic bending of rice leaf sheath bases. Physiol Plant 130: 560–571.

14. Xu J, Tian J, Belanger FC, Huang B (2007) Identification and characterization of an expansin gene *AsEXP1* associated with heat tolerance in C3 *Agrostis* grass species. J Exp Bot 58: 3789–3796.

15. Huang J, Takano T, Akita S (2000) Expression of alpha-expansin genes in young seedlings of rice (*Oryza sativa* L.). Planta 211: 467–473.

16. Wu Y, Thorne ET, Sharp RE, Cosgrove DJ (2001) Modification of expansin transcript levels in the maize primary root at low water potentials. Plant Physiol 126: 1471–1479.

17. Li F, Han Y, Feng Y, Xing S, Zhao M, et al. (2013) Expression of wheat expansin driven by the RD29 promoter in tobacco confers water-stress tolerance without impacting growth and development. J Biotechnol 163: 281–291.

18. Wu Y, Sharp RE, Durachko DM, Cosgrove DJ (1996) Growth maintenance of the maize primary root at low water potentials involves increases in cell-wall extension properties, expansin activity, and wall susceptibility to expansins. Plant Physiol 111: 765–772.

19. Li F, Xing S, Guo Q, Zhao M, Zhang J, et al. (2011) Drought tolerance through over-expression of the expansin gene *TaEXPB23* in transgenic tobacco. J Plant Physiol 168: 960–966.

20. Yang KA, Lim CJ, Hong JK, Park CY, Cheong YH, et al. (2006) Identification of cell wall genes modified by a permissive high temperature in Chinese cabbage. Plant Sci 171: 175–182.

21. Han YY, Li AX, Li F, Zhao MR, Wang W (2012) Characterization of a wheat (*Triticum aestivum* L.) expansin gene, *TaEXPB23*, involved in the abiotic stress response and phytohormone regulation. Plant Physiol Biochem 54:49–58.

22. Queitsch C, Hong SW, Vierling E, Lindquist S (2000) Heat shock protein 101 plays a crucial role in thermotolerance in *Arabidopsis*. Plant Cell 12: 479–492.

23. Yokotani N, Ichikawa T, Kondou Y, Matsui M, Hirochika H, et al. (2008) Expression of rice heat stress transcription factor *OsHsfA2e* enhances tolerance to environmental stresses in transgenic *Arabidopsis*. Planta 227: 957–967.

24. Tang L, Kwon SY, Kim SH, Kim JS, Choi JS, et al. (2006) Enhanced tolerance of transgenic potato plants expressing both superoxide dismutase and ascorbate peroxidase in chloroplasts against oxidative stress and high temperature. Plant Cell Rep 25: 1380–1386.

25. Yang X, Liang Z, Lu C (2005) Genetic engineering of the biosynthesis of glycinebetaine enhances photosynthesis against high temperature stress in transgenic tobacco plants. Plant Physiol 138: 2299–2309.

26. Zhang M, Barg R, Yin M, Gueta-Dahan Y, Leikin-Frenkel A, et al. (2005) Modulated fatty acid desaturation via overexpression of two distinct omega-3 desaturases differentially alters tolerance to various abiotic stresses in transgenic tobacco cells and plants. Plant J 44: 361–371.

27. Chiang C-M, Chen S-P, Chen L-F, Chiang M-C, Chien H-L, et al. (2013) Expression of the broccoli catalase gene (*BoCAT*) enhances heat tolerance in transgenic *Arabidopsis*. J Plant Biochem Biotechnol DOI 10.1007/s13562-013-0210-1

28. Tamura K, Peterson D, Peterson N, Stecher G, Nei M, et al. (2011) MEGA5: molecular evolutionary genetics analysis using maximum likelihood, evolutionary distance, and maximum parsimony methods. Mol Biol Evol 28: 2731–2739.

29. Christie JM, Swartz TE, Bogomolni RA, Briggs WR (2002) Phototropin LOV domains exhibit distinct roles in regulating photoreceptor function. Plant J 32: 205–219.

30. Gallois P, Marinho P (1995) Leaf disk transformation using *Agrobacterium tumefaciens*-expression of heterologous genes in tobacco. In: Jones H, editor. Plant Gene Transfer and Expression Protocols: Springer New York. pp. 39–48.

31. Murashige T, Skoog F (1962) A revised medium for rapid growth and bioassays with tobacco tissue cultures. Physiol Plant 15: 473–497.

32. Hoagland DR, Arnon DI (1950) The water-culture method for growing plants without soil. Cir CA Agric Exp Sta 347: 32.

33. Blum A, Ebercon A (1981) Cell membrane stability as a measure of drought and heat tolerance in wheat. Crop Sci 21: 43–47.

34. Barrs H, Weatherley P (1962) A re-examination of the relative turgidity technique for estimating water deficits in leaves. Aust J Biol Sci 15: 413–428.

35. Arnon DI (1949) Copper enzymes in isolated chloroplasts.Polyphenoloxidase in beta vulgaris. Plant Physiol 24: 1–15

36. Dhindsa RS, Plumb-Dhindsa P, Thorpe TA (1981) Leaf senescence: correlated with increased levels of membrane permeability and lipid peroxidation, and decreased levels of superoxide dismutase and catalase. J Exp Bot 32: 93–101.

37. Heath RL, Packer L (1968) Photoperoxidation in isolated chloroplasts: I. Kinetics and stoichiometry of fatty acid peroxidation. Arch Biochem Biophys 125: 189–198.

38. Patterson Bd, MacRae EA, Ferguson IB (1984) Estimation of hydrogen peroxide in plant extracts using titanium(IV). Anal Biochem 139: 487–492.

39. Zhang J, Kirkham MB (1996) Antioxidant responses to drought in sunflower and sorghum seedlings. New Phytol 132: 361–373.

40. Yaneshita M, Ohmura T, Sasakuma T, Ogihara Y (1993) Phylogenetic relationships of turfgrasses as revealed by restriction fragment analysis of chloroplast DNA. Theor Appl Genet 87: 129–135.

41. Mathews S, Tsai RC, Kellogg EA (2000) Phylogenetic structure in the grass family (Poaceae): evidence from the nuclear gene phytochrome B. Am J Bot 87: 96–107.

42. Sampedro J, Cosgrove DJ (2005) The expansin superfamily. Genome Biol 6: 242.

43. Nilsen ET, Orcutt DM (1996) Physiology of plants under stress. Abiotic factors. New York: John Wiley and Sons. xii + 689 pp.

44. Cosgrove DJ (1997) Relaxation in a high-stress environment: the molecular bases of extensible cell walls and cell enlargement. Plant Cell 9: 1031–1041.

45. Rachmilevitch S DM, and Huang B (2006) Physiological and biochemical indicators for abiotic stress tolerance. In: Huang B, editor. Plant-Environment Interaction. New York: CRC Press.

46. Jespersen D, Meyer W, Huang B (2013) Physiological traits and genetic variations associated with drought and heat tolerance in creeping bentgrass. Int Turfs Soc Res J 12: 459–464.

47. Xu J, Belanger F, Huang B (2008) Differential gene expression in shoots and roots under heat stress for a geothermal and non-thermal *Agrostis* grass species contrasting in heat tolerance. Environ Exp Bot 63: 240–247.

48. Keller E, Cosgrove DJ (1995) Expansins in growing tomato leaves. Plant J 8: 795–802.

49. Li Z, Durachko DM, Cosgrove DJ (1993) An oat coleoptile wall protein that induces wall extension in vitro and that is antigenically related to a similar protein from cucumber hypocotyls. Planta 191: 349–356.

50. Cho HT, Kende H (1997) Expansins in deepwater rice internodes. Plant Physiol 113: 1137–1143.

51. Zhou P, Zhu Q, Xu J, Huang B (2011) Cloning and characterization of a gene, *AsEXP1*, encoding expansin proteins inducible by heat stress and hormones in creeping bentgrass. Crop Sci 51: 333–341.

17

Optimized Scorpion Polypeptide LMX: A Pest Control Protein Effective against Rice Leaf Folder

Xiuzi Tianpei, Yingguo Zhu, Shaoqing Li*

State Key Laboratory of Hybrid Rice; Key Laboratory for Research and Utilization of Heterosis in Indica Rice of Ministry of Agriculture; Engineering Research Center for Plant Biotechology and Germplasm Utilization of Ministry of Education; College of Life Sciences, Wuhan University, Wuhan, China

Abstract

Lepidopteran insect pests are the main class of pests causing significant damage to crop plant yields. Insecticidal scorpion peptides exhibit toxicity specific for insects. Here, we report that a peptide LMX, optimized from the insect-specific scorpion neurotoxin LqhIT2, showed high levels of activity against rice leaf folder *in vitro* and *in planta*. Oral ingestion of LMX protein led to a significant decrease in feeding on rice leaves, repression of larval growth and development, delay in molting, and increase in larval lethality. Compared with LqhIT2 protein, the stability and insecticidal efficacy of LMX was better. Meanwhile, biochemical analysis showed that LMX protein ingestion dramatically decreased ecdysone content in rice leaf folder larvae, and down-regulated enzymatic activities of the detoxification system (α-naphthyl acetate esterase and glutathione S-transferase), the digestive system (tryptase and chymotrypsin), and the antioxidant system (catalase). These changes were tightly correlated with the dosage of LMX protein. Transgene analysis showed that the rate of leaf damage, and the number of damaged tillers and leaves in the transgenic line were greatly reduced relative to wild type plants and empty vector plants. Based on these observations, we propose that the insect-specific scorpion neurotoxin peptide LMX is an attractive and effective alternative molecule for the protection of rice from rice leaf folder.

Editor: Guy Smagghe, Ghent University, Belgium

Funding: This work was supported by the National High Technology Program (2012AA10A304), Transgenic Research and Development Program (2011ZX08001-004). The funders had no role in study design, data collection and analysis, decision to publish, or preparation of the manuscript.

Competing Interests: The authors have declared that no competing interests exist.

* Email: shaoqingli@whu.edu.cn

Introduction

Rice (*Oryza sativa*) acts as a staple food for much of the global population [1], particularly in Asia where about 90% of people live on rice [2]. Thus, rice plays a critical role in safeguarding the food security of the world. However, in commercial production, rice productivity and quality are adversely affected by many biotic stresses, particularly insect pests.

Rice leaf folder (*Cnaphalocrocis medinalis* Guenee) is a migratory insect which causes significant damage to rice yields [3]. Rice leaf folder larvae damage rice crops by folding leaf blades and feeding inside the rolled leaves. A single larva can damage a number of rice leaves, disturbing photosynthesis and reducing the rice yield [4]. Rice leaf folder is widely distributed around the rice-growing areas of the world. Since the middle of the 1960s, rice leaf folder has become one of the most severe paddy field pests and outbreaks of rice leaf folder have been reported in many Asian countries [5]. In China, since 2003, the annual average area damaged by rice leaf folder was more than 20 million hm^2 and the grain yield loss was up to 760 million kg every year. In 2011, rice leaf folder damaged 22 million hm^2 of rice in China [5].

To avoid rice yield loss caused by leaf folder infestation, conventional means rely on the extensive use of chemical pesticides. However, chemical pest control is expensive, environmentally unfriendly and pest-resurgence often occurs [6]. With wide use of modern biotechnology in rice breeding programs, development of pest-resistant plants through the introduction of foreign insect-resistance genes offers a potentially desirable and effective way to accomplish rice pest control [7]. To date, the most widely used pest-resistant gene is *Cry*, isolated from *Bacillus thuringiensis (Bt)*, which has shown good performance in public and private sector rice breeding programs [8]. However, long-term use of *Bt* genes will increase the risk of narrowing the insect-resistant spectrum of transgenic plants [9]. Thus, to meet this challenge, there is an urgent need to explore economically and ecologically sound alternatives so as to enrich genetic diversity.

Scorpion venom contains a variety of polypeptides with diverse biological activities [10]. According to their targets, scorpion venom polypeptides are divided into three categories: mammal neurotoxins, crustacean neurotoxins and insecticidal toxins [11]. The insecticidal toxins are further divided into two groups based on molecular size and activity. One is the short group with 30–40 amino acid residues and 3–4 disulphide bridges, which mainly affect conductance of potassium channels [12]. The other is the long group with 60–70 amino acid residues cross-linked by 4 disulphide bridges, which principally affect sodium channels in excitable cells [13]. Recently, a series of scorpion insecticidal peptides including *LqhIT2* [14,15], *AaIT* [16,17], *ButaIT* [18,19], *BmkIT* [20] and *AaHIT* [21] have been isolated and tested for enhancement of protection against lepidopteran insect pests. It is notable that these scorpion toxins are neuromuscular system-specific in insects and are safe for other animals [22–25]. The specificity to insect pests offers significant potential for the development of safe insecticides with a broad spectrum. However, none of these have yet been employed as an insecticide defense against rice leaf folder by introducing the genes into rice.

LqhIT2, an insect-specific scorpion toxin, is purified from the venom of *Leiurus quinquestriatus hebraeus* [26]. It is confirmed that having the effective toxic to lepidopteran insect pests [14,15], however, the insecticide efficacy still have some distance to meet our expection in the practical use. Here, we designed an optimized peptide named LMX which has more than 85% amino acid and 92% nucleotide identity with LqhIT2 (Figure S1). After optimizing the nucleotide sequence of *LMX* based on rice codon preference, we introduced the gene into rice to improve resistance to rice leaf folder. Our results demonstrated that the *in vitro* expressed LMX fusion protein had high biological activity and toxicity against rice leaf folder, and the toxicity was better than that of LqhIT2 fusion protein. The *LMX* transgenic rice lines showed high resistance to rice leaf folder. LMX is therefore revealed as a good alternative tool for the further improvement of insect-pest resistance in rice.

Materials and Methods

Ethics Statement

The experimental field in the town of Huashan, Wuhan city, did not require any specific permission for use in this study. This study did not involve endangered or protected species. The specific location of the experimental field was as follows: $114°47'\sim114°60'$ E; $30°51'\sim30°59'$N.

Plant materials

The Indica rice variety Yuetai (*Oryza sativa* L.), which was supplied by our laboratory, was used in this research. Rice was planted in the experimental field in the town of Huashan, Wuhan city, during the summer season from 2010 to 2012.

Insect culture

The third instar larvae of the rice leaf folder were collected from the experimental field in Huashan town, Wuhan city. Rice leaf folder larvae were reared in controlled chambers at $27\pm1°C$ and $75\pm5\%$ relative humidity under a photoperiod of 16-h-light/8-h-dark using fresh leaves of Yuetai [27]. The larvae were fed individually from day 1 of 3rd instar.

Construction of the expression vector and the binary vector

The *LMX*-encoding nucleotide sequence was synthesized (Genscript, China) based on plant codon usage bias [28] and cloned into the vector pUC57 (Genscript, China). To construct a plasmid expressing LMX in *E. coli*, the complete coding region of the *LMX* gene was amplified using PCR primer pairs: LMX-F$_1$ (5-CGG-GATCCATGGACGGCTACATCCGCAAG-3) and LMX-R$_1$ (5-CGGGATCCTTAGCCGCAGGTGTTGGTCTC-3). The *LMX* gene was introduced into the expression vector pGEX-6P-1, digested with *Bam*HI, to construct the plasmid pGEX-6P-1-LMX (Figure 1A). A strain of BL21 was used for protein expression.

Plasmid pCAMBIA1301 was used to construct the binary vector for the transgene. The *LMX* gene driven by the rice green-tissue specific promoter rbcS was cloned into the polylinker of the plasmid. The constructed pCAMBIA1301 vector expressing the *LMX* gene contains a rbcS promoter, a fragment of *LMX* cDNA, a NOS terminator and the hygromycin phosphotransferase (*hpt*) gene as a marker. A strain of Agrobacterium tumefaciens (*EHA105*) was used for the transformation experiment.

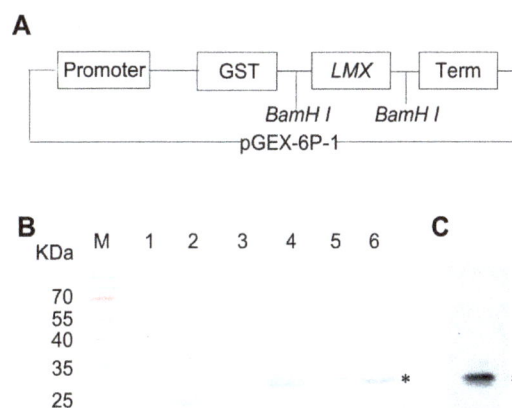

Figure 1. Expression and purification of the LMX protein in *E. coli*. A. Schematic representation of the expression vector pGEX-6P-1-LMX. **B.** Coomassie stained polyacrylamide gel (12%). M, molecular weight marker (SM0671, Fermentas); line 1–2, total protein extracts of pGEX-6P-1 without IPTG induction and with IPTG induction; line 3, total protein extracts of pGEX-6P-1-*LMX* without IPTG induction; line 4–5, supernatant and inclusion bodies of pGEX-6P-1-*LMX* with IPTG induction; line 6, purified sample of supernatant of pGEX-6P-1-LMX with IPTG induction. * represents the target band. **C.** Western blot probed with an antiserum raised against the purified GST-LMX fusion protein. The Molecular mass of polypeptides expressed by the GST-LMX recombinant construct was about 32.8 kDa.

Scorpion LMX protein expression and purification in *E. coli*

The pGEX-6P-1-LMX construct (Figure 1A) was transformed into *E. coli* strain BL21 competent cells and cultured overnight in LB medium at 37°C with 50 µg/ml ampicillin. The culture was diluted 1000-fold in 10 ml of LB medium and allowed to grow to $OD_{600} = 0.8$. The culture was induced with 1 mM IPTG and incubated with shaking for an additional 22 h at 18°C. The IPTG-induced culture was concentrated by centrifugation for 8 minutes at $2,000\times g$ and the bacteria was resuspended in PBS buffer for ultrasonication by the Ultrasonic cell disruption system (SONICS, USA), and then centrifuged at $14,000\times g$ for 30 min at 4°C. The soluble cell extracts and cell pellets were confirmed by 12% SDS-PAGE and western blot.

For large-scale preparation of active LMX protein, the soluble cell extracts of 3 L IPTG-induced culture were collected and purified from clarified bacterial lysate by affinity chromatography on GSTrap FF 1 ml columns (Amersham Biosciences, USA). A polyclonal antiserum was prepared in rabbits against LMX (Newst Biotechnology, China).

Analysis of protein characters and homologous modeling

The basic physic-chemical characters of LMX and LqhIT2 were performed by ProtParam online tools in the ExPASy database (Physico-chemical parameters of a protein sequence, http://expasy.org/tools/protparam.html). The homologous modeling of LMX was performed by SWISS-MODEL which used the three-dimensional structure of LqhIT2 as a template.

Analyses of circular dichroism (CD) spectroscopy

The CD spectra for LMX and LqhIT2 were carried out by using CD spectropolarimeter Jasco J-810 (Japan). The acquisition parameters were 0.1 cm optical path, 50 nm/min scan rate, 25°C temperature, 0.1 nm resolution, 2.0 nm bandwidth, 0.5 nm data pitch, 1 s response and 250~190 nm wavelength range. All treatments were performed in triplicate.

Analysis of the biological toxicity of the LMX protein

To investigate toxin effects of LMX protein on rice leaf folders, the leaf disc method [29] was used as a biological toxicity assay. Freshly collected rice leaves from the Yuetai were cut into 1 cm ×10 cm pieces weighing about 0.2 g, placed in petri dishes padded with humid filter paper and pressed with wet absorbent cotton on the opposite ends of the leaf blade. For the biological toxicity assay, supernatants containing 2.5 μg, 5 μg, 7.5 μg and 10 μg LMX fusion protein were dried under reduced pressure and the dried residues were dissolved in 100 μl PBS buffer. The different dosages of LMX fusion protein were individually daubed on one piece of the leaves. Larva on day 1 of 3rd instar were released and reared individually with one piece of leaf fragment placed in dishes. The dishes were placed in controlled chambers at $27\pm1°C$ and $75\pm5\%$ relative humidity under a photoperiod of 16-h-light/8-h-dark. 100 μl PBS buffer and 10 μg purified GST protein dissolved in 100 μl PBS buffer were used as controls. After 72 h, we recorded the average consumption rates of leaves, the average weight of surviving larvae, and the larvae lethality. Thirty larvae and thirty leaves were used in each treatment group. Three repeats were performed for each treatment, and calculations were done based on the following formula:

$$Comsumed\ rate\ (\%) = \frac{The\ area\ of\ leaf\ by\ larva\ feeding}{The\ total\ area\ of\ leaf} \times 100$$

$$The\ average\ weight\ of\ surviving\ larva\ (mg) = \frac{Total\ weight\ of\ surviving\ larvae}{Number\ of\ surviving\ larvae}$$

$$Larvae\ lethality\ (\%) = \frac{Number\ of\ dead\ larvae}{Number\ of\ total\ larvae} \times 100$$

Measurement of enzyme activities of rice leaf folder

Larvae on day 1 of 3rd instar were released and reared individually with one piece of leaf fragment, which were each applied with a different protein dosage of 2.5, 5 or 7.5 μg of LMX fusion protein. After 72 h, ten larvae as a group were homogenized with PBS buffer, and centrifuged for 30 min at 4,000×g. The suspension of every group of larvae was collected as a crude enzyme extract for measurement. The measurement of the activities of α-naphthyl acetate esterase (α-NAE), glutathione S-transferase, superoxide dismutase (SOD), catalase (CAT), tryptase and chymotrypsin were performed with insect enzyme activity assay kits (Nanjing Jiancheng Bioengineering Institute, China) according to the manufacturer's instructions. PBS buffer and 7.5 μg GST protein were used as controls. Thirty larvae were used in each treatment group. Three repeats were performed in each treatment.

Measurement of rice leaf folder ecdysteroids

Briefly, larvae on day 1 of 3rd instar were individually fed with one piece of leaf fragment coated with different protein dosages of 2.5, 5 or 7.5 μg LMX fusion protein. After 72 h, ten larvae as a

group were weighed and homogenized in 2 ml methanol. Ecdysteroids were extracted according to the method described by Hägele et al. [30]. The ecdysteroid measurement was carried out using High Performance Liquid Chromatography (HPLC) (More Biotechnology, China) based on the following program: Twenty microliters of the solution was injected into a LC-MS instrument (Waters e2695/2489, USA), separations were performed on a Waters Symmetry C_{18} column (4.6×250 mm i.d., 5 μm, Waters, USA) at a flow rate of 0.8 ml/min at 30°C, eluting with acetonitrile, methanol and water in a ratio of 1:2:4. α-ecdysteroid and β-ecdysteroid were used as standards. Thirty larvae were used in each treatment. All treatments were performed in triplicate. PBS buffer and 7.5 μg GST protein were used as controls.

Genetic transformation and molecular identification of LMX transformants

Transformation of Yuetai plant by *A. tumefaciens* was performed following the method described by Peng et al. [31]. The T_0 generation plants were first screened for *LMX* positive lines using specific PCR primer pairs: LMX-F_1 (5-GCTCTAGAATG-GACGGCTACATCCGCAAG-3) and NOS-R_1 (5-GGAATTCG GTTTACCCGCCAATATATCC-3). PCR was performed based on the following program: 94°C for 10 min followed by 35 cycles of 94°C, 1 min; 58°C, 1 min; 72°C, 1 min; and a final extension of 10 min at 72°C. PCR products were separated by 1% (w/v) agarose gel electrophoresis. The PCR positive transgenic lines were further confirmed by Southern blot. Briefly, 10 μg of genomic DNA per sample was digested overnight with *Hind* III and separated on a 0.8% agarose gel, then transferred to a nylon membrane and probed with *LMX* cDNA labeled with biotin-11-dUTP, using the North2South Biotin Random Prime Labeling Kit (Thermo, USA). The blots were hybridized with the labeled probe overnight at 55°C and washed at 55°C for 15 min in 2×SSC and 0.1% SDS three times. The membrane was visualized using the Chemiluminescent Nucleic Acid Detection Module (Thermo, USA).

Quantitative real time PCR

For quantitative real time PCR analysis, total RNA was extracted from approximately 100 mg fresh rice leaves from control and positive transgenic rice plants using the Trizol reagent (Invitrogen) according to the manufacturer's instructions. First-strand cDNA was synthesized using oligo(dT)$_{20}$ primers with the ReverTra Ace transcriptase kit (Toyobo, Japan) according to the manufacturer's instructions.

Quantitative real time PCR was performed in a 20 μl volume using SYBR Green Master Mix (Roche, Switzerland) and data was analyzed on the iCycler iQ5 Real Time PCR Detection System (Bio-Rad, USA) according to the manufacturer's instructions. The cDNA was diluted 10-fold and amplified for quantitative real time PCR analysis with the specific primers LMX-F_1 (5-ATG-GACGGCTACATCCGCAAG-3) and LMX-R_1 (5-TTAGCCG-CAGGTGTTGGTCTC-3). Quantitative real time PCR reactions containing 1 μl of cDNA were carried out by the following the program: 1 cycle at 94°C for 5 min followed by 35 cycles each consisting of 10 s at 94°C, 10 s at 58°C and 10 s at 72°C. Each sample was run in triplicate. The endogenous *ubiquitin* gene was used as the internal control with specific primers (UBI-F_1:5-GTTCGCCAGTTGACATCTC-3, and UBI-F_2: 5-CAGATT-GTTGAGGTTAGTATTGC-3) [32].

Detection of the LMX protein in transgenic rice

The homologous T_2 transgenic plants were used for Western blotting. Approximately 20 mg of protein extracted from fresh rice leaves was loaded in each lane, separated on a 15% Tricine-SDS-PAGE gel, and transferred onto a PVDF membrane (Amersham). Membranes were blocked with TBST buffer (20 mM Tris-HCl pH 8.0, 150 mM NaCl, 0.05% Tween 20) containing 5% fat-free milk, and incubated 1 h at 37°C. The blot was then incubated with LMX rabbit anti-serum diluted 1:2000 overnight at 4°C, and washed three times with TBST at room temperature for 15 min each time. Next the blot was incubated with a secondary antibody goat antirabbit IgG diluted 1:5000 for 1 h at 37°C and washed three times with TBST at room temperature. Immune complexes were detected by Supersignal West Pico assay kit (Thermo, USA).

Evaluation of transgenic rice insect-resistance

To investigate the resistance of transgenic plants expressing *LMX* to rice leaf roller, we used a leaf disc method to analyze the transgenic plants. Freshly collected rice leaves from the 2-month old homozygous transgenic rice plants expressing *LMX* were used. The leaves of transgenic plants were cut into fragments about 10 cm², and placed in petri dishes padded with humid filter paper. Rice leaf folder larvae on day 1 of 3rd instar were released individually with one piece of leaf fragment placed in each dish for 72 h. The plates were kept at a temperature of 27±1°C and relative humidity of 75±5% under a photoperiod of 16-h-light/8-h-dark. The leaves of wild-type plants and transgenic plants expressing the empty vector pCAMBIA1301-rbcS-Nos (rbcS plants) were used as the controls. Ten larvae and ten leaves were placed in each dish, three dishes were used in each treatment, and all treatments were performed in triplicate. The consumption rate of leaves was counted 72 h after release of the insects. The consumption rate was calculated using the following formula:

$$\textit{Comsumed rate } (\%) = \frac{\textit{The area of leaf by larva feeding}}{\textit{The total area of leaf}} \times 100$$

For whole plant analysis of pest-resistance, each homozygous transgenic rice plant was placed in a plastic bucket and infested with five rice leaf folder larvae on day 1 of 3rd instar. Then, the plastic buckets were covered with a net. The severity of tillers and leaves damage to the whole plant was scored on day 7 post-infestation. The wild-type plants and the plants transformed with the empty vector (rbcS plants) were used as controls. Twenty-five larvae and five plants were used in each treatment and each treatment was repeated three times. Further, we also tested resistance of the whole transgenic plant to rice leaf roller under natural conditions. Each plot consisted of 30 plants grown in three rows. No insecticides were applied during the entire growing season. The damage rate to tillers and leaves of the whole plants were calculated using the following formula:

$$\textit{Damage rate of tillers } (\%) = \frac{\textit{Number of tillers by larvae feeding}}{\textit{Number of tillers of the whole plant}} \times 100$$

$$\textit{Damage rate of leaves } (\%) = \frac{\textit{Number of leaves by larvae feeding}}{\textit{Number of leaves of the whole plant}} \times 100$$

Statistical analysis

All experimental data are represented as the mean over three independent replicates. Statistical analysis was done using GraphPad Prism 5.0 software. The values shown in the table and figures represent means ± SD of triplicate. Statistical significance was determined as P≤0.05.

Results

Molecular optimization of LMX polypeptide

In order to enhance the insecticide efficacy of LqhIT2, we changed the amino acid in the non-conservative regions of LqhIT2 to construct the optimized peptide named LMX. The result of amino acid alignment showed that there were nine amino acid sequences differences between LMX and LqhIT2 (Figure S1). The nine amino acid was follow: the lysine, arginine, aspartic acid, valine, alanine, aspartic acid, alanine, tyrosine and glycine at the 5, 6, 8, 12, 13, 22, 27, 28 and 30 were replaced by arginine, lysine, asparagine, isoleucine, serine, asparagine, glycine, phenylalanine and alanine, respectively (Figure S1). The optimized LMX shared has more than 85% amino acid and 92% nucleotide identity with LqhIT2 (Figure S1).

Expression and purification of scorpion polypeptide LMX

To investigate the activity of LMX against rice leaf folder, a GST-LMX recombinant construct was produced using the codon-optimized *LMX* gene (Figure 1A), and expressed in *E. coli*. Coomassie staining revealed that an additional band about 32.8 kDa (LMX 6.8 kDa plus GST 26 kDa) was observed both in soluble and insoluble cell fractions of GST-LMX recombinant *E. coli*, and the GST-LMX fusion protein mainly existed in the soluble cell fraction (Figure 1B). After affinity purification, two bands were visible on a 12% polyacrylamide gel, the biggest one being about 32.8 kDa was GST-LMX fusion protein, and the smaller one around 26 kDa was GST tag protein (Figure 1B). This was further confirmed by Western blot using an antibody specific to the LMX protein (Figure 1C).

Structural analysis of the polypeptide LMX

In order to know the structure of LMX, we analyzed the physico-chemical parameters of LMX protein by using ProtParam online tools in the ExPASy database. The result showed that the instability index and hydropathicity of LMX protein was 30.29 and −0.621 (Table S1). Meanwhile, the instability index and hydropathicity of LqhIT2 protein was 41.89 and −0.651, respectively (Table S1), which means that the hydrophobicity and stability of LMX protein was better than the LqhIT2 protein. Furthermore, we analyzed the secondary structure of LMX and LqhIT2 protein by CD spectra, and the result showed that the spectral curves of LMX and LqhIT2 were highly similar (Figure 2A). Additionally, the homologous modeling showed the three-dimensional structures of LMX and LqhIT2 were also highly similar (Figure 2B). All of these indicated that although there were some differences in physico-chemical parameters between LMX and LqhIT2, the core spatial structure of LMX had no significant change in comparison with LqhIT2 protein.

A

B

LqhIT2

LMX

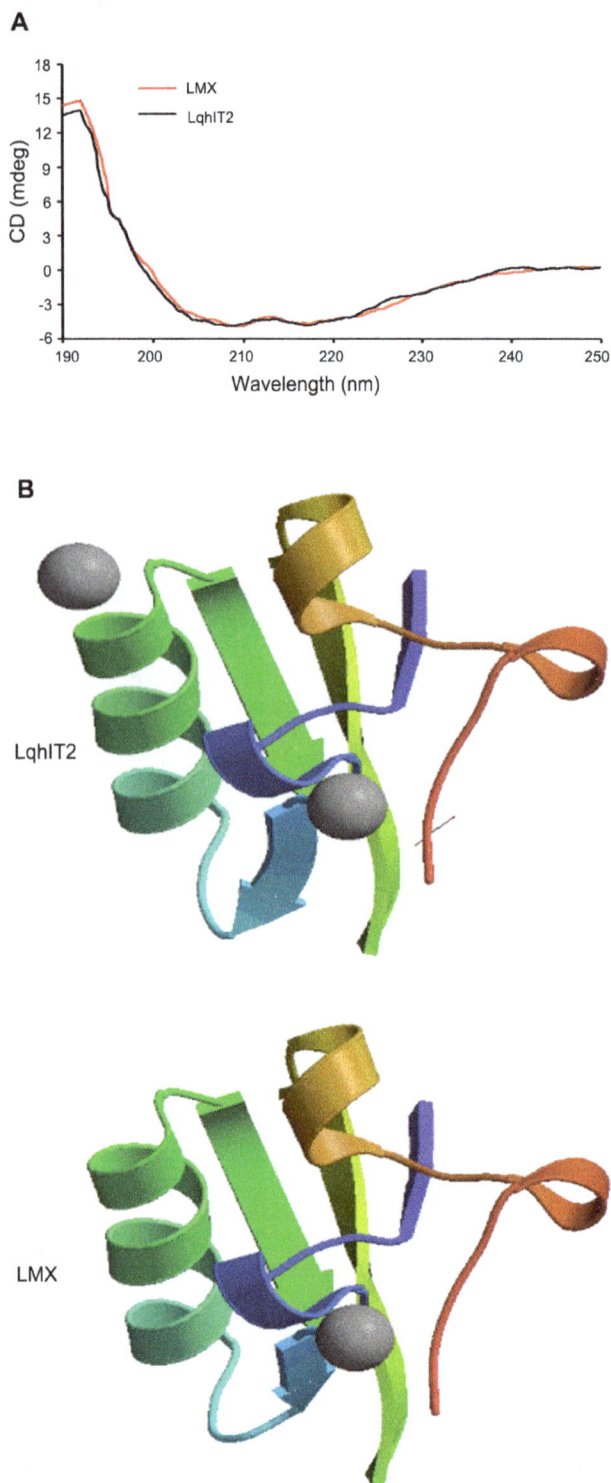

Figure 2. Structural analysis of LMX and LqhIT2 protein. A.
Comparison of the secondary structure of LMX and LqhIT2 protein by
CD spectra. **B**. The homologous modeling of LMX was performed by
SWISS-MODEL. The homologous modeling used the three-dimensional
structure of LqhIT2 as a template.

LMX protein showed high effectively than LqhIT2 in decreases feeding on rice leaves and increase lethality of the rice leaf folder

Some reports have shown that oral ingestion of exogenous
neuropeptides can lead to effective toxicity against pests [19,33].
To investigate the toxicity of the LMX fusion protein against rice
leaf folder by oral ingestion, 2.5~10 µg dose gradients of LMX
fusion protein were applied to Yuetai rice leaves. Compared to a
91.9% damage rate for PBS buffer treated leaves and an 86.6%
damage rate for the 10 µg GST tag protein control, LMX fusion
protein treatments led to significant reductions in damage to rice
leaves on the 3rd day post-infestation. The damage rates of leaves
in the 2.5 µg and 5 µg LMX samples were 45.0% and 20.5%,
respectively; and almost no damage was observed on the leaves
with 7.5 µg and 10 µg of protein (Figure 3A-B). The damage rate
was negatively correlated with the dosage of LMX fusion protein,
and the correlation coefficient reached −0.957 (Figure 3B).
Furthermore, we also detected the toxicity of the LqhIT2 fusion
protein against rice leaf folder by oral ingestion. The damage rates
of leaves in the 2.5~10 µg dose gradients of LqhIT2 protein were
54.0%, 34.5%, 15.2% and 8.6%, respectively. The comparison
result of LMX and LqhIT2 fusion protein showed that the damage
rates of leaves in 2.5~10 µg dose gradients of LqhIT2 samples
were both higher than those in LMX samples (Figure S2A).

Meanwhile, we investigated the effects of LMX fusion protein
on the survival and development of the leaf folder. Results showed
that the lethality rates of insects fed on the rice leaves applied with
LMX protein were much higher than those of the controls on the
3rd day post-infestation (Figure 3C). The lethality rates of larvae
fed with 2.5 µg, 5 µg and 7.5 µg LMX fusion proteins were
12.2%, 45.5% and 68.9%, respectively. When the dosage reached
10 µg, insect mortality was as high as 94.4%. Insect lethality had a
strong positive correlation with the dosage of LMX fusion protein,
and the correlation coefficient was 0.982. There was no significant
difference in insect lethality of controls (larvae ingesting PBS buffer
or 10 µg GST tag protein). Control lethality rates were 7.8% and
8.9%, respectively (Figure 3C). Whereas, the lethality rates of
larvae fed with 2.5 µg, 5 µg, 7.5 µg and 10 µg LqhIT2 fusion
proteins were 11.1%, 35.6%, 58.9% and 78.9%, respectively.
Compared with the insect lethality of 2.5~10 µg dose gradients of
LMX protein, we also found that the lethality rates of gradients of
LqhIT2 protein were much lower (Figure S2B). The results of the
damage rates of leaves and the insect lethality rates suggested that
the insecticidal efficacy of optimized peptide LMX was better than
that of LqhIT2.

LMX protein represses growth and disrupts the equilibrium of ecdysone in larvae of rice leaf folder

Further, we investigated the growth status of the surviving
larvae treated with LMX fusion protein. Compared with the
controls, average weight of the surviving rice leaf folder larvae in
the LMX fusion protein treatment groups was much lower on
both the 3rd day and the 5th day post-infestation (Figure 4). As
shown in Table 1, the average weight of the surviving larvae
ingesting 2.5 µg, 5 µg and 7.5 µg LMX increased 6.1 mg, 3.8 mg
and 2.4 mg after 3 days of treatment, respectively. However, the
average weight of the larvae in the PBS and GST protein control
groups increased 7.4 mg and 6.5 mg, respectively. At the 5th day
of treatment, the larval weight in the GST tag protein control
group increased 10.4 mg, but the LMX (2.5 µg, 5 µg and 7.5 µg)
treatment groups only increased 8.3 mg, 4.7 mg and 2.7 mg,
respectively. This reflects a strong inhibition of LMX fusion
protein on the growth and development of rice leaf folder.

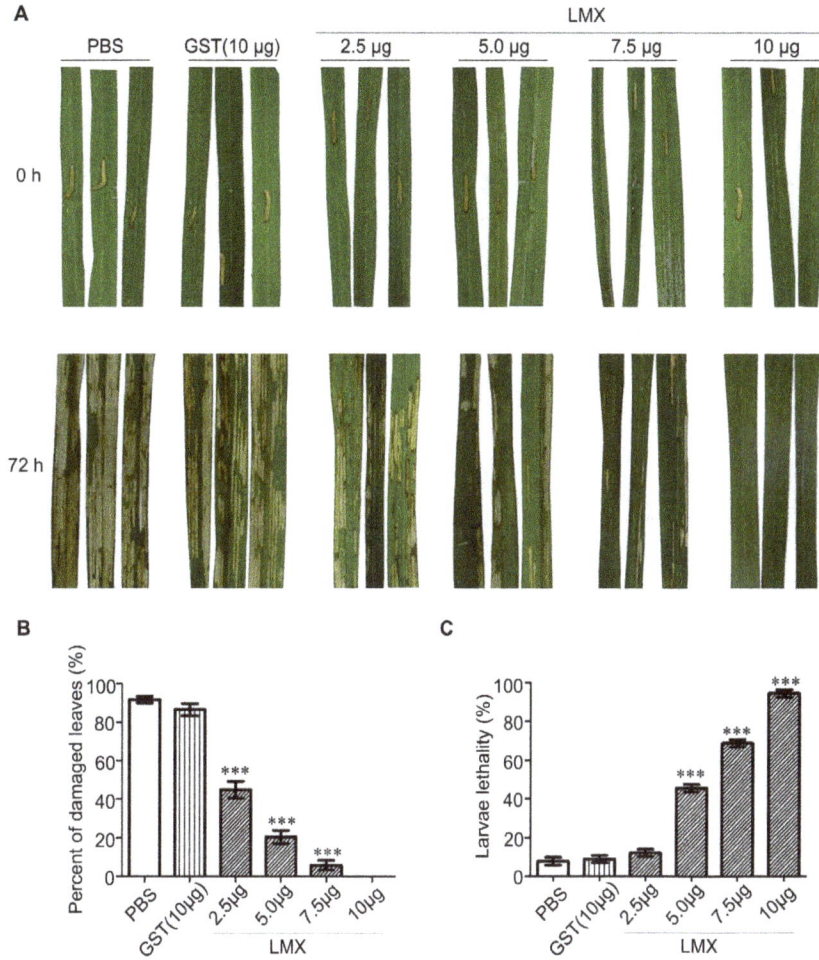

Figure 3. The LMX protein confers resistance to the rice leaf folder. A. Damage to rice leaves daubed with 2.5 µg~10 µg LMX fusion protein, caused by rice leaf folder. **B.** Ratio of the damaged leaf area to the whole leaf. **C.** Larvae lethality of the rice leaf folder fed with 2.5 µg~10 µg bacterial-expressed LMX protein. For the control, the rice leaves were daubed with 1 × PBS buffer and 10 µg GST tag protein. Pictures were taken and ratios of the damaged leaf area were measured after 72 h incubation with the rice leaf folder. Values are means ± SD. *** denotes 0.001 significant difference.

It is well known that growth and development of the larvae of lepidopteran insects are tightly related to molting. Prohibition of the growth of rice leaf folder fed with LMX fusion protein suggests that molting characteristics of the larvae may also change. Interestingly, compared to the normal molting and pupation of the surviving larvae in controls, we observed that molting of the surviving larvae both in the 5 µg and 7.5 µg LMX fusion protein treatment groups was retarded (Figure 5). This indicates that LMX fusion protein could postpone and impede molting and development in the rice leaf folder. α-ecdysone and β-ecdysone are two major insect hormones which play a critical role in controlling insect growth and development [34]. Hence, we investigated the ecdysone content of larvae in the control and treatment groups fed with LMX fusion protein after 72 h of ingestion. Results showed α-ecdysone in the 3rd instar larvae fed with 2.5 µg, 5 µg and 7.5 µg LMX decreased 15.8%, 35.2% and 42.8% respectively, compared to the 7.5 µg GST protein control. Meanwhile, β-ecdysone in larvae of the treatment groups was reduced 5.0%, 31.4% and 50.3% relative to the 7.5 µg GST tag protein control (Figure 5). The changes in α- and β-ecdysone content were both negatively correlated with the dosage of LMX fusion protein, and the correlation coefficients were −0.989 and −0.968, respectively.

This reflects that oral ingestion of LMX fusion protein disturbs the metabolism of α-ecdysone and β-ecdysone in larvae of rice leaf folder.

LMX protein disturbs digestive, detoxification and antioxidant systems in rice leaf folder larvae

In insects, tryptase and chymotrypsin, α-naphthyl acetate esterase (α-NAE) and glutathione S-transferase (GST), superoxide dismutase (SOD) and catalase (CAT) are the key enzymes of the digestive, detoxification and antioxidant systems, respectively. The activities of these enzymes can directly reflect the physiological state of the rice leaf folder. We measured the enzyme activities in larvae of rice leaf folder from the 1st to the 3rd day after infestation. Similar trends were observed for each of the enzymes during the three days (Table S2 and Table 2). According to these trends, we focused on the enzyme activities in larvae on the 3rd day after infestation. After 72 h of treatment, the activities of α-NAE, GST, CAT, POD, tryptase and chymotrypsin in the larvae fed with LMX fusion protein (2.5 µg, 5 µg and 7.5 µg) were reduced. Enzyme activity of α-NAE was decreased 17.5~25.42%, glutathione S-transferase decreased 65.66~76.42%, CAT decreased 28.71~58.63%, tryptase decreased 28.54~40.35%, and chymo-

Figure 4. Phenotype of the rice leaf folder larvae fed with LMX protein. The larvae fed on 2.5 ~7.5 μg LMX protein-applied rice leaves were treatment groups, while the larvae fed on PBS buffer-applied rice leaves and 7.5 μg GST tag protein-applied rice leaves were control groups. Pictures were taken at the start (0 days), 3 days, and 5 days.

trypsin decreased 43.34~63.89% compared to the GST tag protein control (Table 2). In contrast, the activity of SOD was activated, and the increase was 139~149.66% compared to the GST protein control (Table 2).

Expression of the LMX gene improves rice resistance to rice leaf folder

In order to evaluate whether *LMX* can effectively function at the plant level, a rice genetic transformation experiment was

Table 1. Weight of surviving larvae of rice leaf folder on the 3rd and 5th day post-infestation (mg/larvae).

Culture duration	PBS	GST protein (7.5 μg)	LMX (2.5 μg)	LMX (5.0 μg)	LMX (7.5 μg)
0d	21.1±0.31	21.4±0.42	21.0±0.38^{ns}	21.6±0.25^{ns}	21.3±0.32^{ns}
3d	28.5±1.20	27.9±0.60	27.1±0.86^{ns}	25.4±0.89*	23.7±0.87**
5d	32.4±0.45	31.8±0.55	29.3±0.44***	26.3±0.60***	24.0±0.21***

Weight of surviving larvae was measured after 72 h feeding on 2.5 μg~7.5 μg LMX protein-applied rice leaves. The larvae fed on PBS buffer-applied rice leaves and 7.5 μg GST tag protein-applied rice leaves were used as control groups. Values are presented as means of triplicates ±SD. * denotes p<0.05; ** denotes p<0.01; *** denotes p<0.001. ns denotes no significant difference.

A

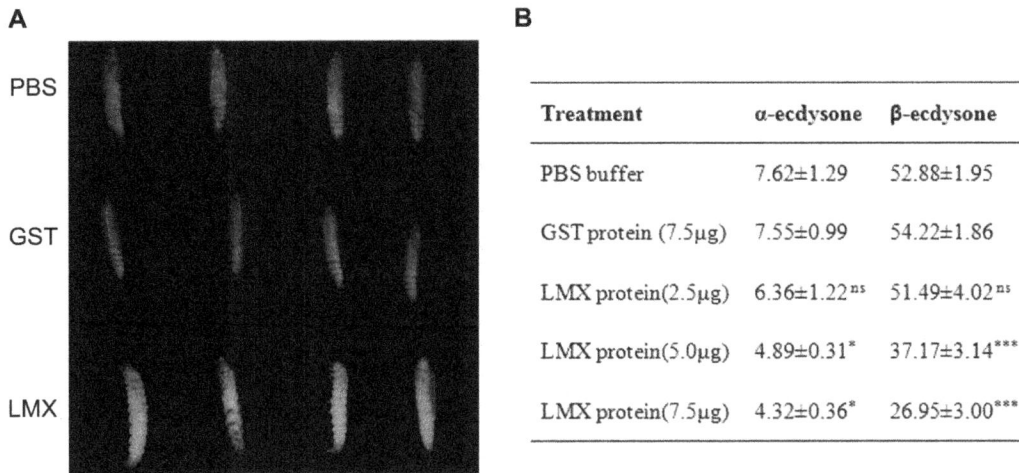

B

Treatment	α-ecdysone	β-ecdysone
PBS buffer	7.62±1.29	52.88±1.95
GST protein (7.5μg)	7.55±0.99	54.22±1.86
LMX protein(2.5μg)	6.36±1.22[ns]	51.49±4.02[ns]
LMX protein(5.0μg)	4.89±0.31[*]	37.17±3.14[***]
LMX protein(7.5μg)	4.32±0.36[*]	26.95±3.00[***]

Figure 5. Molting of rice leaf folder larvae fed with LMX protein. A. Larvae fed with LMX-GST fusion protein delay molting. **B**. Ecdysone content in the larvae of rice leaf folder feeding on LMX fusion protein on the 3rd day post-infestation (ng/mg). The larvae fed on 2.5~7.5 μg LMX protein-applied rice leaves were treatment groups, while the larvae fed on PBS buffer-applied rice leaves and 7.5 μg GST tag protein-applied rice leaves were control groups. Pictures were taken and ecdysone was measured after 72 h feeding. Values are presented as means ± SD of triplicates. * denotes $p<0.05$; *** denotes $p<0.001$. ns denotes no significant difference.

performed. In total, 73 T_0 independent transgenic plants were generated, of which 63 were identified as positive by PCR amplification (Figure 6B). Furthermore, Southern blot analysis showed that three transgenic plants (LMX-5, LMX-10 and LMX-22) harbored a single copy of the insertion (Figure 6C). Western blot and qPCR analysis indicated successful expression of the *LMX* gene under the *rbcS* promoter in these three transgenic rice lines, with the LMX-10 plant having the highest expression level (Figure 6D-E).

Further, we studied the insect resistance of *LMX* transgenic rice plants using wild type and empty vector transgenic plants (rbcS plants) as controls. Larvae on day 1 of the 3rd instar of rice leaf folder were individually fed on one piece of transgenic rice leaf. This showed that the average damage rate to the LMX transgenic rice leaf was about 58.5% after 72 h, whereas the wild type and the rbcS plants were both more than 90% damaged (Figure 7). Meanwhile, we found that *LMX* transgenic rice plants had significantly less damage than the wild type and rbcS plants after a 7 d infestation in an artificial infestation setting (Figure 7 and Table 3). Under natural field conditions, the average damage rates to plant tillers and leaves of wild type plants over two years (2011~2012) were about 86.1% and 68%. The damaged tillers

and leaves of transgenic lines decreased 30.6%~48.3% and 19.1%~36.9% relative to those of the wild type group, respectively (Table 3). However, no significant differences were observed in the damage rates of tillers and leaves between wild type and rbcS plants (Table 3). These data reveal that LMX can efficiently improve the resistance of rice plants against the rice leaf folder.

Discussion

LMX has oral activity against rice leaf folder

It has been proved that scorpion toxin LqhIT2 has high preference for insect voltage-gated sodium channels. In the present work, we designed an optimized protein LMX. The CD spectra and homologous modeling analysis showed that substitution amino acids did not affect the structure, and the core spatial structures of LMX and LqhIT2 were highly similar (Figure 2). It is consistent with previous study that most substitutions in non-conservative regions of LqhIT2 could not affect the CD signature [34]. It was indicated that LMX and LqhIT2 has the similar or same model of action and LMX also affects the insect voltage-gated sodium channels.

Table 2. Enzyme activities in larvae feeding on LMX fusion protein on the 3rd day post-infestation.

Enzymes	PBS	GST protein (7.5 μg)	LMX (2.5 μg)	LMX (5.0 μg)	LMX (7.5 μg)
α-NAE (U/L)	79.59±8.13	81.88±4.39	67.56±6.35[ns]	65.23±5.74[*]	61.07±3.78[*]
GST (U/mg.prot)	452.42±25.25	445.31±11.47	152.91±8.11[***]	114.75±12.79[***]	104.99±3.71[***]
SOD (U/mg.prot)	13.98±0.35	14.54±0.57	34.75±0.31[***]	33.63±0.43[***]	36.30±0.62[***]
CAT (U/mg.prot)	43.95±0.96	44.02±0.35	31.38±0.42[***]	21.56±0.24[***]	18.21±0.28[***]
TPS (IU/ml)	20.58±2.46	21.09±1.05	15.07±1.98[**]	14.79±1.05[**]	12.58±1.27[***]
CTP (U/L)	36.99±1.22	37.61±1.63	21.31±1.13[***]	17.3±0.75[***]	13.58±1.91[***]

For the control, the third larvae were fed with PBS buffer and 7.5 μg GST tag protein. α-NAE: α-naphthyl acetate esterase, GST: glutathione S-transferase, SOD: superoxide dismutase, CAT: catalase, TPS: tryptase, CTP: chymotrypsin. Values are means ± SD of triplicates. * denotes $p<0.05$; ** denotes $p<0.01$; *** denotes $p<0.001$. ns denotes no significant difference.

A

B ... **E**

C ... **D**

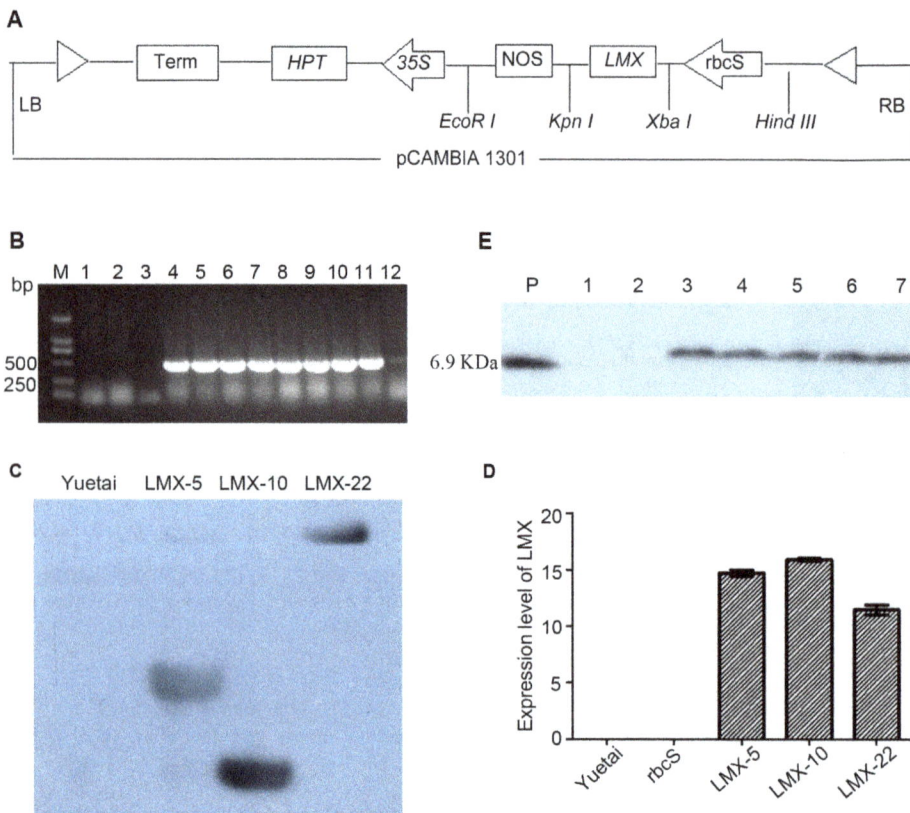

Figure 6. Molecular analysis of the *LMX* transgenic lines. A. Schematic representation of the transgenic binary vector pCAMBIA 1301-LMX. **B**. Analysis of T_0 independent transgenic plants expressing *LMX* by PCR. M, DL2000 marker; 1, Yuetai; 2, ddH$_2$O; 3–12, T_0 generation of transgenic plants expressing *LMX* gene. **C**. Southern-blot analysis of T_0 transgenic lines with single copy insert. The DNA samples were hybridized with the prepared biotin-11-dUTP probe. **D**. qRT-PCR analysis of LMX expression in transgenic lines L-5, L-10, and L-22. The *ubiquitin* gene was used as the internal control. **E**. Western blot assay of T_2 homozygous transgenic line L-10 using LMX antibody. P, purified LMX protein; 1, Yuetai; 2, empty vector transgenic line rbcS; 3–7, homozygous T_2 LMX-10 transgenic lines.

It has been suggested that scorpion insecticidal neurotoxins function by affecting conductance of ion channels by direct injection [35,36]. However, very few studies have investigated the oral activity of scorpion peptides. In the present work, oral ingestion of LqhIT2 protein could affect larval feeding and kill larval (Figure S2), and oral ingestion of LMX protein was also shown to decrease larval feeding on leaves, retard larval development and increase larvae lethality (Figure 3–5; Table 1). The insect-feeding trials clearly demonstrated that bacteria-expressed LMX protein was toxic to rice leaf folder larvae, and the effect of insect-resistance was positively correlated with the dosage of LMX protein. Furthermore, the comparison results demonstrated that the hydrophobicity, stability and toxicity of optimized LMX is better than that of LqhIT2 protein (Figure 3 and Figure S2, Table S1). It it also proved the previous studies that the hydrophobicity and stability of protein are crucial for the insecticidal potency of LqhIT2 [35]. The amino acid Alanine of LqhIT2 which is at 13 site has a unique role in toxin function of LqhIT2 [35]. It could explain that why the toxicity of LMX is better than LqhIT2. It also may be relative with substitution of alanine at 13 site. Ingestion of the LMX protein into the insect gut system is different from the mechanism of direct injection. Though we know little about the exact mechanism of LMX toxicity to the larvae by oral ingestion, we propose two possibilities that may partly explain the toxicity of LMX: (1) The stability of LMX makes LMX peptide have high-levels of resistance to protease. It is

likely to be hyper-stable and has a long residence time in the gut of leaf folder larvae. Based on online prediction by the DiANNA 1.1 web server and DISULFIND web server (https://www.predictprotein.org/), LMX has 4 disulfide bonds; C_{10}-C_{60}, C_{14}-C_{35}, C_{21}-C_{42} and C_{25}-C_{44} (Figure S1C). Disulfide bonds can enforce protein structural stability and resistance to proteases [37]. Hence, absorption of a minority of ingested protein may exert some level of oral activity. This is consistent with the observation that the spider peptide ω-HXTX-Hv1a is orally active against lepidopteran pests when expressed in cotton, poplaf and tobacco plants [38]. (2) The stable LMX protein impedes the physiological and chemical metabolism of leaf folder larvae, which results in disrupted enzyme activities of digestive, antioxidant and detoxification systems and the disequilibrium of ecdysones involved in the process of insect molting, causing the observed delay in larva development (Table 2 and Figure 5). This further inhibits the growth and development of insect larva, and therefore decreases feeding and damage to rice leaves.

LMX is a valuable insecticidal candidate for rice improvement

In the last few decades, some valuable resistance genes have been successfully identified. Combined with transgenic techniques, this will make it possible to breed environmentally friendly rice by improving rice resistance to insect pests through bioengineering techniques. This engineered rice will potentially reduce the use of

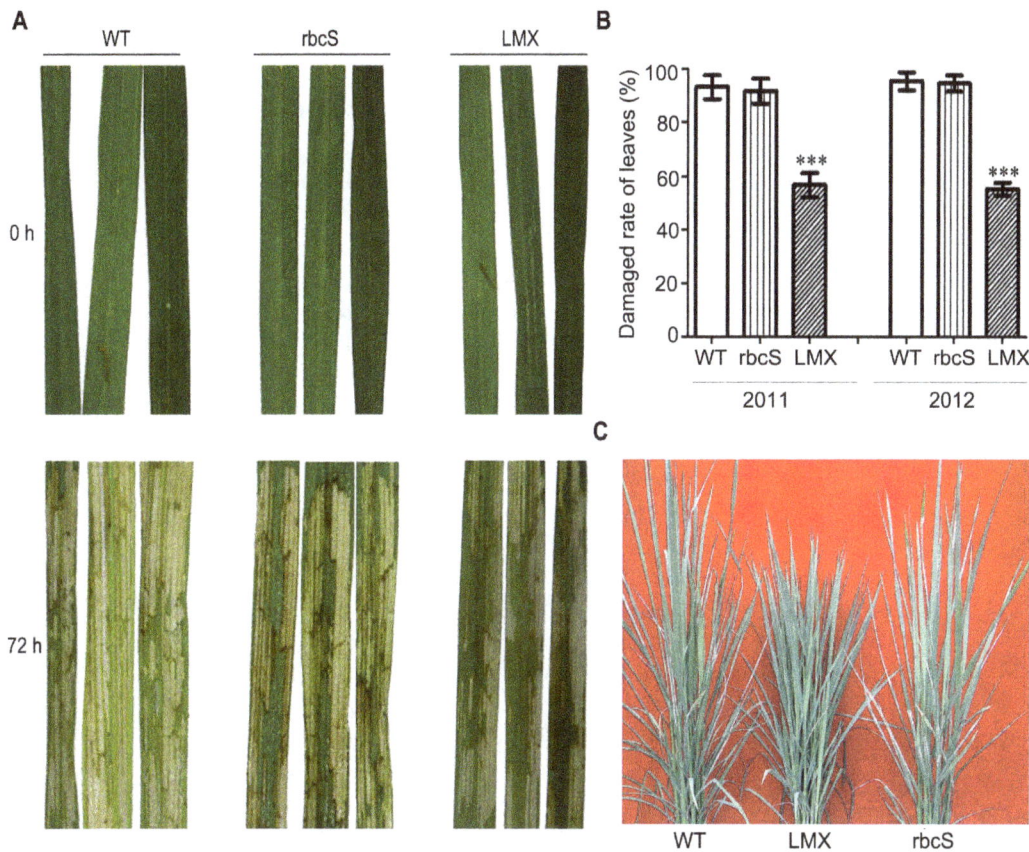

Figure 7. Leaves of the *LMX* transgenic rice plants demonstrate insect resistance. A. Damage caused by rice leaf folder on the leaves of *LMX* transgenic rice plants. **B**. Ratio of the damaged leaf area to the whole leaf. **C**. Whole transgenic rice plants expressing LMX demonstrated improved resistance to rice leaf roller. Wild type plants and empty vector transgenic plants (rbcS plants) were used as the controls and homozygous LMX transgenic rice plants were used as the experimental group. Day 1 of 3rd instar rice leaf roller larvae were utilized in the bioassay. Leaves were photographed and measured after 72 h incubation with the rice leaf folder. Plants were photographed and measured after 7 days of incubation with the rice leaf folder larvae. Values are mean ± SD, *** denotes 0.001 significance value.

pesticide in commercial rice production [39,40]. *Bacillus thuringiensis* (*Bt*) genes are the most prominent examples used for rice genetic improvement in resistance to insects, especially lepidoptera [41]. However, widespread use of a single exogenous *Bt* insecticidal protein will lead to the adaptation of insect pests to this protein. In fact, a number of *BT*-resistant insect strains have been observed under laboratory, greenhouse, and even field conditions [42–44]. This reflects the urgency in exploring new useful insecticidal genes to arm crop plants against challenges from insect pests. Recently, some insect toxin genes, such as proteases [45], protease inhibitors [46–48], and the spider venom toxin gene *Hvt* [49] etc. have been tested for the improvement of plant resistance to insect pests, but the number of available genes conferring effective pest resistance remains limited.

It has been suggested that ectopic expression of insect-selective neurotoxin peptides would be highly advantageous in coping with lepidopteran insect larvae, which attack over 60 crop plants ranging from monocotyledon to dicotyledon [50]. As discussed above, many insecticidal scorpion peptides have been suggested as a promising alternative for insect pest control [17,18,21,51]. Of these, LqhIT2 has been demonstrated to confer resistance to lepidoptera insects [14] without toxicity to mammals [23,24]. Its efficacy in transgenic rice against leaf folder has also been confirmed in our laboratory (data in preparation for subsequent publication). In our study, LMX shares more than 85% amino

acid homology with LqhIT2 (Figure S1). Both nurse chamber feeding and natural field investigations confirmed that *LMX* can effectively improve rice plant resistance to rice leaf folder (Figure 7; Table 3). Therefore, the *LMX* gene is critical for improving resistance to rice leaf folder.

After oral ingestion, we observed the toxicity of LMX protein to larvae to be positively correlated with protein dosage. 10 μg of LMX protein almost completely repressed larvae feeding on rice leaves. Practically, LMX content in transgenic rice leaves will be far lower than this level. As larvae living on LMX transgenic rice can implement the transit from larvae to pupa, this implies that to some extent these larvae are able to endure the toxicity of LMX on the *in-planta* level. If we use only LMX as a resistance gene in rice improvement, it will lead to the leaf folder's swift adaptation to this toxin [15]. Thus, comprehensive strategies such as combining scorpion peptide LMX with other insecticidal genes like spider venom peptide [49] and plant insecticidal proteins [52], are necessary to prevent the occurrence of pest resistance. In the interest of increasing the security of transgenic plants, as well as the safety of mammals, we propose that insecticidal scorpion peptide LMX may play a useful role in the genetic modification of rice for insect pest resistance.

Table 3. Damage analysis of LMX transgenic rice caused by rice leaf folders from 2011 to 2012.

Years	Treatments	Genotypes	Damage rate of tillers (%)	Damage rate of leaves (%)
2011				
		Wild type	95.0±2.35	95.2±1.12
		rbcS	95.3±1.76[ns]	95.5±1.36[ns]
	Artificial infestation	LMX-5	71.2±1.59***	62.8±1.61***
		LMX-10	65.6±3.16***	56.6±3.20***
		LMX-22	78.6±2.0***	74.2±1.58***
		Wild type	95.9±4.40	86.5±3.10
		rbcS	95.6±4.70[ns]	86.2±3.70[ns]
	Natural infestation	LMX-5	54.0±3.68***	59.7±4.46***
		LMX-10	45.8±6.50***	50.7±4.00***
		LMX-22	62.9±2.51***	67.5±1.99***
2012				
		Wild type	93.0±0.93	94.0±2.08
		rbcS	93.8±2.28[ns]	93.0±1.49[ns]
	Artificial infestation	LMX-5	73.1±5.45***	66.5±2.59***
		LMX-10	64.3±2.33***	54.5±2.68***
		LMX-22	80.3±4.01***	76.5±2.46***
		Wild type	76.3±11.80	49.5±10.30
		rbcS	79.0±7.50[ns]	50.5±7.30[ns]
	Natural infestation	LMX-5	34.5±1.66***	21.7±2.44***
		LMX-10	29.8±5.70***	11.5±2.60***
		LMX-22	48.2±4.65***	30.3±1.13***

Values are means ± SD of ten different plants. *** denotes $P<0.001$.

Conclusions

The results of *in vitro* and *in planta* experiments provide evidence that insect-specific scorpion neurotoxin peptide LMX has high levels of activity against rice leaf folder. LMX represses larval growth and molting, affects enzyme activities, and increases larval lethality *in vitro*. In addition, transgenic rice plants expressing *LMX* demonstrate enhanced resistance to rice leaf folders. Therefore *LMX* could be considered as a candidate gene to improve plant resistance to lepidopteran insects.

Supporting Information

Figure S1 LMX disulfide bonds and sequence alignment. A. Alignment of the amino acid sequences of the mature polypeptide LMX and the LqhIT2 polypeptide. The gray frames indicate the same amino acid for the LMX and LqhIT2 polypeptides in non-conservative region. The purple frames highlight the conservative region between LMX and LqhIT2 polypeptides. B. Alignment of the full-length LMX and LqhIT2 nucleic acid sequence according to plant codon usage. The gray frames indicate the same nucleic acid sequence, shared between LMX and LqhIT2. C. Disulfide bonds of the mature polypeptide LMX.

Figure S2 The LqhIT2 protein confers resistance to the rice leaf folder. A. Ratio of the damaged leaf area to the whole leaf applied with 2.5 µg~10 µg bacterial-expressed LqhIT2 protein. B. Larvae lethality of the rice leaf folder fed with 2.5 µg~10 µg bacterial-expressed LqhIT2 protein. For the control, the rice leaves were daubed with 1 × PBS buffer and 10 µg GST tag protein. Pictures were taken and ratios of the damaged leaf area were measured after 72 h incubation with the rice leaf folder. Values are means ± SD. *** denotes 0.001 significant difference.

Table S1 The physic-chemical parameters of LMX and LqhIT2 protein.

Table S2 Enzyme activities in larvae feeding on LMX fusion protein on 1st and 2nd day post-infestation.

Acknowledgments

The authors would like to thank Dr. Zhijian Cao for his help in optimizing the LMX sequence.

Author Contributions

Conceived and designed the experiments: SQL YGZ XZTP. Performed the experiments: XZTP. Analyzed the data: XZTP. Contributed reagents/materials/analysis tools: XZTP. Wrote the paper: XZTP SQL.

References

1. Food and Agriculture Organization of the United Nations (FAO) (2004) The state of food security in the world. pp. 30–31.
2. Khush GS, Brar DS (2002) Biotechnology for rice breeding: progress and potential impact. In: Proc. 20th Session of the Int. Rice Commission: 23rd–26th July 2002; Thailand: Bangkok.
3. Luo SJ (2010) Occurrence of rice leaf roller in China and its identification and prevention. Plant Diseases and Pests 1: 13–18.
4. Alvi SM, Ali MA, Chaudhary S, Iqbal S (2003) Population trends and chemical control of rice leaf folder, Cnaphalocrocis medinalis on rice crop. Int J Agri Biol 5: 615–617.
5. Li SW, Yang H, Liu YF, Liao QR, Du J, et al. (2012) Transcriptome and gene expression analysis of the rice leaf folder, Cnaphalocrosis medinalis. PLoS One 7: e47401.
6. Ho NH, Baisakh N, Oliva N, Datta K, Frutos R, et al. (2006) Translation fusion Hybrid Bt genes confer resistance against yellow stem borer in transgenic elite Vietnamese rice (Oryza sativa L.) cultivars. Crop Sci 46: 781–789.
7. Deka S, Barthakur S (2010) Overview on current status of biotechnological interventions on yellow stem borer Scirpophaga incertulas (Lepidoptera: Crambidae) resistance in rice. Biotechnol Adv 28: 70–81.
8. Kumar S, Chandra A, Pandey KC (2008) Bacillus thuringiensis (Bt) transgenic crop: an environment friendly insect-pest management strategy. J Environ Biol 29: 641–653.
9. Shelton AM, Zhao JZ, Roush RT (2002) Economic, ecological, food, safety and social consequences of the deployment of Bt transgenic plants. Annu Rev Entomol 47: 845–881.
10. Froy O, Sagiv T, Poreh M, Urbach D, Zilberberg N, et al. (1999) Dynamic diversification from a pupative common ancestor of scorpion toxins affecting sodium, potassium and chloride channels. J Mol Evol 48: 187–196.
11. Gurevitz M, Karbat I, Cohen L, Ilan N, Kahn R, et al. (2007) The insecticidal potential of scorpion β-toxins. Toxicon 49: 473–489.
12. Bergeron ZL, Bingham JP (2012) Scorpion toxins specific for potassium (K+) channels: a historical overview of peptide bioengineering. Toxins 4: 1082–1119.
13. Rodríguez de la Vega RC, Possani LD (2005) Overview of scorpion toxins specific for Na+-channels and related peptides: biodiversity, structure-function relationships and evolution. Toxicon 46: 831–844.
14. van Beek N, Lu A, Presnail J, Davis D, Greenamoyer C, et al. (2003) Effect of signal sequence and promoter on the speed of action of a genetically modified Autographa californica nucleopolyhedrovirus expressing the scorpion toxin LqhIT2. Biol Control 27: 53–64.
15. Regev A, Rivkin H, Inceoglu B, Gershburg E, Hammock BD, et al. (2003) Further enhancement of baculovirus insecticidal efficacy with scorpion toxins that interact cooperatively. FEBS Lett 537: 106–110.
16. Wang CS, St Leger RJ (2007) A scorpion neurotoxin increases the potency of a fungal insecticide. Nat Biotechnol 25: 1455–1456.
17. Pava-Ripoll M, Posada FJ, Momen B, Wang CS, St Leger R (2008) Increased pathogenicity against coffee berry borer, Hypothenemus hampei (Coleoptera: Curculionidae) by Metarhizium anisopliae expressing the scorpion toxin (AαIT) gene. J Invertebr Pathol 99: 220–226.
18. Rajendra W, Hackett KJ, Buckley E, Hammock BD (2006) Functional expression of lepidopteran-selective neurotoxin in baculovirus: potential for effective pest management. Biochim Biophys Acta 1760: 158–163.
19. Pham Trung N, Fitches E, Gatehouse JA (2006) A fusion protein containing a lepidopteran-specific toxin from the South Indian red scorpion (Mesobuthus tamulus) and snowdrop lectin shows oral toxicity to target insects. BMC Biotechnol 6: 28–43.
20. Fan XJ, Zheng B, Fu YJ, Sun Y, Liang AH (2008) Baculovirus-mediated expression of a Chinese scorpion neurotoxin improves insecticidal efficacy. Chinese Sci Bull 53: 1855–1860.
21. Wu JH, Luo XL, Wang ZA, Tian YC, Liang AH, et al. (2008) Transgenic cotton expressing synthesized scorpion insect toxin AaHIT gene confers enhanced resistance to cotton bollworm (Heliothis armigera) larvae. Biotechnol Lett 30: 547–554.
22. de Dianous S, Hoarau F, Rochat H (1987) Re-examination of the specificity of the scorpion Androctonus australis Hector insect toxin towards arthropods. Toxicon 25: 411–417.
23. Herrmann R, Moskowitz H, Zlotkin E, Hammock BD (1995) Positive cooperativity among insecticidal scorpion neurotoxins. Toxicon 33: 1099–1102.
24. Benkhalifa R, Stankiewicz M, Lapied B, Turkov M, Zilberberg N, et al. (1997) Refined electrophysiological analysis suggests that a depressant toxin is a sodium channel opener rather than a blocker. Life Sci 61: 819–830.
25. Wudayagiri R, Inceoglu B, Herrmann R, Derbel M, Choudary PV, et al. (2001) Isolation and characterization of a novel lepidopteran-selective toxin from the venom of South Indian red scorpion, Mesobuthus tamulus. BMC Biochem 2: 16–23.
26. Zlotkin E, Eitan M, Bindokas VP, Adams ME, Moyer M, et al. (1991) Functional duality and structursal uniqueness of depressant insect-selective neurotoxins. Biochemistry 30: 4814–4821.
27. Ramachandran R, Khan ZR (1991) Mechanisms of resistance in wild rice Oryza brchyantha to rice leaffolder Cnaphalocrocis medinalis (Guenée) (Lepidoptera: Pyralidae). J Chem Ecol 17: 41–65.
28. Murray EE, Lotzer J, Eberle M (1989) Codon usage in plant genes. Nucleic Acids Res 17: 477–498.
29. Smigocki AC, Ivic-Haymes S, Li HY, Savić J (2013) Pest protection conferred by a Beta vulgaris serine proteinase inhibitor gene. PLoS One 8: e57303.
30. Hägele BF, Wang FH, Sehnal F, Simpson SJ (2004) Effects of crowding, isolation, and transfer from isolation to crowding on total ecdysteroid content of eggs in Schistocerca gregaria. J Insect Physiol 50: 621–628.
31. Peng XJ, Wang K, Hu CF, Zhu YL, Wang T, et al. (2010) The mitochondrial gene orfH79 plays a critical role in impairing both male gametophyte development and root growth in CMS-Honglian rice. BMC Plant Biol 10: 125–135.
32. Hu J, Zhou JB, Peng XX, Xu HH, Liu CX, et al. (2011) The Bphi008a gene interacts with the ethylene pathway and transcriptionally regulates MAPK genes in the response of rice to brown planthopper feeding. Plant Physiol 156: 856–872.
33. Fitches E, Audsley N, Gatehouse JA, Edwards JP (2002) Fusion proteins containing neuropeptides as novel insect control agents: snowdrop lectin delivers fused allatostatin to insect haemolymph following oral ingestion. Insect Biochem Mol Biol 32: 1653–1661.
34. Tawfik AI, Vedrova A, Sehnal F (1999) Ecdysteroids during ovarian development and embryogenesis in solitary and gregarious schistocerca gregaria. Arch Insect Biochem Physiol 41: 134–143.
35. Karbat I, Turkov M, Cohen L, Kahn R, Gordon D, et al. (2007) X-ray structure and mutagenesis of the scorpion depressant toxin LqhIT2 reveals key determinants crucial for activity and anti-insect selectivity. J Mol Biol 366: 586–601.
36. Strugatsky D, Zilberberg N, Stankiewicz M, Ilan N, Turkov M, et al. (2005) Genetic polymorphism and expression of a highly potent scorpion depressant toxin enables refinement of the effects on insect Na-channels and illuminates the key role of Asn-58. Biochemistry 44: 9179–9187.
37. Fass D (2012) Disulfide bonding in protein biophysics. Annu Rev Biophys 41: 63–79.
38. King GF, Hardy MC (2012) Spider-venom peptides: structure, pharmacology, and potential for control of insect pests. Annu Rev Entomol 58: 475–496.
39. Benedict JH, Altman DW (2001) Commercialization of transgenic cotton expressing insecticidal crystal protein. In: Jenkins JJ, Saha S, editors. Genetic improvement of cotton: Emerging technologies. USA: Science Publishers; pp. 137–201.
40. Christou P, Capell T, Kohli A, Gatehouse JA, Gatehouse AMR (2006) Recent developments and future prospects in insect pest control in transgenic crops. Trends Plant Sci 11: 302–308.
41. Perlak FJ, Oppenhuizen M, Gustafson K, Voth R, Sivasupramaniam S, et al. (2001) Development and commercal use of Bollgard cotton in the USA—early promises versus today's reality. Plant J 27: 489–501.
42. Ferré J, Van Rie J (2002) Biochemistry and genetics of insect resistance to Bacillus thuringiensis. Annu Rev Entomol 47: 501–533.
43. Tabashnik BE, Carrière Y, Dennehy TJ, Morin S, Sisterson MS, et al. (2003) Insect resistance to transgenic Bt crops: lessons from the laboratory and field. J Econ Entomol 96: 1031–1038.
44. Tabashnik BE, Brévault T, Carrière Y (2013) Insect resistance to Bt crops: lessons from the first billion acres. Nat Biotechnol 31: 510–521.
45. Sun XL, Wu D, Sun XC, Jin L, Ma Y, et al. (2009) Impact of Helicoverpa armigera nucleopolyhedroviruses expressing a cathepsin L-like protease on target and nontarget insect species on cotton. Biol Control 49: 77–83.
46. Dunse KM, Stevens JA, Lay FT, Gaspar YM, Heath RL, et al. (2010) Coexpression of potato type I and II proteinase inhibitors gives cotton plants protection against insect damage in the field. Proc Natl Acad Sci USA 107: 15011–15015.
47. Schlüter U, Benchabane M, Munger A, Kiggundu A, Vorster J, et al. (2010) Recombinant protease inhibitors for herbivore pest control: a multitrophic perspective. J Exp Bot 61: 4169–4183.
48. Alvarez-Alfageme F, Maharramov J, Carrillo L, Vandenabeele S, Vercammen D, et al. (2011) Potential use of a serpin from Arabidopsis for pest control. PLoS one 6: e20278.
49. Khan SA, Zafar Y, Briddon RW, Malik KA, Mukhtar Z (2006) Spider venom toxin protects plants from insect attack. Transgenic Res 15: 349–357.
50. Christian P (1994) Recombinant baculovirus insecticides: Catalysts for a change of heart? In: Monsour CJ, Reid S, Teakle RE, editors. Proc. Symp. Biopesticides.Brisbane: Opportunities for Australian industry. pp. 40–50.
51. Wang JX, Chen ZL, Du JZ, Sun Y, Liang AH (2005) Novel insect resistance in Brassica napus developed by transformation of chitinase and scorpion toxin genes. Plant Cell Rep 24: 549–555.
52. Koundal KR, Rajendran P (2003) Plant insecticidal proteins and their potential for developing transgenics resistant to insect pests. Indian J Biotechnol 2: 110–120.

18

A Remorin Gene *SiREM6*, the Target Gene of SiARDP, from Foxtail Millet (*Setaria italica*) Promotes High Salt Tolerance in Transgenic *Arabidopsis*

Jing Yue[9], Cong Li[9], Yuwei Liu, Jingjuan Yu*

State Key Laboratory of Agrobiotechnology, College of Biological Sciences, China Agricultural University, Beijing, China

Abstract

Remorin proteins (REMs) form a plant-specific protein family, with some REMs being responsive to abiotic stress. However, the precise functions of REMs in abiotic stress tolerance are not clear. In this study, we identified 11 remorin genes from foxtail millet (*Setaria italica*) and cloned a remorin gene, *SiREM6*, for further investigation. The transcript level of *SiREM6* was increased by high salt stress, low temperature stress and abscisic acid (ABA) treatment, but not by drought stress. The potential oligomerization of SiREM6 was examined by negative staining electron microscopy. The overexpression of *SiREM6* improved high salt stress tolerance in transgenic *Arabidopsis* at the germination and seedling stages as revealed by germination rate, survival rate, relative electrolyte leakage and proline content. The *SiREM6* promoter contains two dehydration responsive elements (DRE) and one ABA responsive element (ABRE). An ABA responsive DRE-binding transcription factor, *SiARDP*, and an ABRE-binding transcription factor, *SiAREB1*, were cloned from foxtail millet. SiARDP could physically bind to the DREs, but SiAREB1 could not. These results revealed that *SiREM6* is a target gene of SiARDP and plays a critical role in high salt stress tolerance.

Editor: Jin-Song Zhang, Institute of Genetics and Developmental Biology, Chinese Academy of Sciences, China

Funding: This work was supported by the National Basic Research Program of China (2012CB215301). The funders had no role in study design, data collection and analysis, decision to publish, or preparation of the manuscript.

Competing Interests: The authors have declared that no competing interests exist.

* Email: yujj@cau.edu.cn

9 These authors contributed equally to this work.

Introduction

Plant growth and development are constrained by environmental stress conditions. Salt stress is one of the major environmental stresses in agriculture worldwide and affects productivity and crop quality [1]. High salinity stress causes hyperosmotic stress, ion toxicity and nutrient deficiency, and can lead to molecular damage and even plant death. To respond and adapt to high salinity stress, plants have developed many strategies, such as selective ion uptake and exclusion, efficient detoxification by the antioxidant system [2], and the accumulation of osmotically protective matter [3]. Numerous salt tolerance- relevant genes are induced in response to salt stress [4].

The remorin protein family exists in all land plants, including angiosperms, gymnosperms, pteridophytes and bryophytes [5]. The first remorin was discovered in potato in 1989 and named pp34 for its 34 kD molecular mass position in protein gels [6]. The protein was renamed as remorin to indicate its ability to attach to the plasma membrane [7]. Recently, more remorin genes have been identified from different plants [8,9,10,11,12,13]. Remorins contain a conserved C-terminal region and a variable N-terminal region. The coiled-coil structure exists in the C-terminal region of remorin and is considered the family's signature. The variable N-terminal region of remorin suggests different structures and functions [14]. Based on the phylogenetic trees analysis and the different N-terminal domains, remorins are divided into six groups. While groups 1, 2 and 3 were not clearly separated by phylogeny, their domain features allowed them to be subdivided further [8]. In addition, many remorins could oligomerize *in vitro* [15].

Transcriptome and proteome analyses suggest that remorins play very important roles in plants in response to biotic and abiotic stresses [16,12,17,18,19,20,21]. A *Medicago truncatula* remorin protein, MtSYMREM1, induced during nodulation, interacts with symbiotic receptors, such as NFP, LYK3 and DMI2 that are important for the perception of bacterial signaling molecules. Oligomeric MtSYMREM1 attaches to the host plasma membrane surrounding the rhizobium, and controls the release of rhizobia into the host cytoplasm. Thus, *MtSYMREM1* has an important role during the plant-bacteria interaction [22]. Remorin gene *LjSYMREM1* was cloned from *Lotus japonicus*. The overexpression of *LjSYMREM1* increased root nodulation in transgenic plants. Functional analysis revealed that the C-terminal region of LjSYMREM1, especially the coiled-coil domain, was very important for protein interactions and remorin oligomerization. The RLK kinase interacted with the LjSYMREM1 protein *in vivo* and phosphorylated a residue in the N-terminal region *in vitro*. The molecular mechanisms of the LjSYMREM1 protein showed a new function and the importance of scaffold proteins during rhizobial infection [23]. *MiREM* from mulberry (*Morus indica*) was the first reported remorin gene involved in abiotic stress. Heterologous expression of MiREM in *Arabidopsis* enhanced drought and high

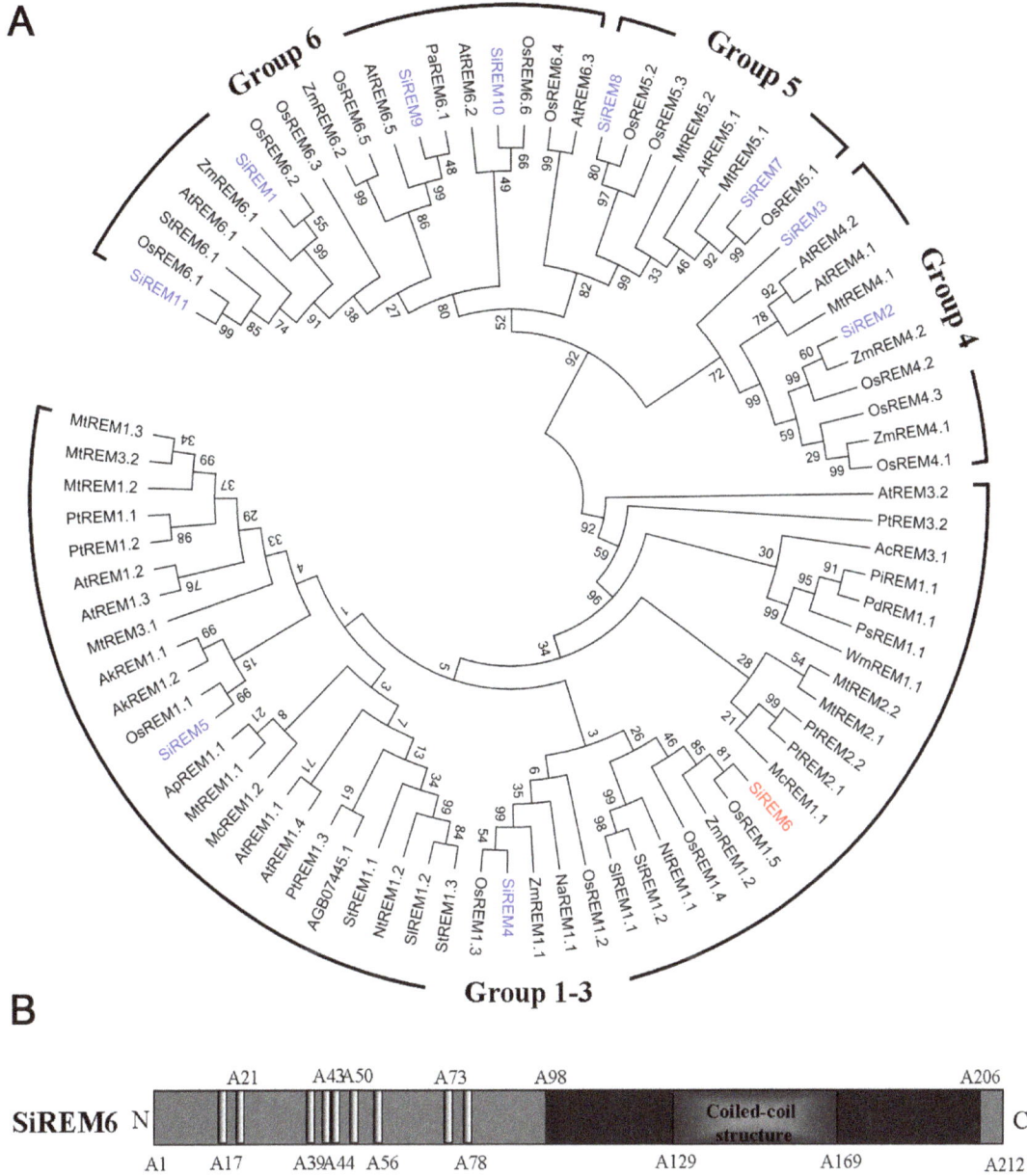

Figure 1. Phylogenetic tree of remorin proteins and predicted domains the SiREM6 protein. (A) Phylogenetic tree of remorin proteins from various plants. The multiple alignments were generated by MUSCLE and the phylogenetic tree was constructed by MEGA5.2.2 using a bootstrap test of phylogeny and the Neighbor Joining test with default parameters. The proteins belonged to four brackets: groups 1 to 3, group 4, group 5 and group 6. The nomenclature is based on Raffaele et al. (2007). The Genbank numbers of the remorin proteins used for the phylogenetic tree are shown in Table S2. (B) Predicted domains in the SiREM6 protein. The nine prolines in the N-terminal region are indicated as amino acid residue numbers. The black box indicates the C-terminal conserved region, and the coiled-coil structure is in this region between A129 and A169.

salinity tolerance during the germination and seedling stages [5]. The study of abiotic stress-response functions for remorins in plants was novel.

The dehydration responsive element binding (DREB)-type transcription factors are a subfamily of the APETALA2 (AP2)/ ethylene responsive factor (ERF) protein family, and play an important role in the responses to various stresses. Since the first DREB gene was cloned using the yeast one-hybrid screening system in *Arabidopsis* [24,25], many DREB genes have been identified from rice, maize and barely [26,27,28]. Most DREB genes were responsive to abiotic stresses. The DREB proteins bind to the DRE core sequence in the promoter region of target genes

and regulate their transcription. The overexpression of these DREB genes enhanced transgenic plant tolerance to abiotic stresses and accumulated osmoprotectants, such as proline and sugars [29]. Past studies indicate that DREB transcription factors regulated downstream gene expression through the abscisic acid (ABA)-independent signal pathway. However, increasing evidence shows that some DREB transcription factors are also responsive to ABA signals and are involved in the ABA signal pathway [30,31].

Foxtail millet, an important crop in China, can grow in marginal soils and has a high tolerance of hostile environments [32]. It is important to identify new stress-relevant genes from foxtail millet. Although there is little data to confirm the function

A

B

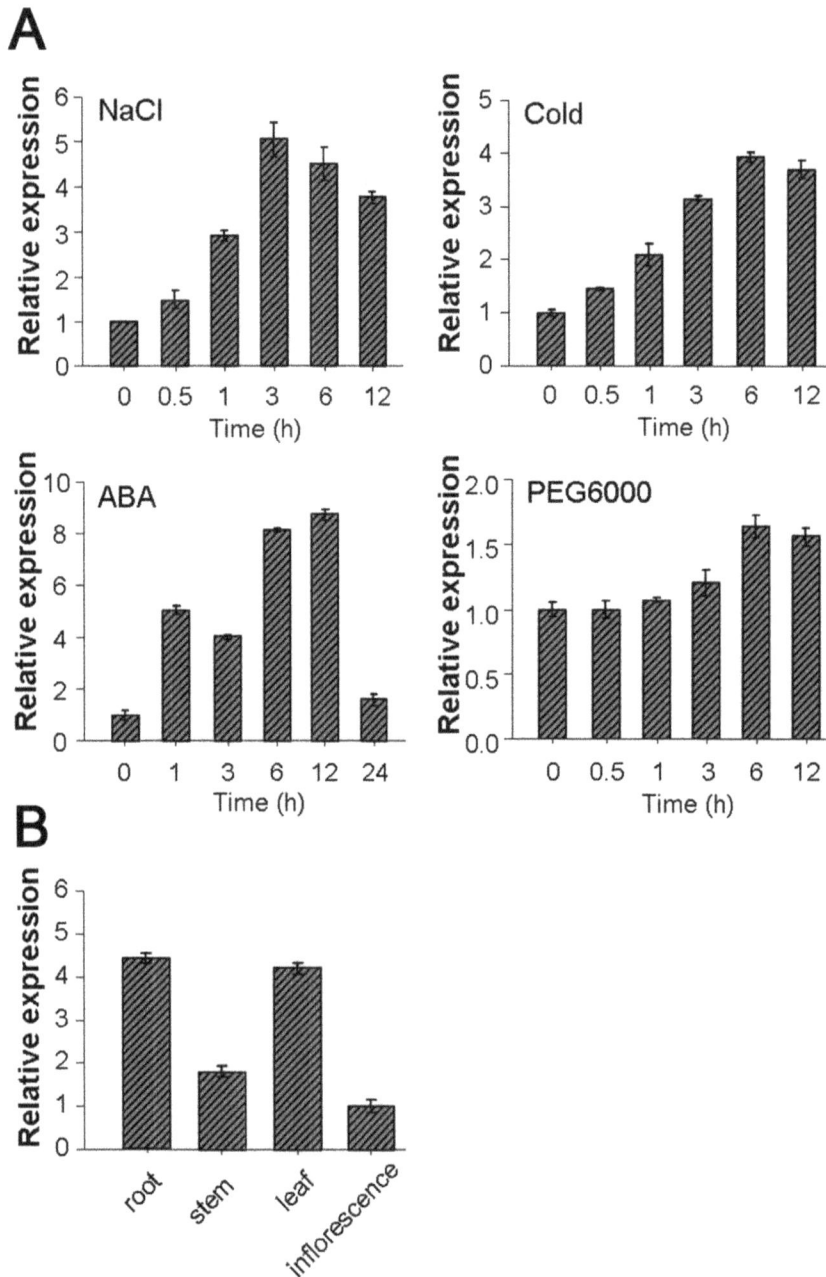

Figure 2. Expression patterns of the SiREM6 under different treatments and in different tissues of foxtail millet (*Setaria italica*). The 17-day-old foxtail millet seedlings subjected to NaCl (150 mM), cold (4°C), ABA (100 μM) and PEG (20% v/v) treatments for selected time periods. (A) Transcription levels of *SiREM6* in response to various stresses in foxtail millet seedlings as demonstrated by qRT-PCR. (B) Transcription levels of *SiREM6* in different tissues of foxtail millet seedlings as demonstrated by qRT-PCR. Foxtail millet *actin* (GenBank: AF288226) was amplified as a normalization control.

of the remorins in abiotic stresses, several expression analyses suggest that remorin genes are responsive to abiotic stresses and involved in signal transduction pathways [33,34]. In the present study, we found 11 remorin genes, based on the C-terminal conserved domain of remorin proteins, in the foxtail millet transcriptome. We cloned them from foxtail millet cDNA and named them *SiREM1* to *SiREM11*. *SiREM6* was induced by high salinity, low temperature and ABA treatment. The overexpression of *SiREM6* in *Arabidopsis* enhanced the tolerance to high salt stress during seed germination and seedling development stages.

SiARDP, an ABA responsive DREB transcription factor, can bind to DRE core elements in the promoter region of *SiREM6*. These results suggest that *SiREM6* is involved in salt tolerance under the control of the SiARDP transcription factor in the ABA-dependent signal pathway.

Materials and Methods

Plant materials and stress treatments

Foxtail millet (*Setaria italica*, cultivar Jigu 11) seeds were germinated on distilled water and grown in growth chambers

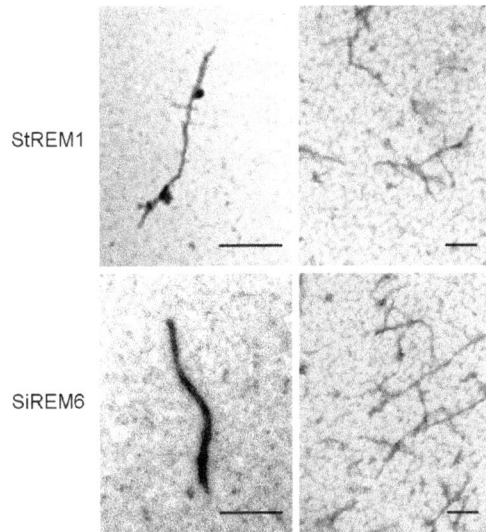

Figure 3. Electron microscopic images of remorin filaments. The protein samples were stained with uranyl acetate as described in the section "Materials and Methods". StREM1: *Solanum tuberosum* remorin 1; SiREM6: *Setaria italica* remorin 6. Bar = 100 nm.

(16 hour: 8 hour, light: dark cycle) at 28°C and 60% relative humidity. Two-week-old seedlings were transferred to 1/3 Hoagland solution, grown for 3 days, and then subjected to various stress treatments. Polyethylene glycol (PEG), NaCl and ABA treatments were conducted by transferring seedlings to 1/3 Hoagland solution containing 20% PEG6000, 150 mM NaCl and 100 μM ABA, respectively, and letting them grow for the indicated time. For the low temperature treatment, 17-day-old seedlings were transferred to a cold chamber and maintained at 4°C for the indicated time. For tissue expression analyses, the roots, stems and leaves were collected from 14-day-old untreated seedlings. The inflorescences were collected during the heading-stage from foxtail millet. These samples were frozen in liquid nitrogen, and then stored at −80°C.

RNA extraction and RNA analysis

Total RNA from foxtail millet and *Arabidopsis* was extracted using TRIzol reagent (Invitrogen, Carlsbad, CA, USA). After digestion with RNase-free DNase I (Takara, Dalian, China), 2 μg of total RNA was converted into cDNA by M-MLV Reverse Transcriptase (Promega, Madison, WI, USA).

The reverse transcription polymerase chain reaction (RT-PCR) was performed using 2× Taq PCR StarMix with Loading Dye (GenStar, Beijing, China). PCR reactions were 95°C for 3 min, followed by 95°C for 30 sec, 60°C for 30 sec, 72°C for 30 sec for 25 cycles and 72°C for 5 min. Primers are listed in Table S1.

A quantitative real-time PCR (qRT-PCR) assay was performed using a LightCycler 480 II RT-PCR detection system (Roche, USA) with the UltraSYBR reagent mixture (CWBIO, Beijing, China). The PCR conditions were 95°C for 10 min, followed by 40 cycles of 95°C for 15 sec, 60°C for 1 min. The relative expression levels of mRNA were calculated using the $\Delta\Delta C_T$ method.

Negative staining electron microscopy

The *SiREM6* and *StREM1* genes were cloned into the pET-28a vector containing a His tag. The recombinant vectors were independently transformed into *Escherichia coli* BL21 cells and then the cells were induced by 1 mM isopropyl-β-D-thiogalactoside (IPTG) for 4 h at 28°C. The fusion proteins were purified by nickel NTA (Qiagen, Germany).

For the negative-staining assay, recombinant remorins were dialyzed against 10 mM Tris (pH 7.5). The final protein concentrations were 80 μg/ml. Then, the recombinant remorins were adsorbed on formvar-coated copper grids for 10 min, stained with 2% uranyl acetate for 4 min, and air-dried. The samples were visualized at a magnification of 80000× using a Hitachi 7500 electron microscope (Japan). Photographs were taken using iTEM (OSIS, Germany).

Generation of transgenic *Arabidopsis* plants

The full-length sequence of *SiREM6* was constructed in the modified binary vector pS1300 at the *Hin*dIII and *Xba*I sites controlled by the cauliflower mosaic virus (CaMV) 35S promoter. The constructed plasmid was introduced into *Agrobacterium tumefaciens* strain LBA4404 competent cells by the freeze-thaw method [35]. *Arabidopsis* plants were transformed by the vacuum infiltration method [36]. The transformed *Arabidopsis* seeds were screened on Murashige and Skoog (MS) medium containing 50 mg/L hygromycin. Three independent homozygous T3 seedling lines were chosen for subsequent experiments.

Phenotypic analysis of transgenics

For the salt stress, approximately 80 seeds from the WT and each T3 generation of *Arabidopsis* lines were used for germination analysis. Surface sterilized seeds were sown on MS medium containing 0, 100, 150 and 175 mM NaCl for 7 days at 22°C. Then, the germination rate was scored and fresh/dry weights of the WT and transgenic *Arabidopsis* seedlings were measured.

For the early growth assay, 5-day-old WT and transgenic *Arabidopsis* seedlings were grown on MS medium and then transferred to MS medium containing 0, 150, 200 and 220 mM NaCl at 22°C. At 5 days, the survival rate was calculated.

For the growth assay, 7-day-old WT and transgenic *Arabidopsis* seedlings were grown on MS medium and then transferred to pots filled with soil and vermiculite (1:1, v/v) for an additional 2 weeks. They were grown for 3 weeks using water containing 400 mM NaCl and then the survival rate was calculated.

Relative electrolyte leakage and proline content were measured as described by Zhao et al [37].

For the ABA treatment, approximately 6 seeds from the WT and each T3 generation of *Arabidopsis* lines were used for analysis. Surface sterilized seeds were sown on MS medium containing 0, 0.5, 0.75 and 1 μM ABA, and grown for 10 days at 22°C. Then the phenotype was observed.

For the dehydration stress, approximately 80 seeds from the WT and each T3 generation of *Arabidopsis* lines were used for analysis. Surface sterilized seeds were sown on MS medium containing 0 and 300 mM mannitol, and grown for 10 days at 22°C. Then the phenotype was observed.

The germination rate, survival rate, fresh/dry weights, relative electrolyte leakage and proline content data were subjected to Student's t-test analysis using GraphPad Prism 5. All experiments were repeated three times.

Electrophoretic mobility shift assay (EMSA)

Using their predicted sequences, *SiARDP* (SiPROV014314m) and *SiAREB1* (SiPROV013188m) were cloned from foxtail millet cDNA. The gene-specific primer pairs are listed in Table S1.

The *SiARDP* and *SiAREB1* genes were cloned into the pGEX-TEV vector containing a GST tag. The recombinant vectors were transformed into *E. coli* BL21 cells, and the cells were induced by

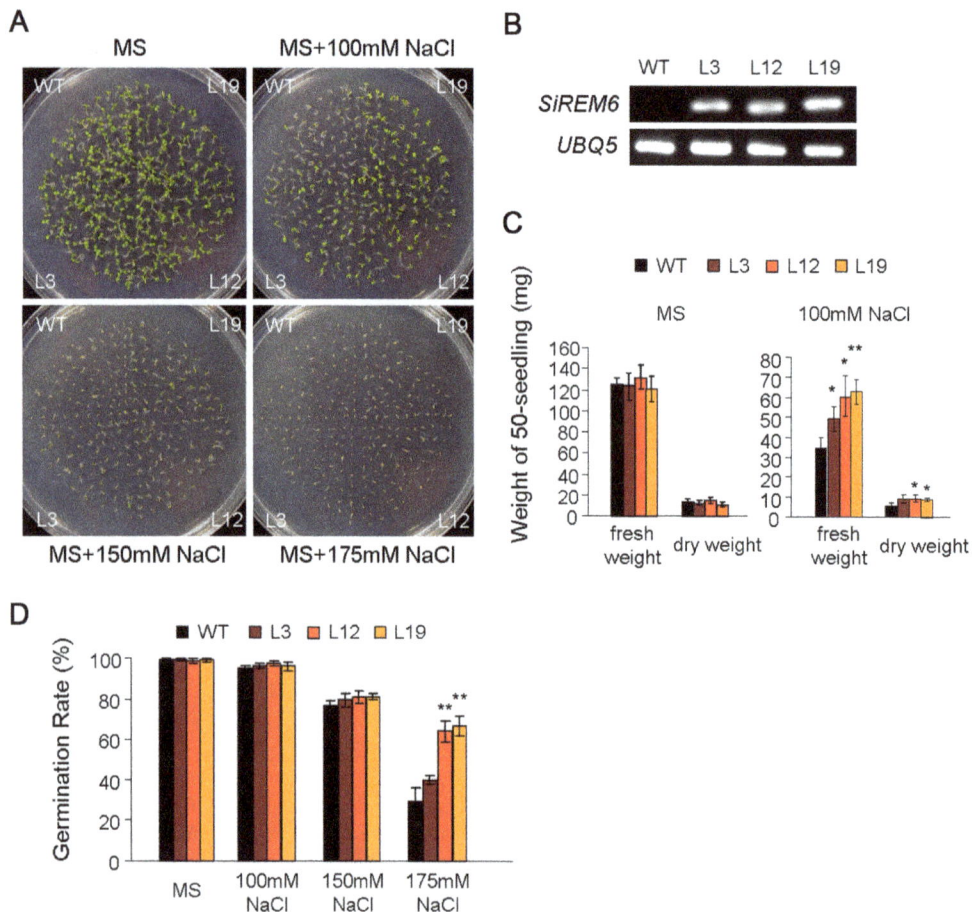

Figure 4. Overexpression of *SiREM6* improves salt stress tolerance in *Arabidopsis* during the germination stage. (A) NaCl stress tolerance of WT and transgenic plants. Seeds of WT and transgenic lines were germinated on medium containing 0 (control), 100, 150 and 175 mM NaCl for 7 days. (B) Transcription levels of SiREM6 in transgenic *Arabidopsis* and WT plants (negative control). *Arabidopsis UBQ5* (GenBank: AT3G62250) was amplified as a normalization control. (C) Fresh/dry weights of 7-day-old seedlings grown on medium containing 0 and 100 mM NaCl. (D) The germination rate of WT and transgenic lines on medium containing 0, 100, 150 and 175 mM NaCl for 7 days. Each data point had three replicates. For C and D, error bars indicate + SD. * and ** indicate statistically significant differences with $P<0.05$ and $P<0.01$ (Student's *t*-test), respectively.

1 mM IPTG for 4 h at 28°C. Then, the fusion proteins were purified using Glutathione Sepharose 4B (GE, USA). Oligonucleotides and their reverse complementary oligonucleotides, which were labeled with biotin, were synthesized. Double-stranded DNA was obtained by heating oligonucleotides at 92°C for 30 sec, and annealing at 30°C. The gel-shift assay was performed following the manufacturer's protocol for the LightShift Chemiluminescent EMSA Kit (Thermo, USA).

Results

The sequence characteristics of SiREM6

To isolate remorin genes from foxtail millet, the C-terminal conserved sequence of remorin was used as a query to search the foxtail millet transcriptome in the Phytozome and the *Setaria italica* databases. Based on the search results, 11 remorin genes were identified and named *SiREM1* to *SiREM11*. A phylogenetic analysis showed that these remorin could be divided into four subgroups. SiREM4, 5 and 6 belonged to groups 1 to 3, SiREM2 and 3 belonged to group 4, SiREM7 and 8 belonged to group 5, and SiREM1, 9, 10 and 11 belonged to group 6 (Fig. 1A). SiREM4, 5 and 6 were then classified into group 1 based on their N-terminal domain features [8]. It had been reported that the

remorin proteins in group 1 might be involved in abiotic stresses [5]. The transcription levels of *SiREM4, 5* and *6* in response to various stresses were analyzed by RT-PCR (data not shown). *SiREM6*, which responded to multiple treatments, was chosen for further analysis.

The open reading frame of *SiREM6* (SiPROV019639m) was cloned by PCR using sequence-specific primers as determined by the database analysis (Table S1). It contains 639 bp and encodes a protein of 212 amino acids with a predicted molecular mass of 23.1 kD. The isoelectric point is 5.40. SiREM6 contains the conserved C-terminal coiled-coil structure, a signature of remorin. Similar to OsREM1.5 and ZmREM1.2, which belong to group 1a, there are nine prolines in the N-terminal region of SiREM6 (Fig. 1B).

Expression analysis of *SiREM6* under abiotic stresses

To analyze the expression patterns of *SiREM6* under different abiotic stresses and ABA treatment, qRT-PCR was conducted. The results showed that *SiREM6* expression was induced by ABA treatment, high salt and cold stresses, but not by drought stress (Fig. 2A). The transcription levels of *SiREM6* reached the highest level of 5.2-fold at 3 h, and maintained a similarly high level in the following 9 hours under 150 mM NaCl treatment. Under ABA

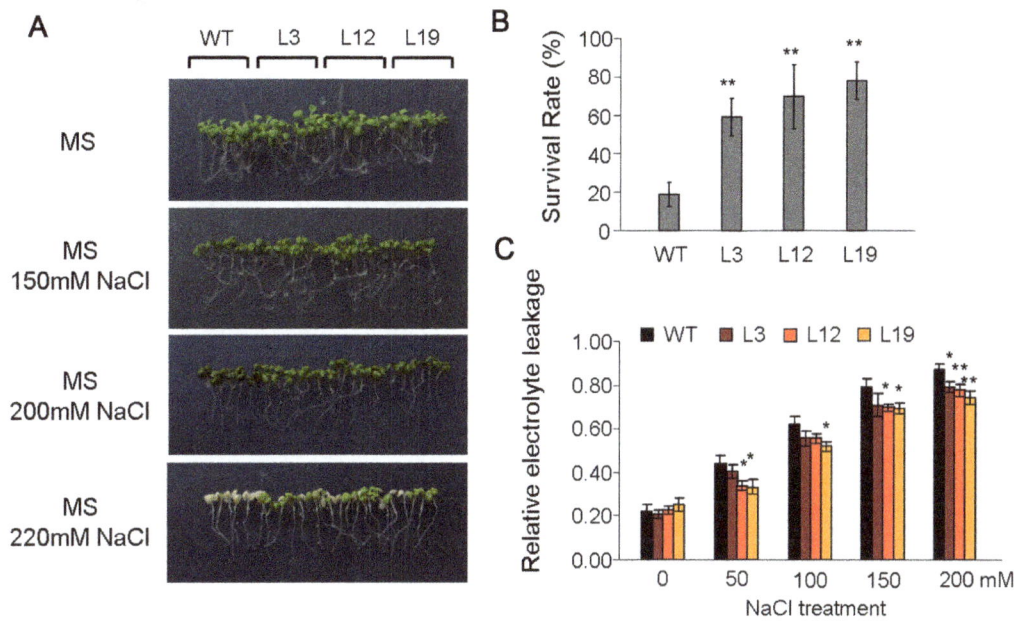

Figure 5. Overexpression of *SiREM6* improves salt stress tolerance in *Arabidopsis* during young seedling stage. (A) NaCl stress tolerance of WT and transgenic lines. Five-day-old seedlings were transferred to medium containing 0, 150, 200 and 220 mM NaCl for 5 days. (B) The survival rates of plants were analyzed after growing on medium containing 220 mM NaCl for 5 days. This experiment had three replicates, and each experiment comprised at least 30 plants. (C) The relative electrolyte leakage in WT and transgenic lines after exposure to different salt stress levels. Each data point had three replicates. For B and C, error bars indicate + SD. * and ** indicate statistically significant differences with $P<0.05$ and $P<0.01$ (Student's *t*-test), respectively.

treatment, the mRNA of *SiREM6* accumulated and reached 9.1-fold at 12 h, then decreased dramatically at 24 h. During cold treatment, *SiREM6* mRNA levels increased gradually and peaked at 6 h. To analyze the expression patterns in different tissues, total RNA isolated from different foxtail millet tissues were reverse transcribed as the templates for qRT-PCR. The results showed that *SiREM6* was expressed in root, stem, leaf and inflorescences (Fig. 2B).

Oligomerization of SiREM6 *in vitro*

To examine the potential oligomerization of SiREM6, negative staining electron microscopy was performed. StREM1 from potato which could oligomerize *in vitro* was used as the positive control [15]. StREM1-His and SiREM6-His fusion proteins were purified and analyzed. Filamentous structures were clearly visible under electron microscopy (Fig. 3). These results suggest that, like other remorin proteins, SiREM6 could oligomerize.

Functional analysis of *SiREM6* in transgenic *Arabidopsis*

To analyze the function of *SiREM6* in stress tolerance, transgenic *Arabidopsis* plants expressing SiREM6 under the control of the CaMV 35S promoter were generated. A total of 30 independent transgenic *Arabidopsis* plants were obtained using a vacuum infiltration method. After RT-PCR analysis, three independent homozygous *SiREM6* overexpression T3 lines (L3, L12 and L19) were selected for further functional analyses (Fig. 4B).

To evaluate the influence of salt stress during the germination stage, seeds of WT and transgenic lines were sown on MS medium containing 0, 100, 150 and 175 mM NaCl. The WT and three transgenic lines did not show any difference on normal MS medium (Fig. 4A); however, the transgenic lines grew better than the WT on MS medium containing 100, 150 and 175 mM NaCl.

The fresh/dry weight of seedlings indicated that the salt stress had a weaker influence on growth in the transgenic lines than in the WT plants (Fig. 4C). The germination rate was calculated for seeds on MS medium containing NaCl after 7 days. The germination rate showed no obvious differences between WT and transgenic lines on MS medium containing 100 and 150 mM NaCl. However, when the MS medium contained 175 mM NaCl, the germination rates of the transgenic lines were much higher than that of WT (Fig. 4D).

To evaluate the influence of salt stress during the seedling stage, the 5-day-old WT and transgenic seedlings grown under normal conditions were transferred to MS medium containing 0, 150, 200 and 220 mM NaCl, and maintained for 5 days. The growth rates of WT and transgenic lines were not obviously different on normal MS medium. When grown on the MS medium containing NaCl, the growth of the WT was affected more seriously than that of the transgenic lines, and more WT seedlings were bleached (Fig. 5A). The survival rates of the transgenic lines grown on medium containing 220 mM NaCl were significantly higher than those of the WT lines (Fig. 5B). The results of electrolyte leakage analysis showed that WT seedlings were more seriously damaged than transgenic seedlings under salt stress (Fig. 5C).

To further test the function of *SiREM6* during salt tolerance, 2-week-old WT and transgenic seedlings, grown under normal conditions, were treated with water containing 400 mM NaCl for three weeks. The WT seedlings were more significantly damaged than the transgenic seedlings (Fig. 6A). The survival rates of the transgenic lines were higher than 60%, while the survival rates of the WT were lower than 50% (Fig. 6B). The proline contents of WT and transgenic seedlings were measured after 14 days under the 400 mM NaCl stress treatment. The result showed that more proline had accumulated in the transgenic lines than in the WT (Fig. 6C).

A

B

C

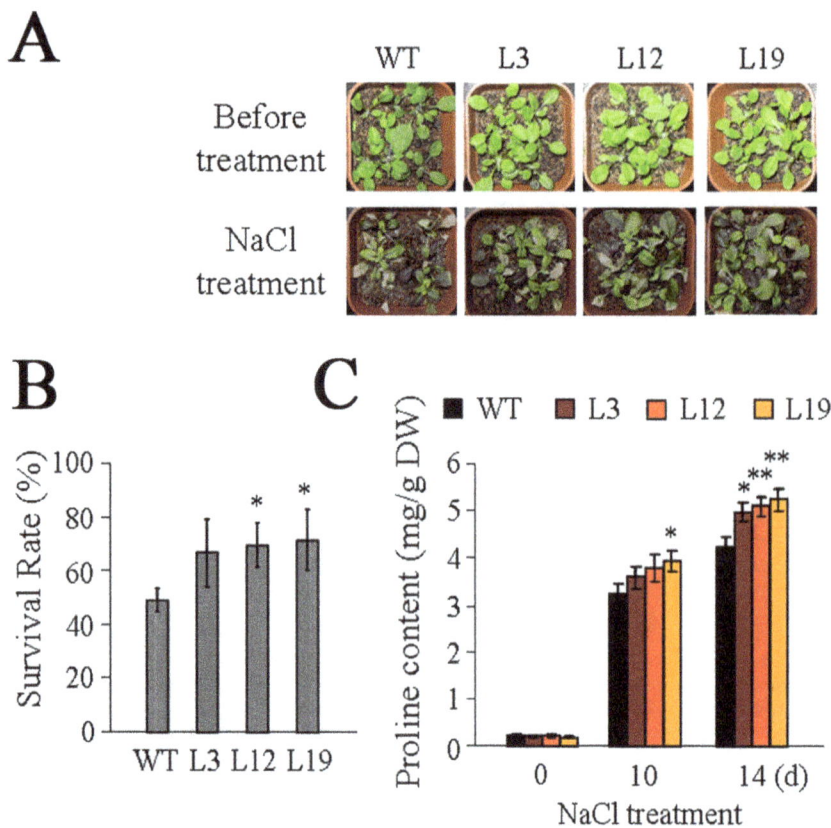

Figure 6. Overexpression of *SiREM6* improves salt stress tolerance in *Arabidopsis* during the seedling stage. (A) NaCl stress tolerance of WT and transgenic *Arabidopsis*. Two-week-old WT and transgenic *Arabidopsis* plants were treated with water containing 400 mM NaCl for three weeks. (B) The survival rates of plants were analyzed after treatment with water containing 400 mM NaCl for 3 weeks. This experiment had three replicates, and each experiment comprised at least 36 plants. (C) The proline content was analyzed in WT and transgenic plants after exposure to salt stress for 10 and 14 days. Each data point had three replicates. Error bars indicate + SD, and * and ** indicate statistically significant differences with $P<0.05$ and $P<0.01$ (Student's *t*-test), respectively.

As *SiREM6* expression was also induced by ABA treatment (Fig. 2A), the transgenic plants were tested for response to ABA treatment. The seeds of WT and trangenic lines were sown on MS medium containing 0, 0.5, 0.75 and 1 μM ABA, and grown for 10 days. The results were shown in Fig. S3. The WT and transgenic lines did not show any difference on normal MS medium, whereas the transgenic lines showed higher sensitivity to ABA treatment than the WT. The response of transgenic lines to drought was further detected even though the transcript level of *SiREM6* was weakly induced by drought stress. The seeds of WT and transgenic lines were sown on MS medium containing 0 and 300 mM mannitol, and grown for 10 days. As shown in Fig. S4, the WT and transgenic lines did not show any difference on MS containing 0 and 300 mM mannitol.

Identification of *SiARDP* and *SiAREB1*

Because *SiREM6* was responsive to both salt stress and ABA treatment, we analyzed the promoter of *SiREM6* and found two DRE and one ABRE core elements (Fig. S1). The DREB transcription factors and AREB transcription factors bind the DRE and ABRE elements, respectively. To analyze the regulation of SiREM6, we cloned a DREB and an AREB transcription factor from foxtail millet, *SiARDP* (SiPROV014314m) and *SiAREB1* (SiPROV013188m), respectively. The SiARDP and SiAREB1 were located in the nucleus and had the ability of transcriptional activity in yeast (data not shown). The transcription levels of

SiARDP and *SiAREB1* were also induced by salt stress and ABA treatment (Fig. S2).

SiARDP binds to the promoter region of *SiREM6*

To assess the consequences of SiARDP and SiAREB1 binding to the elements in the promoter region of *SiREM6*, an EMSA was performed. The sequences that defined DRE1, ACCGAC, and DRE2, GCCGAC, were used as probe 1 (P1) and probe 2 (P2), respectively. The sequence that defined ABRE, ACGTGCG, was used as probe 3 (P3) (Fig. 7A). SiARDP and SiAREB1 were expressed as glutathione S-transferase (GST) fusion proteins in *E. coli*. These fusion proteins were purified and used in the EMSA. The results showed that SiARDP could bind to P1 and P2, but not to P3. The binding affinity of SiARDP for P2 was weaker than for P1. With additional unlabeled probes (competitors 1 and 3), the SiARDP binding signals for P1 and P2 were reduced, but the addition of the mutant probes (competitors 2 and 4) did not obviously reduce the signals. SiAREB1 could not bind to any probes. These results showed that SiARDP specifically binds to two DRE elements in the promoter of *SiREM6*, but SiAREB1 does not.

Discussion

In the past, most studies on the functions of remorins focused on plant-microbe interactions and biotic stresses. However, the

A

Probe: labeled by biotin

P1: GCGATGGTCCTCACCGACGACGCCGCCCGC

P2: ATGGGTTCCTTGCCGACTCTTCTTTTAACC

P3: GCCGCGCGGGTACGTGCGAGCCCCGCGGTC

DRE1: ACCGAC

DRE2: GCCGAC

ABRE: ACGTGCG

Competitor: unlabeled by biotin

Competitor1: GCGATGGTCCTCACCGACGACGCCGCCCGC

Competitor2: GCGATGGTCCTCAAAAACGACGCCGCCCGC

Competitor3: ATGGGTTCCTTGCCGACTCTTCTTTTAACC

Competitor4: ATGGGTTCCTTGAAAAATCTTCTTTTAACC

B

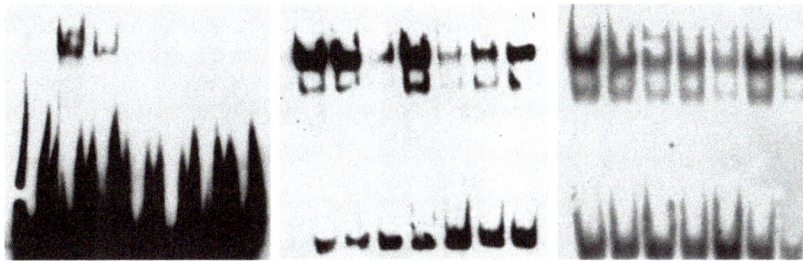

Figure 7. DNA binding abilities of dehydration responsive element (DRE)-binding transcription factor (TF), SiARDP, and abscisic acid responsive element (ABRE)-binding TF, SiAREB1, to the promoter of *SiREM6*. (A) The probes (P) were labeled with biotin, while the competitors were unlabeled. P1, 2 and 3 contained the dehydration responsive element 1 (DRE1), DRE2 and abscisic acid responsive element (ABRE), respectively. Competitors 1 and 2 contained DRE1 and mutant DRE1, in which ACCGAC was replaced with AAAAAC, respectively. Competitors 3 and 4 contained DRE2 and mutant DRE2, in which GCCGAC was replaced with GAAAAA, respectively. (B) SiARDP and SiAREB1 bind to elements. SiARDP binds to the elements in the presence of changing competitor concentrations.

precise functions of remorins are not certain. Additionally, compared with their functions during biotic stress, less data has been reported on the functions of remorins during abiotic stress. Foxtail millet is an important crop in China. It is nutritionally rich and adapts well to stress [38]. However, there is less research on foxtail millet than on other crops, such as rice, maize and wheat. In the present study, 11 remorin genes were identified and cloned from foxtail millet, and the function of *SiREM6* during abiotic stress was analyzed.

The coiled-coil structure in the conserved C-terminal of proteins is a typical remorin signature. Proteins containing the coiled-coil structure usually interact with other coiled-coil proteins and can be oligomerized [39]. The SiREM6 protein contains the signature coiled-coil domain in the C-terminal and could be oligomerized *in vitro* (Fig. 3B). The signature domain features and similarities between remorin proteins combined with a phylogenetic analysis indicated that the remorin family is subdivided into six separate groups, and that SiREM6 is classified as belonging to group 1. The remorins in group 1 are subdivided into groups 1a and 1b according to the number of prolines in their N-terminal region [8]. SiREM6 contains nine prolines in its N-terminal region, thus SiREM6 belongs to the 1a subgroup. Many remorins in group 1a respond to abiotic stress and ABA treatment, and are involved in abiotic stress [5].

The expression levels of certain stress responsive genes may be associated with stress tolerance [40]. The transcription levels of *SiREM6* increased under NaCl, cold stress and ABA treatment (Fig. 2A). That this response to salt stress occurred rapidly, and

was maintained at a high level, strongly implies that the function of *SiREM6* may involve the adaptation to salt stress. The germination rate and live weight are typical physiological parameters for evaluating plant resistance during the germination stage. Relative electrolyte leakage is a relevant index for measuring the cell damage of plants under stresses, and the accumulation of free proline plays a protective role in plants under various stresses. *SiREM6* overexpressing transgenic lines have a higher germination rate and live weight (Fig. 4C and D), and, after high salinity treatments, these transgenic lines had low relative electrolyte leakage and high levels of proline content (Figs. 5C and 6C). These results indicated that *SiREM6* expression in *Arabidopsis* reduced cellular injuries and made the transgenic lines more adaptable to salt stress during the germination and seedling stages. The NaCl stress includes ionic (Na$^+$-specific) and osmotic stresses. The SiREM6 expression was weakly induced by drought, and SiREM6 transgenic lines could not improve the drought tolerance. This implied that the function of remorin proteins may be in resist ion stress.

The plant hormone ABA plays an important role in plants under abiotic stresses [41,42,43], and many remorin genes were induced by ABA treatment [33]. The promoter region of *SiREM6* contained two DRE elements and one ABRE element (Fig. S1). The results of qRT-PCR showed that *SiREM6* was also rapidly induced by ABA treatment (Fig. 2A) and the SiREM6 transgenic lines showed higher sensitive to ABA (Fig. S3). The DREB and AREB transcription factors are important regulatory factors in plants and regulate the expression of target genes during abiotic stress. We cloned the ABA response DRE-binding transcription factor, *SiARDP* and ABA response element (ARE)-binding transcription factor, *SiAREB1*, from foxtail millet. The transcription levels of *SiARDP* and *SiAREB1* were induced by salt stress and ABA treatments (Fig. S2). SiARDP binds to two DRE elements in the promoter region of *SiREM6*, while SiAREB1 did not bind to the ABRE element. These results suggest that SiARDP, but not SiAREB1, regulate the *SiREM6* gene in foxtail millet. SiARDP had a higher affinity for the DRE1 element in P1 than for the DRE2 element in P2. The difference in the core sequences, which occurs at the first base pair (A/G), may explain the difference in the binding affinity. These results indicate that *SiREM6* may be regulated by SiARDP in foxtail millet when under salt stress, and may be involved in the ABA-dependent pathway.

In addition, phosphorylation is a very important process in many abiotic stress signaling pathways. Remorin proteins have been reported to be phosphorylated *in vivo* [44,45,46]. The conserved C-terminal of remorin proteins could provide a stable structure for phosphorylation. Phosphorylation may change the conformation of remorin proteins, and then the changed remorins could interact with other proteins [23] to response the stresses. Further phosphorylation analysis of the SiREM6 will be helpful to deeply understand the molecular mechanism of SiREM6 in response to the stress.

Remorin genes exist extensively in plants, and have different functions in plants. We focused on the function of *SiREM6* in salt stress tolerance in foxtail millet. The expression of *SiREM6* is regulated by transcription factors under salt stress, including SiARDP. Overexpression of *SiREM6* could enhance salt stress tolerance in transgenic *Arabidopsis* plants. These processes rely on the accumulation of protective materials, such as proline, thereby reducing the damage to plant cells. Although the precise mechanism involving *SiREM6* during salt stress is not clear, our results demonstrated that *SiREM6* is involved in salt stress tolerance in plants.

Supporting Information

Figure S1 The *cis*-elements, dehydration responsive element (DRE) and abscisic acid responsive element (ABRE), identified in the *SiREM6*'s promoter. The DRE1 (blue bar), DRE2 (purple bar), AREB (red bar) and TATA box (yellow bar) are shown.

Figure S2 Expression pattern assay of dehydration responsive element (DRE)-binding transcription factor (TF), *SiARDP*, and an abscisic acid responsive element (ABRE)-binding TF, *SiAREB1*, under salt stress and abscisic acid (ABA) treatment in foxtail millet (*Setaria italica*). The 17-day-old foxtail millet seedlings were treated with NaCl (150 mM) and ABA (100 μM) for selected time periods. (A) Transcription levels of *SiARDP* in response to NaCl stress and ABA treatment as demonstrated by qRT-PCR. (B) Transcription levels of *SiAREB1* in response to NaCl stress and ABA treatment as demonstrated by qRT-PCR.

Figure S3 Overexpression of *SiREM6* enhances sensitivity to ABA treatment. Seeds of WT and transgenic lines were sown on MS medium containing 0 (control), 0.5, 0.75 and 1 μM ABA, and grown under normal condition for 10 days.

Figure S4 No difference between *SiREM6* transgenic lines and wild type under dehydration stress. Seeds of WT and transgenic lines were sown on MS medium containing 0 (control) and 300 mM mannitol, and grown under normal condition for 10 days.

Table S1 Gene specific primers used in this study.

Table S2 The GenBank accession numbers of proteins used to develop the remorin phylogenetic tree.

Acknowledgments

We would like to thank Prof. Xianmin Diao (Chinese Academy of Agricultural Sciences) for providing Jigu 11 foxtail millet seeds. We also thank Prof. Zhizhong Gong (China Agricultural University) for providing the pS1300 vector and Shuhua Yang (China Agricultural University) for providing the pUC vector.

Author Contributions

Conceived and designed the experiments: JJY JY CL. Performed the experiments: JY CL. Analyzed the data: JY CL YWL. Contributed reagents/materials/analysis tools: CL YWL. Wrote the paper: CL JY. Revised the manuscript and finalized the manuscript: JJY.

References

1. Tuteja N (2007) Mechanisms of high salinity tolerance in plants. Methods Enzymol 428: 419–438.
2. Yeo AR, Flowers TJ (1984) Nonosmotic effects of polyethylene glycols upon sodium transport and sodium-potassium selectivity by rice roots. Plant Physiol 75: 298–303.
3. Waditee R, Bhuiyan MN, Rai V, Aoki K, Tanaka Y, et al. (2005) Genes for direct methylation of glycine provide high levels of glycinebetaine and abiotic-stress tolerance in *Synechococcus* and *Arabidopsis*. Proc Natl Acad Sci USA 102: 1318–1323.
4. Xiong L, Schumaker KS, Zhu JK (2002) Cell signaling during cold, drought and salt stress. Plant Cell 14: S165–S183.
5. Checker VG, Khurana P (2013) Molecular and functional characterization of mulberry EST encoding remorin (MiREM) involved in abiotic stress. Plant Cell Rep 32: 1729–1741.
6. Farmer EE, Pearce G, Ryan CA (1989) In vitro phosphorylation of plant plasma membrane proteins in response to the proteinase inhibitor inducing factor. Proc Natl Acad Sci USA 86: 1539–1542.
7. Jacinto T, Farmer EE, Ryan CA (1993) Purification of potato leaf plasma membrane protein pp34, a protein phosphorylated in response to oligogalacturonide signals for defense and development. Plant Physiol 103: 1393–1397.
8. Raffaele S, Mongrand S, Gamas P, Niebel A, Ott T (2007) Genome-wide annotation of remorins, a plant-specific protein family: evolutionary and functional perspectives. Plant Physiol 145: 593–600.
9. Fedorova M, van de Mortel J, Matsumoto PA, Cho J, Town CD, et al. (2002) Genome-wide identification of nodule-specific transcripts in the model legume *Medicago truncatula*. Plant Physiol 130: 519–537.
10. Coaker GL, Willard B, Kinter M, Stockinger EJ, Francis DM (2004) Proteomic analysis of resistance mediated by Rcm 2.0 and Rcm 5.1, two loci controlling resistance to bacterial canker of tomato. Mol Plant Microbe Interact 17: 1019–1028.
11. Sánchez-Morán E, Mercier R, Higgins JD, Armstrong SJ, Jones GH, et al. (2005) A strategy to investigate the plant meiotic proteome. Cytogenet Genome Res 109: 181–189.
12. Kistner C, Winzer T, Pitzschke A, Mulder L, Sato S, et al. (2005) Seven *Lotus japonicus* genes required for transcriptional reprogramming of the root during fungal and bacterial symbiosis. Plant Cell 17: 2217–2229.
13. Ofosu-Anim J, Offei SK, Yamaki S (2006) Pistil receptivity, pollen tube growth and gene expression during early fruit development in sweet pepper (Capsicum annum). Int J Agric Biol 8: 576–579.
14. Marin M, Ott T (2012) Phosphorylation of intrinsically disordered regions in remorin proteins. Front Plant Sci 3: 86.
15. Bariola P, Retelska D, Stasiak A, Kammerer R, Fleming A, et al. (2004) Remorins fom a novel family of coiled coil-foming oligomeric and filamentous proteins associated with apical, vascular and embryonic tissues in plants. Plant Mol Biol 55: 579–594.

16. Wienkoop S, Saalbach G (2003) Proteome analysis. Novel proteins identified at the peribacteroid membrane from *Lotus japonicus* root nodules. Plant Physiol 131: 1080–1090.

17. El Yahyaoui F, Kuster H, Ben Amor B, Hohnjec N, Puhler A, et al. (2004) Expression profiling in *Medicago truncatula* identifies more than 750 genes differentially expressed during nodulation, including many potential regulators of the symbiotic program. Plant Physiol 136: 3159–3176.

18. Widjaja I, Naumann K, Roth U, Wolf N, Mackey D, et al. (2009) Combining subproteome enrichment and rubisco depletion enables identification of low abundance proteins differentially regulated during plant defense. Proteomics 9: 138–147.

19. Bray EA (2002) Classification of genes differentially expressed during water-deficit stress in *Arabidopsis thaliana*: an analysis using microarray and differential expression data. Ann Bot 89: 803–811.

20. Kreps JA, Wu Y, Chang HS, Zhu T, Wang X, et al. (2002) Transcriptome changes for *Arabidopsis* in response to salt, osmotic, and cold stress. Plant Physiol 130: 2129–2141.

21. Malakshah SN, Rezaei MH, Heidari M, Salekdeh GH (2007) Proteomics reveals new salt responsive proteins associated with rice plasma membrane. Biosci Biotechnol Biochem 71: 2144–2154.

22. Lefebvre B, Timmers T, Mbengue M, Moreau S, Hervé C, et al. (2010) A remorin protein interacts with symbiotic receptors and regulates bacterial infection. Proc Natl Acad Sci USA 107: 2343–2348.

23. Tóth K, Stratil TF, Madsen EB, Ye J, Popp C, et al. (2012) Functional domain analysis of the remorin protein LjSYMREM1 in *Lotus japonicas*. PloS One 7: e30817.

24. Stockinger EJ, Gilmour SJ, Thomashow MF (1997) *Arabidopsis thaliana* CBF1 encodes an AP2 domain-containing transcriptional activator that binds to the C-repeat/DRE, a cis-acting DNA regulatory element that stimulates transcription in response to low temperature and water deficit. Proc Natl Acad Sci USA 94: 1035–1040.

25. Liu Q, Kasuga M, Sakuma Y, Abe H, Miura S, et al. (1998) Two transcription factors, DREB1 and DREB2, with an EREBP/AP2 DNA binding domain separate two cellular signal transduction pathways in Drought- and Low-Temperature-Responsive gene expression, respectively, in *Arabidopsis*. Plant Cell 10: 1391–1406.

26. Dubouzet JG, Sekuma Y, Ito Y, Kasuga M, Dubouzet EG, et al. (2003) OsDREB genes in rice, *Oryza sativa* L., encode transcription activators that function in drought-, high-salt- and cold-responsive gene expression. Plant J 33: 751–763.

27. Qin F, Kakimoto M, Sakuma Y, Maruyama K, Osakabe Y, et al. (2007) Regulation and functional analysis of ZmDREB2A in response to drought and heat stresses in *Zea mays* L. Plant J 50: 54–69.

28. Xue GP (2003) The DNA-binding activity of an AP2 transcriptional activator HvCBF2 involved in regulation of low-temperature responsive genes in barley is modulated by temperature. Plant J 33: 373–383.

29. Gilmour SJ, Sebolt AM, Salazar MP, Everard JD, Thomashow MF (2000). Overexpression of the *Arabidopsis* CBF3 transcriptional activator mimics multiple biochemical changes associated with cold acclimation. Plant Physiol 124: 1854–1865.

30. Wang Q, Guan Y, Wu Y, Chen H, Chen F, et al. (2008) Overexpression of a rice *OsDREB1F* gene increases salt, drought, and low temperature tolerance in both *Arabidopsis* and rice. Plant Molecular Biology 67: 589–602.

31. Kizis D, Pagès M (2002) Maize DRE-binding proteins DBF1 and DBF2 are involved in *rab17* regulation through the drought-responsive element in an ABA-dependent pathway. Plant J 30: 679–689.

32. Barton L, Newsome SD, Chen FH, Wang H, Guilderson TP, et al. (2009) Agricultural origins and the isotopic identity of domestication in northern China. Proc Natl Acad Sci USA 106, 5523–5528.

33. Lin F, Xu SL, Ni WM, Chu ZQ, Xu ZH, et al. (2003) Identification of ABA-responsive genes in rice shoots via cDNA macroarray. Cell Res 13: 59–68.

34. Kaplan B, Davydov O, Knight H, Galon Y, Knight MR, et al. (2006) Rapid transcriptome changes induced by cytosolic Ca^{2+} transients reveal ABRE-related sequences as Ca^{2+}-responsive cis elements in *Arabidopsis*. Plant Cell 18: 2733–2748.

35. Chen L, Tu Z, Hussain J, Cong L, Yan Y, et al. (2010) Isolation and heterologous transformation analysis of a pollen-specific promoter from wheat (*Triticum aestivum* L.). Mol Biol Rep 37: 737–744.

36. Bechtold N, Pelletier G (1998) In planta *Agrobacterium* Mediated transformation of adult *Arabidopsis thaliana* plants by vacuum infiltration. Methods in Molecular Biology 82: 259–266.

37. Zhao LN, Liu FX, Xu WY, Di C, Zhou SX, et al. (2009) Increased expression of *OsSPX1* enhances cold/subfreezing tolerance in tobacco and *Arabidopsis thaliana*. Plant Biotechnol J 7: 550–561.

38. Bettinger RL, Barton L, Morgan C (2010) The origins of food production in north China: A different kind of agricultural revolution. Evolutionary Anthropology 19: 9–21.

39. Burkhard P, Stetefeld J, Strelkov SV (2001) Coiled coils: a highly versatile protein folding motif. Trends Cell Biol 11: 82–88.

40. Chen L, Song Y, Li S, Zhang L, Zou C, et al. (2012) The role of WRKY transcription factors in plant abiotic stresses. BBA-Gene Regul 1819: 120–128.

41. Busk PK, Pagès M (1998) Regulation of abscisic acid-induced transcription. Plant Mol Biol 37: 425–435.

42. Rock C (2000) Pathways to abscisic acid-regulated gene expression. New Phytol 148: 357–396.

43. Yamaguchi-Shinozaki K, Shinozaki K (2006) Transcriptional regulatory networks in cellular responses and tolerance to dehydration and cold stresses. Ann Rev Plant Biol 57: 781–803.

44. Farmer EE, Moloshok TD, Saxton MJ, Ryan CA (1991) Oligosaccharide signaling in plants. Specificity of oligouronide-enhanced plasma membrane protein phosphorylation. J Biol Chem 266: 3140–3145.

45. Reymond P, Kunz B, Paul-Pletzer K, Grimm R, Eckerskorn C, et al. (1996) Cloning of a cDNA encoding a plasma membrane-associated, uronide binding phosphoprotein with physical properties similar to viral movement proteins. Plant Cell 8: 2265–2276.

46. Widjaja I, Naumann K, Roth U, Wolf N, Mackey D, et al. (2009) Combining subproteome enrichment and Rubisco depletion enables identification of low abundance proteins differentially regulated during plant defense. Proteomics 9: 138–147.

19

Chinese Wild-Growing *Vitis amurensis ICE1* and *ICE2* Encode *MYC*-Type *bHLH* Transcription Activators that Regulate Cold Tolerance in *Arabidopsis*

Weirong Xu[1,2,3]*, **Yuntong Jiao**[4], **Ruimin Li**[4], **Ningbo Zhang**[1,2,3], **Dongming Xiao**[1,2,3], **Xiaoling Ding**[1,2], **Zhenping Wang**[1,2,3]*

1 School of Agronomy, Ningxia University, Yinchuan, Ningxia, P.R. China, 2 Engineering Research Center of Grape and Wine, Ministry of Education, Ningxia University, Yinchuan, Ningxia, P.R. China, 3 Ningxia Engineering and Technology Research Center of Grape and Wine, Ningxia University, Yinchuan, Ningxia, P.R. China, 4 College of Horticulture, Northwest A & F University, Yangling, Shaanxi, P.R. China

Abstract

Winter hardiness is an important trait for grapevine breeders and producers, so identification of the regulatory mechanisms involved in cold acclimation is of great potential value. The work presented here involves the identification of two grapevine *ICE* gene homologs, *VaICE1* and *VaICE2*, from an extremely cold-tolerant accession of Chinese wild-growing *Vitis amurnensis*, which are phylogenetically related to other plant *ICE1* genes. These two structurally different ICE proteins contain previously reported ICE-specific amino acid motifs, the bHLH-ZIP domain and the S-rich motif. Expression analysis revealed that *VaICE1* is constitutively expressed but affected by cold stress, unlike *VaICE2* that shows not such changed expression as a consequence of cold treatment. Both genes serve as transcription factors, potentiating the transactivation activities in yeasts and the corresponding proteins localized to the nucleus following transient expression in onion epidermal cells. Overexpression of either *VaICE1* or *VaICE2* in *Arabidopsis* increase freezing tolerance in nonacclimated plants. Moreover, we show that they result in multiple biochemical changes that were associated with cold acclimation: *VaICE1/2*-overexpressing plants had evaluated levels of proline, reduced contents of malondialdehyde (MDA) and decreased levels of electrolyte leakage. The expression of downstream cold responsive genes of *CBF1*, *COR15A*, and *COR47* were significantly induced in *Arabidopsis* transgenically overexpressing *VaICE1* or *VaICE2* upon cold stress. *VaICE2*, but not *VaICE1* overexpression induced *KIN1* expression under cold-acclimation conditions. Our results suggest that *VaICE1* and *VaICE2* act as key regulators at an early step in the transcriptional cascade controlling freezing tolerance, and modulate the expression levels of various low-temperature associated genes involved in the C-repeat binding factor (*CBF*) pathway.

Editor: Jin-Song Zhang, Institute of Genetics and Developmental Biology, Chinese Academy of Sciences, China

Funding: This work was supported by the National Natural Science Foundation of China (Grant no. 31101512) and Ningxia Innovation Program for the Returned Overseas Scholars. The funders had no role in study design, data collection and analysis, decision to publish, or preparation of the manuscript.

Competing Interests: The authors have declared that no competing interests exist.

* Email: xuwr@nxu.edu.cn (WX); wzhenping@yahoo.com (ZW)

Introduction

Grapevine (*Vitis* L.) is one of the most widely cultivated fruit crops worldwide, and is of great economic importance. However, grape production is often severely limited by various biotic and abiotic stresses [1–5]. For example, low temperature stress greatly restricts the geographic range of grapevine cultivation, and decreases berry yield and quality. Significantly, the grapevine cultivars that currently dominate the market in terms of acreage and production of premium wines are derived from the species *Vitis vinifera*, but they tend to be very sensitive to low winter temperatures [6]. Thus, enhancing low temperature tolerance of grapevine is of great practical importance.

In this context, the current study focuses on *Vitis amurensis* Rupr., a wild grape species that is native to China and is extremely cold-tolerant [7], withstanding freezing temperature as low as −40°C [8]. *V. amurensis* therefore has great potential as an experimental system to identify mechanisms of cold tolerance and as a germplasm resource for grapevine cold-resistance breeding. Some highly cold-resistant *Vitis* cultivars have previously been produced using classical breeding methods; however, resistance to cold stress is a multigenic trait, which limits the effectiveness of using traditional breeding [9]. Therefore, understanding the mechanisms underlying tolerance and adaptation to cold stress could potentially lead to the development of new strategies for improving the yield of cold sensitive agronomic plants and expanding the geographic areas of production.

Cold acclimation has been extensively studied, resulting in considerable evidences at the molecular level that cold stress triggers a multitude of physiological responses [10]. Cold responses are complex and highly regulated via activation of signaling pathways and numerous genes encoding proteins that act directly in stress tolerance. To date, numerous cold-regulated (*COR*) genes have been functionally identified, such as antifreeze proteins, later embryo abundant (*LEA*) proteins, molecular chaperones and enzymes involved in detoxification and biosynthesis of osmopro-tectants [9], [11–13]. Ectopic expression of some *COR* genes have been reported to result in improved cold tolerance, including studies with tobacco [14–16], rice [17], [18], strawberry [19] and *Arabidopsis* [20], [21]. Conversely, the expression of *Craterostigma plantagineum* or spinach *LEA* proteins in tobacco did not induce any significant changes in freezing tolerance [15], [22]. Such findings are not surprising since a clear increase in freezing tolerance is rarely obtained by expressing a single cold-induced gene, even if the end product is directly related to development of freezing tolerance.

Recent studies have shown that *Arabidopsis* cold stress tolerance can be enhanced by modulating the signaling pathways triggered by low temperature stress [23]–[25]. Such pathways include the *CBF* (C-repeat binding factor), also known as *DREB* (dehydration responsive element binding)/*ICE* (inducer of CBF expression) signaling pathway, which have also been characterized in other plant species [25]–[27]. It was reported that increased expression of the entire battery of *COR* genes resulted from overexpressing the *Arabidopsis* transcriptional activator *CBF1* [23], [24], [28], [29]. The underlying mechanism involves the CBF1 protein binding to the CRT/DRE regulatory element located in the promoters of the target genes, thereby inducing cold response metabolic pathways and enhancing cold tolerance. Jaglo-Ottosen et al. [23] further found that constitutive overexpression of *CBF1* induces expression of the *COR* genes in non-acclimated *Arabidopsis* plants and increased freezing tolerance at a whole plant level, an effect that was not observed by expressing *COR15A* alone. This further suggests that freezing tolerance is a multigenic trait involving genes with additive effects. Overexpression of *Arabidopsis CBF* genes, which reside at the nodes of regulatory networks in cold responses, were able to improve chilling/freezing tolerance in different plant species, or homologs from other plant species could enhance the freezing tolerance of transgenic *Arabidopsis* [30], [31]. Other known regulatory components include an upstream transcription factor, *ICE1* (inducer of *CBF* Expression 1), which encodes a *MYC*-type transcription factor that positively regulates *CBF3*, and which plays a critical role in cold acclimation [32]. Additionally, the *Arabidopsis* gene *AtICE2*, a positive regulator belonging to the *bHLH* family, has been shown to activate *AtCBF1* [33].

Since the discovery of the *Arabidopsis ICE* genes, homologs have subsequently been found in a variety of crop species, including wheat, rice, banana, tea, trifoliate orange and grapevines[34]–[44]. In addition, *ICE*-like proteins have been overexpressed in transgenic plants and shown to increase stress tolerance [34], [35], [37]. For example, overexpression of an *ICE*-like gene from *V. amurensis* in tobacco was reported to result in a increased cold tolerance [42], and *Arabidopsis* has been shown to become more tolerant of cold, drought and salt stresses when expressing the *V. vinifera ICE1a* or *-1b* genes [43]. In this current study, we investigated whether two members of the *ICE* gene family from a highly cold-tolerant accession of *V. amurensis* differ functionally from those previously identified from other *Vitis* species, or whether they play similar roles in activating multiple components of cold acclimation responses.

Materials and Methods

Plant materials, growth conditions, and cold treatment

Two-year-old Chinese wild-growing *Vitis amurensis* accession 'Heilongjiang Seedling' potted plants developing from stem cuttings were grown in a greenhouse under natural photoperiod. For the cold treatment, plants with a uniform growth status were transferred to a chamber (LT-BIX120L, LEAD-Tech (SHANG-HAI) SCIENTIFIC INSTRUMENT CO., LTD, China) at 0°C with a 16 h photoperiod (200 µmol m^{-2} s^{-1} light) and 8 h dark. Unstressed plants were used as controls (0 h). Leaves from untreated (control) and cold-treated *V. amurensis* plants were harvested at time points 0, 1, 3, 6, 12, 24, and 48 h and immediately frozen in liquid nitrogen, prior to RNA extraction. More than 3 plants were collected and pooled for each time point, and the sampling was in triple for biological replicates.

Structural features, phylogenetic tree and expression analysis

Total RNA derived from grapevine leaves were isolated using the plant RNA kit (Omega Bio-tek, Doraville, GA, USA), and first-strand cDNA synthesis was synthesized from 2 µg of total RNA using PrimeScript RT Reagent Kit according to the manufacturer's manual (TaKaRa, Dalian, China). Two partial lengths of *V. amurensis ICE* cDNA fragments were amplified by PCR using degenerate primers 5′-TGGACTSSTCSTCKTCGTGYTCK-CC-3′ and 5′-TCATCMGCCTCRTCWWTCAAACCSGA-3′, 5′-CCTYCAGTKGGBKCACAGCCMACKCT-3′ and 5′-GC-AAACCYATTGAARCAGCTGATRACMGC-3′, which were designed based on the known nucleotide sequence of *ICE* homologs from other plant species. Isolation of the full length cDNA sequences was carried out using the SMART RACE Kit (Clontech, Palo Alto, CA, USA). Amplification was performed at 94°C for 3 min; 27 cycles of 94°C for 30 s, 55 to 58°C for 30 s, and 72°C for 1 min; followed by 5 min at 72°C. The cDNA pools for 3′ and 5′ RACE were generated from total RNA extracted from cold-stressed leaves of *V. amurensis*, using the RNAprep Pure Plant Kit (Omega). Subsequently, a nested PCR was performed with the prepared cDNA pool using the adaptor primer UPM and *VaICE1* gene-specific primer (GSP) 5′-CATCCA-CATGTTCTGTGGCCGCAGACCAGG-3′ for 3′RACE, 5′-CTGATGACCGCTTGCTGAATGTCTAGCCCAAGGC-3′ for 5′RACE; and the *VaICE2* gene-specific primer: 5′-GCCGGCGTCAGACAGTACTGGAAGCTTAGG-3′ for 3′RACE, 5′-ATCGCTCGCACTGGAGCTTTGTC-GAAGCGC-3′ for 5′RACE. Amplicons were cloned into the pMD19-T vector (TakaRa) for sequencing and the full-length cDNAs of *ICE* homologs were predicted by comparing and aligning the 5′- and 3′-RACE amplified sequences, using BioEdit software (Version 7.0.1). The putative full-length cDNAs were amplified using primers designed from the extreme 5′ and 3′ ends and 5′RACE-Ready cDNA as template, cloned to pMD-19-T (TaKaRa) to generate pMD-*VaICE1*, and pMD-*VaICE2*, and verified by sequencing. Chromosomal location prediction was performed using the BLAT server (http://www.genoscope.cns.fr/cgi-bin/vitis/webBlat) at the Genoscope Genome Browser. The molecular weights (MW) and isoelectric points (pI) of the corresponding proteins were predicted with the ProParam tool (http://www.expasy.ch/tools/pi_tool.html). Nuclear localization signals were predicted based on the predicted protein sequence using the online server (http://www.predictprotein.org/) and homolog searches were conducted with the NCBI BLAST server (http://blast.ncbi.nlm.nih.gov/Blast.cgi). Molecular model building of bHLH-ZIP domain of VaICE1 or VaICE2 was carried out using SWISS-MODEL server (http://swissmodel.expasy.org). Sequence alignment was performed using the DNAMAN software (Version 6.0, Lynnon Biosoft) and the phylogenetic tree was constructed with MEGA 5.1 software [45] using the neighbor-joining method.

To determine the expression profiles of *VaICE1* and *VaICE2* in *V. amurensis* and *Arabidopsis* plants, total RNAs were extracted as described above. After treatment with RNase-free DNase, the first-strand cDNA was synthesized using the PrimeScript First Strand cDNA Synthesis Kit (TaKaRa). Semi-quantitative RT-PCR was performed to assess the expression of *VaICE1* and *VaICE2* over a cold stress time course. The *VaICE1* cDNA was amplified using the primers 5′-ATGTTACCCAGGTCGAACGACGT-3′ and 5′- CTACAGCATACCGTGGAAGCC TG--3′, and the *VaICE2* cDNA was amplified using the primers 5′-ATGCTGTCCA-GAGTGAACGGCGTC-3′ and 5′-CTACAGCATACCGTG-GAAGCCTG-3′. Grapevine *GAPDH* (GenBank accession no. CB973647) was used as a loading control using the following primers: 5′-TTCTCGTTGAGGGCTATTCCA-3′ and 5′-CCACAGACTTCATCGGTGACA-3′. *Arabidopsis Actin2* (AT3G18780) was served as reference control using the following primers: 5′-CTTGCACCAAGCAGCATGAA-3′ and 5′-CCGATCCAGACACTGTACTTCCTT-3′. Three replicates were performed for each semi-quantitative RT-PCR reaction.

Subcellular localization and transactivation activity assay

To construct green fluorescent protein (GFP) translational fusion vectors, full-length *VaICE1* and *VaICE2* cDNAs were PCR-amplified with the primers 5′-ggaattcCATATGATGTTACC-CAGGTCGAACGACGT-3′ (*EcoR* I) and 5′-gcGTCGACCTA-CAGCATACCGTGGAAGCCTG-3′(*Sal* I), 5′-catgCCATG-GATGCTGTCCAGAGTGAACGGCGTC-3′ (*Nco* I) and 5′-cgGGATCCCTACAGCATACCGTGGAAGCCTG-3′(*BamH* I), and fused in-frame upstream of the of GFP reporter gene in the pBI221 vector. The empty vector containing only GFP sequence was used as a positive control. The isolated plasmids were concentrated to ~1 µg/µl, and used to coat one set of gold particles for bombardment experiments. *VaICE1::GFP* and *VaICE2::GFP* fusion proteins were transiently expressed in onion (*Allium cepa*) epidermal cells using particle bombardment and subsequent localization of the proteins was performed as previously described [46]. Images were collected from the transiently transformed onion epidermal cells (n≥20) that experienced a 24-h dark culture (22°C), using a confocal laser-scanning microscope (LAM510, Carl Zeiss GmbH, Jena, Germany).

For the transcriptional activation assay, the sequences corresponding to the predicted open reading frames (ORFs) of *VaICE1*, amplified with 5′-gcTCTAGAATGTTACCCAGGTCGAAC-GACGT-3′ (*Xba* I) and 5′-gcTCTAGACAGCATACCGTG-GAAGCCTGCCG-3′ (*Xba* I), and *VaICE2*, amplified with 5′-ccgCTCGAGATGCTGTCCAGAGTGAACGGCGTC-3′ (*Xho* I) and 5′-ggGGTACCCAGCATACCGTGGAAGCCTGCCG-3′(*Kpn* I), were fused in frame with the GAL4 DNA binding domain in the pGBKT7 vector, to make pGBKT7-*VaICE1* and pGBKT7-*VaICE2*, respectively. These constructs, as well as pCL1 (positive control) and pGBKT7 (negative control), were transformed into yeast strain AH109 cells and transformants were streaked on plates containing SD/Trp- and SD/Trp-/His-/Ade-medium. After incubation at 28°C for 3 days, the growth of the transformants were evaluated. A β-galactosidase assay was carried out according to the manufacturer's instructions (Clontech).

Generation *Arabidopsis* plants overexpressing *VaICE1* or *VaICE2*

The cDNA sequences corresponding to the ORFs of *VaICE1* or *VaICE2* was amplified from 5′RACE-Ready cDNA using the following gene-specific primers containing sequences for restriction endonucleases: 5′-ccATCGATATGTTACCCAGGTCGAAC-GACGT-3′ (*Cla* I) and 5′-gcTCTAGACTACAGCATACCGTG-GAAGCCTG-3′ (*Xba* I) for *VaICE1* and 5′-ccATCGA-TATGCTGTCCAGAGTGAACGGCGTC-3′ (*Cla* I) and 5′-

ccgCTCGAGCTACAGCATACCGTGGAAGCCTG-3′ (*Xho* I) for *VaICE2*. The PCR products were cloned immediately adjacent to the 3′ end of the CaMV35S promoter in the pART-CAM-S vector [47]. The resulting construct was confirmed by sequencing and transferred into *Agrobacterium tumefaciens* strain GV3101 by electroporation. *Arabidopsis* transformation was performed by the floral dip method [48]. T1 seeds were collected from individual lines and screened on 50 mg/mL kanamycin-contained medium to analyze the segregation of the resistant phenotype. T2 kanamycin-resistant seeds were harvested from non-segregating families and confirm the kanamycin-resistant phenotypes. T3 homozygous lines were validated by RT-PCR using the primers as above described in grapevine expression analysis, and further used for all experiments.

Stress tolerance assay

For the cold treatment, surface-sterilized T3 homozygous seeds of vector-carrying control and *VaICE1/2*-overexpressing *Arabidopsis* lines were germinated and grown on MS medium supplemented with 50 mg/L kanamycin for one week, then transferred to pots containing a mixture of perlite: sand: peat (1:1:1, v/v) for two weeks. The three-week-old seedlings of the transgenic plants overexpressing *VaICE1* and *VaICE2* (n = 50) and control plants (n = 50) were transferred to the chamber (LT-BIX120L) at 0°C under the condition as above-described in grapevine for 0, 6, 12, 24, or 48 h. The freezing tolerance assay was performed by transferring 3-week-old seedlings (n = 50) grown in pots to the pre-chilled chamber (LT-BIX120L) at −6°C for 8 h under continuous dim light conditions (2.5 µmol m^{-2} s^{-1}) and subsequently returning them to normal conditions. Survival rates of plants were evaluated after 7 days. The freezing treatment experiment was performed in triplicate.

Determination of electrolyte leakage, and malondialdehyde and proline content

Three-week-old seedlings from the empty vector control *Arabidopsis* plants (n = 20), *VaICE1* and *VaICE2* T3 transgenic lines (n = 20) were subjected to the cold stress treatment described above and leaves were harvested to determine three biochemical features associated with cold tolerance: electrolyte leakage was assessed by ion leakage analysis as previously described [49]; proline content was determined using acid-ninhydrin reagent and acetic acid [50]; and malondialdehyde (MDA) content was measured using the thio-barbituric acid (TBA) method as previously described [51].

Analysis of cold-responsive genes regulated by *VaICE1* or *VaICE2*

Total RNA was extracted from vector-carrying control and T3 transgenic *Arabidopsis* leaves treated at 0°C for 0, 3, 12, 24 and 48 h using the plant RNA kit (Omega) following the manufacturer's instructions. cDNA was synthesized from 2 µg of total RNA using the PrimeScript RT Reagent Kit with the oligo (dT)18 primer, according to the manufacturer's instructions (TakaRa) and quantitative real-time PCR was performed using the SYBR Premix Ex ^{TM} TaqII kit (TakaRa) on an iCycler iQ5 thermal cycler (Bio-Rad). The reactions were carried out in triplicate in 96-well plates (25 µl/well) in a mixture containing 12.5 µl 2×SYBR Premix Ex ^{TM} TaqII, 1 µl each of primer (10 µM in stock), 1 µl template cDNA and 9.5 µl ddH2O. Two-step real-time PCR reactions were performed under the following conditions: 95°C for 10 s, followed by 40 cycles of denaturation at 94°C for 15 s, annealing and extension at 57°C for 30 s, and data acquisition at

57°C for 15 s. Three replicate PCR amplifications were performed for each sample. Transcript levels of AtICE1, AtICE2, AtCBF1, AtCBF2, AtCBF3, AtCOR15A, AtCOR47, AtRD29A and AtKIN1 were measured in VaICE1 or VaICE2 overexpressing or empty vector control plants. The amount of transcript for each gene, normalized to the internal reference Atactin2, was analyzed using the 2-$\triangle\triangle$Ct method [52]. Oligos used for real-time PCR were: 5′- CTTCCATCCGTTGACACCTAC -3′ and 5′-CTCTAGCTTGCTGGCCTTTAG-3′ for AtICE1; 5′-TCCACAAACGCTGTCTTACC-3′ and 5′-GTTCACTGCC-TTTCCTTCTCT-3′ for AtICE2; 5′-GAGACGATGGTGGA-AGCTATTT-3′ and 5′-AGCATGCCTTCAGCCATATTA-3′ for AtCBF1; 5′-GACCTTGGTGGAGGCTATTT-3′ and 5′-ATCCCTTCGGCCATGTTATC-3′ for AtCBF2; 5′-GACGT-TGGTGGAGGCTATTT-3′ and 5′-AGCATCCCTTCTGC-CATATTAG-3′ for AtCBF3; 5′-GGCGTATGTGGAGGAGA-AAG-3′ and 5′-CCCTACTTTGTGGCATCCTTAG-3′ for AtCOR15A; 5′-GGCTGAGGAGTACAAGAACAA-3′ and 5′-ACAATCCACGATCCGTAACC-3′ for AtCOR47; 5′-GCAA-TGTTCTGCTGGACAAG-3′ and 5′- TCCTTCACGAAGT-TAACACCTC-3′ for AtKIN1; 5′-GCTTTCTGGAACAGAG-GATGTA-3′ and 5′-CGACTCTTCCTCCAACGTTATC-3′ for AtRD29A. Data analysis shown was obtained from three biological replicates.

Results

Characterization of two ICE homologs from V. amurensis

Two ICE-orthologs, designated VaICE1 (GenBank accession no. KC815984) and VaICE2 (GenBank accession no. KC815985), were isolated from leaves of the cold-tolerant Chinese wild-growing V. amurensis accession 'Heilongjiang Seedling'. The complete cDNA of VaICE1 is 1,949 bp, comprising a 97 bp 5′UTR, 301 bp 3′UTR and a 1,551 bp ORF that was predicted to encode a 516 amino acid protein with a MW of 55.5 kDa and a pI of 5.30. The complete cDNA of VaICE2 is 2,108 bp, comprising a 214 bp 5′UTR, 277 bp 3′UTR and a 1,617 bp ORF that was predicted to encode a 538 amino acid protein with a MW of 58.38 kDa and a pI of 5.05. A sequence alignment showed that the two proteins differ significantly from each other (60% amino acid similarity). Chromosome location of the two ICE homologs in the V. vinifera cv. Pinot Noir clone P40024 genome suggested that VaICE1 is mapped on chromosome 1, spanning 2714 bp, while VaICE2 is mapped on chromosome 14 and spans 4234 bp (Fig. 1A). Analysis of structural properties revealed that the predicted VaICE1 or VaICE2 protein possesses the typical features of ICE proteins, including a serine-rich region (S-rich), a basic helix-loop-helix (bHLH) domain, an ICE-specific domain [38], a zipper region (ZIP) and an ACT_UUT-ACR-like domain (Fig. 1B). An interPro scan suggested that both proteins belong to the MYC-like bHLH family of transcription factors, and in support of this, they contain nuclear location signals (NLSs) at position 12–46 aa for VaICE1 and 339–362 aa for VaICE2 (Fig. 1B). The predicted three dimensional structures of the bHLH-ZIP domain of VaICE1 and VaICE2 were distinct, and were more similar to the structures of DIMER (PDB ID: 1r05) and HETERO DIMER (PDB ID: 2ql2), respectively (Fig. 1C).

Phylogenetic analysis of various ICE amino acid sequences indicated the existence of two major groups (Fig. 2A). VrICE4 is phylogenetically distinct from the other groups in which they could be clearly classified into dicot and monocot specific subgroups. Our phylogetic data demonstrated VaICE1 and VaICE2 fell into different clades of the dicot subgroup, with VaICE1 being closely related to the previously reported VaICE14 (98.7% identity) from

V. amurensis [42], VrICE1 (97.7% identity) from V. ripara [44], whereas VaICE2 being closely similar to VrICE2 (99.4% identity) from V. ripara [44], VvICE1 (98.9% identity) and VvICE1a (98.7% identity) from V. vinifera [43]. In addition, VvICE1b shared 98.72% identity with VrICE3, which belongs to another clade of dicot subgroup. Comparison of VaICE1,2 and their homologs AtICE1 and AtICE2 from Arabidopsis, as determined by 49~60% sequence similarity, revealed they share highly conserved regions in the bHLH DNA binding domain and ACT domain in their C-terminal regions (Fig. 2B). In contrast, only a moderate sequence conservation was found in the ZIP domains between Vitis and Arabidopsis ICE proteins. A potential sumoylation site previously identified in Arabidopsis ICE proteins [53] was also observed in VaICE2, and the S-rich region, which has been suggested to be a site of phosphorylation [32], [54] was present in either VaICE1 or VaICE2. Finally, the R236 residue that substituted with H236 in the Arabidopsis ice1 mutant [32], causing a loss of ICE function, was shown to be present in the two V. amurensis ICE proteins (Fig. 2B).

To gain insight into the biological functions of VaICE1 and VaICE2, semi-quantitative RT-PCR was used to examine their expression profiles in leaves of V. amurensis over a cold stress time-course (Fig. 2C). VaICE2 transcript levels were constant over the time course, while a rapid induction of VaICE1 expression was observed after 1 h, followed by a gradual decline from 3 h to the minimum level after 12 h, and finally a return to background levels at the end of the time course. These results suggest that VaICE1 and VaICE2 may be involved in cold stress responses.

VaICE1/VaICE2 is nuclear-localized and can act as transcriptional activators in transient assays

Sequence analysis showed that VaICE1 and VaICE2 have putative NLSs, suggesting that they target the mature proteins to the nucleus. To test this, VaICE1 and VaICE2 were transiently expressed as translational fusions at the N terminus of GFP in onion (Allium cepa) epidermal cells. Confocal imaging showed that GFP alone (control) was present in the cytoplasm and nucleus, as expected, whereas cells transformed with either VaICE1::GFP or VaICE2::GFP showed strong fluorescence exclusively in the nucleus (Fig. 3A).

Yeast one hybrid assays were carried out to determine whether VaICE1 and VaICE2 possess transactivation activity. As shown in Fig. 3B, yeast transformants with pCL1 (positive control), pGBKT7-VaICE1 and pGBKT7-VaICE2 grew well on the SD/Trp- medium, but also grew normally on the SD/Trp-/His-/Ade-medium, and exhibited fairly strong β-galactosidase activity. In contrast, transformants carrying the negative control (pGBKT7) did not grow on the SD/Trp-/His-/Ade-medium and could not show β-galactosidase activity. Thus, VaICE1 and VaICE2 have transactivation activities in yeasts.

Over-expression of VaICE1 or VaICE2 in Arabidopsis increases cold tolerance

To determine whether VaICE1 or VaICE2 enhances cold stress tolerance, Arabidopsis WT-type Col-0 was transformed with constructs carrying empty vector, VaICE1- or VaICE2-coding regions under the control of constitutive 35S promoter (Fig. 4A). Expression of VaICE1 or VaICE2 transgens in these T2 lines was confirmed by semi-quantitative RT-PCR, and two VaICE1 (L2 and L3) and two VaICE2 transgenic T3 homozygous lines (L1 and L6) with similar transcript levels (Fig. 4B) were selected for further analyses. These two independent transgenic lines over-expressing VaICE1 and VaICE2, together with the corresponding empty vector-transformed control lines, were subjected to a whole plant

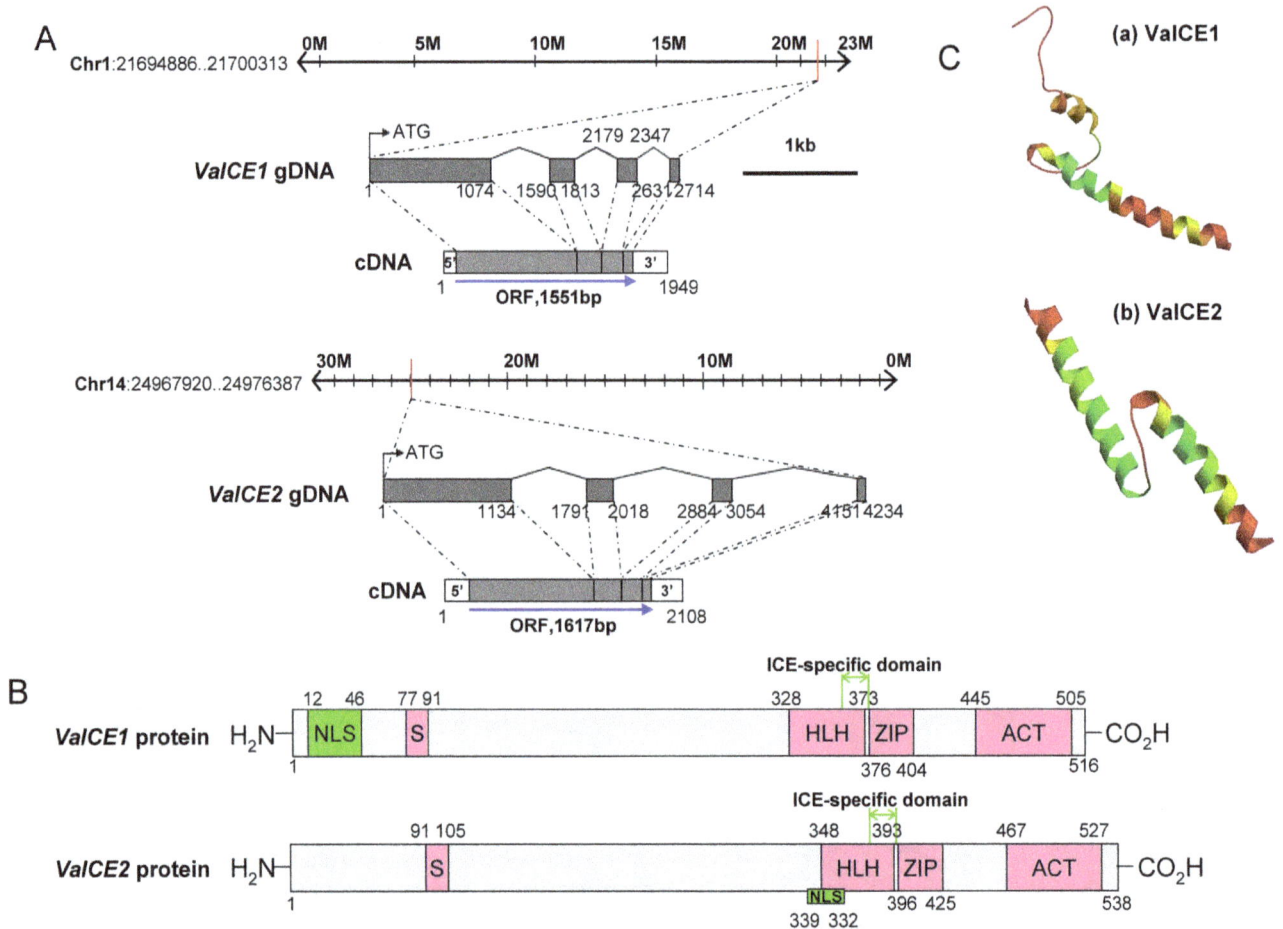

Figure 1. Location and structure of the _ValCE1,2_ genes. (A) Gene structure and locus. Chromosomal localization of _ValCE1,2_ genes were predicted in _V. vinifera_ cv. Pinot Noir clone P40024 genome. _ValCE1_ comprises four exons spanning 2.7 kb with a predicted 1,551 bp ORF, mapping to Chromosome 1 and position 21694886–21700313. _ValCE2_ contains four exons spanning 4.2 kb and has a predicted 1,617 bp ORF, which is mapped to Chromosome 14 and position 24967920–24976387. Nucleotide numbers are indicated above the gene structure. (B) Schematic protein structures of ValCE1 and ValCE2 with the N-terminal domain containing a S-rich motif and the C-terminal domain represented by helix-loop-helix (HLH), ICE-specific domain, Zipper region (ZIP), and ACT_UUR_ACR-like (ACT) domain. The putative nuclear localization signals (NLS) are indicated as green boxes. Codon numbers are indicated above the protein structures. (C) The predicted tertiary structures of the bHLH-ZIP domains of the (a) ValCE1 and (b) ValCE2 proteins evaluated by the SWISS-MODEL sever.

freezing assay to evaluate plant survival after freezing treatment (Fig. 4C). The freezing test consisted of exposing 3- week-old transgenic and control plants to a temperature of $-6°C$ for 8 h, before recovery at normal temperatures. The _ValCE1_ and _ValCE2_ transformed plants exhibited less freezing damage and increased survival rates compared with empty vector control plants after a 7 d recovery period (Fig. 4D). Only ~5% of the control plants (L1 and L2) survived the treatment, while the _ValCE1_-overexpressing plants exhibited survival rates of 91% for L2 and 81% for L3, and 85% for L1 and 95% for L6 in _ValCE2_-overxpressing plants (Fig. 4D). No obvious differences in phenotype were observed between the empty vector control plants and _ValCE1_ or _ValCE2_ transgenic plants grown under normal conditions. These results suggest that _ValCE1_ or _ValCE2_ overexpression leads to enhanced freezing tolerance.

Over-expression of _ValCE1_ or _ValCE2_ affects electrolyte leakage as well as proline and MDA metabolism

To investigate the physiological and biochemical factors that might contribute to the improved cold tolerance of the transgenic

plants, MDA and proline content were evaluated over a time course of exposure to cold stress ($0°C$), as well as electrolyte leakage, which is commonly used as an index of membrane injury [23], [55]. A representative experiment comparing the empty vector control L1 and _35S::ValCE1_ L2 and _35S::ValCE2_ L6 transgenic plants is shown in Fig. 5. Under non-acclimation conditions, electrolyte leakage of all tested lines did not vary greatly i.e. between 22% in the control (L1) and 21% in the _ValCE1_-(L2) and _ValCE2_-(L6) overexpressing lines (Fig. 5A). Electrolyte leakage increased with the prolonged cold stress and reached 47% in the control and 40% and 39% in _ValCE1_ and _ValCE2_ lines, respectively, after 48 h.

Proline accumulation in plants has been associated with a wide variety of environmental stresses, and confers stress tolerance by facilitating osmotic adjustment, protecting proteins and membranes and quenching reactive oxygen species [56], [57]. Significantly, for this study, increases in proline levels have been well-documented in _Arabidopsis_ and other plants during cold acclimation [58], [59]. Similar changes were observed in both _ValCE1_ and _ValCE2_ overexpressing _Arabidopsis_ lines (Fig. 5B), since

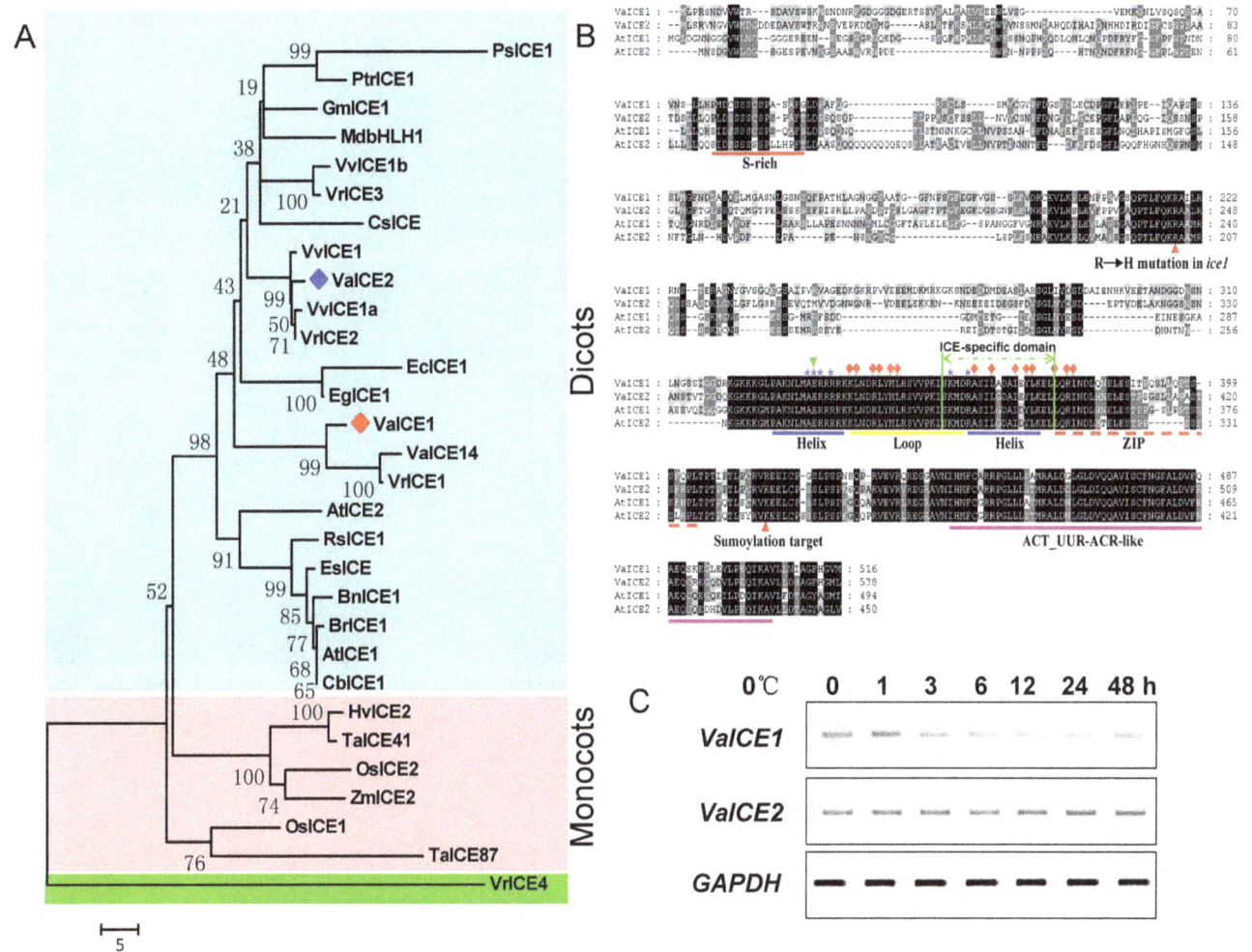

Figure 2. Phylogenetic tree, alignment and expression profiles of *ValCE1,2*. (A) Phylogenetic tree based on the deduced amino acid sequences of ICE from a range of plant species. A neighbor-joining tree, with the following predicted ICE protein sequences, with GenBank accession numbers listed in parentheses: *V. amurensis* (ValCE1, AGP04217; ValCE2, AGP04218; ValCE14, ADY17816), *V. vinifera* (VvICE1, AFI49627; VvICE1a, AGQ03810; VvICE1b, AGQ03811), *V. riparia* (VrICE1, AGG34704; VrICE2, AIA58705; VrICE3, AIA58706; VrICE4, AIA58707), *Arabidopsis thaliana* (AtICE1, NP_189309; AtICE2, NP_172746), *Brassica napus* (BnICE1, AEL33687), *Brassica rapa* subsp. Chinensis (BrICE1, ACB70963), *Capsella bursa-pastoris* (CbICE1, AAS79350), *Camellia sinensis* (CsICE, ACT90640), *Eucalyptus camaldulensis* (EcICE1,ADY68776), *Eucalyptus globules* (EgICE1, AEF33833), *Eutrema salsugineum* (EsICE, ACT68317), *Glycine max* (GmICE1, ACJ39211), *Hordeum vulgare* (HvICE2, ABA25896), *Malus x domestica* (MdbHLH1, ABS50251), *Oryza sativa* (OsICE1, Os11g0523700; OsICE2, Os01g0928000), *Populus suaveolens* (PsICE1, ABF48720), *Populus trichocarpa* (PtrICE1, ABN58427), *Raphanus sativus* (RsICE1, ADY68771), *Triticum aestivum* (TaICE41, ACB69501; TaICE87, ACB69502), *Zea mays* (ZmICE2, ACG46593), was produced by ClustalX 2.0 alignment followed by tree construction using MEGA 5.0 with 100 bootstrap tests. The branch support values are indicated. The length of the scale bar corresponds to 5 substitutions per site. (B) Comparison of ICE amino acid sequences from *V. amurensis* (ValCE1,2) and *Arabidopsis thaliana* (AtICE1,2). Deduced amino acid sequences were aligned using ClustalW. Sequences and accession numbers are shown for the following: *Vitis amurensis* (ValCE1, KC815984; ValCE2, KC815985) and *A. thaliana* (AtICE1, AAP14668; AtICE2, AAO63441). Residues in black and gray regions indicate identical and similar residues, respectively, between isoforms. Four predicted domains are labeled: a S-rich motif, a basic-helix-loop-helix-leucine zipper (bHLH-ZIP) region, *ICE*-specific domain [38] and a ACT-UUR-ACR-like domain. The red triangles indicate the position of the mutation isolated by Chinnusamy et al. [32] and Kanaoka et al. [54], and the residue targeted for sumoylation by SIZ1[53]. Green triangle indicates E-box/N-box specificity site [69], blue asterisks indicate core residues for DNA binding sites [70], and red diamonds indicate dimerization interface/polypeptide binding sites [70]. (C) Expression profiles of *ValCE1* and *ValCE2* during a cold stress time-course experiment. Semi-quantitative RT-PCR was used to determine *ValCE1* and *ValCE2* transcript levels in cold treated grapevine leaves at indicted times. Grapevine *GAPDH* was used as a loading control.

similar levels of proline were detected in the transgenic and wild type plants without cold stress, but substantial proline accumulation occurred in cold-treated *ValCE1* and *ValCE2* transgenic lines. After 48 h of cold stress, the proline contents in the leaves of *ValCE1* or *ValCE2* transgenic lines were ~2.3-fold higher than those of the control plants (Fig. 5B).

MDA is also considered an indicator of plant oxidative stress and structural integrity of the membranes in response to low

temperature [60], [61]. Under non-acclimation conditions, all tested lines had similar MDA levels ranging from 1.76–2.11 µg/g FW (Fig. 5C). However, MDA accumulation was considerably less in the *ValCE1* or *ValCE2* transgenic lines than in the controls during the cold stress time-course. These results indicated that *ValCE1* and *ValCE2* overexpression resulted in increased levels of proline, but decreased levels of MDA and reduced levels of electrolyte leakage.

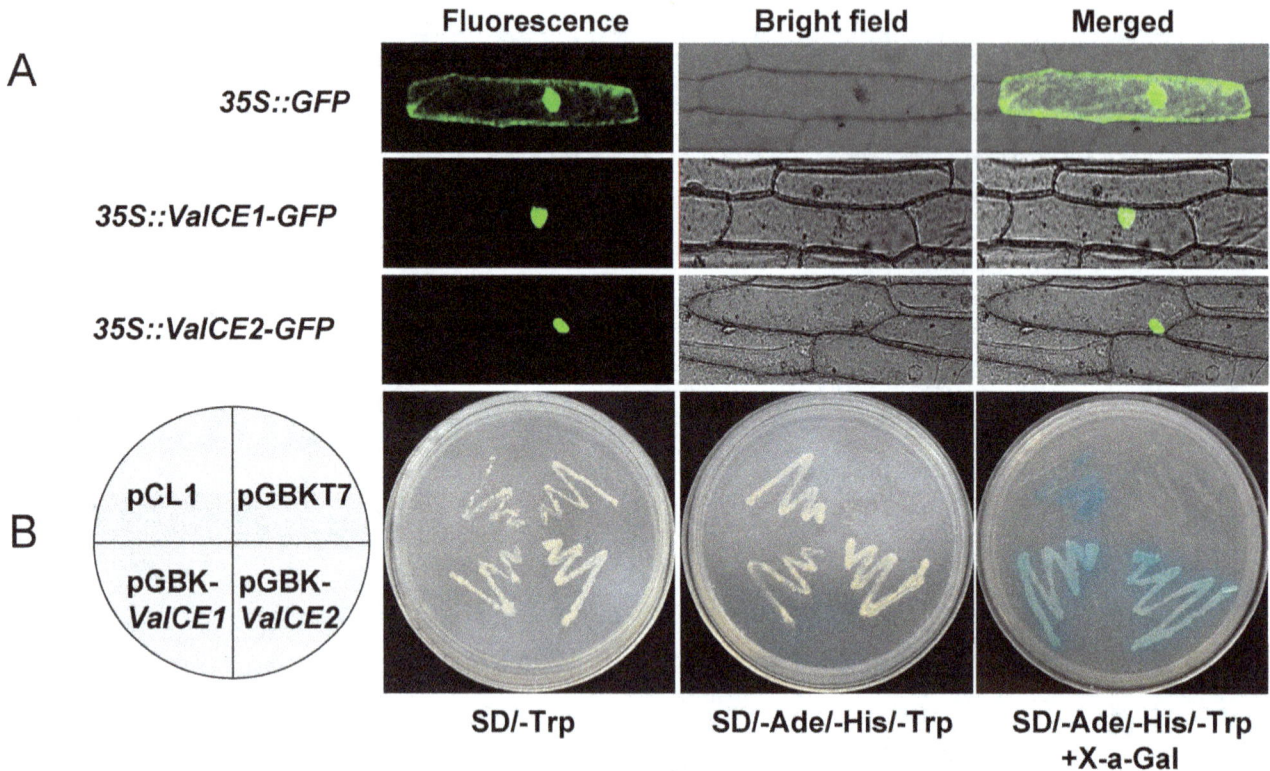

Figure 3. Nuclear localization and transcriptional activation assay of *VaICE1,2.* (A) Confocal imaging of the VaICE1 and VaICE2 transiently expressed in onion epidermal cells as GFP fusion proteins, driven by the 35S promoter. GFP alone was used as a control. Left hand panels show dark field images of green fluorescence, middle panels show the morphology of the cells in bright field and the right hand panels show the merged images. (B) Fusion proteins of pGBKT7-*VaICE1*, pGBKT7-*VaICE2*, pCL1 (positive control) and pGBKT7 (negative control) were expressed in the yeast strain AH109. Transformants were incubated on SD/Trp- and SD/Trp-/His-/Ade- to assess their growth and tested for β-galactosidase activity.

VaICE1 or *VaICE2* positively regulate cold-induced gene expression

To further identify molecular components associated with the *VaICE1* and *VaICE2* mediated stress tolerance that we observed in the *35S::VaICE1* L2 and *35S::VaICE2* L6 transgenic *Arabidopsis* lines, the expression patterns of genes in the *ICE-CBF* pathway were evaluated by real-time PCR (Fig. 6). Under control conditions, *AtICE1* or *AtICE2* expression was relatively low in either the empty vector control or *VaICE1*- and *VaICE2*-overexpressing plants, respectively. However, overexpression of *VaICE1* or *VaICE2* resulted in a reduction in transcript levels compared with the control during the cold stress time course. The possible effect of *VaICE1* or *VaICE2* on the expression of downstream genes through transcriptional regulation was also investigated (Fig. 6). Under control conditions, transcripts of none of the three *CBF* genes was detected in the control, but higher or similar level *CBF* transcript levels were present in the *VaICE1* or *VaICE2* overexpression lines. Expression of the *CBF* genes showed a general decline in the control plants starting at 12 h of cold stress. Starting at 3 h, *CBF1* and *CBF2* transcript levels were considerably greater in the transgenic lines than in the control plants, while a substantially greater expression in the transgenic lines was not observed for *CBF3* until 24 h. Expression of the *CBF* downstream target genes (*COR15A, COR47, KIN1* and *RD29A*) were also investigated. *COR47* had extremely high expression in both transgenic lines and the vector control compared with the other genes, and *COR15A* exhibited higher expression in the transgenic lines compared with the control after 48 h. The *KIN1*

gene was expressed at considerably higher levels in the *VaICE2* overexpression than that in the *VaICE1* overexpression line and the vector control, while no substantial difference in *RD29A* expression was seen among the three different genotypes. These results suggest that *VaICE1* and *VaICE2* positively regulate the expression of the *CBF* genes by differentially controlling downstream genes in response to cold stress, which in turn likely contributes to freezing tolerance.

Discussion

In the current study, two putative *ICE*-orthologs, *VaICE1* and *VaICE2*, from the highly cold-tolerant species *V. amurensis*, were structurally analyzed and functionally tested. We report that these two genes encode *MYC*-type *bHLH* transcription factors that are nuclear-localized, and are candidate regulators of the *CBF* gene expression during cold stress. Reports on *ICE*-orthologs in three genotypes of *Vitis* species [42]–[44] might indicate that there are multiple *ICE*-like genes in grapevine. However, sequence comparisons revealed that none of the mRNA sequences that were previously reported [42], [43] completely matched the two sequences identified here (Fig. S1 in File S1). Thus, *Vitis ICEs* were categorized on the basis of sequence similarity (Fig. S2A in File S1), phylogenetic clustering (Fig. S2B in File S1) and genetic locus (Table S1 in File S1). Our results provide evidence of the presence of 4 polymorphic loci for the identified ICE-like genes in grapevine. *VaICE1*, *VaICE14* and *VrICE1* are alleles from which they are located in grapevine chromosome 1, whereas *VaICE2*, *VrICE2*, *VvICE1* and *VvICE1a* are alleles, mapping to chromosome

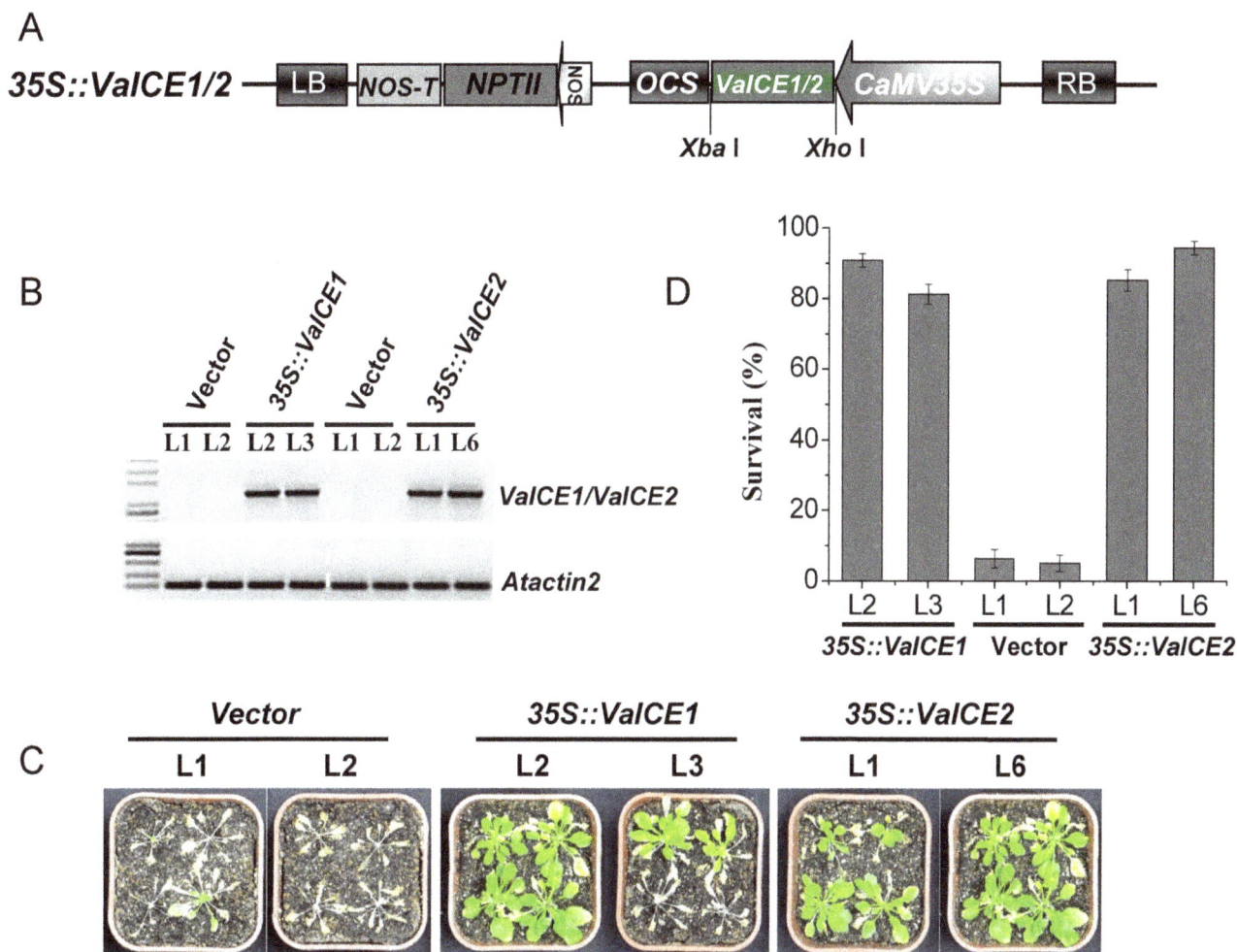

Figure 4. Freezing tolerance evaluation of *35S::ValCE1,2* transgenic *Arabidopsis* plants. (A) A schematic map of the T-DNA region of *35S::ValCE1,2* fusion constructs employed for *Arabidopsis* transformation. RB, right border; LB, left borders; CaMV35S, Cauliflower mosaic virus 35S promoter; OCS, octopine synthetase terminator; NOS, nopaline synthase promoter; NPTII, Kanamycin resistance gene; NOS-T, nopaline synthase terminator. (B) RT-PCR was used to assess the transcript abundance of *ValCE1* or *ValCE2* in the transgenic *Arabidopsis* plants. *Atactin2* was used as a reference control (C) 3-week-old plants were treated at −6°C for 8 h and then transferred back to normal conditions for recovery. Photographs were taken after 7 d of recovery. (D) Survival rates of plants exposed to −6°C. Average survival rates and standard errors were calculated using the results of three separate experiments with 50 seedlings per line for each freezing stress.

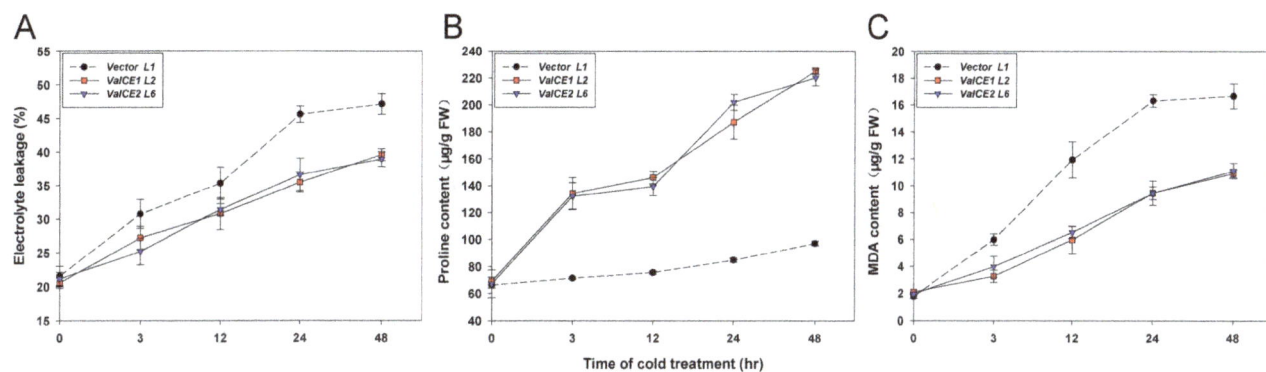

Figure 5. Effect of *ValCE1, 2* expression in *Arabidopsis* on levels of (A) electrolyte leakage, (B) proline and (C) malondialdehyde (MDA). Three-week-old *Arabidopsis* plants of vector-carrying control, *ValCE1* L2, and *ValCE2* L6 were grown at 0°C for the time indicated. Leaves were collected to assess electrolyte leakage and free proline and MDA. Each value is the mean ± SD of three replicates in case of electrolyte leakage, proline and MDA content (6 seedlings each) for the indicated time points.

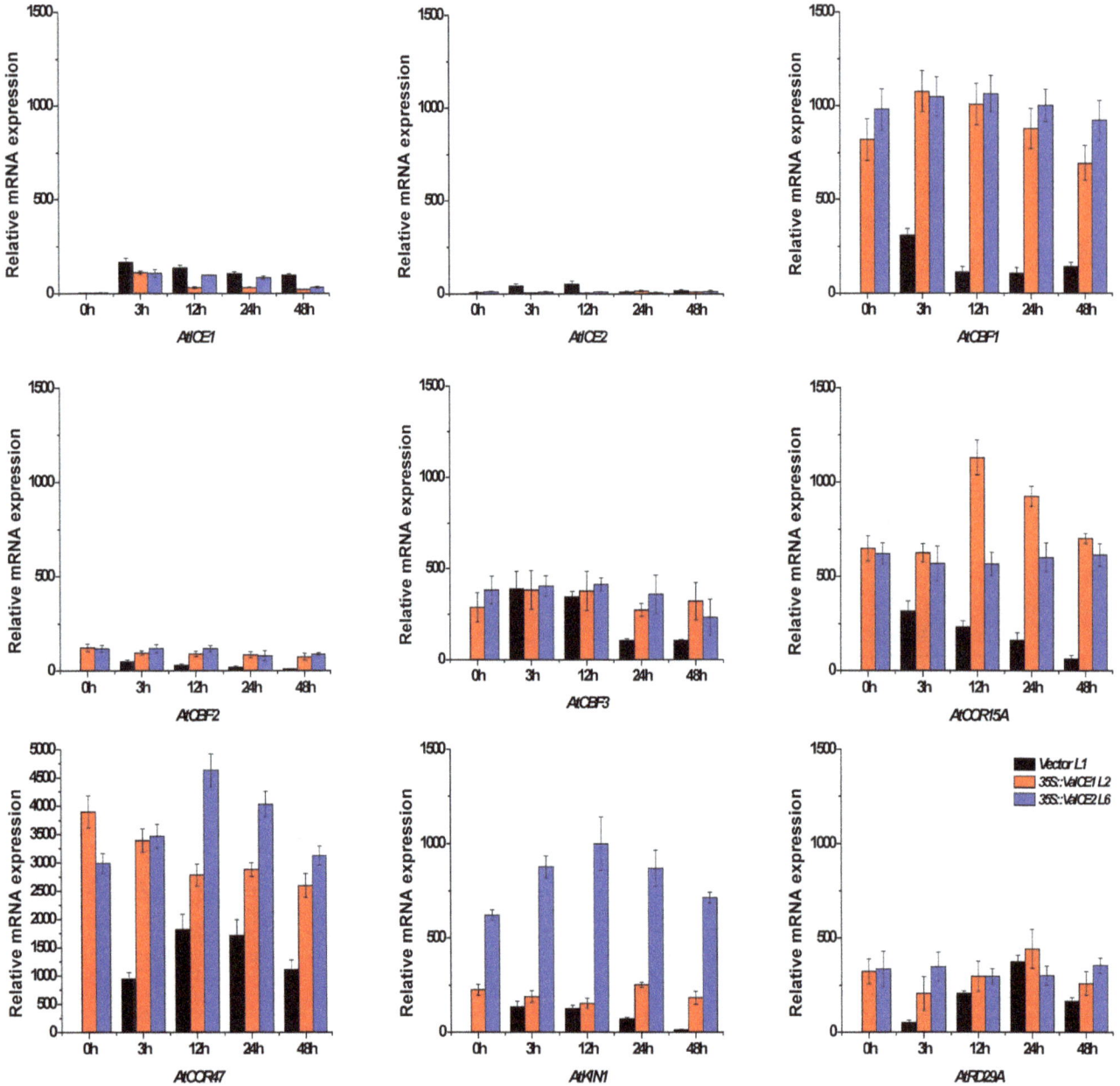

Figure 6. Effect of *VaICE1,2* overexpression in *Arabidopsis* on transcript levels of genes involved in the cold stress pathway. qRT-PCR analysis using leaves from control *Arabidopsis* L1, *35S::VaICE1* L2 and *35S::VaICE2* L6 plants. Three-week-old plants were grown at 0°C for the time indicated. The tested genes were *AtICE1*, *AtICE2*, *AtCBF1*, *AtCBF2*, *AtCBF3*, *AtCOR15A*, *AtCOR47*, *ATKIN1* and *AtRD29A*. *Atactin2* was used as a reference control. Each value is the mean of three replicates.

14. Additionally, the other independent cluster that *VvICE1b* and *VrICE3* involved are alleles, which are located in chromosome 17. Of these three distinct categories, very few, and mostly conserved, substitutions were observed for each category, but it appears that *VrICE4* is phylogenetically distinct from the above-mentioned three categories and located in chromosome 18. A detailed sequence assay on *VrICE4* revealed that it lacked the typical S-rich motif of ICE proteins, presented 3 amino acid mutations (A-T, A-T, and E-D) in ICE-specific domain, and exhibited remarkable sequence difference in ACT_UUR_ACR-like (ACT) domain (Fig. S2A in File S1). Moreover, result from FLAGdb++ database prediction indicated that *VrICE4* has the best match with

GSVIVG01009234001 with functional annotation as DNA binding protein rather than inducer of *CBF* expression 2 (Table S1 in File S1) when compared with other grapevine ICE proteins. Therefore, further investigation on *VrICE4* is still required to confirm its role as *ICE*-like transcription factor.

Evidence from structural and phylogenetic analyses further suggested that *VaICE1* and *VaICE2* may have different properties when involved in cold stress. This is also suggested by their expression profiles (Fig. 2C), which showing that they are both constitutively expressed, but that the expression of *VaICE1* is affected by cold treatments. With the availability of the reported expression data on grapevine *ICE* genes [42–44], a comparison of cold responses of these genes on different genotypes revealed that *VaIC1, 2, VvICE1a, b* and *VrICE1-4* were all detected before and after cold treatment, which is most consistently observed to be constitutively expressed in other plants [32], [34], [40]. However, *VaICE14* is a notable exception where the transcript was not detected under non-stress condition but under low temperature [43]. One possible explanation for such difference has been proposed [44], in which this discrepancy might be attributed to the different genotypes and sampling materials derived from different cultural ways. An additional effort in validation of the expression profile of this gene before and after cold stress in repeated trials is required to clarify its expression property.

Many of the physiological and biochemical changes that occur during cold acclimation are directly correlated to an up-regulation of *COR* gene expression, which is activated by the constitutive expression of the *ICE* transcription factor. Liu et al. [62] found that overexpression of *AtICE1* in cucumber was sufficient to increase chilling tolerance and to simultaneously alter the levels of several cold responses associated factors (e.g., free proline, MDA and soluble sugars). Li et al. [43] discovered that the survival rates of *VvICE1a* and *VvICE1b*-overexpressing *Arabidopsis* lines were significantly higher than those of the wild type under cold stress. Similarly, overexpression of *Solanum lycopersicum ICE1* was reported to improve tolerance of chilling stress in tomato plants, as indicated by differences in electrolyte leakage [37]. In our study, overexpression of either *VaICE1* or *VaICE2* in transgenic *Arabidopsis* plants increased freezing tolerance at the whole-plant level, based on survival rates (Fig. 4B, C), which correlated with increased accumulation of proline, and a reduction in MDA and electrolyte leakage (Fig. 5A-C). These results suggest that although an endogenous *ICE-CBF* pathway is present in *Arabidopsis* system and that enables the ectopic *VaICE1,2* transgene expression to influence the plants' freezing-tolerance capacity. There are also considerable evidences to suggest that proline [63]–[65], MDA [14], [66], [67], and electrolyte leakage [25], [68], contribute to an enhancement of freezing tolerance. Our results confirmed that the greater tolerance of *VaICE1* or *VaICE2*-overexpressing plants to cold stress positively linked to elevated proline levels, and negatively correlated to the reduction of MDA content as well as electrolyte leakage. However, these data did not reveal a difference between the *VaICE1* and *VaICE2* transgenic lines. It is important to note that plants overexpressing *VaICE1* or *2* had similar levels of proline and MDA contents relative to the vector-carry controls under normal growth conditions (0 h), but upregulated in proline level and downregulated in MDA content under time-course cold stresses. We therefore assume that the commonly observed accumulation of free proline and MDA under non-acclimation are partially determined by proline biosynthesis gene (e.g. *P5CS*, d-1-pyrroline-5-carboxylate synthetase) or MDA reductase gene itself property with a relatively low expression, but also induced by the expression of a transcription factor under cold-stressed conditions. Additionally, ionic leakage from the cells is considered as an indicator that the semipermeable nature of the plasma membrane has been lost, at least transiently, in response to freezing. This is probably the main cause for the similar electrolyte leakage observed in controls and *VaICE1-/VaICE2*-overexpressing lines under non-stressed conditions.

We propose that our analyses of the *VaICE1* or *VaICE2* overexpressing *Arabidopsis* plants do not contradict the general cold acclimation model [32], [43], but rather suggest a more integrated circuitry involved in cold acclimation signaling. Chinnusamy et al. [32] reported that over-expression of *AtICE1* increased the expression of *AtCBF2, AtCBF3* and cold-regulated genes (*CORs*) under cold stress. In contrast, Fursova et al. [33] found that over-expression of *AtICE2* only resulted in dominant changes in *AtCBF1* transcription levels after cold acclimation, while Badawi et al. [34] showed that over-expression of wheat *ICE* genes in *Arabidopsis* induced a higher expression of *AtCBF2, AtCBF3* and some *COR* genes only after cold acclimation. Our data suggest that the expression of *AtICE1* or *AtICE2* was slightly lower in the *VaICE1* and *VaICE2* overexpressing *Arabidopsis* plants. One possible explanation for this is that co-suppression of the endogenous gene occurred in the target plant, but we suggest that the *V. amurensis* and *Arabidiopsis ICE1* or *ICE2* cDNA nucleotide sequences are not highly homologous (64.6–69.9%). The observation that several downstream genes were significantly induced in *VaICE1-* or *VaICE2*-overexpressing lines might suggest that the constitutive expression of the transgenes affects the expression of the endogenous gene. Thus, as *VaICE1* and *VaICE2* mRNA levels are abundant in the transgenic plants, competition at a post-transcriptional level could explain this slight decrease in endogenous *AtICE1* or *AtICE2* transcript levels. We speculate that another signal transduction pathway may exist in the *CBF/DREB1* gene regulatory network that is stimulated by *VaICE1* or *VaICE2*. It should be noted that the activation of the expression of these downstream genes differed between the *VaICE1* and *VaICE2* overexpressing lines (Fig. 6). The enhanced tolerance of *35S::VaICE1* and *35S::VaICE2* plants coincides with an up-regulation of the stress-responsive genes *CBF1, COR15A* and *COR47*. However, some differences were seen in the expression patterns between the transgenic lines, and particular in the case of *KIN1*, which suggests functional differences between *VaICE1* and *VaICE2*. In the case of *VvICE1a* and *VvICE1b* transgenic *Arabidopsis* lines, both contribute to the modulation of *AtRD29A* and *AtCOR47* in response to cold stress [43]. Together with the above-mentioned examples, these differences in targeting to the downstream genes are likely due to a varying number of *ICE* homologs in different plant species [32], [33], [40–44]. However, it has not yet been established which of the identified grapevine *ICE* genes specifically control the different sets of cold-responsive genes. A detailed comparison of the expression of these *ICE* genes and their target genes through microarray analysis might help address this issue.

Taken together, *VaICE1* and *VaICE2* are two previously unreported *ICE*-like transcription factors from *V. amurensis*, whose regulatory roles in cold acclimation are suggested by the results presented here. Both genes may act as positive regulators to increase the levels of cold-responsive genes in transgenic lines under cold stress. Moreover, *VaICE1* and *VaICE2* influence cold stress-related factors such as electrolyte leakage, and levels of proline and MDA, thereby alleviating damage by ROS and enhancing osmotic protection. Our data may help elucidate the cold-acclimation pathways of *Vitis* species and more ultimately guide the design of strategies for improving the stress tolerance of agricultural crops.

Supporting Information

File S1   Contains the files: Figure S1 Multiple alignment of the mRNA sequences of *ICE*-homologous from different *Vitis* species. Identical nucleotide sequences are highlighted on a black background while white boxes indicate at least three identical nucleotides. The GenBank accession numbers are reported as follows: *VaICE1* (KC815984), *VaICE2* (KC815985), *VaICE14* (HM231151), *VvICE1* (JQ707298), *VvICE1a* (KC831748), *VvICE1b* (KC831749), *VrICE1* (KF994961), *VrICE2* (KF994962), *VrICE3* (KF994963) and *VrICE4* (KF994964). **Figure S2 Protein sequence similarity and phylogenetic clustering of ICE from three different *Vitis***

species. (A) Protein sequence alignment of 10 grapevine ICEs. Identical residues are outlined in black. Amino acids are numbered on the right. (B) Phylogenetic tree based on the deduced amino acid sequences of ICEs from three different *Vitis* genotypes. A maximum-likelihood phylogenetic tree of the amino acid sequences of ICE from *V. amurensis* (VaICE1, AGP04217; VaICE2, AGP04218; VaICE14, ADY17816), *V. vinifera* (VvICE1, AFI49627; VvICE1a, AGQ03810; VvICE1b, AGQ03811), and *V. riparia* (VrICE1, AGG34704; VrICE2, AIA58705; VrICE3, AIA58706; VrICE4, AIA58707) is constructed by MEGA 5.0 with 1000 bootstrap tests. The branch support values are indicated. The length of the scale bar corresponds to 0.05 substitutions per site. **Table S1 Feature lists of BLAST results or queries of ten grapevine *ICE* genes available in FLAGdb⁺⁺.** A Blastp search with an E-Value of 1.E-50 was

performed on *V. vinifera* using the ten grapevine ICE proteins from *V. amurensis* (VaICE1, AGP04217; VaICE2, AGP04218; VaICE14, ADY17816), *V. vinifera* (VvICE1, AFI49627; VvICE1a, AGQ03810; VvICE1b, AGQ03811), and *V. riparia* (VrICE1, AGG34704; VrICE2, AIA58705; VrICE3, AIA58706; VrICE4, AIA58707) as query.

Author Contributions

Conceived and designed the experiments: WRX ZPW. Performed the experiments: WRX YTJ NBZ RML DMX. Analyzed the data: WRX ZPW RML. Contributed reagents/materials/analysis tools: YTJ NBZ RML DMX XLD. Contributed to the writing of the manuscript: WRX.

References

1. Christen D, Susan S, Jermini M, Strasser RJ, Défago G (2007) Characterization and early detection of grapevine (*Vitis vinifera*) stress responses to esca disease by in situ chlorophyll fluorescence and comparison with drought stress. Environmental and Experimental Botany 60: 504–514.
2. Daldoul S, Guillaumie S, Reustle GM, Krczal G, Ghorbel A, et al. (2010) Isolation and expression analysis of salt induced genes from contrasting grapevine (*Vitis vinifera* L.) cultivars. Plant Science 179: 489–498.
3. C FAaL (2013) Abiotic stress effects on grapevine (*Vitis vinifera* L.): Focus on abscisic acid-mediated consequences on secondary metabolism and berry quality. *Environmental and Experimental Botany.*
4. Kadioglu A, Terzi R, Saruhan N, Saglam A (2012) Current advances in the investigation of leaf rolling caused by biotic and abiotic stress factors. Plant Science 182: 42–48.
5. Kobayashi M, Katoh H, Takayanagi T, Suzuki S (2010) Characterization of thermotolerance-related genes in grapevine (*Vitis vinifera*). Journal of Plant Physiology 167: 812–819.
6. Zhang Y, Dami IE (2012) Foliar Application of Abscisic Acid Increases Freezing Tolerance of Field-Grown *Vitis vinifera* Cabernet franc Grapevines. American Journal of Enology and Viticulture 63: 377–384.
7. Liu L, Li H (2013) Review: Research progress in amur grape, *Vitis amurensis* Rupr. Canadian Journal of Plant Science 93: 565–575.
8. Wan Y, Schwaninger Heidi, Li D, Simon CJ, Wang Y, et al. (2008) The eco-geographic distribution of wild grape germplasm in China. VITIS 47: 77–80.
9. Thomashow MF (1999) PLANT COLD ACCLIMATION: Freezing Tolerance Genes and Regulatory Mechanisms. Annu Rev Plant Physiol Plant Mol Biol 50: 571–599.
10. Lee B-h, Henderson DA, Zhu J-K (2005) The *Arabidopsis* Cold-Responsive Transcriptome and Its Regulation by *ICE1*. The Plant Cell Online 17: 3155–3175.
11. Cushman JC, Bohnert HJ (2000) Genomic approaches to plant stress tolerance. Curr Opin Plant Biol 3: 117–124.
12. Fowler S, Thomashow MF (2002) *Arabidopsis* transcriptome profiling indicates that multiple regulatory pathways are activated during cold acclimation in addition to the *CBF* cold response pathway. Plant Cell 14: 1675–1690.
13. Shinozaki K, Yamaguchi-Shinozaki K, Seki M (2003) Regulatory network of gene expression in the drought and cold stress responses. Curr Opin Plant Biol 6: 410–417.
14. Hara M, Terashima S, Fukaya T, Kuboi T (2003) Enhancement of cold tolerance and inhibition of lipid peroxidation by citrus dehydrin in transgenic tobacco. Planta 217: 290–298.
15. Kaye C, Neven L, Hofig A, Li QB, Haskell D, et al. (1998) Characterization of a gene for spinach *CAP160* and expression of two spinach cold-acclimation proteins in tobacco. Plant Physiol 116: 1367–1377.
16. Nakatsuka T, Haruta KS, Pitaksutheepong C, Abe Y, Kakizaki Y, et al. (2008) Identification and characterization of R_2R_3-MYB and *bHLH* transcription factors regulating anthocyanin biosynthesis in gentian flowers. *Plant and Cell Physiology* 49: 1818–1829.
17. Garg AK, Kim JK, Owens TG, Ranwala AP, Choi YD, et al. (2002) Trehalose accumulation in rice plants confers high tolerance levels to different abiotic stresses. Proc Natl Acad Sci U S A 99: 15898–15903.
18. Pramanik MH, Imai R (2005) Functional identification of a trehalose 6-phosphate phosphatase gene that is involved in transient induction of trehalose biosynthesis during chilling stress in rice. Plant Mol Biol 58: 751–762.
19. Houde M, Dallaire S, N'Dong D, Sarhan F (2004) Overexpression of the acidic dehydrin *WCOR410* improves freezing tolerance in transgenic strawberry leaves. Plant Biotechnol J 2: 381–387.
20. Dai X, Xu Y, Ma Q, Xu W, Wang T, et al. (2007) Overexpression of an $R_1R_2R_3$ MYB gene, *OsMYB3R-2*, increases Tolerance to Freezing, Drought, and Salt stress in Transgenic *Arabidopsis*. Plant Physiol 143: 1739–1751.
21. Gong Z, Lee H, Xiong L, Jagendorf A, Stevenson B, et al. (2002) RNA helicase-like protein as an early regulator of transcription factors for plant chilling and freezing tolerance. Proc Natl Acad Sci U S A 99: 11507–11512.
22. Iturriaga G, Schneider K, Salamini F, Bartels D (1992) Expression of desiccation-related proteins from the resurrection plant *Craterostigma plantagineum* in transgenic tobacco. Plant Mol Biol 20: 555–568.
23. Jaglo-Ottosen KR, Gilmour SJ, Zarka DG, Schabenberger O, Thomashow MF (1998) *Arabidopsis CBF1* overexpression induces *COR* genes and enhances freezing tolerance. Science 280: 104–116.
24. Kasuga M, Liu Q, Miura S, Yamaguchi-Shinozaki K, Shinozaki K (1999) Improving plant drought, salt, and freezing tolerance by gene transfer of a single stress-inducible transcription factor. Nat Biotechnol 17: 287–291.
25. Qin F, Sakuma Y, Li J, Liu Q, Li YQ, et al. (2004) Cloning and functional analysis of a novel *DREB1/CBF* transcription factor involved in cold-responsive gene expression in *Zea mays* L. Plant Cell Physiol 45: 1042–1052.
26. Jaglo KR, Kleff S, Amundsen KL, Zhang X, Haake V, et al. (2001) Components of the *Arabidopsis* C-repeat/dehydration-responsive element binding factor cold-response pathway are conserved in *Brassica napus* and other plant species. Plant Physiol 127: 910–917.
27. Dubouzet JG, Sakuma Y, Ito Y, Kasuga M, Dubouzet EG, et al. (2003) *OsDREB* genes in rice, *Oryza sativa* L., encode transcription activators that function in drought-, high-salt- and cold-responsive gene expression. Plant J 33: 751–763.
28. Gilmour SJ, Sebolt AM, Salazar MP, Everard JD, Thomashow MF (2000) Overexpression of the *Arabidopsis CBF3* transcriptional activator mimics multiple biochemical changes associated with cold acclimation. Plant Physiol 124: 1854–1865.
29. Liu Q, Kasuga M, Sakuma Y, Abe H, Miura S, et al. (1998) Two transcription factors, *DREB1* and *DREB2*, with an EREBP/AP2 DNA binding domain separate two cellular signal transduction pathways in drought- and low-temperature-responsive gene expression, respectively, in *Arabidopsis*. Plant Cell 10: 1391–1406.
30. Chinnusamy V, Zhu J-K, Sunkar R (2010) Gene regulation during cold stress acclimation in plants. Plant Stress Tolerance: Springer. pp. 39–55.
31. Yamaguchi-Shinozaki K, Shinozaki K (2006) Transcriptional regulatory networks in cellular responses and tolerance to dehydration and cold stresses. Annu Rev Plant Biol 57: 781–803.
32. Chinnusamy V, Ohta M, Kanrar S, Lee BH, Hong X, et al. (2003) *ICE1*: a regulator of cold-induced transcriptome and freezing tolerance in *Arabidopsis*. Genes Dev 17: 1043–1054.
33. Fursova OV, Pogorelko GV, Tarasov VA (2009) Identification of *ICE2*, a gene involved in cold acclimation which determines freezing tolerance in *Arabidopsis thaliana*. Gene 429: 98–103.
34. Badawi M, Reddy YV, Agharbaoui Z, Tominaga Y, Danyluk J, et al. (2008) Structure and functional analysis of wheat *ICE* (inducer of *CBF* expression) genes. Plant Cell Physiol 49: 1237–1249.
35. Feng XM, Zhao Q, Zhao LL, Qiao Y, Xie XB, et al. (2012) The cold-induced basic helix-loop-helix transcription factor gene *MdCIbHLH1* encodes an *ICE*-like protein in apple. BMC Plant Biol 12: 22.
36. Lin Y, Zheng H, Zhang Q, Liu C, Zhang Z (2013) Functional profiling of EcaICE₁ transcription factor gene from *Eucalyptus camaldulensis* involved in cold response in tobacco plants. Journal of Plant Biochemistry and Biotechnology: 1–10.
37. Miura K, Shiba H, Ohta M, Kang SW, Sato A, et al. (2012) *SlICE1* encoding a MYC-type transcription factor controls cold tolerance in tomato, *Solanum lycopersicum*. Plant Biotechnology 29: 253–260.
38. Nakamura J, Yuasa T, Huong TT, Harano K, Tanaka S, et al. (2011) Rice homologs of inducer of *CBF* expression (*OsICE*) are involved in cold acclimation. Plant Biotechnology 28: 303–309.
39. Wang X, Sun X, Liu S, Liu L, Liu X, et al. (2005) Molecular cloning and characterization of a novel ice gene from *Capsella bursa-pastoris*. Mol Biol (Mosk) 39: 21–29.

40. Wang Y, Jiang CJ, Li YY, Wei CL, Deng WW (2012) *CsICE1* and *CsCBF1*: two transcription factors involved in cold responses in *Camellia sinensis*. Plant Cell Rep 31: 27–34.

41. Xiang D, Man L, Yin K, Song Q, Wang L, et al. (2013) Overexpression of a *ItICE1* gene from *Isatis tinctoria* enhances cold tolerance in rice. Molecular Breeding 32: 617–628.

42. Dong C, Zhang Z, Ren J, Qin Y, Huang J, et al. (2013) Stress-responsive gene *ICE1* from *Vitis amurensis* increases cold tolerance in tobacco. Plant Physiol Biochem 71: 212–217.

43. Li J, Wang L, Zhu W, Wang N, Xin H, et al. (2014) Characterization of two *VvICE1* genes isolated from 'Muscat Hamburg' grapevine and their effect on the tolerance to abiotic stresses. Scientia Horticulturae 165: 266–273.

44. Rahman MA, Moody MA, Nassuth A (2014): Grape contains 4 ICE genes whose expression includes alternative polyadenylation, leading to transcripts encoding at least 7 different ICE proteins. Environmental and Experimental Botany.

45. Tamura K, Peterson D, Peterson N, Stecher G, Nei M, et al. (2011) MEGA5: molecular evolutionary genetics analysis using maximum likelihood, evolutionary distance, and maximum parsimony methods. Molecular biology and evolution 28: 2731–2739.

46. Xu X, Chen C, Fan B, Chen Z (2006) Physical and functional interactions between pathogen-induced *Arabidopsis WRKY18*, *WRKY40*, and *WRKY60* transcription factors. Plant Cell 18: 1310–1326.

47. Xu W, Zhang N, Jiao Y, Li R, Xiao D, Wang Z (2014). The grapevine basic helix-loop-helix (*bHLH*) transcription factor positively modulates CBF-pathway and confers tolerance to cold-stress in *Arabidopsis*. Mol Biol Rep DOI: 10.1007/s11033-014-3404-2.

48. Clough SJ, Bent AF (1998) Floral dip: a simplified method for *Agrobacterium*-mediated transformation of *Arabidopsis thaliana*. Plant J 16: 735–743.

49. Weigel RR, Bäuscher C, Pfitzner AJ, Pfitzner UM (2001) NIMIN-1, NIMIN-2 and NIMIN-3, members of a novel family of proteins from *Arabidopsis* that interact with NPR1/NIM1, a key regulator of systemic acquired resistance in plants. Plant molecular biology 46(2):143–160

50. Bates L, Waldren RP, Teare I (1973) Rapid determination of free proline for water-stress studies. Plant and soil 39: 205–207.

51. Draper HH, Hadley M (1990) Malondialdehyde determination as index of lipid peroxidation. Methods Enzymol 186: 421–431.

52. Livak KJ, Schmittgen TD (2001) Analysis of relative gene expression data using real-time quantitative PCR and the 2(-Delta Delta C(T)) Method. Methods 25: 402–408.

53. Miura K, Jin J, Lee J, Yoo C, Stirm V, et al. (2007) *SIZ1*-mediated sumoylation of *ICE1* controls *CBF3/DREB1A* expression and freezing tolerance in *Arabidopsis*. Plant Cell 19: 1403–1414

54. Kanaoka M, Pillitteri L, Fujii HY, Bogenschutz N, Takabayashi J, et al. (2008) *SCREAM/ICE1* and *SCREAM2* specify three cell-state transitional steps leading to *Arabidopsis* stomatal differentiation. Plant Cell 20: 1775–1785.

55. Dexter ST, Tottingham WE, Graber LF (1932) Investigations of the hardiness of plants by measurement of electrical conductivity. Plant Physiology 7: 63–78.

56. Hare PD, Cress WA (1997) Metabolic implications of stress-induced proline accumulation in plants. Plant Growth Regulation 21: 79–102.

57. Trovato M, Mattioli R, Costantino P (2008) Multiple roles of proline in plant stress tolerance and development. RENDICONTI LINCEI 19: 325–346.

58. Guy CL, Huber JLA, Huber SC (1992) Sucrose Phosphate Synthase and Sucrose Accumulation at Low Temperature. Plant Physiology 100: 502–508.

59. Wanner LA, Junttila O (1999) Cold-induced freezing tolerance in *Arabidopsis*. Plant Physiol 120: 391–400.

60. Deng Y, Chen S, Chen F, Cheng X, Zhang F (2011) The embryo rescue derived intergeneric hybrid between *chrysanthemum* and *Ajania przewalskii* shows enhanced cold tolerance. Plant Cell Rep 30: 2177–2186.

61. Sato Y, Masuta Y, Saito K, Murayama S, Ozawa K (2011) Enhanced chilling tolerance at the booting stage in rice by transgenic overexpression of the ascorbate peroxidase gene, *OsAPXa*. Plant cell reports 30: 399–406.

62. Liu L, Duan L, Zhang J, Zhang Z, Mi G, et al. (2010) Cucumber (*Cucumis sativus* L.) over-expressing cold-induced transcriptome regulator *ICE1* exhibits changed morphological characters and enhances chilling tolerance. Scientia horticulturae 124: 29–33.

63. Carpenter JF, Crowe JH (1988) The mechanism of cryoprotection of proteins by solutes. Cryobiology 25: 244–255.

64. Nanjo T, Kobayashi M, Yoshiba Y, Kakubari Y, Yamaguchi-Shinozaki K, et al. (1999) Antisense suppression of proline degradation improves tolerance to freezing and salinity in *Arabidopsis thaliana*. FEBS Lett 461: 205–210.

65. Rudolph AS, Crowe JH (1985) Membrane stabilization during freezing: the role of two natural cryoprotectants, trehalose and proline. Cryobiology 22: 367–377.

66. Koszo F, Siklosi C, Simon N (1974) Liposome model experiment for the study of assumed membrane damage in porphyria cutanea tarda. Biochim Biophys Acta 363: 182–189.

67. Zhang C, Liu J, Zhang Y, Cai X, Gong P, et al. (2011) Overexpression of *SlGMEs* leads to ascorbate accumulation with enhanced oxidative stress, cold, and salt tolerance in tomato. Plant Cell Reports 30: 389–398.

68. Kasuga M, Miura S, Shinozaki K, Yamaguchi-Shinozaki K (2004) A combination of the *Arabidopsis DREB1A* gene and stress-inducible *rd29A* promoter improved drought- and low-temperature stress tolerance in tobacco by gene transfer. Plant Cell Physiol 45: 346–350.

69. Massari ME, Murre C (2000) Helix-loop-helix proteins: regulators of transcription in eucaryotic organisms. Molecular and cellular biology 20(2):429–440

70. Murre C, McCaw PS, Baltimore D (1989) A new DNA binding and dimerization motif in immunoglobulin enhancer binding, daughterless, MyoD, and myc proteins. Cell 56(5):777–783

20

Metallothionein 2 (*SaMT2*) from *Sedum alfredii* Hance Confers Increased Cd Tolerance and Accumulation in Yeast and Tobacco

Jie Zhang, Min Zhang, Shengke Tian, Lingli Lu, M. J. I. Shohag, Xiaoe Yang*

MOE Key Laboratory of Environment Remediation and Ecosystem Health, College of Environmental and Resource Sciences, Zhejiang University, Hangzhou, China

Abstract

Metallothioneins are cysteine-rich metal-binding proteins. In the present study, *SaMT2*, a type 2 metallothionein gene, was isolated from Cd/Zn co-hyperaccumulator *Sedum alfredii* Hance. *SaMT2* encodes a putative peptide of 79 amino acid residues including two cysteine-rich domains. The transcript level of SaMT2 was higher in shoots than in roots of *S. alfredii*, and was significantly induced by Cd and Zn treatments. Yeast expression assay showed *SaMT2* significantly enhanced Cd tolerance and accumulation in yeast. Ectopic expression of *SaMT2* in tobacco enhanced Cd and Zn tolerance and accumulation in both shoots and roots of the transgenic plants. The transgenic plants had higher antioxidant enzyme activities and accumulated less H_2O_2 than wild-type plants under Cd and Zn treatment. Thus, *SaMT2* could significantly enhance Cd and Zn tolerance and accumulation in transgenic tobacco plants by chelating metals and improving antioxidant system.

Editor: Wagner L. Araujo, Universidade Federal de Vicosa, Brazil

Funding: The present study was supported by the Fundamental Research Funds for the Central Universities (No. 2013FZA6005), National Natural Science Foundation of China (No. 31372128; 21177107). The funders had no role in study design, data collection and analysis, decision to publish, or preparation of the manuscript.

Competing Interests: The authors have declared that no competing interests exist.

* Email: xyang@zju.edu.cn

Introduction

Heavy metals are known to cause toxic effects and inhibition of plant growth. However, rare plant species, which can accumulate and tolerate extremely high concentrations of heavy metals in their shoots without toxicity effects, have been defined as "hyperaccumulators" [1]. The elucidation of the mechanisms underlying metal hyperaccumulation may enable the phytoremediation of metal-contaminated soils and the biofortification of trace elements in food crops [2–4].

Higher plants have evolved various defense mechanisms to detoxify excess metals. These mechanisms contain compartmentalization in inactive tissues, chelation by metal ligands and detoxification by antioxidants [5]. Metal chelators such as organic acids, amino acids, phytochelatins and metallothioneins play important roles in metal detoxification [6]. Metallothioneins (MTs) are low-molecular-mass, cysteine-rich proteins which are broadly distributed in microorganisms, plants and animals [7]. Plant MTs can be divided into four subfamilies based on the distribution of cysteine residues in their amino- and carboxyl-terminal regions [8]. Several MT genes have been isolated and characterized from plants. There are some evidence indicating that plant MTs are involved in metal homeostasis, detoxification and reactive oxygen species (ROS) scavenging [8–11].

Hyperaccumulating ecotype (HE) of *Sedum alfredii* Hance is a Zn/Cd hyperaccumulator discovered from an old Pb/Zn mining area of China [12,13]. It can accumulate up to 9000 $\mu g\ g^{-1}$ Cd and 29000 $\mu g\ g^{-1}$ Zn in its shoots without toxicity symptoms [12,13]. This large amount of metals in plant cells needs a powerful detoxification system to protect plants from the deleterious effects of the metals. Earlier studies have demonstrated that the hyperaccumulating ecotype of *Sedum alfredii* has a more effective antioxidant enzyme system than non-hyperaccumulating ecotype (NHE) [14,15]. However, the mechanism of hypertolerance of metals in this species has not been fully understood. In the present study, a metallothionein gene from hyperaccumulating ecotype of *Sedum alfredii* Hance, named *SaMT2*, was isolated and cloned. The expression pattern of this gene was studied by Real Time-PCR. To analysis the function of *SaMT2*, its full length cDNA was cloned and expressed in yeast and tobacco. The transgenic yeast and tobacco plants were analyzed to evaluate whether *SaMT2* protein played a role in Cd or Zn tolerance and accumulation.

Materials and Methods

Ethics statement

These field studies did not involve any protected species. No specific permits were required for the collection of samples in the study location.

Plant growth

The hyperaccumulating ecotype of *S. alfredii* Hance was collected from an old Pb/Zn mining site in Zhejiang Province, P. R. China. Plants were grown in non-polluted soils for several generations to minimize internal metal concentrations. Similar size shoot branches were cut and cultured hydroponically. After two weeks, rooted seedlings were then subjected to 4 days exposure of one-fourth, half and full strength nutrient solutions containing 2 mM $Ca(NO_3)_2$, 0.7 mM K_2SO_4, 0.1 mM KH_2PO_4, 0.1 mM KCl, 0.5 mM $MgSO_4$, 10 µM H_3BO_3, 0.5 µM $MnSO_4$, 5 µM $ZnSO_4$, 0.2 µM $CuSO_4$, 0.01 µM $(NH_4)_6 \cdot Mo_7O_{24}$, and 20 µM Fe-EDTA. Nutrient solution pH was adjusted daily to 5.8 with 0.1 M NaOH or HCl. Plants were grown under glasshouse conditions with natural light (day/night of 16/8 h), day/night temperature of 26/20°C and day/night humidity of 70/85%. The nutrient solution was aerated continuously and renewed every 3 d. To compare the expression of *SaMT2*, the precultured seedlings were treated with 100 µM $CdCl_2$ and 500 µM $ZnSO_4$ for 8 days.

Cloning of *SaMT2* cDNA and sequence analysis

The cDNA fragment of *SaMT2* was isolated from RNA-Seq data of *Sedum alfredii* Hance (Gao et al. 2013). The full length of *SaMT2* was isolated using 3′ and 5′ RACE methods as described by the supplier (Smart RACE cDNA amplification kit; Clontech Laboratories, Inc. CA, USA). PCR was performed with the following primers: 5′-CTGGGCGTGGCTCCGAAGCAAGT-GTA-3′ for 3′ RACE and 5′-CGCAACCACAGTTTCCACCA-CAGCA-3′ for 5′ RACE. Alignment of *SaMT2* was performed by ClustalW on the internet (http://clustalw.ddbj.nig.ac.jp/). The phylogenetic tree was constructed using the neighbor-joining algorithm by MEGA 5 software (released from http://www.megasoftware.net/) after ClustalW alignment with 1000 bootstrap trials.

Real time RT-PCR analysis

The total RNA was extracted from various tissues by RNAiso plus (Takara Bio, Inc. Shiga, Japan), and then converted to cDNA using Primescript[TM] RT regent kit with gDNA eraser (Takara Bio,

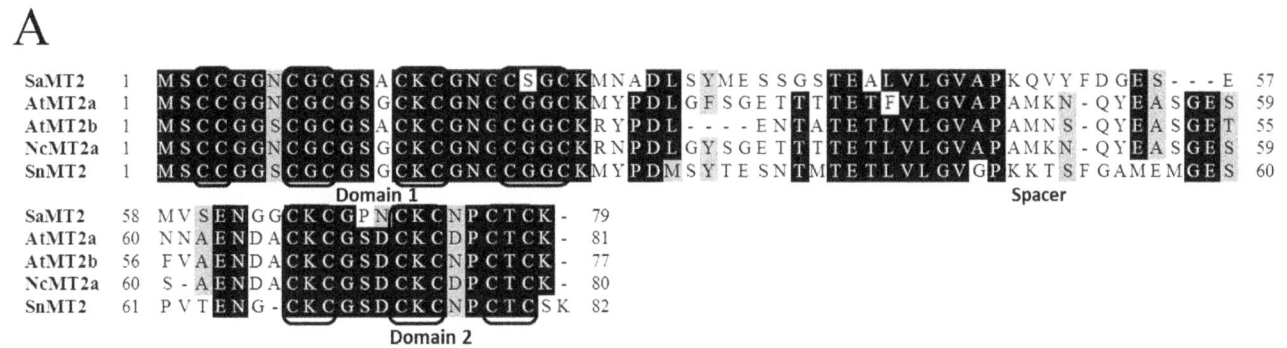

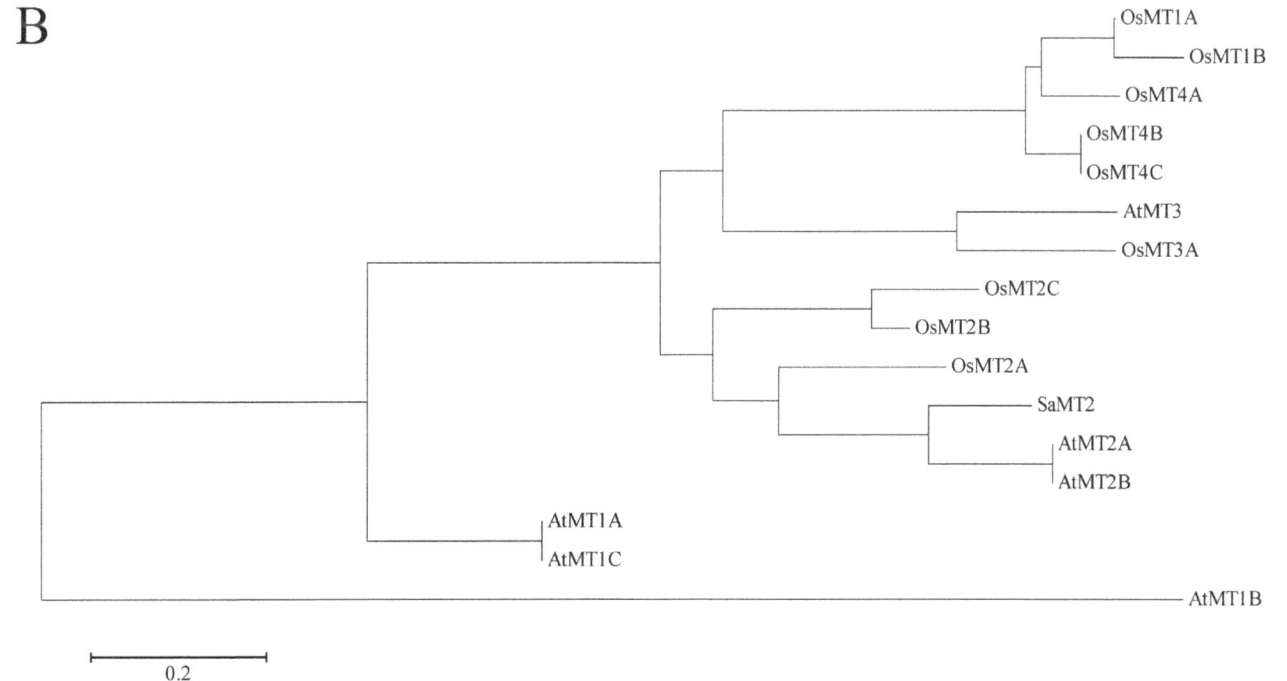

Figure 1. Sequence alignment and phylogenic analysis of *SaMT2* with other MTs. (A) The deduced amino acid sequences encoded by *SaMT2* were aligned with MTs from *Arabidopsis thaliana*, *Noccaea caerulescens* and *Solanum nigrum*. The cysteine-rich domains are boxed. (B) The phylogenic tree of *SaMT2* and MTs from Arabidopsis and rice.

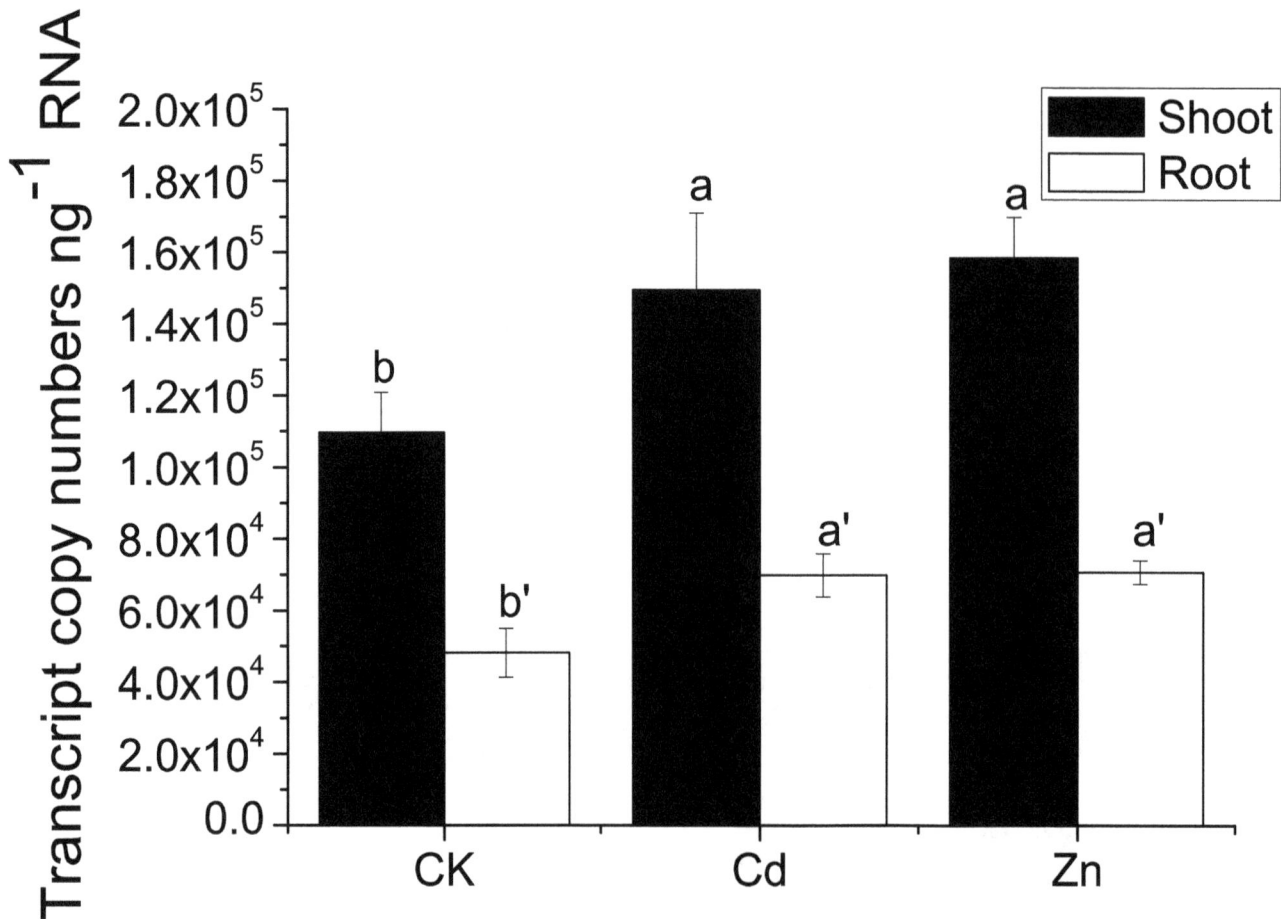

Figure 2. The expression level of *SaMT2* in *Sedum alfredii*. The transcript level of *SaMT2* induced by Cd and Zn treatments. The different letters above the columns indicate the significant difference between the treatments (p<0.05, Tukey's test). CK represents the control group.

Inc. Shiga, Japan). Expression of the *SaMT2* was determined by quantitative RT-PCR with the SYBR Green I reagent (SYBR Premix Ex Taq II; Takara Bio, Inc. Shiga, Japan) on an Eppendorf Mastercycler Epgradient Realplex^2 (Eppendorf AG, Hamburg, Germany). A portion (10 ng) of cDNA was used for the template. The primers used for *SaMT2* were forward 5'-CTGTGGTTGCGGATCTGCTT-3' and reverse 5'- TCCATT-CTCCGACACCATCT-3'. To generate standard curves for the absolute quantification for *SaMT2* copy number, a series of dilutions (from 1×10^{-1} to 1×10^{-6} ng) of plasmids were made and then subjected to real-time PCR.

Plasmids construction

To express *SaMT2* in *Saccharomyces cerevisiae*, the full ORF of *SaMT2* was amplified from the cDNA of *S. alfredii* using primers: 5'-AGCTCGAGATGTCTTGCTGTGGTGGA-3' contains an XhoI site and 5'-GAGGATCCTCATTTGCAAGTGCAGGG-3' contains a BamHI site. The PCR products were then cloned into a pEASY Blunt simple vector (Transgen, Beijing, China) and its sequence confirmed. This vector was double digested with XhoI and BamHI, and the obtained fragment was cloned into pDR195 between XhoI and BamHI sites.

To construct the plant overexpression vector, the full ORF of *SaMT2* was amplified using primers: 5'-AAGATCTGATGTC-TTGCTGTGGTGGA-3' and 5'-AAGGTGACCTCATTTG-CAAGTGCAGG-3', which contained BglII and BstpI restriction

sites, respectively. The obtained fragment was restricted with BglII and BstpI, and then cloned into the BglII and BstpI sites of pCAMBIA 1302 vector and its sequence was confirmed.

Yeast complementation assay

The *S. cerevisiae* strains BY4741 (wild type, *MATα; his2Δ0; met15Δ0; ura3Δ0*), *Δycf1 (MATa; his3Δ1; leu2Δ0; lys2Δ0; ura3Δ0; YDR135c::kanMX4)* and *Δzrc1 (MATα; his3Δ1; leu2Δ0; met15Δ0; ura3Δ0; YMR243c::kanMX4)* mutants were used to investigate the role of *SaMT2* in Cd and Zn tolerance. The yeast transformation was conducted using LiAc/PEG/ssDNA methods, as described by Gietz and Schiestl [16]. To obtain cells for transformation: Inoculate a single colony of the yeast strain with a sterile inoculation loop from a fresh SD (synthetic medium plus dextrose, 0.67% yeast nitrogen base, 2% D-glucose, and amino acids) plate into 5 ml of YPD medium (2% peptone, 1% yeast extracts, 2% D-glucose) and incubate overnight at 30°C. Add 2.5×10^8 cells to 50 ml of YPD medium in a culture flask and incubate until the cell titer is at least 2×10^7 cells ml^{-1}. Cells were harvested by centrifugation at 3,000 g for 5 min and washed twice with sterilized water. Then the cells were re-suspended in 1.0 ml of sterile deionized water and pelleted by centrifugation (13 000×g for 30 sec). The supernatant was discarded and the transformation mixture {containing 240 μl PEG 3350 (50% w/v), 36 μl 1.0 M lithium acetate, 10 μl single-stranded carrier DNA (10 mg ml^{-1}) and plasmid DNA (0.5–1 μg), and sufficient sterile deionized water

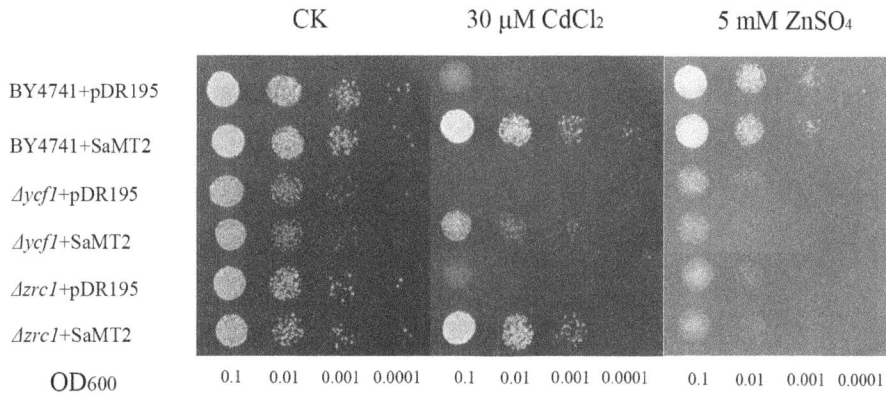

Figure 3. Cd and Zn tolerance of yeast cells expressing *SaMT2*. The *Saccharomyces cerevisiae* BY4741, *Δycf1* and *Δzrc1* yeast cells harboring pDR195 (vector control) or pDR195-*SaMT2* were grown in liquid SD selective medium. Cultures were adjusted to OD_{600nm} of 0.1 and serially 10-fold diluted in water. 10 μl aliquots of each dilution were spotted either on SD selective plates or on plates with 30 μM $CdCl_2$ or 5 mM $ZnSO_4$. After 3 days of incubation at 30°C, plates were photographed. CK represents the control group.

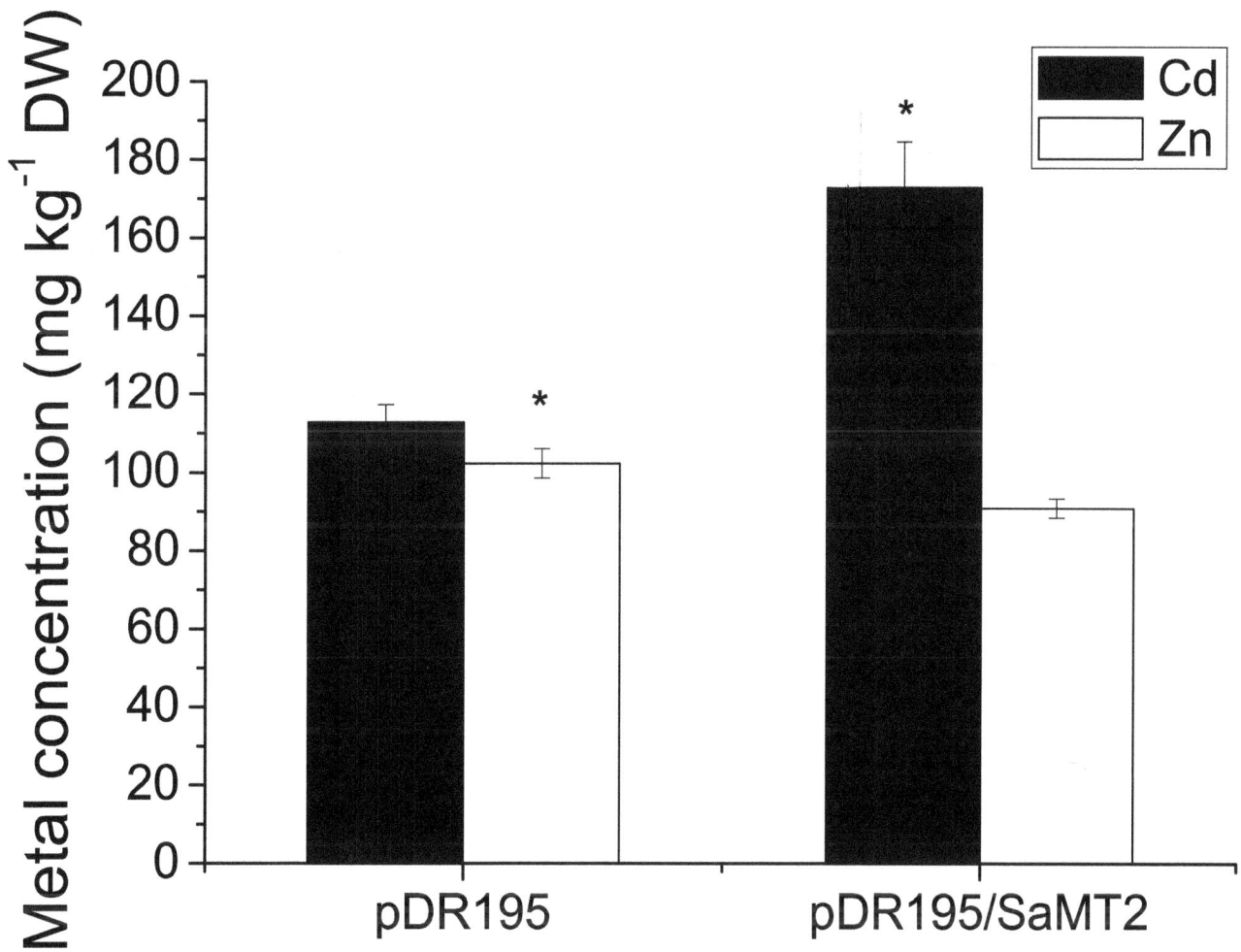

Figure 4. Cd and Zn concentration in *Δycf1* and *Δzrc1* yeast cells expressing *SaMT2*. The yeast transformants containing pDR195 or pDR195-*SaMT2* were grown in liquid SD selective medium with 30 μM $CdCl_2$ and 100 μM $ZnSO_4$ for *Δycf1 and Δzrc1*, respectively. Cells were incubated at 30°C for 48 h and metal contents were measured by ICP-MS. Results are averages (±S.E.) from three independent experiments done with four different colonies. The '*' symbol indicates the mean values were significantly different at p<0.05 (Tukey's test).

Figure 5. Metal tolerance analysis of transgenic tobacco plants over-expressing *SaMT2*. The figure shows the effect of 200 μM ZnSO$_4$ or 100 μM CdCl$_2$ on the growth of WT and transgenic plants on B5 medium. CK represents the control group.

to provide a final volume of 360 μl} were layered over the pellet. The mixture was vortex vigorously for 1 min and subjected to a heat shock at 42°C for 40 min. The transformation mixture was then centrifuged at 13 000×g for 30 sec to pellet the cells. After the supernatant was decanted, the cells were resuspended in 1.0 ml of sterile deionized water. Aliquots of the resuspended cells were plated onto SD-URA media. Plates were incubated for 2–3 days at 30°C until transformants were observed. Single colonies were picked from each transformant plate and established on fresh SD-URA plates.

For the metal tolerance assay, single colonies from SD-URA plates were cultured in liquid SD-URA medium until OD$_{600}$ reached 1.0. After serial dilutions (OD$_{600}$ = 0.1, 0.01, 0.001, 0.0001, respectively) were prepared, each dilution was spotted onto SD-URA medium with or without 5 mM ZnSO$_4$ or 30 μM CdCl$_2$. Plates were photographed after incubation at 30°C for 3 d.

For determination of metal concentration in yeast, transformants were grown in liquid SD-URA medium overnight. Then, cells were adjusted to OD$_{600}$ = 0.2 in the presence of 10 μM

CdCl$_2$ or 100 μM ZnSO$_4$ for Zn determination. After incubation for 48 h, the cells were harvested and washed with distilled water, 20 mM Na$_2$EDTA and distilled water, respectively. Dry weight was determined after 3 days at 60°C. Cells were digested using 5 ml concentrated HNO$_3$, incubated at 95°C for 2 h. The Zn and Cd concentrations were determined by using ICP-MS (Inductively Coupled Plasma Mass Spectrometry, Agilent 7500a, CA, USA).

Heterogeneous expression of *SaMT2* in tobacco

The transformation of tobacco was constructed using the leaf disk method according to Horsch *et al* [17]. Surface-sterilized T$_1$ seeds of two transgenic tobacco lines were germinated on Murashige Skoog (MS) plates containing 40 mg/L hygromycin to select hygromycin-resistance seedlings. For Zn/Cd tolerance analysis, the wild type (WT) and transgenic plants were transferred to MS plates containing 100 μM CdCl$_2$ or 200 μM ZnSO$_4$ for 14 d. To determine the Zn/Cd concentrations in plants, both WT and transgenic plants were transferred to hydroponic culture. One-month old plants were then treated with 50 μM CdCl$_2$ or

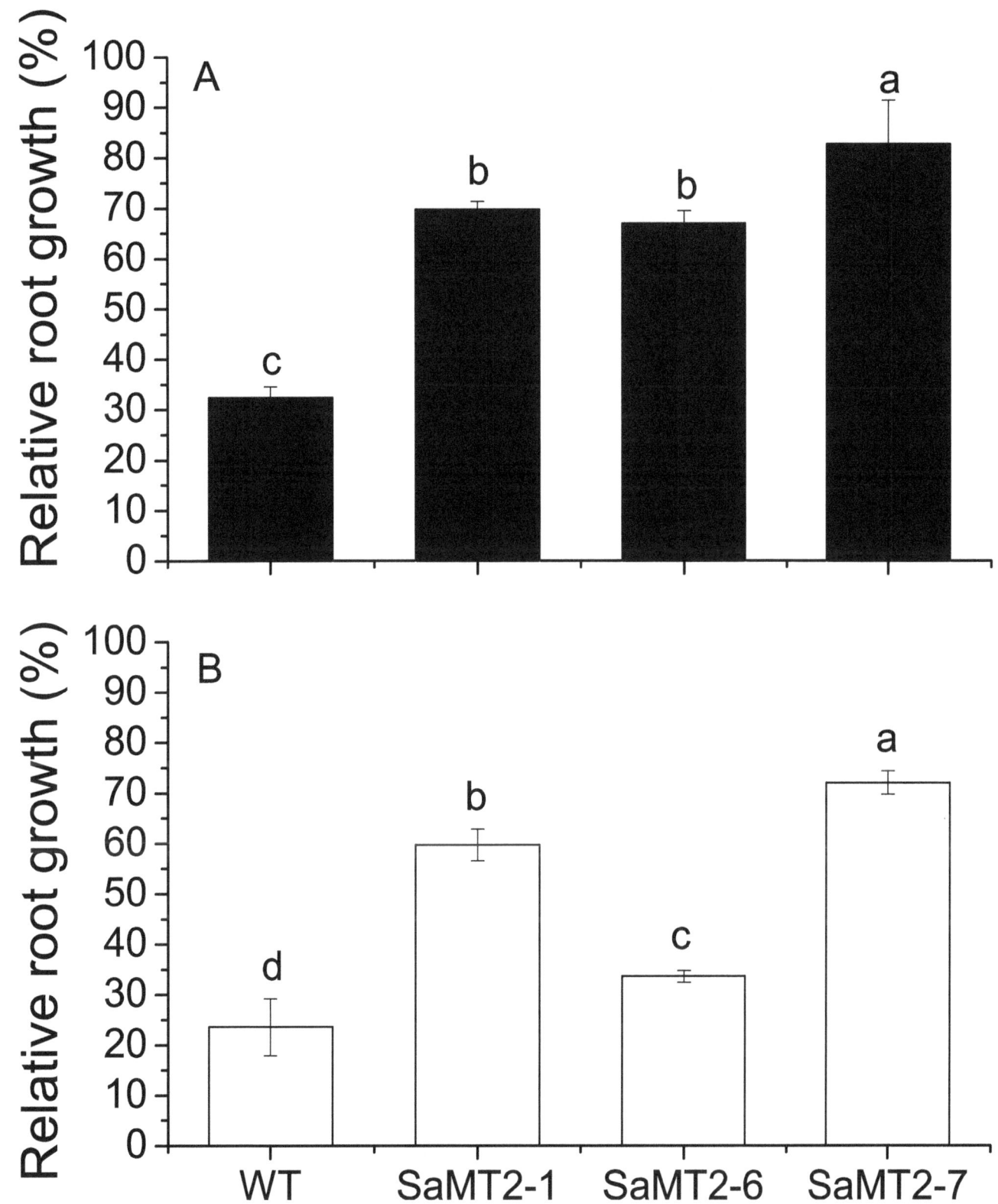

Figure 6. Relative root growth of transgenic tobacco plants. The relative root growth of WT and transgenic tobacco plants under Cd (A) and Zn (B) treatments. Different letters above the columns indicate a significant difference among different plant lines (p<0.05, Tukey's test).

100 μM $ZnSO_4$ for one week. Cadmium and zinc concentrations in plant tissues were measured using ICP-MS as described by Yang *et al.* [13].

Determination of SOD, POD, CAT and H_2O_2

For antioxidant enzyme activity determination, a 0.5-g aliquot of plant sample was homogenized in 5 ml potassium phosphate

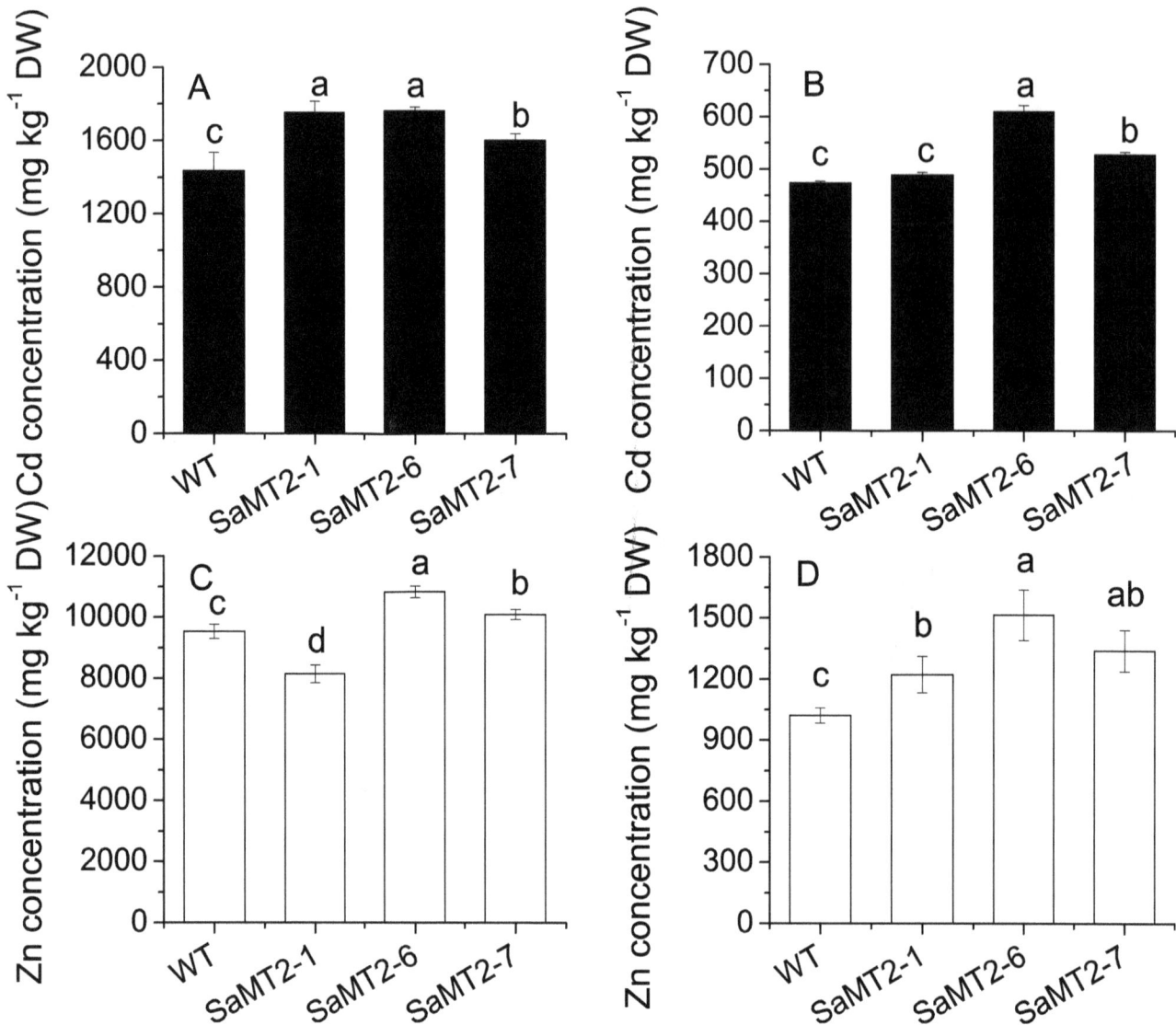

Figure 7. Cd and Zn concentrations in wild type amd transgenic tobacco lines overexpressing *SaMT2*. Three independent *SaMT2* over-expressing lines and wild-type tobacco were grown in nutrient solution containing 50 μM CdCl$_2$, 100 μM ZnSO$_4$ for 1 week. A: Cd concentration in roots, B: Cd concentrantion in shoots, C: Zn concentration in roots, D: Zn concentration in shoots. Results are means ± S.E. (n = 3). Different letter indicate the mean values were significantly different from WT tobacco determined by Tukey's test (p<0.05).

buffer (50 mM, pH 7.8). The homogenates were then centrifuged at 12000×g for 20 min at 4 °C. The supernatants were used for the analysis of enzyme activity. Superoxide dismutase (SOD) activity was determined by the photochemical method described by Giannopotitis and Ries [18]. One unit of the enzyme activity was defined as the amount of enzyme required to result in a 50% inhibition of the rate of nitro blue tetrazolium reduction measured at 560 nm. Catalase (CAT) activity was estimated according to Cakmak *et al.* [19]. The reaction mixture in a total volume of 2 ml contained 25 mM sodium phosphate buffer (pH 7.0), 10 mM H$_2$O$_2$. The reaction was initiated by the addition of 100 μl of enzyme extract and activity was determined by measuring the initial rate of disappearance of H$_2$O$_2$ at 240 nm (E = 39.4 mM^{-1} cm^{-1}) for 30 s. Peroxidase (POD) activity was measured as the increase of absorbance due to guaiacol oxidation [20]. The reaction mixture contained 25 mM phosphate buffer (pH 7.0), 10 mM H$_2$O$_2$, 0.05% guaiacol and 100 μl of enzyme

extract. The reaction was initiated by the addition of H$_2$O$_2$. The oxidation of guaiacol was measured at 470 nm (E = 26.6 mM^{-1} cm^{-1}).

H$_2$O$_2$ was determined according to Loreto & Velikova [21]. Leaf tissues (0.07 g) were homogenized in an ice bath with 5 ml of 0.1% (w/v) trichloroacetic acid (TCA). The homogenate was centrifuged at 12,000×g for 15 min and 0.5 ml of the supernatant was added to 0.5 ml of 10 mM potassium phosphate buffer (pH 7.0) and 1 ml of 1 M KI. The absorbance of the supernatant was measured at 390 nm. The content of H$_2$O$_2$ was calculated by comparison with a standard calibration curve previously made by using different concentrations of H$_2$O$_2$.

Statistical analysis of data

All data were statistically analyzed by using the SPSS package (version 20.0). All values were performed as means of three replicates. Data was tested at significant levels of p<0.05 using

Table 1. The activities of SOD, POD CAT and the content of H_2O_2 in the shoot and roots of wild-type and transgenic tobacco plants.

Metal treatment		SOD (U mg^{-1} Protein)		POD (nanokatals mg^{-1} Protein)		CAT (nanokatals mg^{-1} Protein)		H_2O_2 (μg g^{-1} FW)	
		WT	Transgenic	WT	Transgenic	WT	Transgenic	WT	Transgenic
Shoot	CK	20.1±0.4d	21.4±1.5d	20190.7±205.0d	20045.6±175.0d	36.7±5.0b	35.0±3.3b	154.4±10.0d	156.1±12.1d
	50 μM Cd	23.5±0.3c	28.6±1.2a	29267.5±720.1b	33821.7±688.4a	28.3±3.3c	45.0±1.7a	411.2±20.7a	167.9±10.1d
	100 μM Zn	24.3±0.5c	25.5±0.4b	26777.0±855.1c	30396.0±373.4b	25.0±1.7c	35.0±3.3b	267.3±9.7b	215.2±2.0c
Root	CK	24.2±0.4e	25.7±1.3e	30199.3±338.4e	31156.2±1171.9e	5.0±1.7b	3.3±1.7b	131.6±6.4d	132.0±7.4d
	50 μM Cd	30.2±0.6c	41.6±1.4a	46077.5±606.7b	54069.1±825.1a	5.0±1.7b	11.7±1.7a	400.5±23.0a	210.8±7.7c
	100 μM Zn	28.3±1.0d	34.2±0.7b	34723.6±270.0d	38681.0±1073.5c	6.7±1.7ab	10.0±1.7a	310.1±3.9b	198.6±10.0c

The data presented are mean ± SD of three replicates. Different letters indicate significant differences (p<0.05) among the different treatments and different plant lines.

one-way ANOVA (analysis of variance). The graphical works were made by using Origin software.

Results

Clone and sequence analysis of SaMT2

A cDNA fragment of metallothionein like gene was obtained from RNA-seq of *S. alfredii*. RT-PCR and RACE techniques were used to obtain the full length cDNA of this gene, whose sequence was identified by BLAST search (www.ncbi.nlm.nih.gov/BLAST). The obtained cDNA encoded a 79 amino acids protein, which showed certain similarity to the cDNA of *AtMT2a* or *AtMT2b*. According to the amino acid sequences, it belonged to the Type 2 MTs, and was named *SaMT2* (GeneBank accession number: KJ862538).

Multiple sequence alignment of the deduced amino acid sequences of *SaMT2* with *AtMT2a* (*Arabidopsis thaliana*, NP_187550.1), *AtMT2b* (NP_195858.1), *NcMT2a* (*Noccaea caerulescens*, ACR46966.1) and *SnMT2* (*Solanum nigrum*, ACF10396.1) were conducted (Figure 1A). *SaMT2* shared 62%, 59%, 59% and 66% similarities with *AtMT2a*, *AtMT2b*, *NcMT2a* and *SnMT2*, respectively. Similar to other plant MT proteins, *SaMT2* contained two cysteine-rich domain separated by a large cysteine-free domain [22]. The cysteine-rich domains in the N-terminal region is CCxxxCGCxxxCKCxxxCxGC, which was highly conserved, and that in the C-terminal region contained three CxC motifs. The spacer region between the two terminal regions contained approximately 40 amino acids.

The phylogenetic tree of *SaMT2* and MTs from Arabidopsis and rice was conducted using MEGA software (Figure 1B). These MTs were divided into several groups and *SaMT2* was closely clustered with *AtMT2a* and *AtMT2b*.

Expression analysis of SaMT2 in Sedum alfredii Hance

The expression of *SaMT2* was investigated using absolute quantitative RT-PCR. To investigate whether Cd or Zn were involved in the regulation of *SaMT2*, *S. alfredii* seedlings were treated with 100 μM $CdCl_2$ or 500 μM $ZnSO_4$ and were subjected to determine the transcript level of *SaMT2*. The expression level of *SaMT2* in roots was higher than that in shoots. The expression of *SaMT2* was significantly (p<0.05) increased in both roots and shoots treated with Cd and Zn (Figure 2).

SaMT2 enhanced cadmium but not zinc tolerance in yeast mutants

The *Δycf1* and *Δzrc1* yeast mutants and the wild type strain BY4741 were used to test the Cd and Zn tolerance ability of *SaMT2*. When grown in the control medium, yeasts containing either pDR195 or pDR195-*SaMT2* could grow well. When grown in a medium containing 30 μM $CdCl_2$, the growth of yeast contain pDR195 was significantly inhibited; however, the expression of *SaMT2* could markedly mitigate this growth defect (Figure 3). However, the growth of yeasts in a Zn containing medium was not affected whether *SaMT2* was expressed or not (Figure 3).

Similar trends were found for Cd and Zn concentrations in yeast mutants. Expression of *SaMT2* significantly increased the Cd concentration in yeast *Δycf1* mutant; however, it decreased the concentration of Zn in yeast *Δzrc1* mutant significantly (p<0.05, Figure 4).

Overexpression of SaMT2 in tobacco enhanced Cd and Zn tolerance and accumulation

To evaluate the functions of *SaMT2* in plants, transgenic tobacco plants were generated, ectopically expressing *SaMT2* under the control of CaMV 35S promoter. Three independent transgenic tobacco lines over-expressing *SaMT2* were selected for Cd and Zn tolerance analysis. The wild plants were used as control.

There was no difference in growth between wild type and transgenic plants under control condition. Exposure of the plants to 100 μM $CdCl_2$ or 200 μM $ZnSO_4$ significantly decreased root elongation and plant growth of both wild type and transgenic plants; however, the growth deficiency was less pronounced in transgenic plants (Figure 5). Under 100 μM $CdCl_2$ or 200 μM $ZnSO_4$ treatments, the root growth of wild type plants was decreased by 68% or 76%, compared to the control, respectively. However, the transgenic plants showed a significantly higher resistance to Cd and Zn. Compared to the control, the root growth of the transgenic plants was only decreased by 17%–33% under Cd treatment, and decreased by 28%–66% under Zn treatment (Figure 6).

Over-expression of *SaMT2* gene significantly increased both Cd and Zn concentration in transgenic tobacco plants (p<0.05) (Figure 7). Compared to the wild type plants, the Cd concentration was increased by 11–22% in the roots and by 3–28% in the shoots of the transgenic plants, respectively (Figure 7A, B). Except for the SaMT2-1 line, the Zn concentration was increased by 6–14% in roots and 20–48% in shoots of transgenic plants, respectively (Figure 7C, D). The SaMT2-6 line accumulated the highest amount of Cd and Zn among the three transgenic lines.

Based on the tolerance and accumulation of Cd and Zn in transgenic tobacco lines, the SaMT2-7 line - having similar Cd and Zn tolerance and accumulation level- was selected, to evaluate the reason of elevated tolerance to Cd and Zn of the transgenic plants. The plants treated with different metals were used to determine the activities of SOD, POD, CAT and the content of H_2O_2. The transgenic plant accumulated significantly less H_2O_2 in both roots and shoots than the WT plants under Cd and Zn treatments. The activities of SOD and POD were significantly increased in both roots and shoots of transgenic plants compared to that of wild type plants under Cd and Zn treatments (Table 1). For CAT activity, however, no significant difference was observed between WT and transgenic plants.

Discussion

Metallotheineins (MTs) are cysteine-rich proteins involved in metal tolerance of diverse living organisms. Plant metallothioneins can be divided into four subfamilies based on their sequence similarities and phylogenetic relationships [7,8]. In the present study, the MT gene cloned from *S. alfredii* encoded a protein with two Cys-rich regions, showing high identity with the N- and C-terminal regions of type 2 MTs of other plants; therefore, this MT gene was named as *SaMT2*. *S. alfredii* is a Cd/Zn co-hyperaccumulator, which shows extremely high tolerance to Cd and Zn [12,13]. Thus, It was hypothesized that *SaMT2* cloned from *S. alfredii* might be involved in Cd or Zn tolerance.

It has been reported that different MT genes have distinct tissue specific expression patterns in plants [8]. Generally, MT1s are predominantly expressed in roots, MT2s and MT3s in shoots [22,23]. In the present study, *SaMT2* was more highly expressed in shoots of *S. alfredii* than in roots. Similar results have also been found in Arabidopsis, rice and other plants [22–25]. The expression of MT genes in plants is regulated by many factors, including metal ions, oxidative stress, and stresses such as heat, salt, wounding and so on [11]. Here, the expression of *SaMT2* was significantly increased in both roots and shoots of *S. alfredii* treated with Cd or Zn; in contrast, it has been reported that Cd

and Zn do not induce the expression of *TcMT2* and *TcMT3* in *Thlaspi caerulescens* (now *Noccaea caerulescens*) another Cd/Zn hyperaccumulator [29].

The plant MTs are suggested to be involved in metal homeostasis or tolerance, such as Cu, Cd and Zn. When expressed in yeast or *E. coli*, the plant MTs are able to restore Cu, Cd and Zn tolerance [9,11,26,27]. In the present study, Cd and Zn induced the expression of *SaMT2*, suggesting its possible involvement in Cd and Zn tolerance. This was confirmed in yeast and tobacco plant overexpressing *SaMT2*, which exhibited increased Cd and Zn tolerance and accumulation. Previous studies also reported enhanced tolerance and accumulation of Cd or other heavy metals by over-expressing plant MT genes. For example, the over-expression of *Cajanus cajan* MT1 enhances Cd and Cu tolerance in *E. coli* and Arabidopsis [27]. The expression of *Colocasia esculenta* CeMT2b increases Cd tolerance and accumulation in *E. coli* and tobacco [28]. However, Hassinen et al. [29] have observed that MT expression and Cd accumulation are not correlated among *T. caerulescens* accessions. Furthermore, the overexpression of *TcMT2* and *TcMT3* do not increase Cd accumulation in Arabidopsis shoots. On the other hand, Lv et al. [30] have observed that the ectopic expression of either *BcMT1* or *BcMT2* does increase Cd tolerance, but not the Cd accumulation in Arabidopsis shoots and roots. Thus, the MT genes may have different specific functions, depending on plant species.

The plant MTs are thought to function as metal chelators or ROS scavengers in heavy metal stress [8]. On one hand, plant MT proteins are supposed to have binding activities to heavy metals, such as Cd, Zn and Cu [7,8]. In the present study, the ectopic expression of *SaMT2* in tobacco enhanced Cd tolerance and accumulation, which might be due to reduced activities of free Cd ions in the cytoplasm, by the binding of overexpressed MT protein and Cd. On the other hand, MTs can also function as ROS scavengers which can reduce the ROS induced by Cd or other metals [8]. The plants exposure to heavy metals, such as Cd, can produce ROS and oxidative stress. The present study demonstrated that overexpression of *SaMT2* could significantly reduce H_2O_2 in tobacco exposure to excess Cd. Several studies have demonstrated that MTs can effectively scavenge ROS in plants. Over-expression of *BcMT1*, *BcMT2* [30], *EhMT1* [31], *pCeMT* [9] reduces ROS production in transgenic plants. Using recombinant GhMT3a protein, Xue *et al.* [32] have demonstrated that GhMT3a can scavenge ROS *in vitro*. Plants themselves have developed various antioxidant defense mechanisms to protect from deleterious effects of ROS. One of them is the enzymatic system, which includes SOD, APX, POD, and CAT. Plants overexpressing MT genes show higher antioxidant enzyme activities [8]. The present study demonstrated that tobacco plants overexpressing *SaMT2* showed higher SOD and POD activities than wild type plants, indicating that *SaMT2* might also act as an activator of antioxidant enzyme system.

It has been demonstrated that MTs are not related with Cd or Zn tolerance and accumulation in hyperaccumulator *T. caerulescens*, even though the expression of MT genes varies among *T. caerulescens* accessions [29]. However, in the present study, ectopic expression study in yeast and tobacco revealed that *SaMT2* might play certain roles in Cd and Zn tolerance and accumulation. It is not certain whether *SaMT2* is directly involved in Cd or Zn tolerance in *Sedum alfredii*. There are clear evidence that MTs are not directly related in Zn or Cd tolerance in *T. caerulescens* [29]. Data from the present study demonstrated that *SaMT2* might be involved in the Cd or Zn induced antioxidant stress in *Sedum alfredii*. However, the exact role of *SaMT2* in metal tolerance and accumulation in *Sedum alfredii* needs to be examined by further study in the future.

In conclusion, *SaMT2* is a metallothionein gene cloned from Cd/Zn hyperaccumulator *Sedum alfredii* Hance. Overexpression of this gene could significantly enhance Cd tolerance and accumulation in yeasts and tobacco plants. The mechanism of the elevated Cd tolerance and accumulation by overexpressing of *SaMT2* includes binding of *SaMT2* with Cd and improving the antioxidant system.

Acknowledgments

The *Δzrc1* and *Δycf1* yeast mutant strains and the wildtype strain BY4741 were kindly supplied by Prof. Eide, University of Wisconsin-Madison, USA.

Author Contributions

Conceived and designed the experiments: JZ SKT LLL XY. Performed the experiments: JZ MZ. Analyzed the data: JZ MZ XY. Contributed reagents/materials/analysis tools: JZ MZ XY. Contributed to the writing of the manuscript: JZ MZ MJIS XY.

References

1. Brooks R (1998) Geobotany and hyperaccumulators. In: Robert R. Brooks editor. Plants that Hyperaccumulate Heavy Metals. New York: CAB International. pp. 55–94.
2. Kramer U (2010) Metal hyperaccumulation in plants. Annual Review of Plant Biology 61: 517–534.
3. McGrath SP, Zhao FJ (2003) Phytoextraction of metals and metalloids from contaminated soils. Current Opinion in Biotechnology 14: 277–282.
4. Zhao FJ, McGrath SP (2009) Biofortification and phytoremediation. Current Opinion in Plant Biology 12: 373–380.
5. Verbruggen N, Hermans C, Schat H (2009) Molecular mechanisms of metal hyperaccumulation in plants. New Phytologist 181: 759–776.
6. Hall JL (2002) Cellular mechanisms for heavy metal detoxification and tolerance. Journal of Experimental Botany 53: 1–11.
7. Cobbett C, Goldsbrough P (2002) Phytochelatins and metallothioneins: Roles in heavy metal detoxification and homeostasis. Annual Review of Plant Biology 53: 159–182.
8. Hassinen VH, Tervahauta AI, Schat H, Karenlampi SO (2011) Plant metallothioneins - metal chelators with ROS scavenging activity? Plant Biology 13: 225–232.
9. Kim YO, Jung S, Kim K, Bae HJ (2013) Role of pCeMT, a putative metallothionein from *Colocasia esculenta*, in response to metal stress. Plant Physiology and Biochemistry 64: 25–32.
10. Mir G, Domenech J, Huguet G, Guo WJ, Goldsbrough P, et al. (2004) A plant type 2 metallothionein (MT) from cork tissue responds to oxidative stress. Journal of Experimental Botany 55: 2483–2493.
11. Xia Y, Lv Y, Yuan Y, Wang G, Chen Y, et al. (2012) Cloning and characterization of a type 1 metallothionein gene from the copper-tolerant plant *Elsholtzia haichowensis*. Acta Physiologiae Plantarum 34: 1819–1826.
12. Yang X, Long XX, Ni WZ, Fu CX (2002) *Sedum alfredii* H: A new Zn hyperaccumulating plant first found in China. Chinese Science Bulletin 47: 1634–1637.
13. Yang XE, Long XX, Ye HB, He ZL, Calvert DV, et al. (2004) Cadmium tolerance and hyperaccumulation in a new Zn-hyperaccumulating plant species (*Sedum alfredii* Hance). Plant and Soil 259: 181–189.
14. Jin XF, Yang X, Mahmood Q, Islam E, Liu D, et al. (2008) Response of antioxidant enzymes, ascorbate and glutathione metabolism towards cadmium in hyperaccumulator and nonhyperaccumulator ecotypes of *Sedum alfredii* H. Environmental Toxicology 23: 517–529.
15. Jin XF, Yang XO, Islam E, Liu D, Mahmood Q (2008) Effects of cadmium on ultrastructure and antioxidative defense system in hyperaccumulator and non-hyperaccumulator ecotypes of *Sedum alfredii* Hance. Journal of Hazardous Materials 156: 387–397.
16. Gietz RD, Schiestl RH (2007) High-efficiency yeast transformation using the LiAc/SS carrier DNA/PEG method. Nature Protocols 2: 31–34.
17. Horsch RB, Fry JE, Hoffmann NL, Eichholtz D, Rogers SG, et al. (1985) A simple and general method for transferring genes into plants. Science 227: 1229–1231.
18. Giannopolitis CN, Ries SK (1977) Superoxide dismutases: I. Occurrence in higher plants. Plant Physiology 59: 309–314.
19. Cakmak I, Strbac D, Marschner H (1993) Activities of hydrogen peroxide-scavenging enzymes in germinating wheat seeds. Journal of Experimental Botany 44: 127–132.
20. Zheng X, van Huystee RB (1992) Peroxidase-regulated elongation of segments from peanut hypocotyls. Plant Science 81: 47–56.
21. Loreto F, Velikova V (2001) Isoprene produced by leaves protects the photosynthetic apparatus against ozone damage, quenches ozone products, and reduces lipid peroxidation of cellular membranes. Plant Physiology 127: 1781–1787.
22. Zhou JM, Goldsbrough PB (1995) Structure, organization and expression of the metallothionein gene family in Arabidopsis. Molecular & General Genetics 248: 318–328.
23. Guo WJ, Bundithya W, Goldsbrough PB (2003) Characterization of the Arabidopsis metallothionein gene family: tissue-specific expression and induction during senescence and in response to copper. New Phytologist 159: 369–381.
24. Hsieh HM, Liu WK, Chang A, Huang PC (1996) RNA expression patterns of a type 2 metallothionein-like gene from rice. Plant Molecular Biology 32: 525–529.
25. Hsieh HM, Liu WK, Huang PC (1995) A novel stress-inducible metallothionein-like gene from rice. Plant Molecular Biology 28: 381–389.
26. Roosens NH, Bernard C, Leplae R, Verbruggen N (2004) Evidence for copper homeostasis function metallothionein of metallothionein (MT3) in the hyperaccumulator *Thlaspi caerulescens*. FEBS Letters 577: 9–16.
27. Sekhar K, Priyanka B, Reddy VD, Rao KV (2011) Metallothionein 1 (CcMT1) of pigeonpea (*Cajanus cajan*, L.) confers enhanced tolerance to copper and cadmium in *Escherichia coli* and *Arabidopsis thaliana*. Environmental and Experimental Botany 72: 131–139.
28. Kim YO, Patel DH, Lee DS, Song Y, Bae HJ (2011) High cadmium-binding ability of a novel *Colocasia esculenta* metallothionein increases cadmium tolerance in *Escherichia coli* and tobacco. Bioscience Biotechnology and Biochemistry 75: 1912–1920.
29. Hassinen VH, Tuomainen M, Peraniemi S, Schat H, Karenlampi SO, et al. (2009) Metallothioneins 2 and 3 contribute to the metal-adapted phenotype but are not directly linked to Zn accumulation in the metal hyperaccumulator, *Thlaspi caerulescens*. Journal of Experimental Botany 60: 187–196.
30. Lv YY, Deng XP, Quan LT, Xia Y, Shen ZG (2013) Metallothioneins BcMT1 and BcMT2 from *Brassica campestris* enhance tolerance to cadmium and copper and decrease production of reactive oxygen species in *Arabidopsis thaliana*. Plant and Soil 367: 507–519.
31. Xia Y, Qi Y, Yuan YX, Wang GP, Cui J, et al. (2012) Overexpression of *Elsholtzia haichowensis* metallothionein 1 (EhMT1) in tobacco plants enhances copper tolerance and accumulation in root cytoplasm and decreases hydrogen peroxide production. Journal of Hazardous Materials 233: 65–71.
32. Xue TT, Li XZ, Zhu W, Wu CG, Yang GG, et al. (2009) Cotton metallothionein GhMT3a, a reactive oxygen species scavenger, increased tolerance against abiotic stress in transgenic tobacco and yeast. Journal of Experimental Botany 60: 339–349.

21

A Modular Cloning Toolbox for the Generation of Chloroplast Transformation Vectors

Yavar Vafaee¤, Agata Staniek, Maria Mancheno-Solano, Heribert Warzecha*

Plant Biotechnology and Metabolic Engineering, Technische Universität Darmstadt, Darmstadt, Germany

Abstract

Plastid transformation is a powerful tool for basic research, but also for the generation of stable genetically engineered plants producing recombinant proteins at high levels or for metabolic engineering purposes. However, due to the genetic makeup of plastids and the distinct features of the transformation process, vector design, and the use of specific genetic elements, a large set of basic transformation vectors is required, making cloning a tedious and time-consuming effort. Here, we describe the adoption of standardized modular cloning (GoldenBraid) to the design and assembly of the full spectrum of plastid transformation vectors. The modular design of genetic elements allows straightforward and time-efficient build-up of transcriptional units as well as construction of vectors targeting any homologous recombination site of choice. In a three-level assembly process, we established a vector fostering gene expression and formation of griffithsin, a potential viral entry inhibitor and HIV prophylactic, in the plastids of tobacco. Successful transformation as well as transcript and protein production could be shown. In concert with the aforesaid endeavor, a set of modules facilitating plastid transformation was generated, thus augmenting the GoldenBraid toolbox. In short, the work presented in this study enables efficient application of synthetic biology methods to plastid transformation in plants.

Editor: Mark Isalan, Imperial College London, United Kingdom

Funding: YV and MM-S were funded by the German Academic Exchange Service, https://www.daad.de/en/. The funders had no role in study design, data collection and analysis, decision to publish, or preparation of the manuscript.

Competing Interests: The authors have declared that no competing interests exist.

* Email: warzecha@bio.tu-darmstadt.de

¤ Current address: Department of Horticultural Science, Faculty of Agriculture, University of Kurdistan, Sanandaj, Iran

Introduction

Although the majority of genetically engineered plants today are generated by integrating transgenes into the nuclear genome, engineering of the plastid genome has become a promising technology, both for basic science and applied plant biotechnology [1]. Their potential for successful genetic manipulation stems from the fact that plastids, as relicts of endosymbiotic cyanobacteria, still feature many characteristics of prokaryotes. First, genome modulation can be achieved easily due to the still present and efficiently functioning homologous recombination system. By selecting stretches of plastid DNA to flank any given sequence, transgenes can be integrated into the plastid genome at virtually any location, enabling both mutagenesis of endogenous sequences and incorporation of additional genes with very high efficiency. Second, cells harbor a multitude of plastids, especially chloroplasts; these, in turn, carry multiple genome reprints. This high (trans-) gene copy number per cell, coupled with the utilization of strong promoters, fosters significantly elevated expression rates resulting in unprecedented protein accumulation levels (e.g., 80% TSP for bacteriolysins [2]). Last but not least, in many plant species, plastids are exclusively inherited maternally. This could be considered a built-in genetic containment feature, as the spread of transgenes by pollen is, consequently, largely excluded.

The enumerated traits constitute clear benefits for molecular farming, wherein the expression of one or a few transgenes at maximum levels and the safety of open field applications are major goals. But plastid engineering boasts still more potential that has come into focus only recently: the prospect of multigene stacking and coordinated gene expression for metabolic pathway engineering and the unmitigated access to reducing power from photosynthetic processes, conceivably enabling light-driven generation of metabolites. Polycistronic organization of plastidic operons, affording synchronized expression of multiple genes driven by a single promoter, is a well-established phenomenon [3]. Yet only recently, the group of Ralph Bock has taken advantage of this feature and designed a multigene operon for the concerted expression of three biosynthetic genes leading to the formation of tocochromanols in tomato [4]. The resulting study provides an excellent example of the advantages of plastid engineering for the build-up of metabolic pathways, affording enhanced levels of natural product retrieval. In another very recent report, Lassen *et al.* showed that cytochrome P450 enzymes, requiring electrons (usually delivered from NADPH by an accompanying reductase) for their inherent transformation reactions, can be coupled to the photosynthetic electron translocation machinery within the chloroplasts [5]. Since P450s are important catalysts in numerous biosynthetic routes leading to the formation of valuable natural compounds [6], metabolic engineering within the chloroplasts promises to foster the build-up of efficient pathways fueling light-driven biosynthesis of alternative metabolites.

Despite the numerous advantages of plastid transformation, some persisting bottlenecks and drawbacks still hamper many potential applications. Most discouragingly, not all plant species are amenable to the technique, with the monocotyledons, including agronomically important grasses like rice or maize, proving especially problematic. Although many successful transformation protocols of various plant species have been published (e.g., tomato [7], lettuce [8], or sugar beet [9], to name only a few), only tobacco and, to some extent, tomato and lettuce, as well as the unicellular algae *Chlamydomonas reinhardtii* can be routinely transformed with reasonable effort. Furthermore, identification of a large set of promoters, terminators, and regulatory elements driving the expression of plastid transgenes notwithstanding, the most suitable combination of the aforementioned sequences for any given transgene is hard to predict, as is the stability of the resulting recombinant protein. For example, while the 5'- and 3'-transcript untranslated regions (UTRs) substantively bear upon RNA stability [10], the 5'-segments of coding sequences (CDSs) significantly influence translation efficiency [11,12]. Hence, rational and targeted manipulation of the aforesaid genetic elements can substantially boost expression levels. Moreover, it has been shown that N-terminal fusions of short peptides as well as signal sequences directly affect the stability of recombinant proteins [13]. In light of all the enumerated findings pertaining to the influence of diverse sequence elements on gene expression and protein stability, construction of large sets of transformation vectors becomes a prerequisite for effective modification of the plastid genome. Taken together with the inherent requirement to replace and shuffle flanking sequences necessary for homologous recombination and integration of the transgene cassettes into the genome of a given plant at a specific position, the cited considerations point to extensive and oftentimes cumbersome cloning procedures as the critical hurdle to dynamic development and far-reaching application of plastid genome engineering. Consequently, while an extensive array of expression vectors have been developed and made available to the research community in recent years [14–16], the engineering of novel genetic elements and target plant species still requires tedious redesign and recloning in almost all cases. On the one hand, with DNA synthesis becoming less and less expensive, the challenge can now be addressed through total synthesis of optimized vectors, eliminating repetitive sequences or unfavorable restriction sites by design. On the other, the synthetic approach provides merely case-by-case solutions to the individual experimental objectives and is at odds with the central premise of rational bio-engineering.

Standardization of reusable biological components as a means to efficiently design and engineer biological systems is a paradigm of synthetic biology, as recently reiterated by one of its co-founders [17]. While in many ways the young discipline has been staggeringly successful, with the creation of the minimal cell marking a stepping stone in its ground-breaking advent [18], the very concept of standardization – the driving force of Industrial Revolution and primer of the Information Age shaping modern society [19] – still lacks universal validation in the field of biological engineering.

The first attempt at the development of a standardized strategy for combinatorial manipulation of DNA fragments was reported nearly two decades ago [20]. Although versatile and elegant, NOMAD (nucleic acid ordered assembly with directionality) met with but limited acceptance within the scientific community, while the *ad hoc* experimental design of DNA assembly efforts persisted. In contrast, the BioBrick standard [21], launched in concert with the International Genetically Engineered Machines competition (iGEM), garnered considerable traction and spurred exuberant

development of "standard biological parts" and their applications [22]. While certainly tantalizing, the simplicity of the iterative BioBrick approach turned out to be one of its limitations, as the original design, burdened with the obligatory by-product of residual scarring between individual parts, does not translate into the higher orders of abstraction – beyond genes, into pathways and coordinated circuits. In response to the system constraints, an array of alternative DNA assembly methods have been developed and critically reviewed, addressing their prospective application in both microbial and plant engineering [23,24].

Among others, the expanding toolbox of synthetic biology offers a powerful technology dubbed Golden Gate [25]. Drawing on the distinct properties of type IIs restriction enzymes, the strategy affords multipartite and seamless (or scar-benign) assembly of genetic elements in a "one pot, one step reaction" [26]. In turn, the founding principle of Golden Gate precision cloning proved the corner stone of the concurrent development of two standardized modular cloning systems, MoClo [27] and GoldenBraid, GB [28]. Further coordinated efforts rendered the two compatible and ultimately resulted in the introduction of the common assembly standard for plant synthetic biology, GoldenBraid 2.0 [29], offering its users a starter kit of ready-made genetic modules as well as relevant software tools (https://gbcloning.org/).

The GB 2.0 destination plasmid kit encompasses two complimentary sets of binary vectors based on the pGreenII and pCAMBIA vector backbones, respectively. Thus, the original system solely addresses the *Agrobacterium tumefaciens*-mediated transfer of foreign DNA into the plant cell nucleus. To further establish GoldenBraid as the modular cloning system overarching the full spectrum of plant genetic engineering, we demonstrate its reappropriation for plastid transformation. The proposed comprehensive application of the GoldenBraid grammar will afford straightforward and seamless assembly of coordinated fusions (e.g., promoter-UTR) and multigene operons compatible with the genetic machinery of chloroplasts. It will further allow utilization and effortless shuffling of relevant flanking regions characteristic of not only different parts of a specific plastid genome, but indeed, those of diverse representatives of the plant kingdom, thus enabling easy adjustment to alternative species. Furthermore, bolstering the GB toolbox will foster free exchange of the standardized parts between the nuclear- and plastid-specific transformation vectors. The across-the-board compatibility of the GoldenBraid system thus ensured boasts the potential for prospective establishment of an ever-expanding repository of reusable genetic components and bringing together multiple users within the plant scientific community.

Materials and Methods

Cloning of GB parts (domestication)

All DNA fragments were amplified by PCR using corresponding templates (either plasmid DNA or genomic DNA from tobacco or lettuce) and high fidelity DNA polymerase (Thermo Scientific, St Leon-Roth, Germany) based on the protocol provided by the manufacturer. The DNA sequence encoding griffithsin was ordered as a synthetic gene from Thermo Scientific (St Leon-Roth, Germany) and the primers were obtained from Eurofins MWG GmbH (Ebersberg, Germany). All primers were designed so that they contained the appropriate *Bsm*BI restriction sites and overhangs to be subsequently cloned into the universal domesticator vector (pUPD) [28]. All overhangs released upon *Bsa*I-cleavage of pUPD constructs were designed to give the parts the appropriate identity (e.g., promoter, CDS, etc.). Only pUPD containing left and right targeting regions (LTR and RTR, respectively) were flanked by GGAG at the 5′-end and CGCT at the 3′-end to enable their cloning as single fragments into any α-level pDGB vector.

For templates containing internal type IIs recognition sites (*Bsa*I, *Bsm*BI, and *Bbs*I), additional primers were designed, allowing amplification of the given part in two or more patches (according to [27]). The patch-flanking *Bsm*BI-cleavable over-hangs facilitated in-frame fusion of patches, resulting in parts with point mutations, removing the unfavorable recognition sequences. Ω vectors conferring chloramphenicol resistance were assembled directly from PCR products of backbone parts with compatible overhangs cleaved by *Bbs*I in digestion/ligation GB reactions (see below). The chloramphenicol resistance gene (*cat*), including the appropriate promoter and terminator, was amplified in two patches from the vector pSB1C3 (Biobrick registry part, http://parts.igem.org), while the ori and flanking regions were PCR-synthesized from pICH41306 [27]. GB cassettes with the *lacZ* gene were recloned from the appropriate pDGB1_Ω vectors [28]. The newly assembled vectors were provisionally termed pDGB3_Ω.

PCR products used in the GoldenBraid reactions were purified by QIAquick PCR Purification Kit (Qiagen, Hilden, Germany). Standard GB reactions were set up in 10 µl mixtures containing 75 ng of the target vector, 75 ng of the PCR products (GB parts or patches) or intermediate vectors carrying corresponding fragments, T4 DNA ligase buffer (Promega, Mannheim, Germany), 3 U of the required restriction enzyme (*Bsa*I or *Bsm*BI), and 1 U of T4 DNA ligase (Promega, Mannheim, Germany). The assembly reactions were performed as 25 cycle digestion/ligation reactions (2 min at 37°C, 5 min at 16°C). One µl of each GB reaction mixture was transformed into chemically competent *E. coli* Top10 cells. Positive clones were selected on LB plates containing ampicillin (for the domestication vectors), kanamycin (for α-level destination vectors), and chloramphenicol (for pDGB3_Ω destination vectors). Blue/white selections were performed on plates supplemented with 50 µl X-Gal (2% (w/v) in DMSO) prior to plating. Plasmid DNA preparations were made using the E.Z.N.A. Plasmid Mini Kit I (Omega Bio-Tek, VWR, Darmstadt, Germany). Correct assemblies were confirmed by plasmid restriction analysis using *Eco*RI (pDGB1_α1 and pDGB3_Ω1R), *Bam*HI (pDGB1_α1R and pDGB3_Ω1), *Hind*III (pDGB1_α2 and pDGB3_Ω2R), and *Eco*RV (pDGB1_α2R and pDGB3_Ω2). *Bsa*I was provided by New England Biolabs (Ipswitch, MA, USA). All remaining restriction enzymes were purchased from Fermentas (Thermo Scientific, St Leon-Roth, Germany).

Cloning in α- and Ω-level destination vectors

After assembly of all parts in pUPD vectors, the relevant transcriptional units (TU) were generated in α-level destination vectors (pDGB1). The pDGB1 vectors are derived from pGreenII binary vectors [30], reconstructed and adapted for the GB cloning system by Sarrion-Perdigones *et al.* [28,29].

The 2000 bp stretch of the *Nicotiana tabacum rbcL* gene in pUPD, serving as the left targeting region (LTR) of the expression cassette, was cloned into the pDGB1_α1 vector. In parallel, the aminoglycoside 3′-adenyltransferase (*aadA*) coding sequence was assembled with the *N. tabacum rrn* promoter (*NtPrrn*) and the terminator of the *N. tabacum psbA* gene (*NtTpsbA*) into the pDGB1_α2 vector, yielding the TU*aadA*. The *Bsa*I-GB reactions were performed as 25 cycle digestion/ligation reactions (2 min at 37°C, 5 min at 16°C). After transformation into *E. coli*, positive clones were selected on plates containing kanamycin and extracted plasmids were used as templates for subsequent cloning steps.

TU*aadA* and LTR were then combined in the pDGB3_Ω1 vector in a *Bsm*BI-GB reaction, yielding pDGB3_Ω1:LTR-TU*aadA*. In parallel, the RTR (2000 bp of the *N. tabacum accD* gene) was cloned into the pDGB1_α2 vector and the coding sequence of griffithsin was combined with the *N. tabacum psbA* promoter (*NtPpsbA*) and the *N. tabacum psbA* terminator (*NtTpsbA*) into pDGB1_α1R, yielding TU*GRFT*. RTR and TU*GRFT* were then combined in pDGB3_Ω2, yielding pDGB3_Ω2:RTR-TU*GRFT*. Both pDGB3 (Ω1 and Ω2) constructs were then combined in pDGB1_α2, yielding the final transformation vector.

Verification of the final tobacco chloroplast transformation vector was performed via PCR amplification of its various components (including promoters, CDSs, terminators, and flanking sequences) as well as restriction enzyme digestion.

Chloroplast bombardment and molecular analysis

N. tabacum (cv. petit havana) leaf explants were bombarded with a BioRad PDS1000 (He) gene gun (BioRad, Hercules, CA, USA), as described previously [31] and placed upside down on RMOP medium containing 500 mg×L^{-1} spectinomycin under a 16 h light and 8 h dark photoperiod at 25°C in a cultivation room. To confirm integration of the transgene in the tobacco plastome, DNA preparations of developing green plantlets were tested by PCR with the corresponding primers. To obtain homoplasmy, positive transgenic shoots were subjected to two to three additional rounds of regeneration. Homoplasmic shoots were transferred to the rooting medium (MS containing 500 mg×L^{-1} spectinomycin) under standard cultivation room conditions. Acclimatization was performed by placing the transplastomic plants under a transparent plastic hood for three days in the greenhouse with 16 h of illumination per day.

For Southern blot analysis, total DNA was extracted from transplastomic tobacco lines based on a previously described protocol [32]. The DNA (2 µg) was digested with *Eco*RI for 16 h; the resulting fragments were separated at 25 V on a 1% agarose gel, and transferred to nylon membranes (Roth, Karlsruhe, Germany). A 1 kb DIG-labeled probe was amplified using the PCR-DIG Probe Synthesis Kit (Roche Molecular Biochemicals, Mannheim, Germany). After hybridization at 42°C for 16 h, the membranes were washed with 2×SSC buffer for 15 min and 0.5×SSC buffer for 30 min at room temperature. Probe-target hybrids were detected with alkaline phosphatase conjugated antibody through a color reaction with NBT/BCIP as a substrate (Roche Diagnostics, Mannheim, Germany). For northern blot, RNA extraction was performed as previously described [33]. For each sample, 2 µg of RNA were electrophoretically separated, the gels blotted and hybridized with the DIG-labeled *griffithsin*-specific probe (amplified and detected as before).

Protein detection was carried out by western blotting. Leaf proteins were extracted in 3 volumes (v/w) of 2×SDS sample buffer (100 mM Tris-HCl pH 6.8, 4% SDS, 0.2% bromophenol blue, 20% glycerol), separated by SDS-PAGE (separating gel buffer, 1.5 M Tris-HCl pH 8.8; stacking gel buffer, 0.5 M Tris-HCl pH 6.8; running buffer, 0.025 M Tris base, 0.2 M glycin, 0.1% SDS (w/v); 12.5% acrylamide; PerfectBlue Dual Gel System, PeqLab, Erlangen, Germany), and transferred to a PVDF membrane (Roth, Karlsruhe, Germany) by wet blotting (Perfect-Blue Tank Electro Blotter Web S, PeqLab, Erlangen, Germany) according to the manufacturer's instructions. For the detection of the His-tagged griffithsin, the membrane was incubated with the His-probe mouse monoclonal IgG$_1$ (1:10000 in PBST) and the goat anti-mouse IgG-HRP (1:20000 in PBST) as, respectively, the primary and secondary antibodies (Santa Cruz Biotechnology, Heidelberg, Germany). The blot was developed through application of the CheLuminate-HRP Pico Detect Reagent (AppliChem, Gatersleben, Germany) and visualized by exposure to an X-ray film (Fujifilm Corporation, Tokyo, Japan).

Results and Discussion

The GoldenBraid 2.0 assembly system relies on a series of alpha (α) and omega (Ω) vectors facilitating iterative stacking of transcriptional units. Although the GB vectors are binary T-DNA vectors based on either pGreen (pDGB1) or pCAMBIA (pDGB2) backbones designed for nuclear transformation, they are, in principle, suitable for the assembly of plastid transformation vectors as well. However, the choice of the antibiotic resistance marker in the Ω vectors – the *aadA* gene conferring spectinomycin resistance – counteracts the assembly of standard plastid transformation vectors. Since many plastid expression cassettes contain the *aadA* gene under the control of the *rrn* promoter conjointly functional in *E. coli*, this feature will confer resistance to spectinomycin to all vectors containing the *aadA* transcriptional unit, making it impossible to select assembled Ω vectors from the preceding α vectors. Therefore, we decided to generate a new series of Ω vectors (provisionally dubbed pDGB3_Ω) containing an alternative resistance marker, the chloramphenicol acetyl transferase gene (*cat*). As the new series of Ω vectors for plastid module assembly do not need to contain T-DNA elements, we designed minimal vectors encompassing the *cat* gene, the pMB1 origin of replication, as well as the *lacZ* gene flanked by the appropriate GB 2.0 assembly sites. As the α-level vectors in their original form are amenable for plastid modular cloning, they were not modified.

Next, we wished to include the flanking regions for double homologous recombination into the modular build-up of the expression cassettes to enable maximal flexibility of integration sites within a given plastid genome as well as plant species to be targeted. For this feasibility study, we chose the intergenic region between the *rbcL* and the *accD* genes, proven efficient in previous studies [34], of tobacco and lettuce. Due to the fact that all GB 2.0 (and MoClo) compatible elements have to be devoid of internal type IIs restriction sites (specifically, *Bsa*I, *Bsm*BI, and *Bbs*I), targeting regions needed to be domesticated and the recognition sites removed. Protein encoding sequences, like *rbcL* and *accD*, can be easily mutated (silent mutations) without altering gene functionality. There are other frequently used intergenic regions that might have advantages over this particular site, like the *trnfM* and *trnG* intergenic region [7]. However, mutagenesis of RNA encoding or regulatory sequences might have detrimental effects on the function of these endogenous genes, which needs to be evaluated case-by-case. The selected flanking regions were designed to be incorporated into the universal domesticator plasmid (pUPD) with GB overhangs 1 and 2 [29], enabling subsequent integration into any α vector, depending on the cloning and assembly strategies.

For testing of the GB approach in plastid transformation, we designed and cloned a set of genetic elements to be used in transcriptional unit (TU) assembly. Promoters were designed to encompass 5′-UTRs and terminators, to contain 3′-UTRs and stabilizing elements. Considering the terminology proposed in [29], our strategy resulted in the structure of promoters spanning the GB 2.0 positions 01–12 and terminators, 17–21. But, given the modularity of the approach, any promoter can be combined individually with any 5′-UTR, if parts are designed accordingly (figure 1), with no extra cloning steps. Additionally, we designed superparts: promoter elements including the 5′-segments of coding regions, enabling assembly with coding sequences (CDSs) lacking a start codon and designed to act as fusion partners (spanning GB

2.0 positions 01–13, figure 1A) and providing enhanced stability due to N-terminal amino acid composition determination [11,13]. In general, the GB design allows fusion of any coding sequence of interest to any N-terminal leader as well as to an array of C-terminal extensions. This might further reinforce the stability of the gene product, as previously demonstrated for the HIV fusion inhibitor, cyanovirin-N [35].

After domestication of a basic set of genetic elements for plastid transformation, we proceeded with the design and assembly of an expression construct. As depicted in figure 1A, a common expression cassette is of a generic structure, basically comprised of the two flanking regions (left targeting region and right targeting region, LTR and RTR, respectively), a transcriptional unit (TU) harboring the selection marker, and another TU encompassing the gene of interest. Virtually any transcriptional unit of choice can be assembled from a set of different elements, either parts or superparts (being fusions of parts). Depending on the size of the part collection, a large number of combinations can be easily designed and cloned in one-pot reactions. In course of this study, we started with a basic set of elements that was steadily growing and further included modules characteristic of diverse species.

To test the functionality of GB assembled chloroplast transformation vectors, we used the sequence encoding griffithsin, a viral entry inhibitor and potential topical prophylactic against HIV infection [36]. Griffithsin has been successfully produced in plants via transient expression systems [37]. Our aim was to further evaluate if tobacco chloroplast expression of this algal gene was feasible.

The first TU generated in course of our study was the resistance marker, built from the *aadA* coding sequence together with the *rrn* promoter and the *psbA* terminator from *N. tabacum*. As illustrated in figure 1B, all enumerated modules were taken from the library of standardized parts (in pUPD) and assembled in a GB reaction into pDGB1_α2. In parallel, the LTR was cloned from the pUPD library into pDGB1_α1. The two TUs from the α-level vectors were then combined into pDGB3_Ω1. Similarly, a TU encompassing the griffithsin ORF (open reading frame), P*psbA*, and T*psbA* was assembled in pDGB1_α1R. Using the α1R vector enabled the subsequent combination of relevant TUs in inverse directions, thus preventing the location of two copies of T*psbA* in the same orientation, which might lead to unwelcome homologous recombination events [38]. The griffithsin TU was then combined with the RTR (in pDGB1_α2) into the pDGB3_Ω2 vector. In the final step, the TUs harbored by the pDGB3 Ω1 and Ω2 vectors were combined in the pDGB1_α2 vector, resulting in the ready-to-use transformation vector, shown in figure 2A. Taken together, as depicted in figure 1B, the complete vector was assembled from appropriate parts in seven GB reactions. Since several steps were performed in parallel, only three subsequent reactions were necessary to produce the new vector. As all parts are standardized and reusable, including the TUs and combinations thereof, the presented approach enables generation of a large number of various vectors in only a few additional steps, strongly establishing the power of this modular assembly technology. The fact that the only previously reported endeavors aimed at simplification of plastid transformation vector construction are based on the Gateway recombination cloning system (Life Technologies, Thermo Fisher Scientific) [16,39] further reinforces the superiority of our standardized approach, as it does not require the use of costly proprietary reagents and the persistently substantial (albeit somewhat reduced) array of intermediary vectors and cloning steps.

With the assembled α-level vector, tobacco leaves were bombarded and transplastomic lines selected on spectinomycin

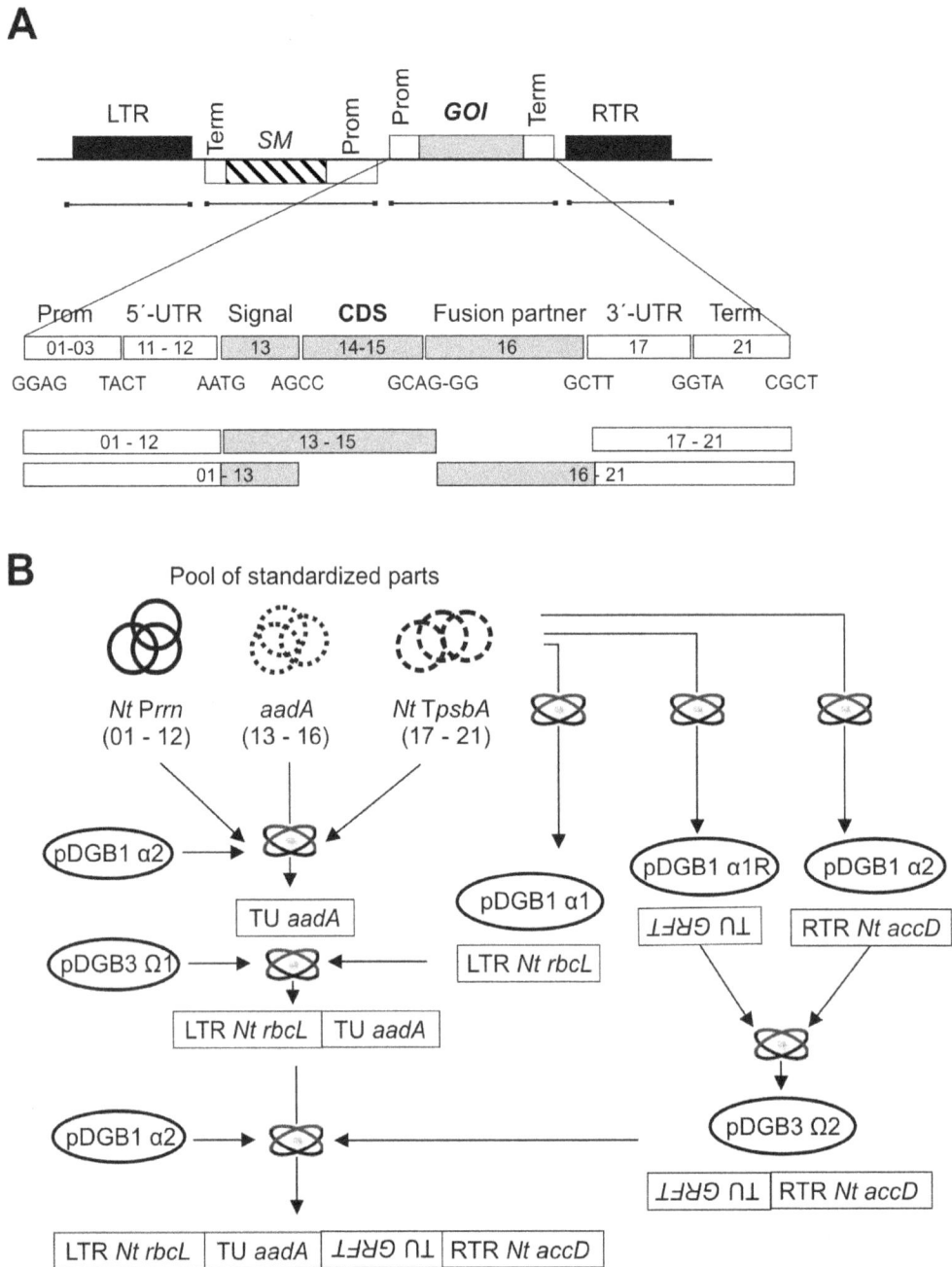

Figure 1. Schematic representation of the modular build-up and the overall GB-based cloning strategy of expression vectors. A) Generic structure of a plastid transformation vector. Magnified details show the modular build-up of a transcriptional unit and how it can be assembled from a set of standardized parts. Numbers within the boxes represent part identity and compatibility. Prom, promoter; Term, terminator; LTR, left targeting region; RTR, right targeting region; UTR, untranslated region; SM, selection marker; GOI, gene of interest; CDS, coding sequence. **B)** Schematic representation of the cloning strategy yielding the expression vector used in this study. In the pool of standardized parts, circles represent pUPD vectors harboring genetic elements (parts). Elliptical structures represent the different α- and Ω-vectors. Two intertwined ellipses on an arrow represent a one-pot restriction/ligation reaction (GB reaction) combining all the relevant parts. Boxes represent the parts and their assembly.

for at least three consecutive regeneration cycles. Two lines were subjected to further investigation. Restriction fragment length polymorphism (RFLP) analysis proved successful integration of the transgene cassette as well as homoplasmy (figure 2B) for both lines (EcoRI, 6572 vs. 4423 bp). Northern blot tests confirmed that specific transcripts were generated (figure 2C). The presence of transcripts of larger size than expected showed that, as described earlier [40], read-through transcripts were also produced with the

GB assembled cassettes. Since the griffithsin construct was designed to encompass a C-terminal hexahistidine tag, we used western blot with antibodies directed against the tag to detect the recombinant protein in plant extracts. Both lines showed appropriate signals (~13 kDa), proving that efficient translation of the griffithsin transcripts took place.

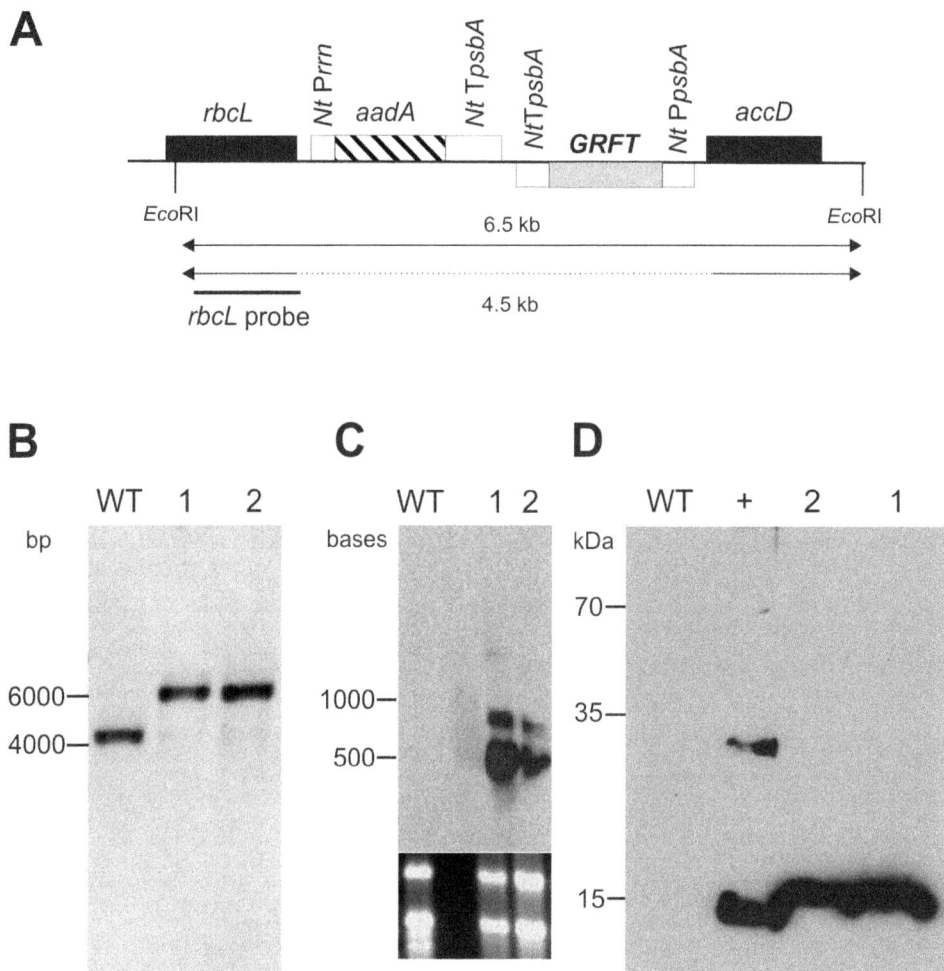

Figure 2. A) Schematic representation of the build-up of the expression construct used in this study (not drawn to scale). **B)** Restriction fragment length polymorphism analysis (RFLP) of the DNA isolated from wild type (WT) plants and two transplastomic lines (1 and 2). **C)** Northern blot analysis of the aforementioned lines. The lower panel shows total RNA, while the upper represents transcripts labeled with the *griffithsin*-specific probe. **D)** Western blot analysis of crude protein extracts of WT plants and two transplastomic lines (1 and 2). The blot was probed with an antibody directed against the hexahistidine tag. + represents an extract of a plant containing thioredoxin with a hexahistidine tag, serving as a positive control.

Conclusions

The postulated comprehensive application of the GoldenBraid modular cloning system now affords straightforward and seamless assembly of transcriptional units, coordinated fusions, and multigene operons compatible with the genetic machinery of chloroplasts. Further, it allows reutilization and effortless shuffling of relevant flanking/targeting regions characteristic of not only different parts of a specific plastid genome, but also of diverse representatives of the plant kingdom, thus facilitating easy swapping of species specificity. Moreover, the demonstrated bolstering of the GB toolbox will foster free exchange of the standardized parts between the nuclear- and plastid-specific transformation vectors. The across-the-board compatibility of the GoldenBraid system thus ensured boasts the potential for prospective establishment of an ever-expanding repository of reusable genetic components and bringing together multiple users within the plant synthetic biology scientific community.

Acknowledgments

YV and MM-S acknowledge the financial assistance of the German Academic Exchange Service (DAAD). The support of COST Action FA1006, *PlantEngine* is gratefully acknowledged. The authors further thank Dr. Houshang Alizadeh for his valuable sugesstions and Sepideh Sadrai for her help with the characterization of transplastomic lines.

Author Contributions

Conceived and designed the experiments: YV HW. Performed the experiments: YV AS MM-S. Contributed to the writing of the manuscript: AS HW YV.

References

1. Bock R, Warzecha H (2010) Solar-powered factories for new vaccines and antibiotics. Trends Biotechnol 28: 246–252.

2. Oey M, Lohse M, Kreikemeyer B, Bock R (2009) Exhaustion of the chloroplast protein synthesis capacity by massive expression of a highly stable protein antibiotic. Plant J 57: 436–445.

3. Zhou F, Karcher D, Bock R (2007) Identification of a plastid intercistronic expression element (IEE) facilitating the expression of stable translatable monocistronic mRNAs from operons. Plant J 52: 961–972.

4. Lu Y, Rijzaani H, Karcher D, Ruf S, Bock R (2013) Efficient metabolic pathway engineering in transgenic tobacco and tomato plastids with synthetic multigene operons. Proc Natl Acad Sci U S A 110: E623–632.

5. Lassen LM, Nielsen AZ, Ziersen B, Gnanasekaran T, Moller BL, et al. (2014) Redirecting photosynthetic electron flow into light-driven synthesis of alternative products including high-value bioactive natural compounds. ACS Synth Biol 3: 1–12.

6. Renault H, Bassard J-E, Hamberger B, Werck-Reichhart D (2014) Cytochrome P450-mediated metabolic engineering: current progress and future challenges. Curr Opin Plant Biol 19: 27–34.

7. Ruf S, Hermann M, Berger IJ, Carrer H, Bock R (2001) Stable genetic transformation of tomato plastids and expression of a foreign protein in fruit. Nat Biotechnol 19: 870–875.

8. Ruhlman T, Verma D, Samson N, Daniell H (2010) The role of heterologous chloroplast sequence elements in transgene integration and expression. Plant Physiol 152: 2088–2104.

9. De Marchis F, Wang Y, Stevanato P, Arcioni S, Bellucci M (2008) Genetic transformation of the sugar beet plastome. Transgenic Res 18: 17–30.

10. Kuroda H, Maliga P (2001) Complementarity of the 16S rRNA penultimate stem with sequences downstream of the AUG destabilizes the plastid mRNAs. Nucleic Acids Res 29: 970–975.

11. Inka Borchers AM, Gonzalez-Rabade N, Gray JC (2012) Increased accumulation and stability of rotavirus VP6 protein in tobacco chloroplasts following changes to the 5′ untranslated region and the 5′ end of the coding region. Plant Biotechnol J 10: 422–434.

12. Herz S, Fussl M, Steiger S, Koop HU (2005) Development of novel types of plastid transformation vectors and evaluation of factors controlling expression. Transgenic Res 14: 969–982.

13. Apel W, Schulze WX, Bock R (2010) Identification of protein stability determinants in chloroplasts. Plant J 63: 636–650.

14. Lutz KA, Azhagiri AK, Tungsuchat-Huang T, Maliga P (2007) A guide to choosing vectors for transformation of the plastid genome of higher plants. Plant Physiol 145: 1201–1210.

15. Verma D, Daniell H (2007) Chloroplast vector systems for biotechnology applications. Plant Physiol 145: 1129–1143.

16. Gottschamel J, Waheed MT, Clarke JL, Lössl AG (2013) A novel chloroplast transformation vector compatible with the Gateway® recombination cloning technology. Transgenic Res 22: 1273–1278.

17. Gardner TS, Hawkins K (2013) Synthetic biology: evolution or revolution? A co-founder's perspective. Curr Opin Chem Biol 17: 871–877.

18. Gibson DG, Glass JI, Lartigue C, Noskov VN, Chuang RY, et al. (2010) Creation of a bacterial cell controlled by a chemically synthesized genome. Science 329: 52–56.

19. Surowiecki J (2002) Turn of the century. Wired 10.01.

20. Rebatchouk D, Daraselia N, Narita JO (1996) NOMAD: a versatile strategy for in vitro DNA manipulation applied to promoter analysis and vector design. Proc Natl Acad Sci U S A 93: 10891–10896.

21. Knight T (2003) Idempotent vector design for standard assembly of BioBricks. MIT Synthetic Biology Working Group Technical Reports. Available: http://hdl.handle.net/1721.1/21168. Accessed 18 June 2014.

22. Smolke CD (2009) Building outside of the box: iGEM and the BioBricks Foundation. Nat Biotechnol 27: 1099–1102.

23. Ellis T, Adie T, Baldwin GS (2011) DNA assembly for synthetic biology: from parts to pathways and beyond. Integr Biol (Camb) 3: 109–118.

24. Patron NJ (2014) DNA assembly for plant biology: techniques and tools. Curr Opin Plant Biol 19: 14–19.

25. Engler C, Gruetzner R, Kandzia R, Marillonnet S (2009) Golden Gate shuffling: a one-pot DNA shuffling method based on type IIs restriction enzymes. PLoS One 4: e5553.

26. Engler C, Kandzia R, Marillonnet S (2008) A one pot, one step, precision cloning method with high throughput capability. PLoS One 3: e3647.

27. Weber E, Engler C, Gruetzner R, Werner S, Marillonnet S (2011) A modular cloning system for standardized assembly of multigene constructs. Plos One 6: e16765.

28. Sarrion-Perdigones A, Falconi EE, Zandalinas SI, Juarez P, Fernandez-del-Carmen A, et al. (2011) GoldenBraid: an iterative cloning system for standardized assembly of reusable genetic modules. PLoS One 6: e21622.

29. Sarrion-Perdigones A, Vazquez-Vilar M, Palaci J, Castelijns B, Forment J, et al. (2013) GoldenBraid 2.0: a comprehensive DNA assembly framework for plant synthetic biology. Plant Physiol 162: 1618–1631.

30. Hellens RP, Edwards EA, Leyland NR, Bean S, Mullineaux PM (2000) pGreen: a versatile and flexible binary Ti vector for Agrobacterium-mediated plant transformation. Plant Mol Biol 42: 819–832.

31. Svab Z, Hajdukkiewicz P, Maliga P (1990) Stable transformation of plastids in higher plants. Proc Natl Acad Sci USA 87: 8526–8530.

32. Glenz K, Bouchon B, Stehle T, Wallich R, Simon MM, et al. (2006) Production of a recombinant bacterial lipoprotein in higher plant chloroplasts. Nat Biotechnol 24: 76–77.

33. Hennig A, Bonfig K, Roitsch T, Warzecha H (2007) Expression of the recombinant bacterial outer surface protein A in tobacco chloroplasts leads to thylakoid localization and loss of photosynthesis. FEBS J 274: 5749–5758.

34. Sinagawa-García SR, Tungsuchat-Huang T, Paredes-López, Maliga P (2009) Next generation synthetic vectors for transformation of the plastid genome of higher plants. Plant Mol Biol 70: 487–498.

35. Elghabi Z, Karcher D, Zhou F, Ruf S, Bock R (2011) Optimization of the expression of the HIV fusion inhibitor cyanovirin-N from the tobacco plastid genome. Plant Biotechnol J 9: 599–608.

36. Mori T, O'Keefe BR, Sowder RC 2nd, Bringans S, Gardella R, et al. (2005) Isolation and characterization of griffithsin, a novel HIV-inactivating protein, from the red alga Griffithsia sp. J Biol Chem 280: 9345–9353.

37. O'Keefe BR, Vojdani F, Buffa V, Shattock RJ, Montefiori DC, et al. (2009) Scalable manufacture of HIV-1 entry inhibitor griffithsin and validation of its safety and efficacy as a topical microbicide component. Proc Natl Acad Sci U S A 106: 6099–6104.

38. Gray BN, Ahner BA, Hanson MR (2009) Extensive homologous recombination between introduced and native regulatory plastid DNA elements in transplastomic plants. Transgenic Res 18: 559–572.

39. Oey M, Ross IL, Hankamer B (2014) Gateway-assisted vector construction to facilitate expression of foreign proteins in the chloroplast of single celled algae. PLoS One 9: e86841.

40. Monde RA, Greene JC, Stern DB (2000) The sequence and secondary structure of the 3′-UTR affect 3′-end maturation, RNA accumulation, and translation in tobacco chloroplasts. Plant Mol Biol 55: 529–542.

PERMISSIONS

The contributors of this book come from diverse backgrounds, making this book a truly international effort. This book will bring forth new frontiers with its revolutionizing research information and detailed analysis of the nascent developments around the world.

We would like to thank all the contributing authors for lending their expertise to make the book truly unique. They have played a crucial role in the development of this book. Without their invaluable contributions this book wouldn't have been possible. They have made vital efforts to compile up to date information on the varied aspects of this subject to make this book a valuable addition to the collection of many professionals and students.

This book was conceptualized with the vision of imparting up-to-date information and advanced data in this field. To ensure the same, a matchless editorial board was set up. Every individual on the board went through rigorous rounds of assessment to prove their worth. After which they invested a large part of their time researching and compiling the most relevant data for our readers.

The editorial board has been involved in producing this book since its inception. They have spent rigorous hours researching and exploring the diverse topics which have resulted in the successful publishing of this book. They have passed on their knowledge of decades through this book. To expedite this challenging task, the publisher supported the team at every step. A small team of assistant editors was also appointed to further simplify the editing procedure and attain best results for the readers.

Apart from the editorial board, the designing team has also invested a significant amount of their time in understanding the subject and creating the most relevant covers. They scrutinized every image to scout for the most suitable representation of the subject and create an appropriate cover for the book.

The publishing team has been an ardent support to the editorial, designing and production team. Their endless efforts to recruit the best for this project, has resulted in the accomplishment of this book. They are a veteran in the field of academics and their pool of knowledge is as vast as their experience in printing. Their expertise and guidance has proved useful at every step. Their uncompromising quality standards have made this book an exceptional effort. Their encouragement from time to time has been an inspiration for everyone.

The publisher and the editorial board hope that this book will prove to be a valuable piece of knowledge for researchers, students, practitioners and scholars across the globe.

LIST OF CONTRIBUTORS

Ramanna Hema, Ramu S. Vemanna, Shivakumar Sreeramulu, Chandrasekhara P. Reddy and Makarla Udayakumar
Department of Crop Physiology, University of Agricultural Sciences, GKVK, Bangalore, India

Muthappa Senthil-Kumar
Department of Crop Physiology, University of Agricultural Sciences, GKVK, Bangalore, India
National Institute of Plant Genome Research (NIPGR), Aruna Asaf Ali Marg, New Delhi, India

WenZhi Jiang and Donald P. Weeks
Department of Biochemistry, University of Nebraska, Lincoln, Nebraska, United States of America

Bing Yang
Department of Genetics, Development and Cell Biology, Iowa State University, Ames, Iowa, United States of America

Shuping Gu, Chao Liu, Cheng Sun, Wenduo Ye and YiPing Chen
Department of Cell and Molecular Biology, Tulane University, New Orleans, Louisiana, United States of America

Weijie Wu
Department of Cell and Molecular Biology, Tulane University, New Orleans, Louisiana, United States of America
Department of Dentistry, ZhongShan Hospital, FuDan University, Shanghai, P.R. China

Ling Yang
Department of Cell and Molecular Biology, Tulane University, New Orleans, Louisiana, United States of America
Guanghua School of Stomatology, Sun Yat-sen University, Guangzhou, Guangdong, P.R. China

Xihai Li
Department of Cell and Molecular Biology, Tulane University, New Orleans, Louisiana, United States of America

Academy of Integrative Medicine,
Fujian University of Traditional Chinese Medicine, Fuzhou, Fujian, P.R. China

Jianquan Chen and Fanxin Long
Department of Internal Medicine, Washington University School of Medicine, St. Louis, Missouri, United States of America

Wanke Zhao, Taleah Farasyn, Wanting Tina Ho and Zhizhuang Joe Zhao
Department of Pathology, University of Oklahoma Health Sciences Center, Oklahoma City, Oklahoma, United States of America,

Kang Zou
Oklahoma School of Science and Mathematics, Oklahoma City, Oklahoma, United States of America

Elisa Santovito and Serena Anna Minutillo
Dipartimento di Scienze del Suolo, della Pianta e degli Alimenti, Universitàdegli Studi di Bari Aldo Moro, Bari, Italy

Donato Gallitelli and Tiziana Mascia
Dipartimento di Scienze del Suolo, della Pianta e degli Alimenti, Universitàdegli Studi di Bari Aldo Moro, Bari, Italy
Istituto di Virologia vegetale del Consiglio Nazionale della Ricerca, UnitàOperativa di Supporto di Bari, Bari, Italy

Shahid A. Siddiqui and Jari P. T. Valkonen
Department of Agricultural Sciences, University of Helsinki, Helsinki, Finland

Wenxian Wu, Zhiwei Cheng, Mengjie Liu, Xiufen Yang and Dewen Qiu
The State Key Laboratory for Biology of Plant Disease and Insect Pests, Institute of Plant Protection, Chinese Academy of Agricultural Science, Beijing, China

Sharad K. Singh
Genetics and Plant Breeding Division, CSIR-Central Institute of Medicinal and Aromatic Plants, Lucknow, Uttar Pradesh, India
Biotechnology Division, CSIR-Central Institute of Medicinal and Aromatic Plants, Lucknow, Uttar Pradesh, India

Ashutosh K. Shukla and Ajit K. Shasany
Biotechnology Division, CSIR-Central Institute of Medicinal and Aromatic Plants, Lucknow, Uttar Pradesh, India

Om P. Dhawan
Genetics and Plant Breeding Division, CSIR-Central Institute of Medicinal and Aromatic Plants, Lucknow, Uttar Pradesh, India

Hoang Trong Phan
Department of Molecular Genetics, Leibniz Institute of Plant Genetics and Crop Plant Research (IPK), Gatersleben, Germany
Department of Plant Cell Biotechnology, Institute of Biotechnology (IBT), Vietnam Academy of Science and Technology (VAST), Hanoi, Vietnam

Bettina Hause
Cell and Metabolic Biology, LeibnizInstitute of Plant Biochemistry (IPB), Halle, Germany

Gerd Hause
Microscopy Unit, Biocenter, University of Halle-Wittenberg, Halle, Germany

Elsa Arcalis and Eva Stoger
Molecular Plant Physiology, University of Natural Resources and Applied Life Sciences, Vienna, Austria

Daniel Maresch and Friedrich Altmann
Department of Chemistry, University of Natural Resources and Applied Life Sciences, Vienna, Austria

Jussi Joensuu
VTT Technical Research Centre of Finland, Espoo, Finland

Udo Conrad
Department of Molecular Genetics, Leibniz Institute of Plant Genetics and Crop Plant Research (IPK), Gatersleben, Germany

Narendra Singh Yadav, Vijay Kumar Singh and Dinkar Singh
Discipline of Marine Biotechnology and Ecology, CSIR-Central Salt and Marine Chemicals Research Institute, Bhavnagar, Gujarat, India

Bhavanath Jha
Discipline of Marine Biotechnology and Ecology, CSIR-Central Salt and Marine Chemicals Research Institute, Bhavnagar, Gujarat, India

Academy of Scientific and Innovative Research, CSIR, New Delhi, India

Shuichi Toyoda, Junya Mizuta and Jun-ichi Miyazaki
Division of Stem Cell Regulation Research, Osaka University Graduate School of Medicine, Osaka, Japan

Takuji Yoshimura
Division of Stem Cell Regulation Research, Osaka University Graduate School of Medicine, Osaka, Japan
Laboratory of Reproductive Engineering, the Institute of Experimental Animal Sciences, Osaka University Medical School, Osaka, Japan

Anne K. Zaiss, Chun Chu, Hidevaldo B. Machado, Jing Jiao, Arthur B. Catapang, Tomo-o Ishikawa, Jose S. Gil and Harvey R. Herschman
Department of Medical and Molecular Pharmacology, David Geffen School of Medicine, University of California Los Angeles, Los Angeles, California, United States of America
Department of Biological Chemistry, David Geffen School of Medicine, University of California Los Angeles, Los Angeles, California, United States of America

Johannes Zuber and Scott W. Lowe
Cold Spring Harbor Laboratory and Howard Hughes Medical Institute, New York, New York, United States of America

w
College of Environment and Plant Protection, Hainan University, Haikou, China
State Key Laboratory for Biology of Plant Diseases and Insect Pests, Institute of Plant Protection, Chinese Academy of Agricultural Sciences, Beijing, China
Qiongtai Teachers College, Haikou, China

Yue Liu and Fengqin Cao
College of Environment and Plant Protection, Hainan University, Haikou, China

Xiuping Chen
State Key Laboratory for Biology of Plant Diseases and Insect Pests, Institute of Plant Protection, Chinese Academy of Agricultural Sciences, Beijing, China

Lisheng Cheng
Qiongtai Teachers College, Haikou, China

Jörg Romeis
State Key Laboratory for Biology of Plant Diseases and Insect Pests, Institute of Plant
Protection, Chinese Academy of Agricultural Sciences, Beijing, China
Agroscope, Institute for Sustainability Science ISS, Zurich, Switzerland

Yunhe Li and Yufa Peng
State Key Laboratory for Biology of Plant Diseases and Insect Pests, Institute of Plant
Protection, Chinese Academy of Agricultural Sciences, Beijing, China

Yi Huang, Jiaqin Shi, Zhangsheng Tao, Xinfa Wang, Qing Yang, Guihua Liu and Hanzhong Wan
Oil Crops Research Institute, Chinese Academy of Agricultural Sciences, Wuhan, Hubei, P. R. China

Lida Zhang
Plant Biotechnology Research Center, School of Agriculture and
Biology, Shanghai Jiao Tong University, Shanghai, P. R. China

Qiong Liu
Oil Crops Research Institute, Chinese Academy of Agricultural Sciences, Wuhan, Hubei, P. R. China
School of Life Science and Technology, Hubei University, Wuhan, P. R. China

Zhao Zhang
Laboratory of Phytopathology, Wageningen University, Wageningen, The Netherlands
Key Laboratory of Plant Molecular Physiology, the Chinese Academy of Sciences, Beijing, China

Yin Song and Bart P. H. J. Thomma
Laboratory of Phytopathology, Wageningen University, Wageningen, The Netherlands

Chun-Ming Liu
Key Laboratory of Plant Molecular Physiology, the Chinese Academy of
Sciences, Beijing, China

Xu-Jing Wang, Xi Jin, Ning Kong, Shi-Rong Jia, Qiao-Ling Tang, and Zhi-Xing Wang
Biotechnology Research Institute, Chinese Academy of Agricultural Sciences, Beijing, China

Bao-Qing Dun
National Key Facility for Crop Gene Resources and Genetic Improvement, Institute of Crop Sciences, Chinese Academy of Agricultural Sciences, Beijing, China

Qian Xu, Xiao Xu, Yang Shi and Jichen Xu
National Engineering Laboratory for Tree Breeding, Beijing Forestry University, Beijing, China

Bingru Huang
Dep. of Plant Biology and Pathology, Rutgers, the State Univ. of New Jersey, New Brunswick, New Jersey, United States of America

Xiuzi Tianpei, Yingguo Zhu and Shaoqing Li
State Key Laboratory of Hybrid Rice; Key Laboratory for Research and Utilization of Heterosis in Indica Rice of Ministry of Agriculture; Engineering Research Center for Plant Biotechology and Germplasm Utilization of Ministry of Education; College of Life Sciences, Wuhan University, Wuhan, China

Jing Yue, Cong Li, Yuwei Liu and Jingjuan Yu
State Key Laboratory of Agrobiotechnology, College of Biological Sciences, China Agricultural University, Beijing, China

Weirong Xu, Ningbo Zhang, Dongming Xiao and Zhenping Wang
School of Agronomy, Ningxia University, Yinchuan, Ningxia, P.R. China
Engineering Research Center of Grape and Wine, Ministry of Education, Ningxia University, Yinchuan, Ningxia, P.R. China
Ningxia Engineering and Technology Research Center of Grape and Wine, Ningxia University, Yinchuan, Ningxia, P.R. China

Yuntong Jiao and Ruimin Li
College of Horticulture, Northwest A & F University, Yangling, Shaanxi, P.R. China

Xiaoling Ding
School of Agronomy, Ningxia University, Yinchuan, Ningxia, P.R. China
Engineering Research Center of Grape and Wine, Ministry of Education, Ningxia University, Yinchuan, Ningxia, P.R. China

Jie Zhang, Min Zhang, Shengke Tian, Lingli Lu, M. J. I. Shohag and Xiaoe Yang
MOE Key Laboratory of Environment Remediation and Ecosystem Health, College of Environmental and Resource Sciences, Zhejiang University, Hangzhou, China

Yavar Vafaee, Agata Staniek, Maria Mancheno-Solano and Heribert Warzecha
Plant Biotechnology and Metabolic Engineering, Technische Universita"t Darmstadt, Darmstadt, Germany

Index

www.ingramcontent.com/pod-product-compliance
Lightning Source LLC
Chambersburg PA
CBHW061249190326
41458CB00011B/3619